Mathematics of Planet Earth

Volume 6

Springer's Mathematics of Planet Earth collection provides a variety of well-written books of a variety of levels and styles, highlighting the fundamental role played by mathematics in a huge range of planetary contexts on a global scale. Climate, ecology, sustainability, public health, diseases and epidemics, management of resources and risk analysis are important elements. The mathematical sciences play a key role in these and many other processes relevant to Planet Earth, both as a fundamental discipline and as a key component of cross-disciplinary research. This creates the need, both in education and research, for books that are introductory to and abreast of these developments.

More information about this series at http://www.springer.com/series/13771

Arkadi Berezovski • Tarmo Soomere
Editors

Applied Wave Mathematics II

Selected Topics in Solids, Fluids, and Mathematical Methods and Complexity

Editors
Arkadi Berezovski
Department of Cybernetics
School of Science
Tallinn University of Technology
Tallinn, Estonia

Tarmo Soomere
Department of Cybernetics
School of Science
Tallinn University of Technology
Tallinn, Estonia

Estonian Academy of Sciences
Tallinn, Estonia

ISSN 2524-4264 ISSN 2524-4272 (electronic)
Mathematics of Planet Earth
ISBN 978-3-030-29953-8 ISBN 978-3-030-29951-4 (eBook)
https://doi.org/10.1007/978-3-030-29951-4

Mathematics Subject Classification (2010): 35-XX, 65-XX, 74-XX, 76-XX, 82-XX

This Springer imprint is published by the registered company Springer Nature Switzerland AG.
The registered company address is: Gewerbestrasse 11, 6330 Cham, Switzerland

There are more things in heaven and earth,
Horatio,
Than are dreamt of in your philosophy.

William Shakespeare, Hamlet 1.5.167-8

Dedicated to Jüri Engelbrecht
on the occasion of his 80th birthday

Preface

This edited volume consists of 16 invited contributions that all address different aspects of theory and applications of linear and nonlinear waves and associated phenomena. The volume is a partial insight to possible answers to the intriguing question "What is a wave?" by Jüri Engelbrecht (2015). His own answer was enthralling: "As surprising as it may sound, there is no simple answer to this question." Indeed, the definition of 'wave' depends on the physical context at hand, remarks one of the contributors to this book (Christov, 2014).

While it seemed in the remote past that wave phenomena are intrinsic to hyperbolic (wave) equations, we recognise today that also certain parabolic (e.g., diffusion) equations allow for the propagation of wavelike structures. Moreover, the variety of environments that may host wave propagation is rapidly increasing in both number and complexity.

The goal of this volume is to recapitulate at least a few developments in a selection of research fields that either focus on wave phenomena or are complementary to the progress of mathematics of wave phenomena. To a large extent, the contributions reflect the above question of (Engelbrecht, 2015) that largely arose from new and emerging problems in solid mechanics. The selection of contributions mirrors worldwide partnership of researchers with the national Centre of Excellence in Nonlinear Studies (CENS) at the Department of Cybernetics, Tallinn University of Technology in Estonia.

The structure of the book reflects the classic line of thinking in wave science from linear harmonic waves over weakly nonlinear environments to strongly nonlinear effects and their numerical representation and mathematical foundations:

Part I Linear Waves
Part II Nonlinear Waves
Part III Modelling and Mathematics

Following the similar collection published a decade ago (Quak and Soomere, 2009), the papers fit the general theme of "mathematics in the analysis of wave phenomena and complexity research". The chapters thus address a wide variety of topics concerning the investigation of wave phenomena in solids and fluids, as well as some of the mathematical methods that form the foundations of the analytical and numerical treatment of complex phenomena.

The papers are written in a tutorial style, having in mind nonspecialist researchers and students as readers. Strictly speaking, they are not just a survey of the work of others, nor a new research paper of the author's own latest results. The authors first describe a problem setting that is currently of interest in the scientific community, and then communicate their own experiences in tackling the problem, including detailed analysis of a few examples. The intention was to describe the approaches taken by the authors in an understandable way so that single chapters can be used for educational purposes, such as material for a course or a seminar. The overall goal was to produce a book, which highlights the importance of applied mathematics for relevant issues in the studies of waves and associated complex phenomena in different media, and motivates the readers to address the open challenges.

To ensure the quality of the contributions, each manuscript was carefully reviewed by one national and at least one international expert. Special thanks go to all the authors and reviewers. Without their efforts making this book would not have been possible. We also thank Prof. Kevin Parnell for valuable consultations on technical English. The friendly collaboration with Springer Verlag is kindly appreciated.

* * *

Even though waves are an intrinsic constituent of our world, understanding of the nature and properties of wave motion requires the use of advanced mathematical methods. While a century ago a wave was mostly a nice, smooth, sinusoidal signal (except the gentle "mountain of water" followed by John Scott Russell), today we see an amazing variety of phenomena in a multitude of environments united under this category.

The production of new materials constantly reinforces the need for analysis of the propagation of the simplest, linear waves under new properties and geometries of the environment. This framework is developed in the first part of the collection. Both contributions in this part focus on the derivation of the dispersion equation that serves as the core element of each wave class. Manfred Braun and Merle Randrüüt address in Chap. 1 approximate models to analyse the wave dynamics in infinite elastic layers or plates. The wave equations in the one-dimensional case are equivalent to those used for the dynamics of certain kinds of microstructured solids. The dispersion curves for the acoustical branch and for the optical branch derived in the framework of an approximate approach fit adequately with their exact counterparts.

Victor Eremeyev discusses in Chap. 2 a new class of antiplane surface waves that may exist in an elastic half space with surface stresses. He considers surface elasticity within so-called stress gradient model. A specific feature of models of this type is that surface stresses are related to surface strains via an integral constitutive dependence. For the kind of motions in question, the problem can be reduced to a variation of the wave that is complemented with a nonclassical dynamic boundary condition. The process of derivation of the dispersion relation makes *inter alia* also clear when such waves may exist.

The analysis of weakly and strongly nonlinear wave systems starts from the classic water wave problems in Chap. 3. Tomas Torsvik, Ahmed Abdalazeez, Denys Dutykh, Petr Denissenko, and Ira Didenkulova first introduce this framework and explain the basic categories for linear surface waves, and focus then on the modelling and measurement of wave runup in the context of long waves and tsunami attack. They present the main results of a series of fascinating experiments in the Large Wave Flume in Hannover, Germany. The experimental data set is complemented by detailed numerical analysis of fully nonlinear shallow water equations in a framework created by Howell Peregrine. While wave dispersion does have a certain impact on the wave propagation, the maximum runup is much less sensitive to dispersive effects and thus the runup height could be evaluated from nondispersive models.

Oxana Kurkina, Andrey Kurkin, Efim Pelinovsky, Yury Stepanyants, and Tatiana Talipova provide in Chap. 4 a consistent and systematic version of the derivation of a higher order evolution equation for internal (interfacial) waves in a two-layer model environment in the weakly nonlinear framework. The derivation is carried on without the classic assumption of irrotational flow in the layers and is thus applicable for situations when surface tension or friction between the layers is no more negligible. The conditions of wave propagation under the resulting fifth order KdV-type evolution equation greatly depend on the particular values of the coefficients at various terms of this equation. The physics of wave motion in real water bodies is such that several coefficients of this equation may become small, vanish, or change their sign.

Andrus Salupere, Martin Lints, and Lauri Ilison focus in Chap. 5 on the propagation and fate of various kinds of signals in a similar equation for wave motion in microstructured media. They consider so-called hierarchical Korteweg–de Vries (KdV) equation that, in essence, is a linear combination of two KdV operators and involves partial derivatives up to fifth order. The goal is to shed some light on the effect the microstructure has on the field of motion driven by sinusoidal waves (that are not valid solutions for the KdV equation) and two classes of solutions to the KdV equation: cnoidal waves and classic KdV solitons. Numerical simulations reveal a variety of different structures that are formed in the system. The most interesting are solitons – waves that propagate at a constant speed, keep their shape and interact with other similar entities elastically.

Karima R. Khusnutdinova and Matthew R. Tranter revert in Chap. 6 back to the problem of waves in new and emerging materials. They focus on strain waves in layered structures that are frequently used in modern engineering constructions. As typical for wave mathematics, equations for such waves are similar to those that are used in radically different environments. They use Boussinesq-type equations that are developed using lattice models. Thus, the entire problem can be framed as an extension of the classic Fermi–Pasta–Ulam problem. The authors apply semi-analytical approaches to construct the solution of an initial value problem.

Alexey V. Porubov and Alena E. Osokina address in Chap. 7 waves that may propagate in graphene that is one of potential materials of the future technologies. They construct a two-dimensional graphene lattice model that takes into account translational and angular interactions between the elements of two sublattices. This model is analysed using an asymptotic procedure. The aim is to derive an equation for weakly transversely perturbed nonlinear longitudinal plane waves. The analysis leads to a two-dimensional nonlinear equation for longitudinal strains.

Tommaso Ruggeri and Shigeru Taniguchi guide the reader into the strongly nonlinear world of shock waves in Chap. 8. Propagation of shock waves is a mathematically and physically extremely interesting phenomenon with numerous applications in various fields of engineering and technology. They start from phenomena associated with ideal shock waves such as the phase transition in real gases. The presence of dissipation leads to a multitude of changes in the behaviour of shocks, including changing the shock thickness and subshock formation.

Nobumasa Sugimoto and Dai Shimizu complete the cycle of mostly theoretical studies in this collection. In Chap. 9 they consider several aspects of linear and nonlinear approaches for the description of thermoacoustic waves. They focus in the environment that occurs in gas filled tubes subject to a temperature gradient. Similarly to the presentation in several earlier chapters, they start from a linear description of the system, focusing on the approximation of the thermoacoustic wave equation in two cases of particular interest, thin and thick diffusion layers. The approximately linear theories are then systematically extended to the weakly nonlinear regime by means of asymptotic methods.

The largest selection of papers in this book tackles various ways of modelling of nonlinear wave phenomena using either powerful numerical methods or advanced mathematical concepts. In Chap. 10, Tanel Peets and Kert Tamm present an insight into contemporary concepts of modelling of propagation and fate of nerve signals. Their starting point is the observation that the nerve function is a much richer phenomenon than a set of electrical action potentials alone. Thus, they develop a model according to which the nerve signal is an ensemble of waves and includes, to a first approximation, the pressure wave in axoplasm, the longitudinal waves in the surrounding biomembrane, transverse displacements of the biomembrane, and temperature changes. A coupled model for the nerve signal is presented in the form of a system of nonlinear partial differential equations.

In the subsequent Chap. 11, Harm Askes, Dario De Domenico, Mingxiu Xu, Inna M. Gitman, Terry Bennett, and Elias C. Aifantis guide the reader into the mathematics of the world of gradient enriched continua. This is an elegant and versatile class of material models that are able to simulate a variety of physical phenomena such as crack tips and dislocations that do not obey classic wave propagation features. An adequate description of such processes usually requires the use of higher order partial differential equations. The authors focus on operator split methods that allow reducing the order of the governing equations. The examples of applications involve problems of elastodynamics and dynamic piezomagnetics.

A better understanding of processes in such complicated environments requires also the development of new, powerful and effective numerical methods that can follow also strongly nonlinear signals. Mihhail Berezovski and Arkadi Berezovski build in Chap. 12 a two-dimensional numerical scheme based on the representation of computational cells as thermodynamic systems. The scheme follows the logic of the finite volume method and provides an accurate implementation of conditions at interfaces and boundaries in terms of explicit ready-to-use expressions.

Chapter 13 introduces the fascinating world of dynamic materials. They are defined as material substances with properties that may change in space and time in a tailored manner. Even though such materials are rare today, Mihhail Berezovski, Stan Elektrov, and Konstantin Lurie introduce and analyse a specific realisation of such a (meta)material. The electrical environment they develop is able to amplify the radio frequency signal and accumulate energy in the system by means of suitable switching of properties of the circuit elements.

Aditya A. Ghodgaonkar and Ivan C. Christov provide in Chap. 14 a massive landscape of novel ways of numerically solving nonlinear parabolic partial differential equations. This is done by means of strongly implicit finite difference scheme that has a second order accuracy in both space and time. The scheme is developed and applied for viscous gravity currents. The treatment is valid for both classic Newtonian fluids as well as for a selection of non-Newtonian fluids. The resulting scheme accurately respects the mass conservation and/or balance constraints.

Another view on the wave propagation offer Domènec Ruiz-Balet and Enrique Zuazua in Chap. 15 in terms of mathematics of control theory and agent based modelling approach. It is amazing that the spreading of opinions in some models follows the classic heat equation. Reaching a desired state of the system, however, is a deeply nontrivial process that encounters a multitude of mathematical problems, especially if the cost of the procedure is taken into account in some manner. As any rich in content problem, the rigorous results presented in this chapter highlight new interesting open problems.

In the final Chap. 16, Peter Ván raises the fundamental question of how entropy flux and entropy production could be described in generic phase field theory. He gives a partial answer using a nonequilibrium thermodynamic framework with weakly nonlocal internal variables by means of the classical method of irreversible thermodynamics by separating full divergences.

Tallinn, Estonia
July, 2019

Arkadi Berezovski
Tarmo Soomere

References

Christov, I.C.: Wave solutions. In: Hetnarski, R.B. (ed.) Encyclopedia of Thermal Stresses, pp. 6495–6506. Springer, Dordrecht (2014). https://doi.org/10.1007/978-94-007-2739-7_33

Engelbrecht, J.: Questions About Elastic Waves. Springer, Cham (2015). https://doi.org/10.1007/978-3-319-14791-8

Quak, E., Soomere, T. (eds.): Applied Wave Mathematics. Selected Topics in Solids, Fluids, and Mathematical Methods. Springer, Heidelberg (2009). https://doi.org/10.1007/978-3-642-00585-5

Jüri Engelbrecht: From Hardcore Science to Great Leadership

Mati Kutser and Tarmo Soomere

After graduating from Tallinn University of Technology as civil engineer in 1964, Jüri Engelbrecht chose to focus his doctoral studies on structural mechanics. His Candidate in Engineering Sciences (today equivalent to PhD) thesis (1968) was devoted to the statics of hanging structures with negative curvature. In 1969 he joined the Institute of Cybernetics of the Estonian Academy of Sciences where he started to address stress waves in elastic and thermoelastic solids.

It was already clear that linear methods and approaches could provide only very limited information about the nature and properties of such waves. It was thus natural to expand their treatment towards considering the influence of various nonlinear effects on the wave motion in such environments. It soon became clear that the physical and geometrical nonlinearities should be accounted for simultaneously, in order to understand what happens in elastic media. Following the principle of equipresence, it was necessary to simultaneously take into account thermal and viscous effects. In other words, in order to progress the understanding of such media, it was necessary to solve several thermodynamic issues as well.

The problem of wave motion in these environments thus required new, much deeper and substantially more complicated mathematical treatment. J. Engelbrecht derived several enhanced mathematical models that took into account a multitude of impacts and phenomena, such as geometrical and physical nonlinearities and several other second order effects caused by viscosity, dispersion, inhomogeneity, etc. He developed a mathematical model of thermoviscoelasticity that took into account both finite deformations and physical nonlinearity. This approach made it

M. Kutser
Department of Cybernetics, School of Science, Tallinn University of Technology, Tallinn, Estonia
e-mail: matik@cs.ioc.ee

T. Soomere (✉)
Department of Cybernetics, School of Science, Tallinn University of Technology, Tallinn, Estonia

Estonian Academy of Sciences, Tallinn, Estonia
e-mail: soomere@cs.ioc.ee

possible significantly progress the comprehension of the properties and evolution of bounded pulses as well as progressing knowledge as to how the finite velocity of heat flux affects the wave motion in this environment. These achievements formed the core of the Doctor of Sciences thesis (1981) devoted to nonlinear deformation waves in solids. The developed principles of modelling of elastic continua have been summarised in the monograph (Engelbrecht, 1983).

Processes in living organisms are usually much more complicated than wave motions in different kinds of solids. The dynamics of nerve pulses drew the attention of J. Engelbrecht in the 1980's. The classic nerve pulse equation (that served as the foundation of the relevant research at that time) was derived from telegraph equations. As this derivation neglected some terms of this equation, the resulting models and equations had certain convergence problems and missed many important properties of signals in our nerves. J. Engelbrecht reexamined this problem and developed a consistent evolution equation for nerve pulses. Solution to this equation reflected all typical properties of a nerve pulse. Moreover, it also offered a mathematically correct explanation for the existence of nonoscillating solutions for the signal propagation in nerve fibres. Solutions of this kind better follow the properties of real nerve pulses.

This direction of research saw the development of a new generation of models that would be able to take into account the entire complexity of signal propagation in nerve fibres. This process incorporates elements of various inorganic media with physics and electrobiochemistry of living tissues. Namely, electrical signals in nerves are coupled to mechanical waves in the internal axoplasmic fluid and in the surrounding biomembrane. The mathematically consistent equations developed by J. Engelbrecht over decades provide the possibility of a much deeper understanding of the role of governing factors in the whole process of propagating the nerve pulse.

Waves in the Earth's core are another field where *in situ* observations are problematic and where sophisticated models and equations are a must. The fact that in seismology, the attenuation process of some waves deviates from the forecast of simple models drew the attention of J. Engelbrecht. This problem motivated him to develop a model that can explain the fracturing of materials. It was based on the concept of dilatons, that are short-lived fluctuations of microdynamical density. Specifically, they are able to absorb energy from the surrounding medium. When the accumulated energy in a dilaton has reached a certain critical value, the dilaton breaks up and releases the stored energy. This concept was generalised from microdilatons (that characterise a fracture in a small body) to macrodilatons that could exist in solids with internal large-scale structure, such as the block structure of the Earth's crust.

A generalisation of this viewpoint is the concept of internal variables. This approach has proved useful far beyond the classic problems of modelling deformation waves in thermoelastic media. It provided new insight into studies of various damage processes, made it possible to better understand the dynamics of liquid crystals, etc. These models and the associated problem of the formation of symmetric and asymmetric solitary waves was generalised in the monograph (Engelbrecht, 1996).

Some time ago the nonlinear theory of continuous media served as the basis of derivation of evolution equations. At the present time, the attention is directed mainly to more complicated microstructured materials where classic theory does not work.

One of the basic results is the analysis of wave propagation in functionally graded materials. In particular, the elaboration of the proper theory for wave propagation in materials with complicated internal structure has given interesting and rich in content results.

The achievements include *inter alia* new results in the analysis of the dependence of wave hierarchy on scaling parameters, in the theory of dual internal variables, derivation of new evolution equations and establishing their asymmetric and solitonic solutions, and have also led to the development of various kinds of numerical methods for their analysis. An overview of numerical methods for analysis such problems is presented together with co-authors A. Berezovski and G.A. Maugin in the monograph (Berezovski et al., 2008).

The necessity to evaluate the properties of microstructure and the impact of the presence of microstructure to wave motion directed the attention of J. Engelbrecht to the world of inverse problems. A wide class of problems in this field is associated with the determination of various parameters of the governing equations from measured data. Problems of this kind are only meaningful if the functional form of the equations is relevant to the problem. Consequently, only advanced mathematical models of waves in microstructured solids can provide the basis for such a kind of analysis.

The problem is much more complicated in environments that support solitary wave propagation. A careful analysis of the properties of solitary waves can be used for solving inverse problems. It also has the potential for opening novel applications to nonlinear differential equations. A selection of results of the cooperation of J. Engelbrecht with mathematician J. Janno highlight several mathematical models and methods for solving inverse problems (Janno and Engelbrecht, 2011).

A never-ending question is how to properly relate mathematical models and physical processes. An associated question is how complicated the model must be to describe the essence of the physical processes. These questions turned the attention of J. Engelbrecht to the detailed analysis of the phenomenon of complexity itself. A number of examples of fascinating complex systems and samples of how the interactions of a few components may determine the behaviour of the whole system are presented in the monographs (Engelbrecht, 1997, 2017).

* * *

J. Engelbrecht has substantially contributed to the development of several Estonian and international institutions and academic organisations. At the end of the 1990s the Institute of Cybernetics became a central node of cooperation of a cluster of excellent scientists who have focused on studies of complex nonlinear processes in various environments. Similar studies also became popular in several other research groups in Estonia. J. Engelbrecht suggested that bringing the scientific potential of Estonia engaged in interdisciplinary studies for complex dynamical nonlinear processes under one umbrella might boost this research and may lead to extra collaborative benefits.

Following this idea, in 1999, a virtual network of research groups under the name tag Centre for Nonlinear Studies (CENS) was established in the Department of Mechanics and Applied Mathematics of the Institute of Cybernetics at Tallinn

University of Technology. It took only a couple of years to develop this virtual unit into a working centre of excellence. CENS was included in the first (2002–2007) and the second (2011–2015) Estonian National Programme for Centres of Excellence in Research. This programme was jointly supported by the state budget and the European Regional Development Fund.

The influence of CENS extended far beyond the Baltic States and led to the formation of the CENS-CMA cooperation (2005–2009) with the similar centre of excellence in the Norway Centre of Mathematics for Applications in the University of Oslo. While the highlights of the research results of the CENS have been published in numerous sources, the CENS-CMA cooperation led to the edited collection (Quak and Soomere, 2009).

J. Engelbrecht was elected to the Estonian Academy of Sciences in 1990, was the president of the Academy in 1994–2004, served as the vice president of the Academy in 2004–2014 and is until now a board member of the Academy. The political process of the restoration of independence of the Republic of Estonia in 1991 enabled significant advances in the status of the Academy of Sciences. J. Engelbrecht basically relaunched the Academy as a contemporary institution. This work promoted the good reputation of Estonia as a whole and the scientific publications published in Estonia. He served this field as a member of the editorial boards of several academic journals and the editor-in-chief of the Proceedings of the Estonian Academy of Sciences in 1991–1995.

Since 1994 Engelbrecht has been involved in designing the science policy for Estonia (funding schemes, research strategies, accreditation, etc.). As a member of the Estonian Science and Development Council (1994–2004) and as a chairman of the Science Competence Council at the Ministry of Education (1994–2004), J. Engelbrecht took an active part in shaping Estonian science policy. He was the chairman of the Estonian State Science Award Committee (1995–2004). J. Engelbrecht had a leading role in the process of establishing the Estonian National Committee for Mechanics (he served as its chair 1991–2008), ensuring it adhered to the rules of the International Union of Theoretical and Applied Mechanics (IUTAM).

J. Engelbrecht also participated actively in international community of scientists. He was invited to the European Mechanics Council Euromech in 1988 and was six years (1989–1993) a full member. In 1993 the European Mechanics Council was transformed into EUROMECH – the European Mechanics Society. J. Engelbrecht served EUROMECH as an elected council member for years 1995–2000. In 2001–2005 he was the chairman of the EUROMECH Advisory Board. Showing great distinction and the trust of the international academic community, J. Engelbrecht was elected to serve for several periods in the General Assembly, the Congress Committee and the Bureau of IUTAM (1996–2008) as member and treasurer (2004–2008).

J. Engelbrecht worked tirelessly with the European Federation of National Academies of Sciences and Humanities (ALLEA). He was a member of the General Assembly, head of a Working Group and was elected president for the period 2006–2011. In ALLEA he initiated an analysis of the research strategies of smaller European countries, research cooperation among the Academies, and reflection of the Academies in the European Research Area.

He has also wide experience in implementing and advising on the European science policy by serving on various *ad hoc* committees of the European Union; *inter alia* as a member of the European Science Foundation Governing Council and a member of the European Academies' Scientific Advisory Council (EASAC) board.

J. Engelbrecht has continually taken care of young generation of scientists. His lecture courses at Tallinn University of Technology as well as single lectures and courses in Aachen, Udine, Helsinki, and Budapest, have always been dedicated to contemporary problems of science such as advanced mathematical modelling, biomechanics, nonlinear dynamics and chaos.

He has published more than 200 scientific papers and several books (published by Pitman, Longman, Springer, and Kluwer) and has held visiting appointments in many European universities (Cambridge, Paris 6, Turin, Aachen, etc.). In his numerous essays and articles published on science policy, science philosophy, and the beauty of science (Engelbrecht, 2015), J. Engelbrecht has pointed out the qualities that are valued in science and in a scientist: ethical behaviour, integrity, exactness in words and deed, not stopping halfway.

The beauty of science, the interplay of its complexity, transparency and simplicity, and the secrets of nonlinearity and chaos have drawn J. Engelbrecht's interest to the wide world of knowledge ranging from solid mechanics to the philosophy of science and to the history of the development of scientific ideas. He has received honours from Estonia, Finland, France and Poland. In all his activities he insists on the excellence and integrity of research, as well as on maintaining the links between science and society.

References

Berezovski, A., Engelbrecht, J., Maugin, G.A.: Numerical Simulation of Waves and Fronts in Inhomogeneous Solids. World Scientific, Singapore (2008)

Engelbrecht, J.: Nonlinear Wave Processes of Deformation in Solids. Pitman Monographs and Surveys in Pure and Applied Mathematics, vol. 16. Pitman Advanced Publishing Program (1983)

Engelbrecht, J.: An Introduction to Asymmetric Solitary Waves. Pitman Monographs and Surveys in Pure and Applied Mathematics, vol. 56. Longman Scientific & Technical, Harlow (1996)

Engelbrecht, J.: Nonlinear Wave Dynamics: Complexity and Simplicity. Kluwer Texts in the Mathematical Sciences, vol. 17. Springer, Dordrecht (1997). https://doi.org/10.1007/978-94-015-8891-1

Engelbrecht, J.: Questions About Elastic Waves. Springer, Cham (2015). https://doi.org/10.1007/978-3-319-14791-8

Engelbrecht, J.: Akadeemilised mõtisklused (Academic thoughts). Ilmamaa, Tallinn (2017) (in Estonian)

Janno, J., Engelbrecht, J.: Microstructured Materials: Inverse Problems. Springer, Berlin (2011)

Quak, E., Soomere, T. (eds.): Applied Wave Mathematics. Selected Topics in Solids, Fluids, and Mathematical Methods. Springer, Heidelberg (2009). https://doi.org/10.1007/978-3-642-00585-5

An Attempt to Classify Brilliance

Tarmo Soomere

Not all scientists are the same. Some can be described as great scientists, but a definite threshold is yet to be set. Various sources place the limit between one hundred and one hundred and ten kilograms (net weight). Among this category of great scientists, visual observation reveals a subcategory of formidable or outstanding scientists. They can be reliably distinguished by viewing the full-length side profile. Whilst interpreting the resulting data is a task not entirely devoid of subjectivity, the shape of the body above the belt tends to impart sufficient information.

Then we have the brilliant scientists. There are many subcategories: from a hair-free patch just above the forehead channelling the owner's enlightenment, down to a nearly completely reflective dome indicative of a particularly illuminated individual. Our esteemed Russian colleagues would round out this list with the eminent status of "сложившийся"[1], which each of us will reach sooner or later, but none of us is in a particular hurry to achieve.

This classical taxonomy lacks a suitable category for our esteemed colleague, Member of the Estonian Academy of Sciences Jüri Engelbrecht, whose passport belies his apparent age because there is no way he is anywhere near his formal age! No way to imagine him as a classic outstanding scientist who advances slowly and carefully like a balloon and requires an extra-wide podium for his or her presentations. A former talented young sportsman whose physical fitness still outranks many colleagues half his age, he has firmly precluded this option of life. This point will be well understood by anyone who has tried to keep up with him as he strides up the

[1]This word is used to illustrate eminent scientists but also has a side meaning of a body with crossed hands in a coffin.

T. Soomere (✉)
Department of Cybernetics, School of Science, Tallinn University of Technology, Tallinn, Estonia

Estonian Academy of Sciences, Tallinn, Estonia
e-mail: soomere@cs.ioc.ee

Superga hill during his annual Torino trip, persistently and determinedly ignoring the hilltop shuttle labouring right next to him.

Those two character traits – persistence and determination – have always served him well. He, in turn, has served as a vector of sorts, passing these traits on to everyone in his scientific vicinity, whether in research, science communication or administration. In this sense, he is undeniably a classic modern scientist.

The paradigm of what it means to be a scientist has undergone a radical change over the last several hundreds of years. Leaving aside the Platonic Academy, science used to be a pastime, if not a hobby, of rich individuals even just a couple of centuries ago. Research became a lifestyle barely a hundred years ago, when society decided that the upkeep of such individuals represented an investment rather than an expense.

The paradigm underwent another change a few decades ago, and scientist became something of a diagnosis. How else can people be categorised, who, absent of any external pressure, adhere to the Kantian ideas of pure reason, as our esteemed colleague Jüri Engelbrecht has stubbornly (that is, with determination) done as long as his colleagues can recall? His public speeches and writings alike carry the 'leitmotif' that researchers make a point of going beyond any existing experiences to make more sense of the world and its laws.

Perhaps the only such socially acceptable group, they are capable of not just establishing, but also applying high-level, regulative, directive, simplifying, and differentiating principles, even if those principles do not refer to anything real or substantive, and even if they clash with prevailing thought or social experience. Principles such as dignity, harmony, research integrity and honesty, which any self-respecting scientist adheres to without thinking, are for science, only a slightly adjusted version of the categorical imperative (only act according to maxims that ought to be universal laws).

Diagnosing a scientist is at least as tricky as distinguishing fouls from diving in football, or telling malingerers from the chronically ill in recruitment to the Soviet army. A sure-fire way is to see whether the affliction is infectious. If it is not, doubt persists. In the case of Jüri Engelbrecht, however, there is not a shadow of a doubt: it is highly contagious. This is evidenced by the many scientists working on the first floors of the Institute of Cybernetics who entered or settled in their field of science due to him, by the long list of degree students whom he has successfully supervised, and by the number of students in his lectures. A few of those that he has infected have passed on the contagion far and wide, so that in two separate cases, the problem has turned into a focus of infection: that is, the collaboration has resulted in a national Centre of Excellence in Research. The contagion has crossed borders and spread internationally: to date, a diagnosis has been confirmed in (at least) Italy, Finland, the USA, Latvia, Sweden, France, Poland, and Bulgaria. Our own Presidents have corroborated Jüri Engelbrecht twice with state decorations.

In the case of such a serious and relatively rare condition, categorising individuals is understandably challenging, and tends to require long-term observation. As is frequently the case with complex cases that stray from at least nominally accepted social norms, a lot of information can be deduced by analysing the communication mechanism of the disease. It varies significantly by scientist. In many cases, interaction with a scientist tends to grant fellow citizens long-term immunity to all scientific phenomena. Such scientists are socially entirely harmless. For some scientists, only a few cases of contagion are documented. In those cases, the target group are relatively defenceless youth full of wonder at the world's complexity, searching for their niche in it or depending on this otherwise relatively innocuous scientist directly or indirectly. In this case, the scientific disease is more akin to a replacement activity, and it is easy enough to cure.

But with some scientists, the contagion spreads faster than the Spanish flu. Observing them tends to lead to a surprisingly straightforward result. Those individuals shine. And not just glow like a glow worm or scintillate like a 'will o' the wisp'. Rather, it is as if a naughty little boy, persistent and determined, reflects a ray of sun in your face with a mirror. After all, sometimes, you'd like to leisurely stretch out in the shade – but you can't, because the light shines through no matter how hard you shut your eyes. And so, brilliant scientists, among whom my colleagues and I finally, as a matter of consensus, categorised Jüri Engelbrecht, are the terror lurking in the worst nightmares of paper-pushers engrossed in their daily routines. They simply will not let stupid decisions prevail – and they will act in a Kantian manner as described above.

And for Jüri Engelbrecht, the shine sparkles on, like a Cheshire Cat's smile, long after he has dashed off to meet his next challenge.

A comment written to the 75th birthday of Jüri Engelbrecht. Originally published in Estonian: Sirp 29 (3499), pp. 29–30 (2014). http://www.sirp.ee/s1-artiklid/c21-teadus/2014-07-24-09-42-42/
Translated into English by Mari Arumäe

Contributors

Ahmed Abdalazeez Department of Marine Systems, School of Science, Tallinn University of Technology, Tallinn, Estonia

Elias C. Aifantis Laboratory of Mechanics and Materials, Aristotle University of Thessaloniki, Thessaloniki, Greece

Harm Askes Department of Civil and Structural Engineering, University of Sheffield, Sheffield, UK

Terry Bennett School of Civil, Environmental and Mining Engineering, University of Adelaide, Adelaide, SA, Australia

Arkadi Berezovski Department of Cybernetics, School of Science, Tallinn University of Technology, Tallinn, Estonia

Mihhail Berezovski Embry-Riddle Aeronautical University, Daytona Beach, FL, USA

Manfred Braun Lehrstuhl für Mechanik und Robotik, Universität Duisburg-Essen, Duisburg, Germany

Ivan C. Christov School of Mechanical Engineering, Purdue University, West Lafayette, IN, USA

Petr Denissenko School of Engineering, University of Warwick, Coventry, UK

Ira Didenkulova Department of Marine Systems, School of Science, Tallinn University of Technology, Tallinn, Estonia

Dario De Domenico Department of Engineering, University of Messina, Messina, Italy

Denys Dutykh University Grenoble Alpes, University Savoie Mont Blanc, CNRS, LAMA, Chambéry, France

Stanislav Elektrov Department of Mathematics, Worcester Polytechnic Institute, Worcester, MA, USA

Victor A. Eremeyev Faculty of Civil and Environmental Engineering, Gdańsk University of Technology, Gdańsk, Poland

Southern Federal University, Rostov on Don, Russia

Southern Scientific Center of RAS, Rostov on Don, Russia

Inna M. Gitman Department of Mechanical Engineering, University of Sheffield, Sheffield, United Kingdom

Aditya A. Ghodgaonkar School of Mechanical Engineering, Purdue University, West Lafayette, IN, USA

Lauri Ilison Department of Cybernetics, School of Science, Tallinn University of Technology, Tallinn, Estonia

Karima R. Khusnutdinova Department of Mathematical Sciences, Loughborough University, Loughborough, UK

Andrey Kurkin Nizhny Novgorod State Technical University n.a. R.E. Alekseev, Nizhny Novgorod, Russia

Oxana Kurkina Nizhny Novgorod State Technical University n.a. R.E. Alekseev, Nizhny Novgorod, Russia

Mati Kutser Department of Cybernetics, School of Science, Tallinn University of Technology, Tallinn, Estonia

Martin Lints Department of Cybernetics, School of Science, Tallinn University of Technology, Tallinn, Estonia

Konstantin Lurie Department of Mathematics, Worcester Polytechnic Institute, Worcester, MA, USA

Alena E. Osokina Peter the Great St. Petersburg Polytechnic University (SPbPU), Saint Petersburg, Russia

Efim Pelinovsky Federal Research Center Institute of Applied Physics of the Russian Academy of Sciences (IAP RAS), Nizhny Novgorod, Russia

Tanel Peets Department of Cybernetics, School of Science, Tallinn University of Technology, Tallinn, Estonia

Alexey V. Porubov Institute for Problems in Mechanical Engineering, Saint-Petersburg, Russia

Merle Randrüüt Lehrstuhl für Mechanik und Robotik, Universität Duisburg-Essen, Duisburg, Germany

Tommaso Ruggeri University of Bologna, Bologna, Italy

Domènec Ruiz-Balet Departamento de Matemáticas, Universidad Autónoma de Madrid, Madrid, Spain

Fundación Deusto Bilbao, Basque Country, Spain

Andrus Salupere Department of Cybernetics, School of Science, Tallinn University of Technology, Tallinn, Estonia

Dai Shimizu Department of Mechanical Engineering, Fukui University of Technology, Fukui, Japan

Tarmo Soomere Department of Cybernetics, School of Science, Tallinn University of Technology, Tallinn, Estonia

Estonian Academy of Sciences, Tallinn, Estonia

Yury Stepanyants University of Southern Queensland, Toowoomba, QLD, Australia

Nobumasa Sugimoto Department of Pure and Applied Physics, Kansai University, Osaka, Japan

Tatiana Talipova Nizhny Novgorod State Technical University n.a. R.E. Alekseev, Nizhny Novgorod, Russia

Kert Tamm Department of Cybernetics, School of Science, Tallinn University of Technology, Tallinn, Estonia

Shigeru Taniguchi Natíonal Institute of Technology, Kitakyushu College, Kitakyushu, Japan

Tomas Torsvik Norwegian Polar Institute, Fram Centre, Tromsø, Norway

Geophysical Institute, University of Bergen, Bergen, Norway

Matthew R. Tranter Department of Mathematical Sciences, Loughborough University, Loughborough, UK

Peter Ván Department of Theoretical Physics, Wigner Research Centre for Physics, RMKI, Budapest, Hungary

Mingxiu Xu Department of Applied Mechanics, University of Science and Technology Beijing, Beijing, China

Enrique Zuazua Department of Mathematics, Friedrich-Alexander-Universität Erlangen-Nürnberg, Erlangen, Germany

Fundación Deusto, Bilbao, Basque Country, Spain

Departamento de Matemáticas, Universidad Autónoma de Madrid, Madrid, Spain

Contents

Chapter 1
An Approximate Theory of Linear Waves in an Elastic Layer and Its Relation to Microstructured Solids

Manfred Braun and Merle Randrüüt

Abstract For the dynamics of an infinite elastic layer or plate, an approximate model is introduced. The governing equations are obtained by Lagrange's method with suitable restrictions imposed to the transverse deformation of the layer. It turns out that the resulting equations, in the one-dimensional case, are of the same kind as those introduced by Engelbrecht and Pastrone (Proc. Estonian Acad. Sci. Phys. Math. **52**(1), 12–20 (2003)) for the dynamics of a microstructured solid in the sense of Mindlin. These equations cover a wide range of possible applications, but usually, there is no validation as to what extent they describe the actual behaviour in reality. The approximate layer model is governed by the same system of equations. It may serve as a test that allows to compare the approximation with the known results of the *exact* theory. To this end, the propagation of harmonic waves in the layer is considered. The dispersion curves of the approximate and exact theories are compared for some values of Poisson's ratio. It is shown that the main, acoustical branch fits well for wavelengths above three times the plate thickness. The optical branches of the approximate model deviate from their exact counterparts but exhibit qualitatively the same behaviour.

1.1 Introduction

The linear theory of waves propagating in an elastic layer is usually attributed to Lord Rayleigh and Horace Lamb. Rayleigh's article on waves in an elastic half space (Rayleigh, 1885), the waves that bear his name today, appeared in 1885. Four years later, he addressed the problem of waves in an infinite plate (Rayleigh, 1889). In the same year, only a few months later, Lamb published his paper "On the flexure of an elastic plate" (Lamb, 1889) which, according to its title, seems not to be related to wave propagation. In the appendix, however, he writes that the paper "as originally

M. Braun (✉) · M. Randrüüt
Lehrstuhl für Mechanik und Robotik, Universität Duisburg–Essen, Duisburg, Germany
e-mail: manfred.braun@uni-due.de

A. Berezovski, T. Soomere (eds.), *Applied Wave Mathematics II*, Mathematics of Planet Earth 6, https://doi.org/10.1007/978-3-030-29951-4_1

drafted" contained a discussion of the propagation of waves, a topic that "has been fully treated by Lord Rayleigh in a paper which has in the meantime appeared". Lamb worked further on the problem, and his presentation (Lamb, 1917) really concludes the theory.

It should be mentioned that the companion problem of waves in an infinite cylindrical rod, which turns out to be formally even more complicated, was solved in 1876 by Pochhammer (1876). In both cases, the original research provides only the abstract theory and lacks the elaboration of the details. The dispersion diagrams of the Rayleigh–Lamb and the Pochhammer theories as they are known nowadays, were not presented by the name-giving authors.

A brief look at the dispersion diagrams explains why they were not available in these early papers. Each curve has to be composed of points satisfying a transcendental equation. It needs a lot of numerical skill to solve them, and without a computer, it would be very hard to generate the diagrams.

Starting around 1950, this challenge was accepted by several researchers. Raymond Mindlin wrote a report in 1955 for the U. S. Army Signal Corps Laboratories, which has been edited and republished by Jiashi Yang on the occasion of Mindlin's one hundredth birthday (Mindlin, 1955). There one finds for the first time a complete dispersion diagram of waves in plates (Mindlin, 1960). Mindlin also contributed to the companion problem of waves in cylindrical rods (Onoe et al., 1962) working out a corresponding dispersion diagram. It is quite characteristic that the main textbooks on elastic waves (Achenbach, 1973; Graff, 1975; Miklowitz, 1978) present Mindlin's original dispersion diagrams. Apparently, Mindlin has realised that, despite the availability of the exact solution, there is a need for approximate models of the plate problem. Together with M.A. Medick, he proposed such a model in 1958 (Mindlin and Medick, 1958, 1959).

Mindlin was active also in quite a different field of continuum mechanics. The idea of microstructured materials has, on the first sight, nothing to do with the theory of elastic plates. It was again Mindlin (1964) who presented a linear theory of elastic materials with embedded microcells that interact with each other and with the surrounding matrix material. The physical nature of the microcells is left open, they may represent, for instance, "a molecule of a polymer, a crystallite of a polycrystal or a grain of a granular material". This arbitrariness makes the theory very general, but on the other hand also rather unspecific. Mindlin's paper contains already a discussion about the propagation of waves in microstructured media. It is explained that there are optical branches emerging from cutoff frequencies, very much like in plate waves. While the cutoff frequencies of plate waves describe thickness vibrations, their counterparts in microstructured media pertain to vibrations of the microcells.

Engelbrecht and Pastrone (2003) have taken up Mindlin's idea, reduced it to one space dimension and augmented it with nonlinearities in both the macro and the micro scale. The nonlinearities may interact with the dispersive effects induced by the microstructure and thus pave the way to solitary waves. Although the emphasis is on this combined effect of nonlinearities and dispersion that eventually leads to solitary waves, the purely linear case has also been studied in subsequent papers and theses as, for instance, (Engelbrecht et al., 2005, 2006; Peets et al., 2006; Randrüüt,

2010; Peets, 2011). The dispersion diagrams of waves in such a microstructured material consist of an acoustical and an optical branch.

A special extension of the theory must still be mentioned in the present context. Engelbrecht et al. (2006) consider the possibility of a hierarchical microstructure where every microcell includes again deformable cells at a smaller scale. In this case, there are two optical branches in addition to the main, acoustical branch.

The present chapter develops an approximate dynamic theory of an infinite elastic layer of constant thickness. The layer or plate[1] is considered as a two-dimensional (2D) object in the sense that the variables describing the deformation depend only on the two coordinates that span the middle surface and, of course, on time. The first order model is based on the assumption that the transverse fibres of the plate remain straight and undergo a uniform extension. It turns out that the governing equations of this model if restricted to one dimension, are of exactly the same kind as those obtained by Engelbrecht and Pastrone for a Mindlin-type microstructured elastic solid.

Our second order model allows the transverse fibres to be extended and bent adopting a symmetric parabolic shape. In this case, the governing equations match those of the two-scale model of Engelbrecht et al. (2005, 2006).

It is surprising that the approximate models of an elastic layer can be interpreted as microstructured solids. What is the nature of the embedded microstructure? Its role is taken by the transverse fibres. They are, however, not embedded into the 2D continuum at a smaller scale but live in the third, transverse dimension.

As mentioned above, Mindlin's theory of a microstructured material is rather unspecific, and this holds also for the one-dimensional (1D) adaption by Engelbrecht and Pastrone. If it is applied, for instance, to granular material, how are the model parameters determined by the size, shape, and distribution of the grains? In our plate example, though not really a microstructured material, the parameters are directly expressed in terms of the four parameters of the elastic layer, namely the density, the two elastic moduli, and the thickness.

Another point might be of interest. If we describe a microstructured material by a Mindlin-type model, we do not know, in general, how good this approximation is. For the layer problem, the exact theory of Rayleigh–Lamb is available, which allows us to compare and assess the quality of the approximation.

This chapter is organised as follows. In the subsequent Sect. 1.2, the governing equations of our approximate models are developed for an elastic layer considered as a 2D continuum. Some details of the derivation are found in the Appendix. In Sect. 1.3 the equations are reduced to one spatial dimension and contrasted with Engelbrecht and Pastrone's equations governing a 1D microstructured material. Section 1.4 is devoted to the propagation of harmonic waves in an elastic layer as

[1]In engineering mechanics, a plate is considered as a flat structural element whose deformation especially allows transverse displacements of the middle surface. Since our model restricts the middle surface to inplane deformations, it might be appropriate to use the more neutral term *layer*, in accordance with (Achenbach, 1973).

described by the approximate models. In Sect. 1.5, the dispersion diagrams of the two approximate models are opposed to the lowest *exact* dispersion curves according to the Rayleigh–Lamb theory. Our main results are restated in Sect. 1.6.

1.2 Governing Equations

We consider an infinite elastic layer of thickness $2h$, equipped with Cartesian coordinates (x_1, x_2, z) such that the free surfaces of the layer are represented by $z = \pm h$. The aim is to describe the time dependent deformation approximately by functions depending on $(\boldsymbol{x}, t)$ only, where $\boldsymbol{x} = (x_1, x_2)$ denotes the 2D position vector in the centre plane of the layer. The analysis is restricted to deformations that leave the deformed layer symmetric to the centre plane $z = 0$. Our approach is based on the assumption that the displacement field is either linear or quadratic in the transverse coordinate z.

A displacement field that is *linear* in z has the general form[2]

$$\mathsf{u}(\boldsymbol{x}, z, t) = \left[\begin{array}{c} \boldsymbol{u}(\boldsymbol{x}, t) \\ \hline z\varphi(\boldsymbol{x}, t) \end{array}\right]. \tag{1.1}$$

According to this model, the deformation is governed by the plane vector field $\boldsymbol{u}(\boldsymbol{x}, t)$ representing the inplane displacement of the transverse fibres, and the scalar strain field $\varphi(\boldsymbol{x}, t)$ that describes the stretching of the fibres. The deformation of a transverse fibre in an (x, z)-plane is visualised in Fig. 1.1 (left).

According to the linear model, any cross sectional fibre remains straight as it is only shifted and elongated or compressed. In a second step, we extended the model by including a quadratic term that additionally bends the fibre to a parabola. The corresponding displacement field is taken in the form

$$\mathsf{u}(\boldsymbol{x}, z, t) = \left[\begin{array}{c} \boldsymbol{u}(\boldsymbol{x}, t) + \frac{1}{2}\big(z^2 - j^2\big)\boldsymbol{w}(\boldsymbol{x}, t) \\ \hline z\varphi(\boldsymbol{x}, t) \end{array}\right]. \tag{1.2}$$

[2]Three-dimensional (3D) vectors and tensors are typeset in a sans-serif boldface font, like u, E. They occur only in the derivation of the model equations and are represented by partitioned column vectors and matrices. The usual, italic boldface font is reserved for 2D vectors and tensors like $\boldsymbol{u}$ or $\boldsymbol{E}$ that finally describe the approximate model.

The value of j^2 is chosen such that on any cross section of the layer, the *average* displacement is

$$\frac{1}{2h}\int_{-h}^{+h} \mathbf{u}(\boldsymbol{x}, z, t)\,\mathrm{d}z = \left[\frac{\boldsymbol{u}(\boldsymbol{x}, t)}{0}\right]. \tag{1.3}$$

This condition is fulfilled if one takes

$$j^2 = \frac{1}{2h}\int_{-h}^{+h} z^2\,\mathrm{d}z = \frac{h^2}{3}. \tag{1.4}$$

In principle, j^2 can be omitted in the displacement *ansatz*, in which case $\boldsymbol{u}(\boldsymbol{x}, t)$ would represent the displacement at the centre plane $z = 0$. It turns out, however, that the subsequent analysis is simplified by taking the *average* horizontal displacement as the main variable $\boldsymbol{u}$ rather than the displacement in the middle of the layer. Figure 1.1 provides a qualitative visualisation of the quadratic model, again restricted to an (x, z)-plane. Since the analysis is based on the linear theory of elasticity, only small deformations are allowed. In the figure, they are magnified to show the principal behaviour of the deformation.

Even higher order displacements could be considered. Symmetric deformations of the layer are obtained by taking even functions in z for the inplane deformation and odd functions for the lateral deformation. However, it makes little sense to extend the order too far. Instead of a more involved approximate model, one could immediately take recourse to Rayleigh and Lamb's *exact* solution. An approximation is justified only if it is remarkably simpler.

Even though we have two different models and want to compare them, we establish the governing equations only for the second order model (1.2). The equations for the

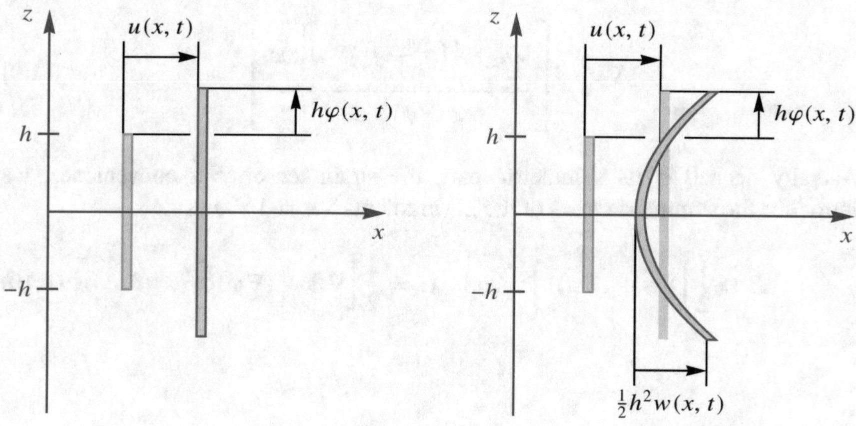

Fig. 1.1 Deformation of a transverse fibre using first order (left) and second order (right) models.

first order model can always be recovered by taking $\boldsymbol{w} \equiv 0$. The equations of motion are derived via Lagrange's method.

The derivative of (1.2) with respect to time t yields the velocity vector

$$\mathbf{v}(\boldsymbol{x}, z, t) = \frac{\partial \mathbf{u}}{\partial t} = \left[\begin{array}{c} \dot{\boldsymbol{u}} + \frac{1}{2}(z^2 - j^2)\dot{\boldsymbol{w}} \\ \hline z\dot{\varphi} \end{array}\right]. \tag{1.5}$$

Here dots indicate differentiation with respect to time.

The kinetic energy per unit area of the layer is

$$\mathcal{K} = \frac{1}{2}\int_{-h}^{+h} \rho\, \mathbf{v} \cdot \mathbf{v}\, \mathrm{d}z. \tag{1.6}$$

We assume the mass density ρ to be constant over the thickness. In evaluating the expression above one encounters the integral

$$l^4 \equiv \frac{1}{2h}\int_{-h}^{+h} \frac{1}{4}\left(z^2 - j^2\right)^2 \mathrm{d}z = \frac{1}{20}h^4 - \frac{1}{6}j^2h^2 + \frac{1}{4}j^4 = \frac{h^4}{45}. \tag{1.7}$$

Using the two length parameters j and l the kinetic energy per unit area of the layer is written as

$$\mathcal{K} = \rho h\left(\dot{\boldsymbol{u}} \cdot \dot{\boldsymbol{u}} + j^2\dot{\varphi}^2 + l^4\dot{\boldsymbol{w}} \cdot \dot{\boldsymbol{w}}\right). \tag{1.8}$$

It contains only the squares of the basic velocities $\dot{\boldsymbol{u}}$, $\dot{\varphi}$, and $\dot{\boldsymbol{w}}$, but no mixed products. This is due to the chosen *ansatz* (1.2) with $\boldsymbol{u}$ denoting the *average* displacement of a cross section.

Turning to the spatial derivatives we start with the displacement gradient[3]

$$\nabla\mathbf{u} = \left[\begin{array}{c|c} \nabla\boldsymbol{u} + \frac{1}{2}(z^2 - j^2)\nabla\boldsymbol{w} & z\boldsymbol{w} \\ \hline z\,(\nabla\varphi)^{\mathsf{T}} & \varphi \end{array}\right]. \tag{1.9}$$

Actually needed is its symmetric part, the strain tensor. For convenience, we introduce the symmetric parts of the 2D gradients $\nabla\boldsymbol{u}$ and $\nabla\boldsymbol{w}$ as

$$\boldsymbol{E} = \frac{1}{2}\left[\nabla\boldsymbol{u} + (\nabla\boldsymbol{u})^{\mathsf{T}}\right] \quad \text{and} \quad \boldsymbol{G} = \frac{1}{2}\left[\nabla\boldsymbol{w} + (\nabla\boldsymbol{w})^{\mathsf{T}}\right], \tag{1.10}$$

[3]We use the same gradient symbol ∇ for 2D and 3D vector and tensor fields.

respectively. Then the 3D strain tensor becomes

$$\mathsf{E} = \left[\begin{array}{c|c} \boldsymbol{E} + \frac{1}{2}\left(z^2 - j^2\right)\boldsymbol{G} & \frac{z}{2}\left(\nabla\varphi + \boldsymbol{w}\right) \\ \hline \frac{z}{2}\left(\nabla\varphi + \boldsymbol{w}\right)^{\mathsf{T}} & \varphi \end{array}\right]. \tag{1.11}$$

Since the analysis is restricted to a linear and isotropic elastic medium, only two strain invariants are needed, namely the trace or first moment

$$\operatorname{tr}\mathsf{E} = \mathsf{E}\cdot\mathsf{I} = \operatorname{tr}\boldsymbol{E} + \frac{1}{2}\left(z^2 - j^2\right)\operatorname{tr}\boldsymbol{G} + \varphi \tag{1.12}$$

and the second moment

$$\operatorname{tr}\mathsf{E}^2 = \mathsf{E}\cdot\mathsf{E} = \boldsymbol{E}\cdot\boldsymbol{E} + \left(z^2 - j^2\right)\boldsymbol{E}\cdot\boldsymbol{G} + \frac{1}{4}\left(z^2 - j^2\right)^2\boldsymbol{G}\cdot\boldsymbol{G} + \\ + \frac{z^2}{2}\left(\nabla\varphi + \boldsymbol{w}\right)\cdot\left(\nabla\varphi + \boldsymbol{w}\right) + \varphi^2. \tag{1.13}$$

These two invariants enter the strain energy density

$$\mathcal{W} = \frac{1}{2}\left[\lambda\left(\operatorname{tr}\mathsf{E}\right)^2 + 2\mu\operatorname{tr}\mathsf{E}^2\right] \tag{1.14}$$

of a linear and isotropic elastic material, where λ and μ denote Lamé's elastic parameters, which are assumed constant throughout the whole layer.

By integrating over the thickness one encounters again integral (1.7). Explicitly the strain energy per unit area becomes

$$W = \frac{1}{2}\int_{-h}^{+h}\left[\lambda\left(\operatorname{tr}\mathsf{E}\right)^2 + 2\mu\operatorname{tr}\mathsf{E}^2\right]\mathrm{d}z = \lambda h\left[\left(\operatorname{tr}\boldsymbol{E} + \varphi\right)^2 + l^4\left(\operatorname{tr}\boldsymbol{G}\right)^2\right] + \\ + \mu h\left[2\left(\boldsymbol{E}\cdot\boldsymbol{E} + \varphi^2\right) + j^2\left(\nabla\varphi + \boldsymbol{w}\right)\cdot\left(\nabla\varphi + \boldsymbol{w}\right) + 2l^4\boldsymbol{G}\cdot\boldsymbol{G}\right]. \tag{1.15}$$

Combining this with the kinetic energy per unit area, expression (1.8), gives the Lagrangian $\mathcal{L} = \mathcal{K} - \mathcal{W}$ as

$$\mathcal{L} = \rho h\left(\dot{\boldsymbol{u}}\cdot\dot{\boldsymbol{u}} + j^2\dot{\varphi}^2 + l^4\dot{\boldsymbol{w}}\cdot\dot{\boldsymbol{w}}\right) - \lambda h\left[\left(\operatorname{tr}\boldsymbol{E} + \varphi\right)^2 + l^4\left(\operatorname{tr}\boldsymbol{G}\right)^2\right] - \\ - \mu h\left[2\left(\boldsymbol{E}\cdot\boldsymbol{E} + \varphi^2\right) + j^2\left(\nabla\varphi + \boldsymbol{w}\right)\cdot\left(\nabla\varphi + \boldsymbol{w}\right) + 2l^4\boldsymbol{G}\cdot\boldsymbol{G}\right], \tag{1.16}$$

which, in principle, determines the equations of motion. It should be noted that by omitting the function $\boldsymbol{w}(\boldsymbol{x}, t)$ and its spatial and time derivatives, $\boldsymbol{G}$ and $\dot{\boldsymbol{w}}$, the Lagrangian of the linear model can be obtained.

Establishing Lagrange's equation is straightforward. The details of the derivation are shifted to the Appendix. The first order model is governed by the system of equations

$$\begin{aligned} \rho \ddot{\boldsymbol{u}} &= (\lambda + \mu)\nabla \operatorname{div} \boldsymbol{u} + \mu \Delta \boldsymbol{u} + \lambda \nabla \varphi\,, \\ \rho j^2 \ddot{\varphi} &= \mu j^2 \Delta \varphi - (\lambda + 2\mu)\varphi - \lambda \operatorname{div} \boldsymbol{u}\,. \end{aligned} \tag{1.17}$$

The equations are coupled through their last terms, $\lambda \nabla \varphi$ and $-\lambda \operatorname{div} \boldsymbol{u}$.

The full system of the second order model, as derived in the Appendix, is

$$\begin{aligned} \rho \ddot{\boldsymbol{u}} &= (\lambda + \mu)\nabla \operatorname{div} \boldsymbol{u} + \mu \Delta \boldsymbol{u} + \lambda \nabla \varphi\,, \\ \rho j^2 \ddot{\varphi} &= \mu j^2 \Delta \varphi - (\lambda + 2\mu)\varphi - \lambda \operatorname{div} \boldsymbol{u} + \mu j^2 \operatorname{div} \boldsymbol{w}\,, \\ \rho l^4 \ddot{\boldsymbol{w}} &= (\lambda + \mu) l^4 \nabla \operatorname{div} \boldsymbol{w} + \mu l^4 \Delta \boldsymbol{w} - \mu j^2 \left(\boldsymbol{w} + \nabla \varphi\right)\,. \end{aligned} \tag{1.18}$$

The first equation is identical to that of the first order model. Also, the coupling between the first and second equations is the same as before. The second and third equations are coupled through the last terms, $\mu j^2 \operatorname{div} \boldsymbol{w}$ and $-\mu j^2 \nabla \varphi$.

1.3 Relation to Mindlin-Type Models of Microstructured Materials

As mentioned in the Introduction there is some correspondence between the presented approximate models of an elastic layer and Mindlin's theory of microstructured materials. To see this in detail we restrict the deformation of the layer to plane strain in the (x_1, z)-plane. Then the x_2-coordinate becomes obsolete and the vectorial quantities $\boldsymbol{u}$ and $\boldsymbol{w}$ are reduced to scalar functions. Thus we have

$$\boldsymbol{u} = u(x, t)\boldsymbol{e}_1\,, \quad \varphi = \varphi(x, t)\,, \quad \text{and} \quad \boldsymbol{w} = w(x, t)\boldsymbol{e}_1\,, \tag{1.19}$$

with $x = x_1$ denoting the single remaining spatial coordinate. Also, the differential operators ∇ and div simplify to partial derivatives $\partial/\partial x$.

The governing equations of the first order model (1.17) reduce to

$$\begin{aligned} \rho \ddot{u} &= (\lambda + 2\mu) u'' + \lambda \varphi'\,, \\ \rho j^2 \ddot{\varphi} &= \mu j^2 \varphi'' - \lambda u' - (\lambda + 2\mu)\varphi\,, \end{aligned} \tag{1.20}$$

where primes indicate differentiation with respect to x. These equations have the same formal appearance as those of a 1D microstructured material in the sense of Mindlin, although the layer does not contain any microstructure.

The microcells in Mindlin's model of a microstructured material undergo homogeneous deformations (Mindlin, 1964) as do the transverse fibres in the first order model of the elastic layer. Engelbrecht and Pastrone's 1D version, in the linear case, is characterised by the governing equations (Engelbrecht et al., 2005, (11))

$$\begin{aligned} \rho u_{tt} &= \alpha u_{xx} + A\varphi_x\,, \\ I\varphi_{tt} &= C\varphi_{xx} - Au_x - B\varphi\,, \end{aligned} \tag{1.21}$$

where subscripts indicate partial differentiation. They have the same mathematical structure as those of the 1D layer model described by (1.20). This means that the analysis of the linear Mindlin–Engelbrecht–Pastrone (MEP) model that has been studied in subsequent papers and theses (Engelbrecht et al., 2005, 2006; Peets et al., 2006; Randrüüt, 2010; Peets, 2011), could be applied directly to the layer problem.

While the MEP model (1.21) has six independent parameters, the layer model (1.20) has only four. The density ρ appears in either model with the same meaning. Also, the radius of inertia, $j = h/\sqrt{3}$, is just another way to describe the moment of inertia, $I = \rho j^2$. Altogether, the parameters of the MEP model can be expressed as

$$I = \rho j^2\,, \quad \alpha = \lambda + 2\mu\,, \quad A = \lambda, \quad B = \lambda + 2\mu\,, \quad C = \mu j^2 \tag{1.22}$$

in terms of the parameters of the first order layer model.

Dividing the governing Eq. (1.20) by density ρ reduces the number of independent parameters by 1. Instead of the two elastic constants, one may introduce the speeds of longitudinal and transverse waves in the unbounded elastic medium by

$$c_0^2 = \frac{\lambda + 2\mu}{\rho} \quad \text{and} \quad c_1^2 = \frac{\mu}{\rho}. \tag{1.23}$$

Incidentally, velocities c_0 and c_1 were introduced also by Engelbrecht et al. (2005) for the MEP model. When applied to the elastic layer these velocities turn out to be the two velocities (1.23) of waves in the isotropic elastic medium. Using c_0 and c_1 as parameters the system of Eq. (1.20) assumes the form

$$\begin{aligned} \ddot{u} &= c_0^2 u'' + \left(c_0^2 - 2c_1^2\right)\varphi', \\ j^2\ddot{\varphi} &= j^2 c_1^2 \varphi'' - c_0^2\varphi - \left(c_0^2 - 2c_1^2\right)u'. \end{aligned} \tag{1.24}$$

The half thickness h of the layer as an additional parameter is included in the radius of inertia j. The propagation of waves according to this system of equations will be analysed in the next section.

Turning now to the second order model described by (1.18), the reduction to plane strain in the (x_1, z)-plane simplifies the system to

$$\begin{aligned} \rho\ddot{u} &= (\lambda + 2\mu)\, u'' + \lambda\varphi' , \\ \rho j^2\ddot{\varphi} &= \mu j^2\varphi'' - \lambda u' - (\lambda + 2\mu)\,\varphi + \mu j^2 w' , \\ \rho l^4\ddot{w} &= (\lambda + 2\mu)\, l^4 w'' - \mu j^2\varphi' - \mu j^2 w . \end{aligned} \tag{1.25}$$

Surprisingly there is again a corresponding system of equations describing a microstructured medium. Engelbrecht et al. (2006) have extended their model and considered a two-scale model, in which every deformable cell of the microstructure includes again deformable cells at a smaller scale. In the original notation, the governing equations are (Engelbrecht et al., 2006, (3.35)–(3.37))

$$\begin{aligned} \rho u_{tt} &= \alpha u_{xx} + A_1\varphi_x , \\ I_1\varphi_{tt} &= C_1\varphi_{xx} - A_1 u_x - B_1\varphi + A_2\psi_x , \\ I_2\psi_{tt} &= C_2\psi_{xx} - A_2\varphi_x - B_2\psi . \end{aligned} \tag{1.26}$$

It is easily seen that our second order model (1.25) is of the same form. The correspondence table of the parameters replacing (1.22) is now enlarged to

$$\begin{aligned} &I_1 = \rho j^2 , && A_1 = \lambda , && B_1 = \lambda + 2\mu , && C_1 = \mu j^2 , \\ &I_2 = \rho l^4 , \quad \alpha = \lambda + 2\mu , && A_2 = \mu j^2 , && B_2 = \mu j^2 , && C_2 = (\lambda + 2\mu) l^4 . \end{aligned} \tag{1.27}$$

The number of independent parameters has increased to 10 in the two-scale model (1.26), while the dynamics of the elastic layer keeps its four parameters even in more refined models. Therefore, several parameters of the two-scale model when applied to the layer problem take the same or related values.

As before the number of independent constants can still be reduced by 1, if the equations are divided by density ρ. This leads to the system

$$\begin{aligned} \ddot{u} &= c_0^2 + \left(c_0^2 - 2c_1^2\right)\varphi' , \\ j^2\ddot{\varphi} &= c_1^2 j^2\varphi'' - c_0^2\varphi - \left(c_0^2 - 2c_1^2\right)u' + c_1^2 j^2 w' , \\ l^4\ddot{w} &= c_0^2 l^4 w'' - c_1^2 j^2 w - c_1^2 j^2\varphi' , \end{aligned} \tag{1.28}$$

generalising (1.24). The refined model of the elastic layer has again no direct relation to microstructure. Especially, there is no equivalent to the cell-in-cell concept that led to the two-scale model (1.26). Nevertheless, the governing equations of the layer problem have the same appearance as those of the two-scale microstructured material. What the models have in common though is the order of scales of the three equations. If the first is assumed $O(h^0)$, the second is $O(h^2)$ and the third $O(h^4)$, recalling that j and l are fixed multiples of the half thickness h.

1.4 Dispersion Analysis

The propagation of harmonic waves in microstructured materials described by both the one-scale and the two-scale model has been studied in (Engelbrecht et al., 2005) and subsequent papers. In principle, the analysis could be taken over and applied to the layer problem by simply inserting the relevant parameters according to the correspondence tables (1.22) and (1.27). We prefer a direct approach for the following reason. Our ultimate goal is to compare the results of the approximate layer models with the exact solutions. Therefore, we should use the normalisation of the equations prevalent in the literature on Rayleigh–Lamb waves, which doesn't coincide with the one used for microstructured materials.

Solutions of the systems (1.24) and (1.28) that represent harmonic waves are obtained by the *ansatz*

$$\begin{bmatrix} u(x,t) \\ \varphi(x,t) \\ w(x,t) \end{bmatrix} = \Re\left\{ \begin{bmatrix} \hat{u} \\ \hat{\varphi} \\ \hat{w} \end{bmatrix} \mathrm{e}^{\mathrm{i}(kx-\omega t)} \right\}. \tag{1.29}$$

Traditionally, the frequency ω is assumed to be real, while the wavenumber k is allowed to be complex, in general. The amplitudes $\hat{u}$, $\hat{\varphi}$, and $\hat{w}$ must be taken complex to allow different phasing of the variables.

Inserting the *ansatz* for u and φ into the governing Eq. (1.24) of the first order layer model leads to the algebraic eigenvalue problem

$$\begin{bmatrix} c_0^2k^2 - \omega^2 & -\mathrm{i}k(c_0^2 - 2c_1^2) \\ \mathrm{i}k(c_0^2 - 2c_1^2) & c_0^2 + j^2(c_1^2k^2 - \omega^2) \end{bmatrix} \begin{bmatrix} \hat{u} \\ \hat{\varphi} \end{bmatrix} = 0. \tag{1.30}$$

Its characteristic equation represents the dispersion law and can be written in the form

$$j^2(\omega^2 - c_0^2k^2)(\omega^2 - c_1^2k^2) = c_0^2\omega^2 - 4c_1^2(c_0^2 - c_1^2)k^2. \tag{1.31}$$

It coincides with the dispersion law for waves in the single-scale microstructured material (Randrüüt, 2010, (3.50)), if the parameters are adjusted to the layer problem according to the correspondence table (1.22). The qualitative appearance of the dispersion diagrams as presented in (Engelbrecht et al., 2005, 2006; Randrüüt, 2010) depends on some velocity introduced as c_R in (Engelbrecht et al., 2005, 2006) and renamed $\bar{c}$ in (Randrüüt, 2010). It is defined in terms of the parameters ρ, α, A, and B of the MEP model and describes the slope of the acoustical branch at the origin. In the case of the layer problem, the correspondence table (1.22) yields

$$\bar{c} = c_1\sqrt{\frac{2}{1-\nu}} > c_1. \tag{1.32}$$

Therefore, only the type $\bar{c} > c_1$ of the dispersion diagrams, like (Engelbrecht et al., 2005, Fig. 2) or (Randrüüt, 2010, Fig. 18), appears in the layer problem.

In the dispersion diagrams mentioned above, the frequency is normalised by a certain time constant p, while for normalising the wavenumber the reference length pc_0 is employed. The layer problem, however, has its own inherent length h, the half thickness of the layer, and it is quite natural to use this for defining a dimensionless wavenumber.

A corresponding time constant for a dimensionless frequency can then be generated by either of the velocities c_0 and c_1. In the exact theory of plate waves, see (Mindlin, 1955, 1960; Achenbach, 1973; Graff, 1975; Miklowitz, 1978; Rose, 2014) the shear wave velocity c_1 is preferred as the basic velocity. The reason is that a general theory of plate waves includes also the SH waves, which are not considered here, and which are governed solely by c_1, while c_0 doesn't appear. In this sense, c_1 is more universal and, therefore, used to define the dimensionless frequency. Conforming with the plate theory we introduce the dimensionless wavenumber and frequency as

$$\xi = kh \quad \text{and} \quad \eta = \frac{\omega h}{c_1}. \tag{1.33}$$

When compared with the corresponding dispersion diagrams in (Engelbrecht et al., 2005; Randrüüt, 2010) our figures are distorted due to the different normalisation.

The eigenvalue problem (1.30) can be reformulated in terms of the dimensionless quantities (1.33). It is reasonable to take the amplitudes as $\hat{u}$ and $h\hat{\varphi}$ having the physical dimension of length. Since $j^2 = h^2/3$, there remains only one dimensionless material parameter in the matrix, namely the ratio

$$\kappa^2 = \frac{c_0^2}{c_1^2} = \frac{2(1-\nu)}{1-2\nu}. \tag{1.34}$$

As Poisson's ratio is restricted by $0 \leq \nu < 1/2$, this parameter varies in the range $2 \leq \kappa^2 < \infty$. The normalised version of the eigenvalue problem (1.30) is

$$\begin{bmatrix} \kappa^2\xi^2 - \eta^2 & -\mathrm{i}(\kappa^2 - 2)\xi \\ \mathrm{i}(\kappa^2 - 2)\xi & \kappa^2 + \frac{1}{3}(\xi^2 - \eta^2) \end{bmatrix} \begin{bmatrix} \hat{u} \\ h\hat{\varphi} \end{bmatrix} = 0 \tag{1.35}$$

and leads to the normalised dispersion relation

$$\left(\eta^2 - \kappa^2\xi^2\right)\left(\eta^2 - \xi^2\right) = 3\left[\kappa^2\eta^2 - 4\left(\kappa^2 - 1\right)\xi^2\right]. \tag{1.36}$$

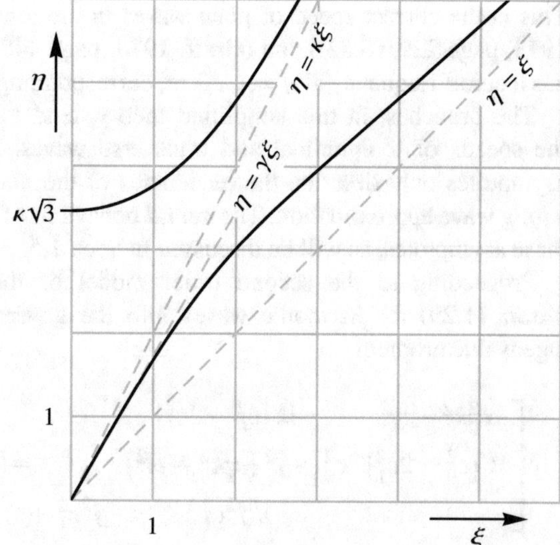

Fig. 1.2 Dispersion diagram of the first order model for $\nu = 1/3$ (real part only).

The factor 3 on the right-hand side originates from the ratio $h^2/j^2 = 3$ according to (1.4). The dispersion diagram is shown in Fig. 1.2 for the Poisson's ratio $\nu = 1/3$ or, equivalently, $\kappa = 2$.

The dispersion diagram consists of two branches,[4] which is typical for a system of two variables each of which is equipped with inertia. The acoustical branch starts with the slope

$$\gamma = \frac{2}{\kappa}\sqrt{\kappa^2 - 1} = \sqrt{\frac{2}{1-\nu}}, \tag{1.37}$$

that corresponds to the speed

$$\bar{c} = \gamma c_1 = \sqrt{\frac{2\mu}{\rho(1-\nu)}} = \sqrt{\frac{E}{\rho(1-\nu^2)}}. \tag{1.38}$$

[4]By analogy with waves in lattices, the two branches are sometimes called *acoustical* and *optical* see Brillouin (1953, page 55). In lattices, there is usually a frequency band gap between the two branches. The nomenclature is motivated by the fact that, in crystals, the frequencies of the lower branch are in the range of acoustical signals, while those of the upper branch reach the magnitude of infrared frequencies. In the case of wave guides, such as a layer, there is usually no upper limit of the frequency of the lower branch. Also, the cutoff frequencies of the higher branches may be in a moderate range. Therefore, the denotations *acoustical* and *optical* might not be appropriate. Nevertheless, we follow the accepted usage and call those branches "optical" that have a nonzero cutoff frequency.

This is the correct speed of plate waves in the long wave limit, see (Achenbach, 1973, page 229, (6.82)) and (Graff, 1975, page 448, (8.1.95)). The optical branch has a cutoff frequency at $\eta = \kappa\sqrt{3}$ or, correspondingly, $\omega = c_0\sqrt{3}/h$.

The branches, in this simplified theory, tend to asymptotes corresponding to the speeds of longitudinal and transverse waves. It should be noted that these asymptotes only describe the tendencies of the underlying model that represents a long wave approximation. The *actual* behaviour of short waves is not reflected by these asymptotes, as will be discussed in Sect. 1.5.

Proceeding to the second order model of the elastic layer, inserting the *ansatz* (1.29) for harmonic waves into the governing system (1.28) yields the eigenvalue problem

$$\begin{bmatrix} c_0^2k^2 - \omega^2 & -\mathrm{i}k\left(c_0^2 - 2c_1^2\right) & \\ \mathrm{i}k\left(c_0^2 - 2c_1^2\right) & c_0^2 + j^2\left(c_1^2k^2 - \omega^2\right) & -\mathrm{i}kj^2c_1^2 \\ & \mathrm{i}kj^2c_1^2 & j^2c_1^2 + l^4\left(c_0^2k^2 - \omega^2\right) \end{bmatrix} \begin{bmatrix} \hat{u} \\ \hat{\varphi} \\ \hat{w} \end{bmatrix} = 0. \tag{1.39}$$

As in the case of the first order model the eigenvalue problem can be reformulated using the dimensionless wavenumber ξ and frequency η according to (1.33). The two characteristic lengths j and l are fixed multiples of the half thickness h. The normalised eigenvalue problem assumes the form

$$\begin{bmatrix} \kappa^2\xi^2 - \eta^2 & -\mathrm{i}\left(\kappa^2 - 2\right)\xi & 0 \\ \mathrm{i}\left(\kappa^2 - 2\right)\xi & \kappa^2 + \frac{1}{3}\left(\xi^2 - \eta^2\right) & -\frac{1}{3}\mathrm{i}\xi \\ 0 & \frac{1}{3}\mathrm{i}\xi & \frac{1}{3} + \frac{1}{45}\left(\kappa^2\xi^2 - \eta^2\right) \end{bmatrix} \begin{bmatrix} \hat{u} \\ h\hat{\varphi} \\ h^2\hat{w} \end{bmatrix} = 0. \tag{1.40}$$

The three amplitudes in this system have the physical dimension of length. They could be taken also as $\hat{u}/h$, $\hat{\varphi}$, and $h\hat{w}$, in which case all of them are dimensionless. It is important that they are of the same dimension in order to keep the matrix Hermitian and without the occurrence of the length h.

Nontrivial solutions of the homogeneous system (1.40) exist if the determinant of its matrix vanishes. This condition delivers the dispersion relation

$$\begin{aligned} &\kappa^4\xi^6 - \kappa^2(\kappa^2 + 2)\xi^4\eta^2 + (2\kappa^2 + 1)\xi^2\eta^4 - \eta^6 + \\ &+ 3\left[4\kappa^2(\kappa^2 - 1)\xi^4 - (\kappa^4 + 9\kappa^2 - 4)\xi^2\eta^2 + (\kappa^2 + 5)\eta^4\right] + \\ &+ 45\left[4(\kappa^2 - 1)\xi^2 - \kappa^2\eta^2\right] = 0. \end{aligned} \tag{1.41}$$

The factors in front of the square brackets can be recognised as the ratios $h^2/j^2 = 3$ and $h^4/l^4 = 45$ as defined in (1.4) and (1.7), respectively. The dimensionless wavenumber ξ and frequency η, as well as the material parameter κ, occur only

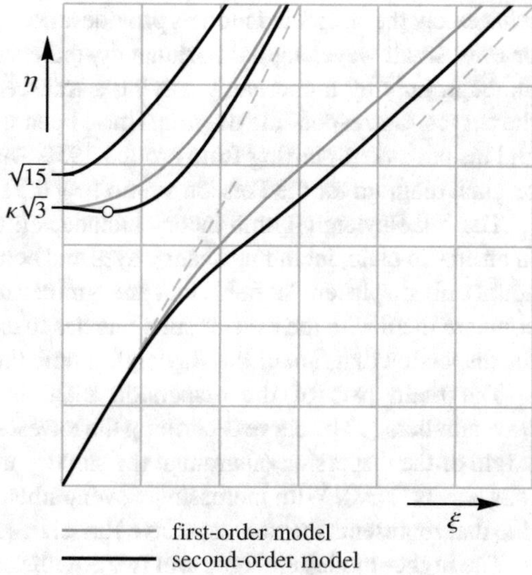

Fig. 1.3 Dispersion diagrams for $\nu = 1/3$ (real part only).

squared, a property that is shared by the characteristic equation of the first order model (1.36).

The dispersion equation is cubic in both the squared wavenumber ξ^2 and frequency η^2. If only the real part of the dispersion diagram is of interest, for each value of the wavenumber ξ, a cubic equation in η^2 whose real coefficients depend on ξ^2 must be solved. The dispersion diagram is shown in Fig. 1.3, again for the Poisson's ratio $\nu = 1/3$ or, equivalently, $\kappa = 2$. For comparison, the curves of the first order model are depicted in grey.

The diagram now has three branches. The acoustical branch starts nearly straight indicating nondispersive behaviour, as in the first order model. Then it turns more sharply away towards the asymptote $\eta = \xi$. The first optical branch starts at the same cutoff frequency as the single optical branch of the first order model. For the chosen parameter $\nu = 1/3$, it starts horizontally but turns downward until it reaches a minimum. Only then does the curve turn up again, tending towards the asymptote $\eta = \kappa\xi$. The second optical branch starts at the fixed cutoff frequency $\eta = \sqrt{15}$ also in horizontal direction but turns upward immediately. For $\kappa = \sqrt{5}$, which corresponds to the Poisson's ratio $\nu = 3/8$, the two cutoff frequencies coincide and the corresponding optical branches start at finite angles. This is shown later (Fig. 1.8) and compared with the exact dispersion diagram.

1.5 Comparison with the Exact Rayleigh–Lamb Theory

The acoustical branch of the dispersion diagram predicts, in the long wave limit $\xi \to 0$, a wave propagating at the velocity $\bar{c}$, which is the correct result for a wave whose wavelength is large compared with the thickness of the layer. It is of interest to what extent the simplified models provide an acceptable approximation for moderate or even small wavelengths. Fortunately, the exact dispersion diagram is available, although only by numerically solving a transcendental equation for each point of the curves. Corresponding diagrams have been worked out and analysed by Mindlin and his coworkers starting from around 1950. Many textbooks reproduce Mindlin's original diagram for the Poisson's ratio $\nu = 0.31$.

The full Rayleigh–Lamb theory includes all types of waves that may propagate in an elastic plate, including shear waves and bending waves. Our simplified models admit only displacement fields that are symmetric to the centre plane of the layer. To compare them with the exact results one has to extract these symmetric modes from the dispersion diagram of the Rayleigh–Lamb theory.

The main part of the dispersion diagram consists of the curves for *real* wavenumbers ξ. The curve describing the lowest symmetric mode starts out from the origin of the dispersion diagram at the slope γ that corresponds to the velocity $\bar{c}$ of long waves (1.38). With increasing wavenumber, the branch approaches the straight line that represents the nondispersive Rayleigh surface waves.

The higher modes emerge from two equidistant sequences of cutoff frequencies, $\eta = m\kappa\pi/2$, $m = 1, 3, 5, \ldots$ and $\eta = n\pi$, $n = 1, 2, 3, \ldots$. Since at the cutoff frequencies the wavenumber vanishes, the corresponding waves do not propagate, rather they represent synchronous vibrations of the whole plate. Vibrations in thickness are related to the speed of longitudinal waves and occur at the frequencies of the m-sequence, while symmetric shear vibrations are governed by the shear wave speed with frequencies of the n-sequence.

Depending on the velocity ratio κ or, correspondingly, on Poisson's ratio ν, the smallest cutoff frequency is

$$\eta = \begin{cases} \kappa\frac{\pi}{2,} & \text{if} \quad \sqrt{2} \le \kappa \le 2 \quad \text{or} \quad 0 \le \nu \le \frac{1}{3}\,, \\ \pi & \text{if} \quad 2 \le \kappa < \infty \quad \text{or} \quad \frac{1}{3} \le \nu < \frac{1}{2}\,. \end{cases} \tag{1.42}$$

In general, all optical branches start with a horizontal tangent. In exceptional cases, the cutoff frequencies of the two sequences may coincide. The two branches then emerge from this "double" cutoff frequency in different, nonhorizontal directions. This happens, if the speed ratio κ assumes an even integer value. Figure 1.4 shows

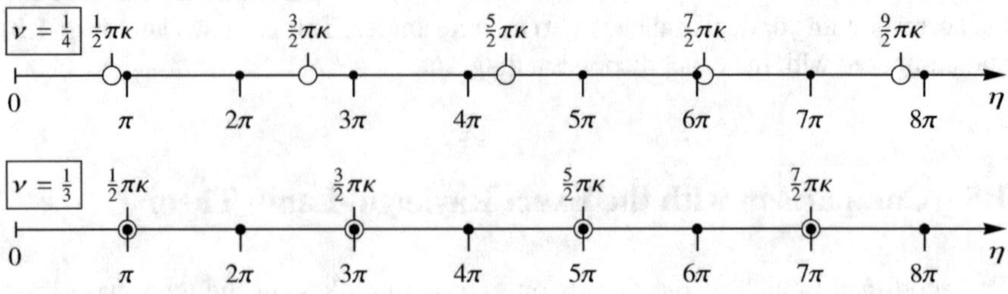

Fig. 1.4 Cutoff frequencies for $\nu = 1/4$ ($\kappa = \sqrt{3}$) and $\nu = 1/3$ ($\kappa = 2$).

the two sequences of cutoff frequencies first for $\nu = 1/4$, which corresponds to the irrational speed ratio $\kappa = \sqrt{3}$ and leads, therefore, to disjoint sequences. For $\nu = 1/4$, however, the speed ratio is $\kappa = 2$, which means that every cutoff frequency of the longitudinal sequence coincides with every second cutoff frequency of the transverse sequence.

With increasing wavenumber all optical branches eventually approach the line $\eta = \xi$ that represents shear waves. The asymptotic behaviour, however, becomes manifest only at wavelengths much below the thickness of the plate. Therefore, it is not recognisable in the long wave section of the diagrams as presented here.

Our approximate models allow either two or three branches in the dispersion diagram. To compare them with the exact results, we shall select just the lowest branches from the two infinite sequences provided by the Rayleigh–Lamb theory.

To fully understand the behaviour of the dispersion curves it is advisable to include their continuation to complex values. In the standard representation of a dispersion relation, the frequency is assumed real and nonnegative, $\eta \geq 0$, while the wavenumber is allowed to be complex with nonnegative real and imaginary parts, i.e., $\Re\{\xi\} \geq 0$ and $\Im\{\xi\} \geq 0$. In the exact dispersion diagram, the acoustical branch of the symmetric modes remains purely real. The first two optical branches are connected by an imaginary arc that joins the two cutoff frequencies. Purely imaginary wavenumbers describe nonpropagating evanescent waves. The first of the higher branches, although starting horizontally, bends down to reach a minimum before turning upward again. From this minimum, a complex branch emerges and leads down to the floor $\eta = 0$. It represents a propagating evanescent wave.

Figure 1.5 contrasts the dispersion diagram of our first order model with the corresponding part of the exact dispersion diagram. As to the acoustical branch, there is a perfect agreement at least up to $\xi \approx 1$, which corresponds to wavelengths bigger than $2\pi h$, i.e., three times the thickness of the layer. For higher wavenumbers, the approximate model does not follow the sharp kink of the exact curve. Its inflection point is located at a much higher wavenumber outside the displayed figure.

The cutoff frequency of the approximate model is in the range of the two lowest exact cutoff frequencies, and the emerging real branch winds somehow between the exact branches. The intricate behaviour for complex wavenumbers cannot be mimicked by our poor model. Since there is only one optical mode, it can be grounded only via an imaginary arc.

The second order model gives quite a different impression. Figure 1.6 shows, for the same value of Poisson's ratio, the model's dispersion diagram, again contrasted with the exact one. The lower branch tends now more towards the exact course. It looks like it is pushed there by the second branch that needs more space. Qualitatively, the two higher branches of the model show the same behaviour as the exact curves, with an imaginary arc connecting the two cutoff frequencies and a complex branch growing out from the minimum point of the second branch. The values of the cutoff frequencies, however, differ considerably. Especially the higher cutoff frequency at $\sqrt{15}$ is clearly above the correct value π. Therefore, also the loop with imaginary wavenumber becomes too large and the uppermost real branch is well above the correct path.

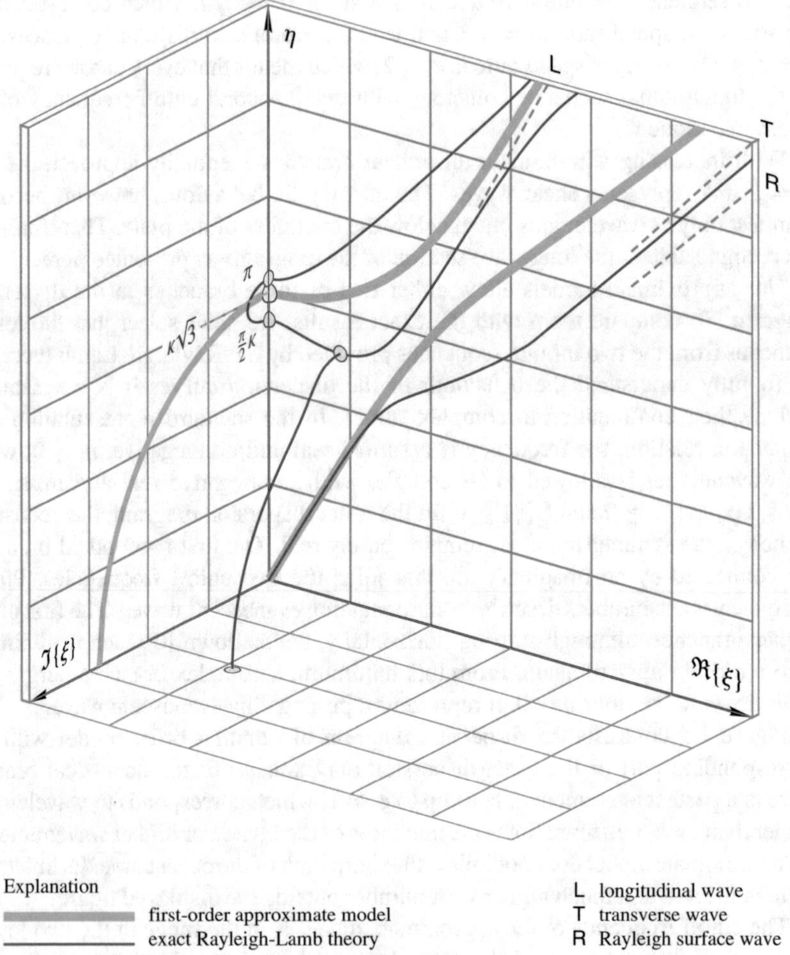

Fig. 1.5 Dispersion diagram of the first order model for $\nu = 1/4$.

The discrepancy between the exact and the approximate dispersion diagrams becomes even more pronounced for the Poisson's ratio $\nu = 1/3$ (Fig. 1.7). In this case, the two cutoff frequencies at $n\pi\kappa/2$ and $n\pi$, $n = 1, 3, 5, \ldots$, coincide as shown already in Fig. 1.4. Thus the two upper branches of the exact dispersion diagram emerge from the same cutoff frequency $\pi\kappa/2 = \pi$. The approximate model, however, still exhibits two different cutoff frequencies both of which are located above the correct one.

The cutoff frequencies of the approximate model eventually also coincide, but this happens only at the Poisson's ratio $\nu = 3/8$ (Fig. 1.8). In the exact theory, at this Poisson's ratio, there are again two distinct cutoff frequencies, just with exchanged

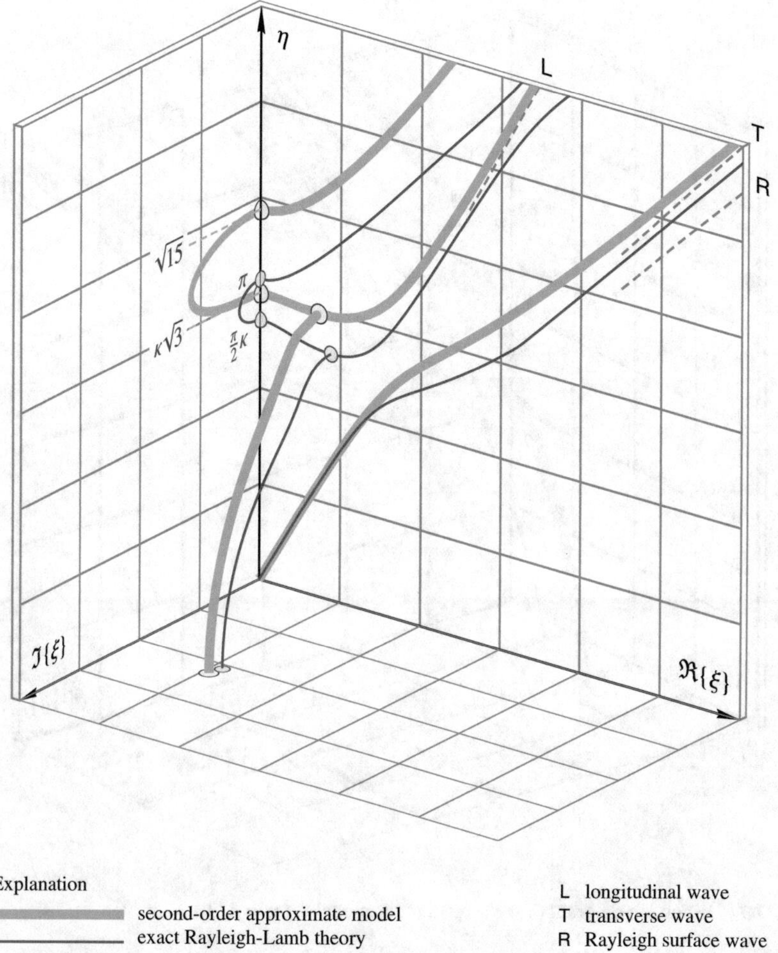

Fig. 1.6 Dispersion diagram of the second order model for $\nu = 1/4$.

roles. The first thickness mode at $\eta = \pi\kappa/2$ has meanwhile passed the first shear mode that remains at $\eta = \pi$.

Our comparison has emphasised those aspects of the approximate models, which deviate from the exact results of the Rayleigh–Lamb theory. This may give the impression that the models are rather poor. In favour of the approximate models, it should be kept in mind that the acoustical branch at low wavenumbers, or wavelengths bigger than three times the thickness, matches the exact curve perfectly.

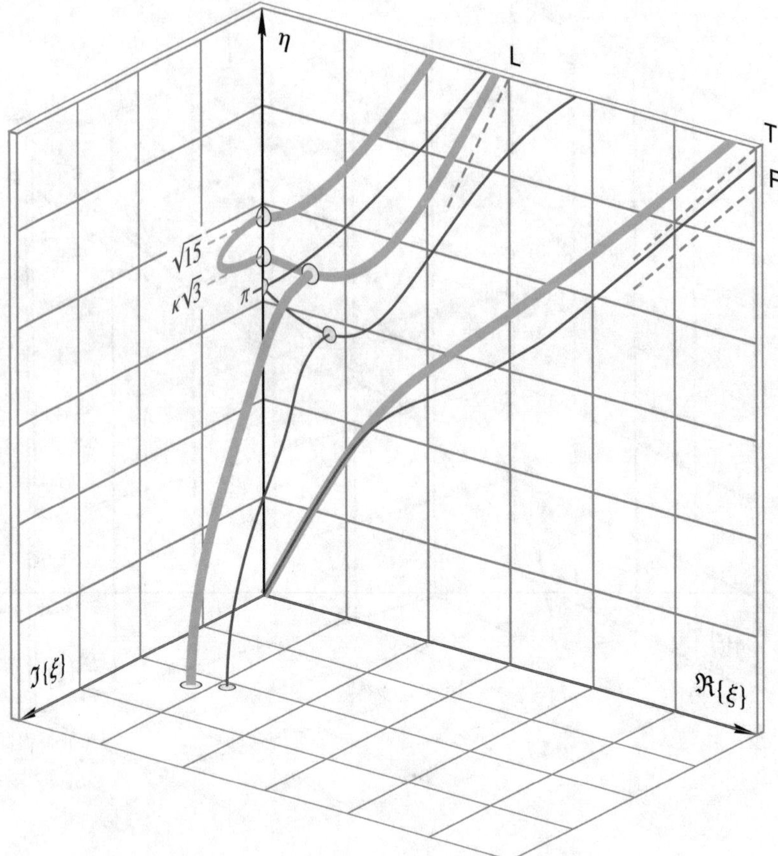

Fig. 1.7 Dispersion diagram of the second order model for $\nu = 1/3$.

1.6 Conclusions

Two approximate dynamic models of an infinite elastic layer of constant thickness have been presented. The governing equations are derived via Lagrange's method, and the approximation consists of restrictions imposed on the deformation of the transverse fibres of the layer. In the first order model, the fibres have to remain straight, but may extend or shrink uniformly. The second order model allows the fibres to bend assuming a parabolic shape.

The governing equations have been derived first for the layer as a 2D continuum and later specialised to the 1D case. The model equations can be combined into the

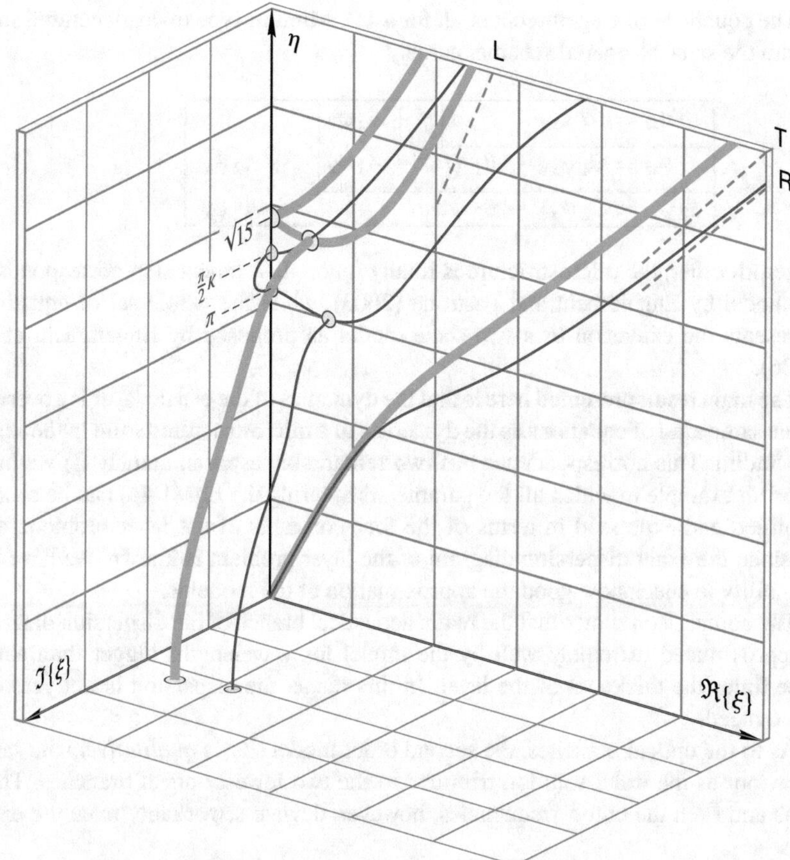

Fig. 1.8 Dispersion diagram of the second order model for ν = 3/8.

staggered scheme

$$\begin{array}{lllll} \rho\,\ddot{u} = & (\lambda + 2\mu)\,u'' & & +\,\lambda\varphi' & \\ \rho j^2\,\ddot{\varphi} = & \mu j^2\,\varphi'' & -\,(\lambda+2\mu)\,\varphi & -\,\lambda u' & +\,\mu j^2 w' \\ \rho l^4\,\ddot{w} = & (\lambda+2\mu) l^4 w'' & -\,\mu j^2\,w & & -\,\mu j^2\,\varphi' \end{array} \quad (1.43)$$

The scheme contains also the "zeroth order" model that restricts the transverse fibres to no deformation. The governing equation is then simply the nondispersive wave equation for the displacement u with longitudinal wave speed, certainly not an acceptable model for a plate. The first order model introduces the transverse strain φ as a second variable and, correspondingly, a second equation that is coupled to the first. Lastly, the second order model brings up a third variable w signifying the curvature of the deformed transverse fibre. The new equation is coupled with the second.

The equations of Engelbrecht et al. for a 1D, Mindlin-type microstructured solid fit into the same staggered scheme, namely

$$\begin{array}{lllll} \rho\, u_{tt} = & \alpha\, u_{xx} & & + A_1 \varphi_x & \\ I_1\, \varphi_{tt} = & C_1\, \varphi_{xx} & - B_1\, \varphi & - A_1 u_x & + A_2 \psi_x \\ I_2\, \psi_{tt} = & C_2\, \psi_{xx} & - B_2\, \psi & & - A_2\, \varphi_x \end{array} . \tag{1.44}$$

At zeroth order, the microstructure is totally ignored. The next step corresponds to the model by Engelbrecht and Pastrone (2003), while the whole set of equations represents the extension to a two-scale model as proposed by Engelbrecht et al. (2006).

The main result presented here is that the dynamics of the elastic layer is governed by the same kind of equations as the dynamics of a microstructured solid in the sense of Mindlin. This correspondence has two remarkable aspects, namely (i) we have found an example in which all the parameters entering the Eq. (1.44) can be clearly identified and expressed in terms of the four constants of the layer problem, and (ii) since the exact dispersion diagram of the layer problem is known, we have the possibility to check how good the approximation of the model is.

The comparison shows that the main, acoustical branch of the dispersion diagram is approximated extremely well by the model for wavelengths bigger than about three times the thickness of the layer. In this range, the dispersion is not yet very pronounced.

As to the optical branches, the second order model shows *qualitatively* the same behaviour as the exact solution trimmed to the two lowest optical branches. Their paths and even the cutoff frequencies, however, deviate remarkably from the exact ones.

It should be mentioned that our approximate model of the elastic layer is somehow related to an approximation proposed by Mindlin and Medick (1958, 1959). They expand the dependence on the z-coordinate by Legendre polynomials. If truncated at second order, it corresponds, in principle, to our *ansatz* (1.29). At a later stage, however, they introduce "additional adjustments to compensate, as well as possible, for the omission of the polynomials of higher degrees". Actually, they include some correction factors κ_r that are not clearly motivated but make the *ansatz* more flexible. These factors are determined then by forcing the cutoff frequencies to the correct values. As a result, also the optical branches of their model match the exact ones better than our model.

Appendix

To work out the details let us switch temporarily to index notation. The Lagrangian is a quadratic form in the variables

$$\left\{ u_{\alpha,t}\ \varphi_{,t}\ w_{\alpha,t}\ u_{\alpha,\beta}\ \varphi_{,\beta}\ w_{\alpha,\beta}\ \varphi\ w_\alpha \right\} . \tag{1.45}$$

First, the corresponding partial derivatives of the Lagrangian are provided. The derivatives with respect to the first group are

$$\frac{\partial \mathcal{L}}{\partial u_{\alpha,t}} = 2\rho h u_{\alpha,t}\,, \qquad \frac{\partial \mathcal{L}}{\partial \varphi_{,t}} = 2\rho h j^2 \varphi_{,t}\,, \qquad \frac{\partial \mathcal{L}}{\partial w_{\alpha,t}} = 2\rho h l^4 w_{\alpha,t}\,. \tag{1.46}$$

For the second group, the auxiliary formulas

$$\begin{aligned} \frac{\partial\, \mathrm{tr}\, \boldsymbol{E}}{\partial u_{\alpha,\beta}} &= \delta_{\alpha\beta}\,, \qquad \frac{\partial \boldsymbol{E}\cdot\boldsymbol{E}}{\partial u_{\alpha,\beta}} = 2E_{\alpha\beta} = 2u_{(\alpha,\beta)}\,,\\ \frac{\partial\, \mathrm{tr}\, \boldsymbol{G}}{\partial w_{\alpha,\beta}} &= \delta_{\alpha\beta}\,, \qquad \frac{\partial \boldsymbol{G}\cdot\boldsymbol{G}}{\partial w_{\alpha,\beta}} = 2G_{\alpha\beta} = 2w_{(\alpha,\beta)} \end{aligned} \tag{1.47}$$

will be useful. They provide the partial derivatives of the Lagrangian as

$$\begin{aligned} \frac{\partial \mathcal{L}}{\partial u_{\alpha,\beta}} &= -2\lambda h \left(u_{\gamma,\gamma} + \varphi\right) \delta_{\alpha\beta} - 4\mu h u_{(\alpha,\beta)}\,,\\ \frac{\partial \mathcal{L}}{\partial \varphi_{,\beta}} &= -2\mu h j^2 \left(\varphi_{,\beta} + w_\beta\right),\\ \frac{\partial \mathcal{L}}{\partial w_{\alpha,\beta}} &= -2\lambda h l^4 w_{\gamma,\gamma}\delta_{\alpha\beta} - 4\mu h l^4 w_{(\alpha,\beta)}\,. \end{aligned} \tag{1.48}$$

The last group reflects the direct dependence on the functions φ and $\boldsymbol{w}$. The corresponding partial derivatives are

$$\frac{\partial \mathcal{L}}{\partial \varphi} = -2\lambda h \left(u_{\gamma,\gamma} + \varphi\right) - 4\mu h \varphi, \qquad \frac{\partial \mathcal{L}}{\partial w_\alpha} = -4\mu h j^2 \left(\varphi_{,\alpha} + w_\alpha\right), \tag{1.49}$$

which complete the list of partial derivatives of the Lagrangian. They have to be inserted into the Lagrange equations of the second kind

$$\begin{aligned} \left(\frac{\partial \mathcal{L}}{\partial u_{\alpha,t}}\right)_{,t} + \left(\frac{\partial \mathcal{L}}{\partial u_{\alpha,\beta}}\right)_{,\beta} - \frac{\partial \mathcal{L}}{\partial u_\alpha} &= 0,\\ \left(\frac{\partial \mathcal{L}}{\partial \varphi_{,t}}\right)_{,t} + \left(\frac{\partial \mathcal{L}}{\partial \varphi_{,\beta}}\right)_{,\beta} - \frac{\partial \mathcal{L}}{\partial \varphi} &= 0,\\ \left(\frac{\partial \mathcal{L}}{\partial w_{\alpha,t}}\right)_{,t} + \left(\frac{\partial \mathcal{L}}{\partial w_{\alpha,\beta}}\right)_{,\beta} - \frac{\partial \mathcal{L}}{\partial w_\alpha} &= 0. \end{aligned} \tag{1.50}$$

At this point, we assume that the half thickness h of the plate is constant. All partial derivatives exhibit the factor $2h$. If this is constant, we can simplify the equations immediately by dividing all ingredients by $2h$. Then the equations become explicitly

$$\begin{aligned} \rho u_{\alpha,tt} &= (\lambda+\mu)u_{\beta,\beta\alpha} + \mu u_{\alpha,\beta\beta} + \lambda\varphi_{,\alpha}\,, \\ \rho j^2\varphi_{,tt} &= \mu j^2\left(\varphi_{,\beta\beta} + w_{\beta,\beta}\right) - (\lambda+2\mu)\varphi - \lambda u_{\beta,\beta}\,, \\ \rho l^4 w_{\alpha,tt} &= (\lambda+\mu)l^4 w_{\beta,\beta\alpha} + \mu l^4 w_{\alpha,\beta\beta} - \mu j^2\left(\varphi_{,\alpha} + w_\alpha\right). \end{aligned} \tag{1.51}$$

Returning to symbolic notation this is the system of Eq. (1.18) that governs the second order model.

The equations of the first order model are obtained by omitting the third equation and discarding the $w_{\beta,\beta}$ term in the second equation. Explicitly, the equations are

$$\begin{aligned} \rho u_{\alpha,tt} &= (\lambda+\mu)u_{\beta,\beta\alpha} + \mu u_{\alpha,\beta\beta} + \lambda\varphi_{,\alpha}\,, \\ \rho j^2\varphi_{,tt} &= \mu j^2\varphi_{,\beta\beta} - (\lambda+2\mu)\varphi - \lambda u_{\beta,\beta}\,. \end{aligned} \tag{1.52}$$

Converted to symbolic notation they correspond to the Eq. (1.17) that govern the first order model.

References

Achenbach, J.D.: Wave Propagation in Elastic Solids. North-Holland, Amsterdam (1973)

Brillouin, L.: Wave Propagation in Periodic Structures. Electric Filters and Crystal Lattices. Second edition, with corrections and additions, Dover Publications, New York (1953)

Engelbrecht, J., Pastrone, F.: Waves in microstructured solids with strong nonlinearities in microscale. Proc. Estonian Acad. Sci. Phys. Math. **52**(1), 12–20 (2003)

Engelbrecht, J., Berezovski, A., Pastrone, M., Braun, M.: Waves in microstructured materials and dispersion. Phil. Mag. **85**(33-35), 4127–4141 (2005). https://doi.org/10.1080/14786430500362769

Engelbrecht, J., Pastrone, F., Braun, M., Berezovski, A.: Hierarchies of waves in nonclassical materials. In: Delsanto, P.P. (ed.), The Universality of Nonclassical Nonlinearity: Applications to Non-destructive Evaluations and Ultrasonics, pp. 29–47. Springer, New York (2006). https://doi.org/10.1007/978-0-387-35851-2_3

Graff, K.F.: Wave Motion in Elastic Solids. Dover Publications, New York (1991)

Lamb, H.: On the flexure of an elastic plate. Proc. Lond. Math. Soc. **21**(1), 70–90 (1889). https://doi.org/10.1112/plms/s1-21.1.70

Lamb, H.: On waves in an elastic plate. Proc. Roy. Soc. A **93**(648), 114–128 (1917)

Miklowitz, J.: The Theory of Elastic Waves and Wave Guides. North-Holland, Amsterdam (1978)

Mindlin, R.D.: An Introduction to the Mathematical Theory of Vibrations of Elastic Plates. J. Yang (ed.) World Scientific, New Jersey (2006). https://doi.org/10.1142/6309

Mindlin, R.D.: Waves and vibrations in isotropic elastic plates. In: Goodier, J.N., Hoff, N.J. (eds.) Structural Mechanics, pp. 199–232. Pergamon Press, New York (1960)

Mindlin, R.D.: Microstructure in linear elasticity. Arch. Ration. Mech. Anal. **16**(1), 51–78 (1964). https://doi.org/10.1007/bf00248490

Mindlin, R.D., Medick, M.A.: Extensional vibrations of elastic plates. Report, Columbia University, Department of Civil Engineering and Engineering Mechanics, New York (1958). https://doi.org/10.21236/ad0200695

Mindlin, R.D., Medick, M.A.: Extensional vibrations of elastic plates. J. Appl. Mech. **26**, 561–569 (1959)

Onoe, M., NcNiven, H.D., Mindlin, R.D.: Dispersion of axially symmetric waves in elastic rods. J. Appl. Mech. **29**(4), 729–734 (1962). https://doi.org/10.1115/1.3640661

Peets, T.: Dispersion Analysis of Wave Motion in Microstructured Solids. Ph.D. Thesis, Tallinn University of Technology, Tallinn (2011)

Peets, T., Randrüüt, M., Engelbrecht, J.: On modelling dispersion in microstructured solids. Wave Motion **45**(4), 471–480 (2008). https://doi.org/10.1016/j.wavemoti.2007.09.006

Pochhammer, L.A.: Über die Fortpflanzungsgeschwindigkeiten kleiner Schwingungen in einem unbegrenzten isotropen Kreiscylinder. J. Reine Angew. Math. **81**, 324–336 (1876)

Randrüüt, M.: Wave Propagation in Microstructured Solids: Solitary and Periodic Waves. Ph.D. Thesis, Tallinn University of Technology, Tallinn (2010)

[Lord] Rayleigh: On waves propagated along plane surfaces of an elastic solid. Proc. Lond. Math. Soc. **17**(1), 4–11 (1885). https://doi.org/10.1112/plms/s1-17.1.4

[Lord] Rayleigh: On the free vibrations of an infinite plate of homogeneous isotropic elastic matter. Proc. Lond. Math. Soc. **20**(1), 225–234 (1889). https://doi.org/10.1112/plms/s1-20.1.225

Rose, J.L.: Ultrasonic Guided Waves in Solid Media. Cambridge University Press, Cambridge (2014). https://doi.org/10.1017/CBO9781107273610

Chapter 2
Antiplane Surface Wave Propagation Within the Stress Gradient Surface Elasticity

Victor A. Eremeyev

Abstract We discuss a new class of antiplane surface waves in an elastic half space with surface stresses. Here we consider a surface elasticity within stress gradient model, that is when the surface stresses relate to surface strains through an integral constitutive dependence. For antiplane motions the problem is reduced to the wave equation with nonclassical dynamic boundary condition. The dispersion relation is derived.

2.1 Introduction

Nowadays the model of surface elasticity and viscoelasticity found various applications in modelling anomalous mechanical properties of materials at the nanoscale, see, e.g., (Duan et al., 2008; Wang et al., 2011; Eremeyev, 2016). Among these models it is worth mentioning those by Gurtin and Murdoch (1975, 1978) and by Steigmann and Ogden (1997, 1999), where surface and interfacial phenomena were discussed. Considering surface related phenomena at the nanoscale the influence of long range interactions that results in nonlocal models can be observed, see, e.g., (de Gennes, 1981; Rowlinson and Widom, 2003; de Gennes et al., 2004; Israelachvili, 2011).

From the mathematical point of view, the presence of surface stresses may significantly change the mathematical properties of solutions of corresponding boundary value problems. In particular, even new types of solutions become possible. Among the latter there are surface antiplane waves in media with surface energy, see, e.g., (Vardoulakis and Georgiadis, 1997; Gourgiotis and Georgiadis, 2015;

V. A. Eremeyev (✉)
Faculty of Civil and Environmental Engineering, Gdańsk University of Technology, Gdańsk, Poland

Southern Federal University, Rostov on Don, Russia

Southern Scientific Center of RAS, Rostov on Don, Russia
e-mail: victor.eremeev@pg.edu.pl

A. Berezovski, T. Soomere (eds.), *Applied Wave Mathematics II*, Mathematics of Planet Earth 6, https://doi.org/10.1007/978-3-030-29951-4_2

Eremeyev et al., 2016, 2019) and the references therein, whereas within the classic linear elasticity theory such waves do not exist (Achenbach, 1973).

Here we discuss the propagation of antiplane surface waves in an elastic half space with surface stresses. For the latter we consider constitutive relations of integral type. Following Maugin's classification of nonlocal models we have here a strongly nonlocal model of surface elasticity (Maugin, 2017).

The paper is organised as follows. In Sect. 2.2 we present the basic equations in the bulk. Then in Sect. 2.3 we introduce the governing equations of integral type surface elasticity. Here we introduce both the integral constitutive equations and their differential counterparts. These constitutive equations are full analogues of Eringen's nonlocal three-dimensional (3D) elasticity (Eringen, 2002). Finally, we derive the dispersion relations in Sect. 2.4.

2.2 Antiplane Motions of an Elastic Half Space

Let us consider an elastic half space $-\infty \le x_3 \le 0$, where x_1, x_2, x_3 are Cartesian coordinates with corresponding base vectors $\mathbf{i}_k$, $k = 1, 2, 3$. Hereinafter we use the direct tensor calculus as described in (Simmonds, 1994; Lebedev et al., 2010; Eremeyev et al., 2018a).

For antiplane motions the displacement field $\mathbf{u}$ takes the form

$$\mathbf{u} = u(x_2, x_3, t)\mathbf{i}_1\,, \tag{2.1}$$

where t is time (Achenbach, 1973). In what follows we consider linear elastic isotropic material in the bulk with the constitutive relations

$$\boldsymbol{\sigma} = 2\mu\mathbf{e} + \lambda\mathbf{I}\,\mathrm{tr}\,\mathbf{e}, \quad \mathbf{e} = \frac{1}{2}\left[\nabla\mathbf{u} + (\nabla\mathbf{u})^T\right], \quad \nabla\mathbf{u} = \frac{\partial u_j}{\partial x_i}\mathbf{i}_i \otimes \mathbf{i}_j\,, \tag{2.2}$$

where $\boldsymbol{\sigma}$ and $\mathbf{e}$ are the stress and strain tensors, respectively, λ and μ are Lamé elastic moduli, tr is the trace operator, the superscript T means the transpose operation, ∇ is the three-dimensional nabla operator, $\otimes$ denotes the dyadic product, and $\mathbf{I}$ is the 3D unit tensor.

For antiplane motions we get in the bulk the wave equation

$$\mu\Delta u = \rho\partial_t^2 u\,, \tag{2.3}$$

where ρ is the mass density, $\Delta = \partial_2^2 + \partial_3^2$ is the two-dimensional (2D) Laplace operator. For brevity, we denote partial derivatives as $\partial_k = \partial/\partial x_k$ and $\partial_t = \partial/\partial t$.

2.3 Stress Gradient Surface Elasticity

Within the surface elasticity models, in addition to constitutive equations in the bulk we independently introduce surface stresses $\boldsymbol{\tau}$. As a result, the classic boundary condition for a free surface

$$\mathbf{n} \cdot \boldsymbol{\sigma} = \mathbf{0} \tag{2.4}$$

should be replaced by the generalised Young–Laplace equation

$$\mathbf{n} \cdot \boldsymbol{\sigma} = \nabla_s \cdot \boldsymbol{\tau} - m \partial_t^2 \mathbf{u}, \tag{2.5}$$

where $\mathbf{n}$ is the unit outward vector normal to the boundary, the dot stands for scalar product, $\nabla_s \equiv \mathbf{P} \cdot \nabla$ is the surface nabla operator, $\mathbf{P} \equiv \mathbf{I} - \mathbf{n} \otimes \mathbf{n}$ is the surface unit second order tensor, and m is the surface mass density defined as in (Gurtin and Murdoch, 1978).

The key point of the surface elasticity model is the choice of constitutive equations for $\boldsymbol{\tau}$. Here we assume the following strongly nonlocal model

$$\boldsymbol{\tau}(\mathbf{x}) = \int_{-\infty}^{\infty} \int_{-\infty}^{\infty} \alpha \left(\|\mathbf{x} - \mathbf{x}'\| \right) \left\{ 2\mu_s \boldsymbol{\epsilon}(\mathbf{x}') + \lambda_s \left[\operatorname{tr} \boldsymbol{\epsilon}(\mathbf{x}') \right] \mathbf{P} \right\} \mathrm{d}x_1' \mathrm{d}x_2', \tag{2.6}$$

where $\mathbf{x} = x_k \mathbf{i}_k$ is the position vector, $s = \|\mathbf{x} - \mathbf{x}'\|$, $\alpha(s)$ is a kernel function, λ_s and μ_s are the surface Lamé moduli, $\boldsymbol{\epsilon}$ is the infinitesimal surface strain tensor defined as in (Gurtin and Murdoch, 1975):

$$\boldsymbol{\epsilon} = \frac{1}{2} \left[\mathbf{P} \cdot (\nabla_s \mathbf{u}) + (\nabla_s \mathbf{u})^T \cdot \mathbf{P} \right].$$

Here for consistency with the constitutive relations in the bulk, we assume linear isotropic behaviour at the surface. So (2.6) describes a homogeneous and isotropic material. For examples of kernel functions we refer to (Eringen, 2002). In particular, it is convenient to use a fundamental solution as a kernel function. Let $\mathcal{L}$ be an elliptic differential operator, so α can be taken as the normalised solution of the problem

$$\mathcal{L}(\partial_1, \partial_2)\alpha = \delta(\mathbf{x}), \qquad \int_{-\infty}^{\infty} \int_{-\infty}^{\infty} \alpha \left(\|\mathbf{x} - \mathbf{x}'\| \right) \mathrm{d}x_1' \mathrm{d}x_2' = 1, \tag{2.7}$$

where $\delta(\mathbf{x})$ is the Dirac delta function. Applying $\mathcal{L}$ to (2.6) we get

$$\mathcal{L}(\partial_1, \partial_2)\boldsymbol{\tau} = 2\mu_s \boldsymbol{\epsilon} + \lambda_s (\operatorname{tr} \boldsymbol{\epsilon})\mathbf{P}. \tag{2.8}$$

As an example of $\mathcal{L}$ we can take $\mathcal{L} = -q^{-2}\Delta + 1$, where the parameter q is a reciprocal length. The corresponding normalised solution of (2.7) is given by

$$\alpha(s) = \frac{1}{2\pi} K_0(qs), \tag{2.9}$$

where K_0 is a modified Bessel function of the second kind. In this case we have the following stress gradient constitutive equation:

$$-q^{-2}\Delta\boldsymbol{\tau} + \boldsymbol{\tau} = 2\mu_s\boldsymbol{\epsilon} + \lambda_s(\operatorname{tr}\boldsymbol{\epsilon})\mathbf{P}. \tag{2.10}$$

If we take $\alpha = \delta(\mathbf{x})$ we get the known linear Gurtin–Murdoch model (Gurtin and Murdoch, 1975):

$$\boldsymbol{\tau} = 2\mu_s\boldsymbol{\epsilon} + \lambda_s(\operatorname{tr}\boldsymbol{\epsilon})\mathbf{P}.$$

In what follows we restrict ourselves to antiplane motions. In this case the surface stresses assume the form:

$$\boldsymbol{\tau} = \tau\,(\mathbf{i}_1 \otimes \mathbf{i}_2 + \mathbf{i}_2 \otimes \mathbf{i}_1), \quad \tau = \tau(x_2, x_3),$$

whereas $\boldsymbol{\epsilon}$ is given by

$$\boldsymbol{\epsilon} = \epsilon(\mathbf{i}_1 \otimes \mathbf{i}_2 + \mathbf{i}_2 \otimes \mathbf{i}_1), \quad \epsilon = \frac{1}{2}\partial_2 u.$$

As $\mathbf{n} = \mathbf{i}_3$ and $\mathbf{n}\cdot\boldsymbol{\sigma} = \sigma_{31}\mathbf{i}_1$, (2.5) transforms into

$$\sigma_{31} = \partial_2\tau - m\partial_t^2 u$$

or, considering the stress strain relation in the bulk, into

$$\mu\partial_3 u = \partial_2\tau - m\partial_t^2 u. \tag{2.11}$$

In the case of antiplane deformations, the integral constitutive equation (2.6) is reduced to

$$\tau = \int_{-\infty}^{\infty}\int_{-\infty}^{\infty} \alpha\left(\|\mathbf{x} - \mathbf{x}'\|\right)\mu_s\partial_2 u(\mathbf{x}')\,\mathrm{d}x_1'\mathrm{d}x_2'. \tag{2.12}$$

Considering the constitutive equation in its differential form (2.8), we have for antiplane motions:

$$\mathcal{L}(0, \partial_2)\tau = \mu_s\partial_2 u. \tag{2.13}$$

For the operator $\mathcal{L} = -q^{-2}\Delta + 1$, this equation becomes

$$-q^{-2}\partial_2^2\tau + \tau = \mu_s\partial_2 u. \tag{2.14}$$

2.4 Dispersion Relation

For steady state behaviour, we assume the solution of (2.3) in the form:

$$u = U(x_2, x_3)\exp(-\mathrm{i}\omega t), \tag{2.15}$$

where U is an amplitude, ω is a circular frequency, and i is the imaginary unit. As a result, (2.3) transforms into

$$\mu(\partial_2^2 + \partial_3^2)U = -\rho\omega^2 U. \tag{2.16}$$

Solutions of this equation that decay for $x_3 \to -\infty$ are:

$$U = U_0\exp(\varkappa x_3)\exp(\mathrm{i}kx_2), \tag{2.17}$$

where k is a wavenumber, U_0 is a constant, and

$$\varkappa = \varkappa(k, \omega) \equiv \sqrt{k^2 - \frac{\rho}{\mu}\omega^2}.$$

A nontrivial solution (2.17) exists if and only if it satisfies the boundary conditions at $x_3 = 0$. The latter will lead to a dispersion relation, i.e., an equation relating the wavenumber k and (angular) frequency ω. In particular, it is known that for (2.4) the antiplane waves do not exist (Achenbach, 1973), whereas within the Gurtin–Murdoch elasticity this type of waves exists (Eremeyev et al., 2016).

The displacement field $u(x_2, x_3, t)$ according to (2.15) and (2.17) leads to a surface stress in the form:

$$\tau = T\exp(\mathrm{i}kx_2)\exp(-\mathrm{i}\omega t),$$

where T is a constant. Solving (2.13) we get:

$$T = \frac{\mathrm{i}k}{L(k)}\mu_s U_0, \quad L(k) = \mathcal{L}(0, \mathrm{i}k).$$

This, in turn, being inserted into (2.11) results in the dispersion relation

$$\mu\varkappa(k, \omega) = m\omega^2 - \frac{\mu_s k^2}{L(k)}. \tag{2.18}$$

For example, for the operator $\mathcal{L} = -q^{-2}\Delta + 1$, one gets $L(k) = k^2/q^2 + 1$, and (2.18) has the form:

$$\mu\varkappa(k,\omega) = m\omega^2 - q^2\frac{\mu_s k^2}{k^2+q^2}\,. \tag{2.19}$$

If $q \to \infty$, then $L(k) \to 1$ and (2.19) coincides with the dispersion relation for the Gurtin–Murdoch model analysed by Eremeyev et al. (2016).

Introducing the phase velocity $c = \omega/k$ we transform (2.18) into

$$c^2 = \frac{c_s^2}{L(k)} + \frac{\mu}{m}\frac{1}{|k|}\sqrt{1-\frac{c^2}{c_T^2}}\,, \tag{2.20}$$

where $c_T = \sqrt{\mu/\rho}$ is the phase velocity of transverse waves and $c_s = \sqrt{\mu_s/m}$ is the shear wave velocity in the thin film associated with the Gurtin–Murdoch model. Obviously, a solution of (2.20) exists if c is in the range

$$\frac{c_s}{\sqrt{L(k)}} \le c \le c_T\,,$$

which significantly depends on k. Equation (2.19) becomes

$$c^2 = c_s^2\left(1+\frac{k^2}{q^2}\right)^{-1} + \frac{\mu}{m}\frac{1}{|k|}\sqrt{1-\frac{c^2}{c_T^2}}\,. \tag{2.21}$$

For the Gurtin–Murdoch model there is a characteristic wavenumber $p = \rho/m$. Using this notation, (2.21) can be written in the dimensionless form:

$$\frac{c^2}{c_T^2} = \frac{c_s^2}{c_T^2}\left(1+\frac{k^2}{q^2}\right)^{-1} + \frac{p}{|k|}\sqrt{1-\frac{c^2}{c_T^2}}\,. \tag{2.22}$$

Typical dispersion curves are shown in Fig. 2.1 for the stress gradient model (2.10) with different values of q/p. All curves start from the point $(0, c_T)$ with a horizontal tangent. Here we assumed that $c_s = 3/4c_T$. Note that $q = \infty$ corresponds to the Gurtin–Murdoch model, whereas other values of q change the behaviour of the curves.

In particular, as $\lim_{k\to\infty} L(k) = \infty$ we have that $c \to 0$ when $k \to \infty$, whereas within the Gurtin–Murdoch model we get the limit $c \to c_s$ when $k \to \infty$. Within a fixed range $0 \le k \le k_1$, the dispersion curve of the stress gradient model for $q \to \infty$ will come arbitrarily close to the dispersion curve of the Gurtin–Murdoch model. For q fixed and $k \to \infty$, however, the curves eventually approach the k-axis, while the Gurtin–Murdoch curve tends to the finite velocity c_s. Moreover, all dispersion curves are enclosed between the lower limiting curve for $q \to 0$, which is given by

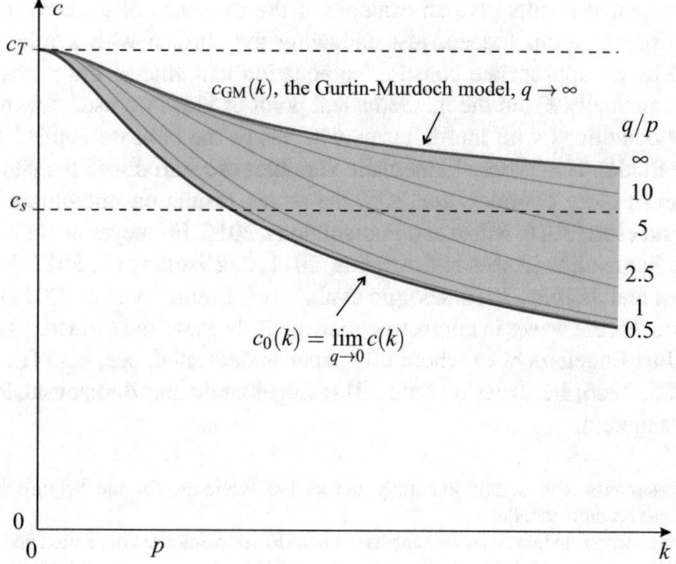

Fig. 2.1 Dispersion relations for stress gradient surface elasticity.

the formula

$$c_0^2 = \frac{c_T^2 p^2}{2k^2}\left(\sqrt{1+\frac{4k^2}{p^2}}-1\right) \tag{2.23}$$

and the dispersion curve of the Gurtin–Murdoch model (Fig. 2.1). So we have the following bounds for c:

$$c_0(k) \le c(k) \le c_{GM}(k), \tag{2.24}$$

where c_{GM} denotes the phase velocity for the Gurtin–Murdoch model.

2.5 Conclusions

The propagation of antiplane surface waves in an elastic half space with surface stresses modelled through the stress gradient model of the surface elasticity has been discussed. Within the model, the surface stresses are expressed through surface strains using integral constitutive relations. Taking into account the surface mass density as described by Gurtin and Murdoch (1978), we proved the existence of antiplane surface waves and derived the dispersion relations.

The presented results give an example of the influence of a microstructure on the wave propagation. Indeed, just endowing the surface with a microstructure governed by an appropriate constitutive equation can change the propagation of waves dramatically. From the mathematical point of view, we used new nonclassic boundary conditions with inertia terms whereas in the bulk we applied the linear elasticity model. If additional kinematic variables are introduced the situation can become even more complex, see, e.g., the recent results on micromorphic solids (Berezovski et al., 2016; Misra and Poorsolhjouy, 2016; Eremeyev et al., 2018b) and strain gradient solids (Askes and Aifantis, 2011; dell'Isola et al., 2012; Rosi et al., 2015; Rosi and Auffray, 2016; Giorgio et al., 2017; Eremeyev et al., 2019).

We note that the waves in microstructured media is one of the main fields of interest of Prof. Jüri Engelbrecht to whom this paper is dedicated, see, e.g., (Engelbrecht et al., 2005, 2006; Berezovski et al., 2011; Engelbrecht and Berezovski, 2015) and references therein.

Acknowledgements The author gratefully thanks the Reviewer for the helpful constructive comments and recommendations.

The author acknowledges financial support from the Russian Science Foundation under the grant "Methods of microstructural nonlinear analysis, wave dynamics and mechanics of composites for research and design of modern metamaterials and elements of structures made on its base" (No 15-19-10008-P).

References

Achenbach, J.D.: Wave Propagation in Elastic Solids. North-Holland, Amsterdam (1973)

Askes, H., Aifantis, E.C.: Gradient elasticity in statics and dynamics: An overview of formulations, length scale identification procedures, finite element implementations and new results. Int. J. Solids Struct. **48**(13), 1962–1990 (2011). https://doi.org/10.1016/j.ijsolstr.2011.03.006

Berezovski, A., Engelbrecht, J., Berezovski, M.: Waves in microstructured solids: a unified viewpoint of modeling. Acta Mechanica **220**(1-4), 349–363 (2011). https://doi.org/10.1007/s00707-011-0468-0

Berezovski, A., Giorgio, I., Corte, A.D.: Interfaces in micromorphic materials: wave transmission and reflection with numerical simulations. Math. Mech. Solids **21**(1), 37–51 (2016). https://doi.org/10.1177/1081286515572244

dell'Isola, F., Madeo, A., Placidi, L.: Linear plane wave propagation and normal transmission and reflection at discontinuity surfaces in second gradient 3D continua. ZAMM **92**(1), 52–71 (2012). https://doi.org/10.1002/zamm.201100022

Duan, H.L., Wang, J., Karihaloo, B.L.: Theory of elasticity at the nanoscale. Adv. Appl. Mech., **42**, 1–68 (2008). https://doi.org/10.1016/S0065-2156(08)00001-X

Engelbrecht, J., Berezovski, A.: Reflections on mathematical models of deformation waves in elastic microstructured solids. Math. Mech. Complex Systems **3**(1), 43–82 (2015). https://doi.org/10.2140/memocs.2015.3.43

Engelbrecht, J., Berezovski, A., Pastrone, F., Braun, M.: Waves in microstructured materials and dispersion. Phil. Mag. **85**(33-35), 4127–4141 (2005). https://doi.org/10.1080/14786430500362769

Engelbrecht, J., Pastrone, F., Braun, M., Berezovski, A.: Hierarchies of waves in nonclassical materials. In: Delsanto P.P. (ed.) Universality of Nonclassical Nonlinearity: Applications to Non-destructive Evaluations and Ultrasonic, pp. 29–47. Springer, New York (2006). https://doi.org/10.1007/978-0-387-35851-2_3

Eremeyev, V.A.: On effective properties of materials at the nano-and microscales considering surface effects. Acta Mechanica **227**(1), 29–42 (2016). https://doi.org/10.1007/s00707-015-1427-y

Eremeyev, V.A., Rosi, G., Naili, S.: Surface/interfacial anti-plane waves in solids with surface energy. Mech. Res. Commun. **74**, 8–13 (2016). https://doi.org/10.1016/j.mechrescom.2016.02.018

Eremeyev, V.A., Cloud, M.J., Lebedev, L.P.: Applications of Tensor Analysis in Continuum Mechanics. World Scientific, New Jersey (2018a). https://doi.org/10.1142/10959

Eremeyev, V.A., Lebedev, L.P., Cloud, M.J.: Acceleration waves in the nonlinear micromorphic continuum. Mech. Res. Commun. **93**, 70–74 (2018b). https://doi.org/10.1016/j.mechrescom.2017.07.004

Eremeyev, V.A., Rosi, G., Naili, S.: Comparison of anti-plane surface waves in strain-gradient materials and materials with surface stresses. Math. Mech. Solids **24**(8), 2526–2535 (2019). https://doi.org/10.1177/1081286518769960

Eringen, A.C.: Nonlocal Continuum Field Theories. Springer, New York (2002). https://doi.org/10.1007/b97697

de Gennes, P.G.: Some effects of long range forces on interfacial phenomena. J. Phys. Lettr. **42**(16), 377–379 (1981). https://doi.org/10.1051/jphyslet:019810042016037700

de Gennes, P.G., Brochard-Wyart, F., Quéré, D.: Capillarity and Wetting Phenomena: Drops, Bubbles, Pearls, Waves. Springer, New York (2004). https://doi.org/10.1063/1.1878340

Giorgio, I., Della Corte, A., dell'Isola, F.: Dynamics of 1D nonlinear pantographic continua. Nonlin. Dyn. **88**(1), 21–31 (2017). https://doi.org/10.1007/s11071-016-3228-9

Gourgiotis, P., Georgiadis, H.: Torsional and {SH} surface waves in an isotropic and homogenous elastic half-space characterized by the Toupin–Mindlin gradient theory. Int. J. Solids Struct. **62**, 217–228 (2015). https://doi.org/10.1016/j.ijsolstr.2015.02.032

Gurtin, M.E., Murdoch, A.I.: A continuum theory of elastic material surfaces. Arch. Rat. Mech. Anal. **57**(4), 291–323 (1975). https://doi.org/10.1007/bf00261375

Gurtin, M.E., Murdoch, A.I.: Surface stress in solids. Int. J. Solids Struct. **14**(6), 431–440 (1978)

Israelachvili, J.N.: Intermolecular and Surface Forces, 3rd edn. Academic Press, Amsterdam (2011). https://doi.org/10.1016/b978-0-12-391927-4.10024-6

Lebedev, L.P., Cloud, M.J., Eremeyev, V.A.: Tensor Analysis with Applications in Mechanics. World Scientific, New Jersey (2010). https://doi.org/10.1142/7826

Maugin, G.A.: Non-Classical Continuum Mechanics: A Dictionary. Springer, Singapore (2017). https://doi.org/10.1007/978-981-10-2434-4

Misra, A., Poorsolhjouy, P.: Granular micromechanics based micromorphic model predicts frequency band gaps. Cont. Mech. Thermodyn. **28**(1-2), 215–234 (2016). https://doi.org/10.1007/s00161-015-0420-y

Rosi, G., Auffray, N.: Anisotropic and dispersive wave propagation within strain-gradient framework. Wave Motion **63**, 120–134 (2016). https://doi.org/10.1016/j.wavemoti.2016.01.009

Rosi, G., Nguyen, V.H., Naili, S.: Surface waves at the interface between an inviscid fluid and a dipolar gradient solid. Wave Motion **53**, 51–65 (2015). https://doi.org/10.1016/j.wavemoti.2014.11.004

Rowlinson, J.S., Widom, B.: Molecular Theory of Capillarity. Dover, New York (2003)

Simmonds, J.G.: A Brief on Tensor Analysis, 2nd edn. Springer, New York (1994)

Steigmann, D.J., Ogden, R.W.: Plane deformations of elastic solids with intrinsic boundary elasticity. Proc. Roy. Soc. A **453**(1959), 853–877 (1997). https://doi.org/10.1098/rspa.1997.0047

Steigmann, D.J., Ogden, R.W.: Elastic surface-substrate interactions. Proc. Roy. Soc. A **455**(1982), 437–474 (1999). https://doi.org/10.1098/rspa.1999.0320

Vardoulakis, I., Georgiadis, H.G.: SH surface waves in a homogeneous gradient-elastic half-space with surface energy. J. Elasticity **47**(2), 147–165 (1997)

Wang, J., Huang, Z., Duan, H., Yu, S., Feng, X., Wang, G., Zhang, W., Wang, T.: Surface stress effect in mechanics of nanostructured materials. Acta Mech. Solida Sin. **24**, 52–82 (2011). https://doi.org/10.1016/s0894-9166(11)60009-8

Part II
Nonlinear Waves

Chapter 3
Dispersive and Nondispersive Nonlinear Long Wave Transformations: Numerical and Experimental Results

Tomas Torsvik, Ahmed Abdalazeez, Denys Dutykh, Petr Denissenko, and Ira Didenkulova

Abstract The description of gravity waves propagating on the water surface is considered from a historical point of view, with specific emphasis on the development of a theoretical framework and equations of motion for long waves in shallow water. This provides the foundation for a subsequent discussion about tsunami wave propagation and runup on a sloping beach, and in particular the role of wave dispersion for this problem. Wave tank experiments show that wave dispersion can play a significant role for the propagation and wave transformation of wave signals that include some higher frequency components. However, the maximum runup height is less sensitive to dispersive effects, suggesting that runup height can be adequately calculated by use of nondispersive model equations.

T. Torsvik (✉)
Norwegian Polar Institute, Fram Centre, Tromsø, Norway

Geophysical Institute, University of Bergen, Bergen, Norway
e-mail: Tomas.Torsvik@uib.no

A. Abdalazeez
Department of Marine Systems, School of Science, Tallinn University of Technology, Tallinn, Estonia
e-mail: ahmed.abdalazeez@taltech.ee

D. Dutykh
University Grenoble Alpes, University Savoie Mont Blanc, CNRS, LAMA, Chambéry, France
e-mail: Denys.Dutykh@univ-savoie.fr

P. Denissenko
School of Engineering, University of Warwick, Coventry, UK
e-mail: P.Denissenko@warwick.ac.uk

I. Didenkulova
Department of Marine Systems, School of Science, Tallinn University of Technology, Tallinn, Estonia

Nizhny Novgorod Technical University n.a. R.E. Alekseev, Nizhny Novgorod, Russia
e-mail: irina.didenkulova@taltech.ee

A. Berezovski, T. Soomere (eds.), *Applied Wave Mathematics II*, Mathematics of Planet Earth 6, https://doi.org/10.1007/978-3-030-29951-4_3

3.1 Introduction

Surface gravity waves propagating on the air–sea interface are categorised as *long waves* when their wave length λ is large compared with the average water depth h in a water basin. Waves of this type include very large scale phenomena such as tides and seiches, and local phenomena such as nonbreaking shoaling waves at the coast. In recent years much attention has been connected with the study of tsunamis, with respect to their propagation over vast distances in the open ocean, their transformation in coastal waters and their resulting inundation of coastal areas.

A range of different model equations have been discussed in connection with long wave propagation, including nonlinear shallow water (NLSW) equations and Boussinesq-type equations, which differ in their ability to represent nonlinear and dispersive effects. While elaborate model equations may provide more accurate representation of the wave propagation and transformation, they are generally more computationally demanding to integrate over time. In practical cases where a prediction of the wave behaviour is needed quickly, such as for a tsunami warning system, it has therefore been common practice to rely on simple NLSW equations rather than Boussinesq-type equations. Questions regarding the tradeoff between accuracy of prediction and efficiency of computation for shallow water model equations remain an active area of research to this day.

In this chapter we consider the problem of shallow water waves in a historical context, introducing some basic concepts of wave propagation. Thereafter we discuss the importance of these factors in the context of tsunami wave propagation and runup on a sloping beach. Finally we consider some examples of different wave types, and assess the suitability of NLSW equations and a Boussinesq-type equation for each of these.

3.2 Historical Background

The description of surface gravity wave propagation at the air–sea interface is one of the truly classical subjects in fluid mechanics, and developed in multiple stages with early contributions from some of the most prominent figures in science history such as Newton, Euler, Laplace, Lagrange, Poisson, Cauchy and Airy (see (Darrigol, 2003; Craik, 2004, 2005) for historical references).

For instance, in 1786 Lagrange demonstrated that small amplitude waves would propagate in shallow water with a velocity of $c = \sqrt{gh}$, where g represents the acceleration of gravity and h represents the water depth. Laplace (1776) was the first to pose the general initial value problem for water wave motion, i.e., *given a localised initial disturbance of the sea surface, what is the subsequent motion?* He was also the first to derive the full linear dispersion relation for water waves. A complete linear wave theory, which included wave dispersion, was later published by Airy (1845). Despite the long and extensive history of investigations into this problem, the study of dispersive surface gravity waves continues to be an active field of research to this day.

3.2.1 Airy Wave Theory

In the following discussion we will restrict our attention to wave propagation in one horizontal dimension x on a surface that can be displaced in the vertical z direction (see Fig. 3.1). It is fairly simple to extend the theory to two horizontal dimensions, but we will not consider any examples where this is necessary, e.g. crossing wave patterns. A more thorough description of these equations can be found in standard fluid mechanics textbooks, e.g., (Kundu, 1990).

A simple model equation for the propagation of surface gravity waves can be derived under the assumption of irrotational fluid motion, ignoring viscous effects, in which case the flow velocity components can be expressed in terms of a *velocity potential* ϕ, defined by

$$u = \frac{\partial \phi}{\partial x} \quad \text{and} \quad w = \frac{\partial \phi}{\partial z}, \tag{3.1}$$

for horizontal and vertical velocity components u and w, respectively. We do not consider effects due to surface tension, which is an important effect for short and steep waves but do not contribute significantly to long crested waves. Lastly, we assume that the water depth does not change very abruptly, i.e., that the water depth is fairly constant over the wave length.

The basic equation of motion is derived from the continuity equation

$$\frac{\partial u}{\partial x} + \frac{\partial w}{\partial z} = 0, \tag{3.2}$$

which is transformed to the Laplace equation

$$\frac{\partial^2 \phi}{\partial x^2} + \frac{\partial^2 \phi}{\partial z^2} = 0, \tag{3.3}$$

with substitution of the velocity potential. The sea bed is traditionally considered to be rigid and nonpermeable, which implies that flow is only permitted along the bed profile. Under this assumption the boundary condition at the sea bed requires a zero

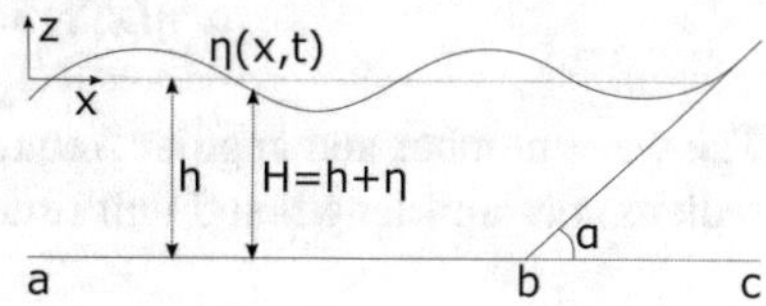

Fig. 3.1 Reference coordinate system for surface gravity waves.

normal velocity to the bed surface itself, which for a flat sea bed simplifies to

$$w = \left.\frac{\partial \phi}{\partial z}\right|_{z=\eta} = 0. \tag{3.4}$$

At the sea surface the fluid parcels are not restricted by any rigid boundary, and in fact the location of this free surface boundary is a variable, we wish to determine by solving the equations of motion. If the wave field is sufficiently smooth we can assume that this boundary is well represented by a material surface, i.e., that fluid parcels at the boundary never leave the surface. Under this assumption the kinematic boundary condition prescribed at $z = \eta$ becomes

$$\frac{\partial \eta}{\partial t} + u \left.\frac{\partial \eta}{\partial x}\right|_{z=\eta} = \left.\frac{\partial \phi}{\partial z}\right|_{z=\eta}. \tag{3.5}$$

Provided the nonlinear term in (3.5) is sufficiently small, this expression can be replaced by the linear equation

$$\frac{\partial \eta}{\partial t} = \left.\frac{\partial \phi}{\partial z}\right|_{z=\eta}. \tag{3.6}$$

Finally, the dynamic boundary condition prescribes that the pressure is continuous across the free surface boundary, which is expressed by the linear form of the Bernoulli equation

$$\frac{\partial \phi}{\partial t} + \frac{P}{\rho} + gz = 0, \tag{3.7}$$

where P is pressure and ρ is density. Assuming that the ambient pressure is zero at the free surface, i.e., $P = 0$ at $z = \eta$, this condition simplifies to

$$\frac{\partial \phi}{\partial t} + g\eta = 0 \quad \text{at} \quad z = \eta. \tag{3.8}$$

The complete boundary value problem is defined by (3.3), (3.4), (3.6), and (3.8).

In order to solve the equations we need to assume an initial wave form. By Fourier analysis it is possible to decompose any continuous disturbance into a sum of sinusoidal components, hence we will assume an initial condition specified by one such component with wave number k and angular frequency ω

$$\eta(x, t) = a\cos(kx - \omega t). \tag{3.9}$$

The wave number and angular frequency represents the number of wave cycles (in radians) per unit length and unit time, respectively, and are defined in terms of the wave length λ and wave period T as

$$k = \frac{2\pi}{\lambda} \quad \text{and} \quad \omega = \frac{2\pi}{T}.$$

Solving the Laplace equation (3.3) with kinematic boundary conditions (3.4) and (3.6) with this initial condition results in a velocity potential

$$\phi = \frac{a\omega}{k} \frac{\cosh k(z+h)}{\sinh kh} \sin(kx - \omega t), \tag{3.10}$$

and the original velocity components become

$$u = a\omega \frac{\cosh k(z+h)}{\sinh kh} \cos(kx - \omega t), \tag{3.11}$$

$$w = a\omega \frac{\sinh k(z+h)}{\sinh kh} \sin(kx - \omega t). \tag{3.12}$$

By combining this solution with the dynamic boundary condition (3.8) we find the *dispersion relation* between k and ω as

$$\omega = \sqrt{gk \tanh kh}, \tag{3.13}$$

and the corresponding *phase velocity*

$$c_p \equiv \frac{\omega}{k} = \sqrt{\frac{g}{k}} \tanh kh. \tag{3.14}$$

The dispersion relation Eq. (3.13) links the wave number k with the angular frequency ω, and represents a necessary condition for consistency of linear wave solutions for the equations of motion. This implies that sinusoidal wave solutions Eq. (3.9) can exist if and only if the wave length and period are strictly linked with each other according to (3.13). Equation (3.14) demonstrates that the speed of propagation for linear waves depends on the wave number (equivalently, wave length), hence an initial disturbance that contains wave components with various wave numbers will tend to separate into clusters of individual components as the waves propagate away from the source.

For waves in a dispersive medium, the energy of wave components does not propagate with the phase velocity c_p (3.14), but with the *group velocity* $c_g = d\omega/dk$. With dispersion relation (3.13), the group velocity becomes

$$c_g \equiv \frac{d\omega}{dk} = \frac{c_p}{2}\left(1 + \frac{2kh}{\sinh 2kh}\right). \tag{3.15}$$

3.2.1.1 Deep and Shallow Water Approximations

As seen in (3.14), the phase velocity for wave components with different wave numbers depends on the hyperbolic tangent function. It is instructive to consider the behaviour of this equation in the deep and shallow water conditions, which is determined by the value of kh (i.e., water depth relative to wave length). A commonly used classification is to consider $kh \geq \pi$ as deep water, $kh \leq \pi/10$ as shallow water, and $\pi/10 < kh < \pi$ as intermediate water depth. This classification should be considered a "rule-of-thumb" rather than a strict rule, as the dispersion relation varies continuously over the range of kh.

For the deep water approximation, the hyperbolic functions in (3.14) and (3.15) can be approximated as

$$\lim_{kh\to\infty} \tanh kh = 1 \quad \text{and} \quad \lim_{kh\to\infty} \frac{2kh}{\sinh 2kh} = 0,$$

in which case the phase and group speed in deep water become

$$c_p = \sqrt{\frac{g}{k}} \quad \text{and} \quad c_g = \frac{c_p}{2}, \tag{3.16}$$

respectively. This implies that short waves in deep water propagate slower than longer waves, and the wave energy propagates slower than the wave phase. Note that (3.16) is derived under the assumption that gravity is the only relevant restoring force, which is not always correct. For instance, at very short wave lengths (cm scale at the air–water interface) surface tension becomes the dominant restoring force, which allows shorter wave components to propagate faster than longer wave components. The velocity components simplify to

$$u = a\omega e^{kz} \cos(kx - \omega t), \tag{3.17}$$

$$w = a\omega e^{kz} \sin(kx - \omega t), \tag{3.18}$$

which are circular orbits with a radius of a at the surface. It should be noted that the linear wave theory assumes that effects due to the finite wave amplitude are negligible. In reality waves have a finite amplitude, which induces a slow drift in the direction of wave propagation, and therefore the orbits of fluid parcels are not perfect circles but display a coil-like behaviour. This effect is called *Stokes drift*.

In shallow water the hyperbolic functions in (3.14) can be replaced by kh because

$$\tanh kh = kh + O\left[(kh)^2\right] \quad \text{as} \quad kh \to 0,$$

and the hyperbolic function in (3.15) can be approximated as

$$\lim_{kh\to 0} \frac{2kh}{\sinh 2kh} = 1\,,$$

hence the simplified expressions for the phase speed and group speed become

$$c_p = \sqrt{gh} \quad \text{and} \quad c_g = c_p\,, \tag{3.19}$$

respectively. In this case the phase velocity is not dependent on the wave number k, hence waves in the shallow water limit are nondispersive. This is also reflected in the group velocity, which becomes identical to the phase velocity in shallow water. In the special case of unidirectional flow, the shallow water wave field therefore becomes stationary in the coordinate system that follows the phase speed c_p. The velocity components for shallow water waves (of small but finite depth) are

$$u = \frac{a\omega}{kh} \cos(kx - \omega t)\,, \tag{3.20}$$

$$w = a\omega\left(1 + \frac{z}{h}\right) \sin(kx - \omega t)\,. \tag{3.21}$$

These are elliptic orbits where the vertical component is much smaller than the horizontal.

3.2.2 Nonlinear Long Waves

The wave theory developed by Airy is a linear system, requiring both the underlying equations of motion and boundary conditions to be linear, and therefore any wave solution to this system must conform to the superposition principle. This means that the net response to the system of two or more stimuli can be established by determining the response of each stimulus separately, and subsequently adding these together. Equivalently, any linear combination or scaling of valid solutions will produce a new valid solution to the problem.

In particular, this means that the wave amplitude, which can be altered by a scalar multiplication, must be an independent variable that cannot have any functional dependence on other wave properties. This property is specific for linear systems, whereas for nonlinear systems the wave amplitude will normally be linked with other wave properties. In fact, waves of this type had already been described at the time when Airy published his account.

A few years prior to the publication of Airy's wave theory, the naval engineer John Scott Russell had published accounts of observations and experiments devoted to surface gravity waves (Russell, 1844). Russell seem to have devoted most of his efforts to explain wave generation and propagation in channels, which was of practical importance for inland waterway transport at that time.

A particularly famous account describes his first observation of a *large, solitary, progressive wave*, which was generated by a boat in a channel and propagated upstream of the boat. Russell was able to follow this wave on horseback for more than a mile, and while it retained its original shape it then gradually subsided.

In a series of subsequent experiments he determined that the wave progressed upstream with a velocity $c = \sqrt{g(h+\eta)}$, and that the wave making resistance against the boat motion was at a maximum when the boat was traveling at this speed. He also proposed that tidal motion could be explained as solitary waves of very large extent, and suggested a mechanism whereby the tidal motion could generate tidal bores in rivers and channels.

Airy devoted some attention to Russell's experiments, but he dismissed Russell's treatment of solitary waves. According to Airy's wave theory, maintaining such a singular disturbance in the absence of any additional force would require the surface slope of the disturbance to be constant, but since the slope should vanish at infinity such a disturbance could not exist.

The existence and importance of solitary waves remained a contested issue for several decades after the initial treatments by Russell and Airy. For example, the prominent scientist Georges Gabriel Stokes first dismissed the possibility of such waves and their relevance to tidal motion in his 1846 hydrodynamic researches review (Stokes, 1846), but later became supportive of the idea after researching finite oscillatory waves.

In 1870 Adhémar Jean Claude Barré de Saint-Venant published an account of tidal bores in rivers (named *mascaret* in French), and the following year (de Saint-Venant, 1871) he presented a set of equations that described the phenomenon

$$\frac{\partial A}{\partial t} + \frac{\partial (Au)}{\partial x} = 0, \tag{3.22}$$

$$\frac{\partial u}{\partial t} + u\frac{\partial u}{\partial x} + g\frac{\partial \eta}{\partial x} = -\frac{P_w}{A}\frac{\tau}{\rho}, \tag{3.23}$$

where $A(x,t)$ is the channel cross section area, $u(x,t)$ is the depth averaged horizontal velocity component, $P_w(x,t)$ is the length of wetted channel perimeter at the cross section, $\tau(x,t)$ is the wall shear stress, and ρ is the water density.

The set of Eqs. (3.22) and (3.23) represent conservation of mass and balance of momentum, respectively, and is possibly the first version of NLSW equations to be presented in a publication. Due to the friction force induced by the shear stress at channel walls, the momentum of an initial disturbance will not be conserved in the model system. The shallow water equations (3.22, 3.23) can describe the propagation of a solitary wave, but the wave will transform over time, with a steepening of the wave front and a decrease in the slope behind the crest

While this behaviour nicely described the transformation of a regular tidal wave to a tidal bore in a channel, it did not provide an adequate framework for describing the solitary waves of constant shape observed by Russell. Although similar shallow water equations had been presented prior to Saint-Venant's treatment of mascarets

(Fenton, 2010), the one-dimensional (1D) version of the shallow water equations are often referred to as *Saint-Venant equations* in honour of his contribution to understand shallow water hydrodynamics.

The same year as Saint-Venant presented the NLSW equations for description of mascarets, one of his disciples, Joseph Boussinesq, presented the first approximate solution of a solitary wave propagating without deformation (Boussinesq, 1871), which finally provided a firm theoretical support for the existence of Russell's wave. The following year (Boussinesq, 1872) he presented a derivation of equations which permitted his wave solution

$$\frac{\partial \eta}{\partial t} + \frac{\partial (H u_b)}{\partial x} = \frac{h^3}{6} \frac{\partial^3 u_b}{\partial x^3}, \tag{3.24}$$

$$\frac{\partial u_b}{\partial t} + u_b \frac{\partial u_b}{\partial x} + g \frac{\partial \eta}{\partial x} = \frac{h^2}{2} \frac{\partial^3 u_b}{\partial t \partial x^2}, \tag{3.25}$$

where $H = h + \eta$ (Fig. 3.1) and u_b is the horizontal velocity at the sea bed $z = -h$. This is the original version of what is now called *Boussinesq equations*. In the absence of higher order derivatives (right-hand side of (3.24), (3.25)) the Boussinesq system becomes equivalent to the Saint-Venant equations (3.22), (3.23) without a friction term.

Boussinesq derived his equations from the Euler equations by eliminating the explicit dependence on the vertical coordinate z in these equations, while retaining nonlinear terms of highest order. This procedure, which is now commonly used when deriving shallow water equations, does not *a priori* stipulate the vertical reference level to be used for the horizontal velocity component or which higher order terms to retain in the derivation.

Numerous variations of Boussinesq-type systems can therefore be derived by selecting different reference variables and forms of nonlinear terms, resulting in equations with slightly different dispersive and nonlinear properties, as well as numerical stability properties. A particularly useful variation was derived by Peregrine (1967)

$$\frac{\partial \eta}{\partial t} + \frac{\partial (H u)}{\partial x} = 0, \tag{3.26}$$

$$\frac{\partial u}{\partial t} + u \frac{\partial u}{\partial x} + g \frac{\partial \eta}{\partial x} - \frac{h}{2} \frac{\partial^3 (hu)}{\partial^2 x \partial t} + \frac{h^2}{6} \frac{\partial^3 u}{\partial x^2 \partial t} = 0, \tag{3.27}$$

which can be applied under gently varying depth conditions.

While the achievement of Boussinesq is widely recognised, his results were not immediately seized upon by his contemporaries. Five years after Boussinesq presented his solitary wave solution, Lord Rayleigh independently derived a long wave equation for the solitary wave of constant shape (Rayleigh, 1876). When Korteweg and de Vries later derived their famous *Korteweg–de Vries (KdV) equation*, they reference to Rayleigh's work but were apparently unaware of the earlier contribution by Boussinesq (Korteweg and de Vries, 1895).

3.2.3 Model Equations for Long Wave Runup on a Beach

In the classical formulations of long wave equations it is usually assumed that the waves propagate in a water basin with small and gentle changes in water depth. However, we would like to apply these model equations to study wave runup on a beach, and this requires some modifications to the standard equation formulations. In the following we consider a depth profile

$$h(x) = \begin{cases} h_0, & \text{if } x \in [a, b] \\ h_0 - (x - b)\tan\alpha, & \text{if } x \in [b, c] \end{cases}, \tag{3.28}$$

with waves approaching the beach from the offshore point a (Fig. 3.1). The modified NLSW equations are defined as

$$\frac{\partial H}{\partial t} + \frac{\partial (Hu)}{\partial x} = 0, \tag{3.29}$$

$$\frac{\partial (Hu)}{\partial t} + \frac{\partial}{\partial x}\left(Hu^2 + \frac{g}{2}H^2\right) = gH\frac{\partial h}{\partial x}, \tag{3.30}$$

where $u(x,t)$ is the depth averaged flow velocity. For comparison, we use a Boussinesq-type equation based on Peregrine's formulation, which we call the modified Peregrine equations.

$$\frac{\partial H}{\partial t} + \frac{\partial Q}{\partial x} = 0, \tag{3.31}$$

$$\begin{aligned} &\left(1 + \frac{1}{3}\frac{\partial H^2}{\partial x} - \frac{H}{6}\frac{\partial^2 H}{\partial x^2}\right)\frac{\partial Q}{\partial t} - \frac{H^2}{3}\frac{\partial^3 Q}{\partial x^2 \partial t} - \\ &- \frac{H}{3}\frac{\partial H}{\partial x}\frac{\partial^2 Q}{\partial x \partial t} + \frac{\partial}{\partial x}\left(\frac{Q^2}{H} + \frac{g}{2}H^2\right) = gH\frac{\partial h}{\partial x}, \end{aligned} \tag{3.32}$$

where $Q = Hu$ represents the horizontal momentum. The modified Peregrine equations have been studied in detail in (Durán et al., 2018).

3.2.4 Numerical Method

In the following discussion we will apply numerical methods for integration of the Boussinesq equations over time. For simple channel geometries it is possible to derive exact solitary wave and periodic wave solutions to the Boussinesq equations (Clarkson, 1990; Chen, 1998; Yan and Zhang, 1999). Furthermore, the runup properties of such general wave solutions can be investigated by analytical methods for some regular beach profiles (Pelinovsky and Mazova, 1992; Didenkulova et al., 2007; Didenkulova and Pelinovsky, 2011). However, such analytical methods are not practical when considering general wave types and variable depth conditions.

The numerical model we use is based on a finite volume method for both the modified NLSW and modified Peregrine equations (Dutykh et al., 2011; Durán et al., 2018). This involves discretisation of the governing equations, and obtaining solutions on a finite mesh covering the model domain. In the finite volume method, the divergence theorem is applied to convert divergence terms in the differential equations to surface integrals, which are evaluated as fluxes at the cell surfaces in the mesh. Finite volume methods are particularly useful for problems where quantities should be preserved, e.g., mass or momentum, since whatever quantity flows out of one grid cell surface will be identical to the inflow into the neighbouring grid cell.

The simplest approximation to a solution in the finite volume formulation is obtained by considering all variables as constant within each grid cell, whereby a piecewise constant solution can be obtained. However, using this approach, the spatial discretisation error will be determined by the grid size. In order to obtain more accurate results, a common method is to replace the piecewise constant data with a piecewise polynomial representation of the solution. In our simulations we have applied the nonoscillatory UNO2 scheme, which is designed to constrain the number of local extrema in the numerical solution at each time step (Harten and Osher, 1987).

Integration of the solution forward in time is achieved by the Bogacki–Shampine time stepping method (Bogacki and Shampine, 1989). This is a version of a Runge–Kutta method, and is a third order method with four stages. An embedded second order method is used to estimate the local error and if necessary adapt the timestep size.

3.3 Tsunami Propagation and Runup

Developing model equations that adequately describe the propagation and runup of tsunamis is a challenging task. Suggested model formulations range from simple nonlinear shallow water (NLSW) theory to the very elaborate fully nonlinear Navier–Stokes theory, with Boussinesq theory occupying an intermediate place in between. NLSW has often been favoured for long wave runup calculations over dispersive wave models represented by Boussinesq-type approximations. Wave runup calculated using dispersive model formulations is prone to numerical instabilities, which make computations more sensitive to numerical parameters (Bellotti and Brocchini, 2001). Furthermore, the Boussinesq terms in the dispersive model tend to zero at the shoreline, so that dispersive equations simplify to NLSW in this region (Madsen et al., 1997).

High accuracy can be achieved by applying the fully nonlinear Navier–Stokes equations, but this approach requires large computational resources and a lengthy integration time, making it unsuitable for operational forecasting in oceanwide or even regional scale applications. Horrillo et al. (2006) studied dispersive effects during the 2004 Indian Ocean tsunami propagation by comparing NLSW with the fully nonlinear Navier–Stokes equations. They concluded that NLSW offered the more suitable framework for hazard assessments, providing an adequate assessment at a very low computational cost. Although the NLSW model tended to overpredict the maximum wave runup, the overprediction was considered to be within a reasonable range for a safety buffer, and hence did not degrade the overall assessment.

For tsunami warning purposes it is of critical importance to determine the time of arrival of the leading wave to different coastal sections. These leading waves are usually well described by NLSW, whereas the trailing wave train may contain shorter wave components that are more sensitive to wave dispersion (Løvholt et al., 2012). For this reason, the NLSW is often considered to be more appropriate than more elaborate Boussinesq-type methods for warning purposes (Glimsdal et al., 2013). Note that biggest wave is often not the first one, at least for tsunamis propagating over a long distance, see, for example, (Candella et al., 2008).

3.3.1 Wave Tank Experiment

Wave tank experiments were carried out at the Large Wave Flume (GWK) located in Hannover, Germany, which is the world's largest publicly available research facility of its kind. It has a length of about 310 m usable for experiments, a width of 5 m, and a maximum depth of 7 m. Access to this facility was granted by the Integrating Activity HYDRALAB IV program, and experiments were carried out over two periods; 10–16 Oct. 2012 and 29 July–9 August 2013. The basic experiment setup consisted of a wave generator at one end of the flume, a 251 m channel of constant depth, and a ramp of 1:6 slope at the opposite end of the flume representing the beach.

The water depth in the channel was kept at a constant $h_0 = 3.5$ m for all the experiments. Wave gauges were placed at 16–18 locations along the channel to measure the waves propagating in the channel and up the slope. The wave runup was measured by a capacitance probe and also recorded by two regular video cameras. A series of experiment runs were performed with different initial wave signals, and with varying roughness of the ramp slope surface. Details of the experiments are described in Didenkulova et al. (2013).

3.4 Measured and Modelled Wave Propagation and Runup

In order to illustrate the wave transformation and runup properties for different wave signals, we consider the four experimental test cases listed in Table 3.1. These consist of a regular sine wave, a biharmonic wave signal, a wave train that resembles a ship wake signal, and a single positive pulse. The wave maker produced waves of period $T = 20$ s, which remained constant for the sine and biharmonic signals, but gradually reduced to $T = 10$ s for the wake-like train. The initial wave amplitude was different for each experiment, with the largest initial wave amplitude $A = 0.20$ m for the sine wave (Fig. 3.2). However, the biharmonic wave signal contained two wave components with amplitude $A = 0.12$ m that could interfere constructively to produce instances of larger amplitude wave peaks than the sine wave (Fig. 3.3).

The wake-like train did not contain waves of equal amplitude. Instead, initially long, low amplitude waves were followed by progressively shorter and larger-amplitude waves (Fig. 3.4). The single positive pulses were generated with $A = 0.15$ m, but were not initiated as stable solitary wave shapes and hence reduced in amplitude to approximately $A = 0.10$ m at an early stage during the wave propagation (Fig. 3.5). Each figure shows a comparison between the experimental record and two model results; the dispersive modified Peregrine model (hereafter mPer) and the NLSW model solutions.

Figure 3.2 shows the sine wave propagation and runup. In this case mPer is fairly close to the measured waves throughout the propagation phase and for the runup, although there is a tendency for mPer to underestimate the runup height. It is noticeable that NLSW has a lower wave height near the wave maker than the measured wave, but increase in amplitude relative to the reference solution, and in the final stage produce significantly larger runup values than the measured values. It is clear that the dispersive properties of mPer in this case balance the nonlinear effect to produce a relatively stable wave train, while this feature is missing for NLSW and therefore results in excessive nonlinear steepening and amplification.

Note that the capacitance runup gauge does not record the wave form correctly in the receding phase. The reason is that the wires are submerged in the thin near surface layer of water when the bulk of the wave is gone. Therefore, only the rising front phase of the experimentally measured runup should be used for comparison with the simulations and the rundown values indicated by the gauge should be ignored.

Table 3.1 Parameters for four experiment runs of different wave types, and the measured runup for each case.

Type of waves	Wave period (s)	Initial wave amplitude (m)	Experimental runup (m)
Sine wave	20	0.20	0.571
Biharmonic wave	20	0.12	0.794
Wake-like train	20→10	~0.10	0.517
Positive pulse	20	0.15	0.438

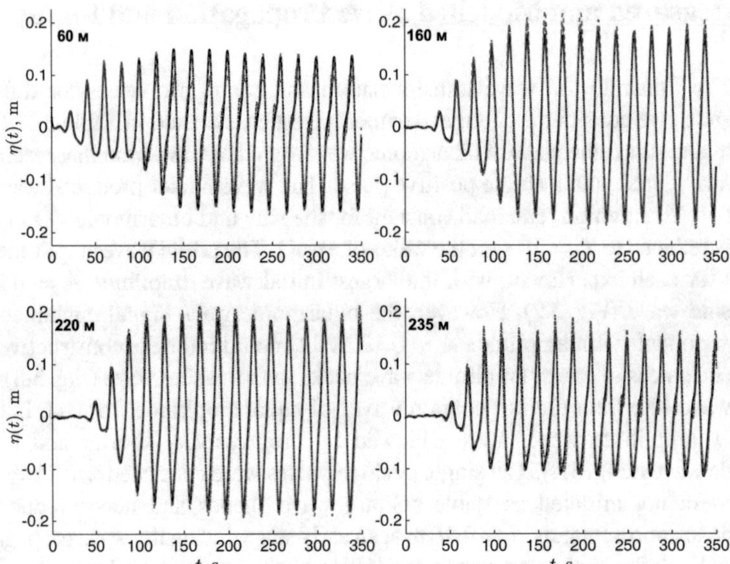

(a) Water surface elevation at different wave gauges (x = 60 m, 160 m, 220 m, and 235 m from the wave maker).

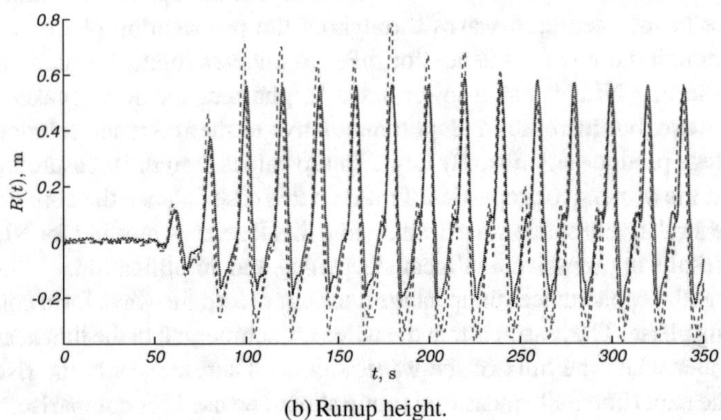

(b) Runup height.

Fig. 3.2 Wave propagation and runup for a sine wave with A = 0.2 m and T = 20 s on a beach slope $\tan\alpha$ = 1:6, mPer is shown with the red dashed line, NLSW solution is shown with blue dash dots line and the experimental record is shown with the black dotted line.

Figure 3.3 shows the biharmonic wave propagation and runup. In this case we again see a reasonably good agreement in wave structure between mPer and the measurements, but there is a clear tendency that mPer underestimates the wave amplitude both in the propagation phase and the runup phase. The NLSW solution looks fairly reasonable in the early stages, but significant discrepancies appear at $x = 180$ m and $x = 230$ m in the later stages of the wave train. The biharmonic

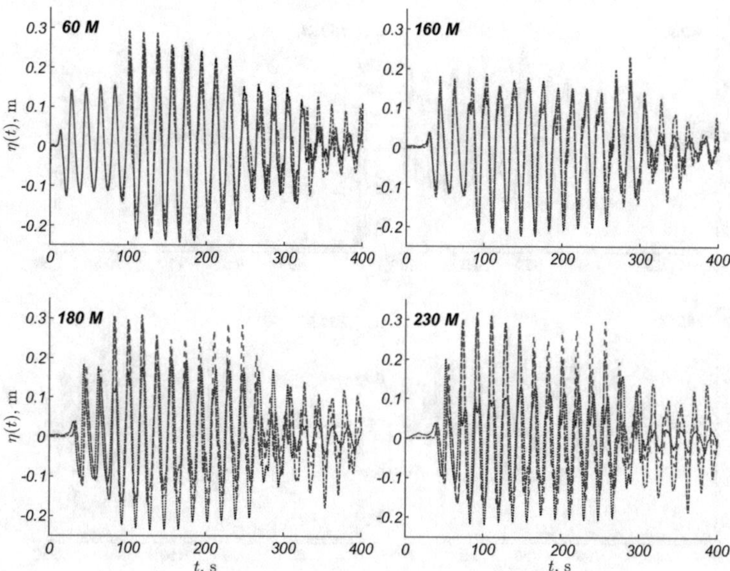

(a) Water surface elevation at different wave gauges (x = 60 m, 160 m, 180 m, and 230 m from the wave maker).

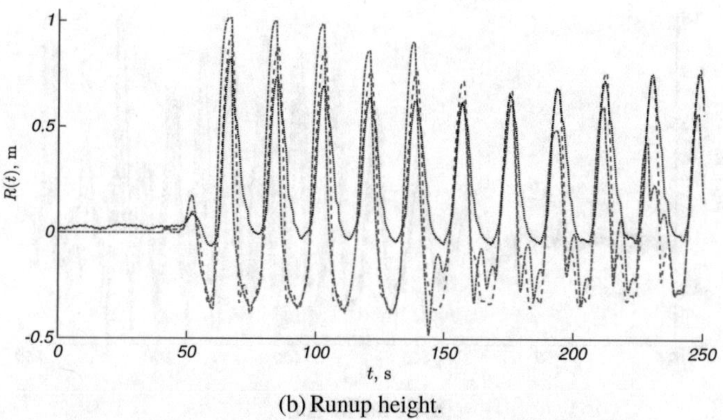

(b) Runup height.

Fig. 3.3 Wave propagation and runup for a biharmonic wave with A = 0.12 m and T = 20 s on a beach slope $\tan\alpha$ = 1:6, mPer is shown with the red dashed line, NLSW solution is shown with Blue dash dots line and the experimental record is shown with the black dots line.

signal is particularly sensitive to the phase speed as wave components may interfere both constructively and destructively at different stages, hence the inclusion of wave dispersion plays a significant role in this case.

Figure 3.4 shows a wave train with a wake-like structure, with a distinct envelope shape created by an initial long, low amplitude wave followed by shorter, higher amplitude waves. The initial phase of the wave train is captured well by both mPer

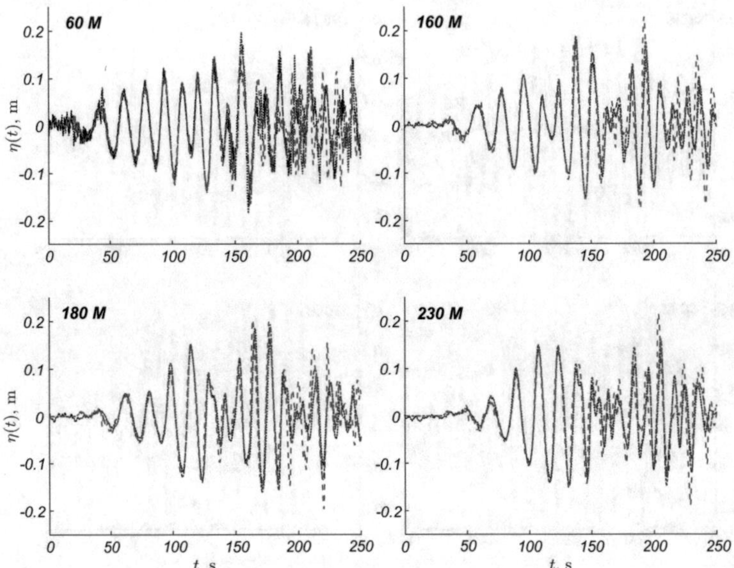

(a) Water surface elevation at different wave gauges ($x = 60$ m, 160 m, 180 m, and 230 m from the wave maker).

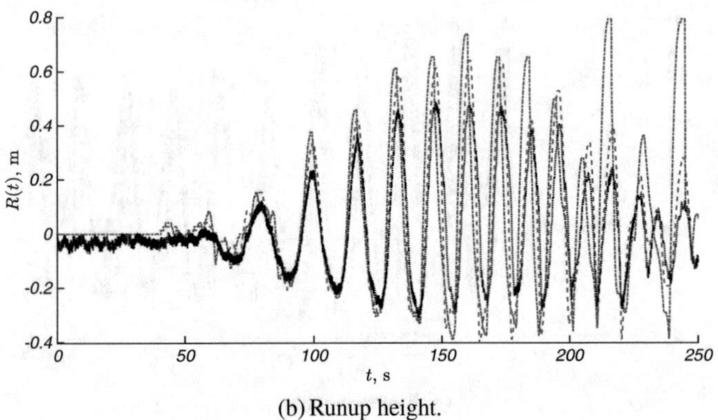

(b) Runup height.

Fig. 3.4 Wave propagation and runup for a wake-like wave train with $A = 0.1$ m and $T \in [10, 20]$ s on a beach slope $\tan\alpha = 1{:}6$, mPer is shown with the red dashed line, NLSW is shown with blue dash dots line and the experimental record is shown with the black dotted line.

and NLSW, although both models struggle to reproduce the later stages of the wave train. Both models also reproduce the runup phase fairly well, although NLSW develops a slight phase shift relative to the reference solution, and both models severely overestimates the runup for the trailing waves.

Figure 3.5 shows the wave propagation and runup for single positive pulse waves. The model results for mPer and NLSW are remarkably similar for the propagation

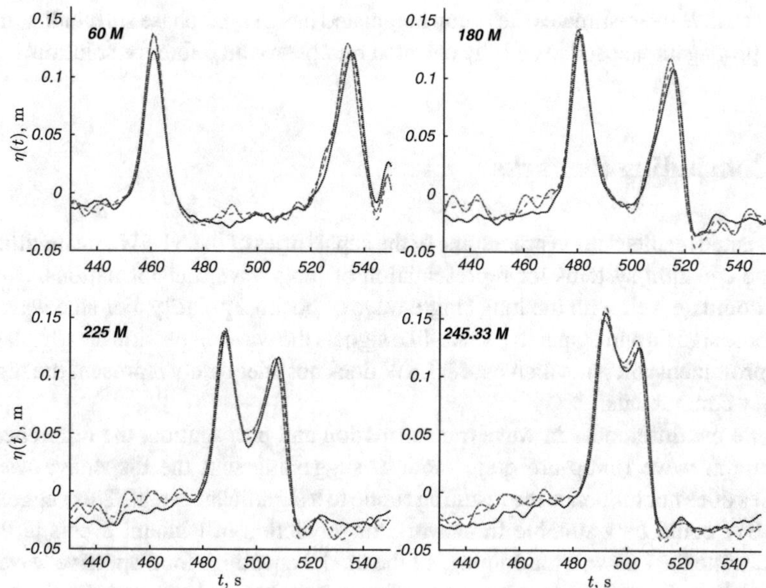

(a) Water surface elevation of a solitary wave at different wave gauges (x = 60 m, 180 m, 225 m, and 245.53 m from the wave maker).

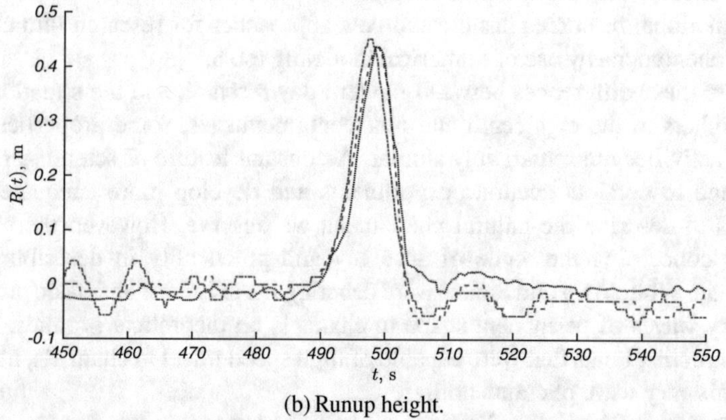

(b) Runup height.

Fig. 3.5 Wave propagation and runup for a single positive pulse (solitary wave) with A = 0.15 m and $T \in [10, 20]$ s on a beach slope $\tan\alpha$ = 1:6, mPer is shown with the red dashed line, NLSW is shown with blue dash dots line and the experimental record is shown with the black dotted line.

phase in this case, although both models tend to overestimate the wave amplitude slightly. This discrepancy can likely be explained by inaccuracies in the initial conditions for the wave, as it deviates slightly from a stable solitary wave form. The runup results are likewise very similar between mPer, NLSW and the reference solution, but again we see the tendency that mPer underestimates the runup height, whereas NLSW overestimates the runup height and has a slight phase shift indicating that the propagation speed is slightly elevated relative to the reference solution.

3.5 Concluding Remarks

The presented results demonstrate some of the capabilities of the NLSW and modified Peregrine equation systems for representation of long wave transformations. Both models compare well with the long single wave of positive polarity. For sine waves, biharmonic signals and dispersive wake-like signals the wave dispersion clearly plays a more prominent role, in which case NLSW does not adequately represent the high frequency components.

Despite the differences in wave transformation and propagation, the differences in maximum wave runup are quite modest, suggesting that the dispersive wave properties does not influence the resulting runup to a significant extent. This suggests that NLSW could be a suitable framework for prediction of tsunami events in the future, despite the known shortcomings of the model equations for dispersive waves.

Research into surface gravity wave phenomena has a long and fascinating history. Modern day researchers benefit greatly by working within a framework where theories for, e.g., Fourier analysis, ordinary and partial differential equations, potential theory, and perturbative methods, are well established. The emergence of computational resources has created new approaches for research into complex physical phenomena by use of numerical modeling tools.

Despite these differences between modern day research and the situation faced by researchers in the eighteenth and nineteenth centuries, some properties of the research activities are remarkably similar. A constant feature of scientific research is the need to conduct accurate experiments and develop more adequate model equations to describe the natural phenomena we observe. However, there is also a debate concerning the value of accuracy and practicality in describing these phenomena. While Airy and Stokes were debating the existence and basic properties of solitary waves of permanent shape in channels on theoretical grounds, Russell was constructing boats that were capable of high speed travel in channels, helped in part by this very wave phenomenon.

To some extent, a similar debate is on-going today within the tsunami research community, where on one side there is a need to develop models that represent fundamental properties of tsunami waves as accurately as possible in order to study the wave transformation and runup processes in detail, and on the other side there is a need to develop tools for operational forecasting of tsunami wave events that are adequate and practical for warning purposes.

Acknowledgements This work was supported by Estonian Research Council (ETAg) grant PUT1378. Authors also thank the PHC PARROT project No 37456YM, which funded the authors' visits to France and Estonia and allowed this collaboration.

References

Airy, G.B.: Tides and waves. Encyclopaedia Metropolitana (1845)

Bellotti, G., Brocchini, M.: On using Boussinesq-type equations near the shoreline: a note of caution. Ocean Eng. **29**(12), 1569–1575 (2001). https://doi.org/10.1016/s0029-8018(01)00092-0

Bogacki, P., Shampine, L.F.: A 3(2) pair of Runge–Kutta formulas. Appl. Math. Lett. **2**(4), 321–325 (1989). https://doi.org/10.1016/0893-9659(89)90079-7

Boussinesq, J.: Théorie de l'intumescence liquide appelée onde solitaire on de translation, se propageant dans un canal rectangulaire. C. R. Acad. Sci. Paris **72**, 755–759 (1871)

Boussinesq, J.: Théorie des ondes et des remous qui se propagent le long d'un canal rectangulaire horizontal, en communiquant au liquide contenu dans ce canal des vitesses sensiblement pareilles de la surface au fond. Journal de Mathématiques Pures et Appliquées **17**, 55–108 (1872)

Candella, R.N., Rabinovich, A.B., Thomson, R.E.: The 2004 Sumatra tsunami as recorded on the Atlantic coast of South America. Adv. Geosci. **14**, 117–128 (2008). https://doi.org/10.5194/adgeo-14-117-2008

Chen, M.: Exact solutions of various Boussinesq systems. Appl. Math. Lett. **11**(5), 45–49 (1998). https://doi.org/10.1016/S0893-9659(98)00078-0

Clarkson, P.A.: New exact solution of the Boussinesq equation. Eur. J. Appl. Math. **1**(3), 279–300 (1990). https://doi.org/10.1017/S095679250000022X

Craik, D.D.: The origins of water wave theory. Annu. Rev. Fluid Mech. **36**, 1–28 (2004). https://doi.org/10.1146/annurev.fluid.36.050802.122118

Craik, D.D .: Georges Gabriel Stokes on water wave theory. Annu. Rev. Fluid Mech. **37**, 23–42 (2005). https://doi.org/10.1146/annurev.fluid.37.061903.175836

Darrigol, O.: The spirited horse, the engineer, and the mathematician: Water waves in nineteenth-century hydrodynamics. Arch. History Exact Sci. **58**(1), 21–95 (2003). https://doi.org/10.1007/s00407-003-0070-5

Didenkulova, I., Pelinovsky, E.: Runup of tsunami waves in U-shaped bays. Pure Appl. Geophys. **168**(6–7), 1239–1249 (2011). https://doi.org/10.1007/s00024-010-0232-8

Didenkulova, I., Pelinovsky, E., Soomere, T., Zahibo, N.: Runup of nonlinear asymmetric waves on a plane beach. In: Kundu, A. (ed.) Tsunami & Nonlinear Waves, pp. 175–190. Springer, Berlin (2007). https://doi.org/10.1007/978-3-540-71256-5_8

Didenkulova, I., Denissenko, P., Rodin, A., Pelinovsky, E.: Effect of asymmetry of incident wave on the maximum runup height. J. Coastal Res. **65**, 207–212 (2013). https://doi.org/10.2112/SI65-036.1

Durán, A., Dutykh, D., Mitsotakis, D.: Peregrine's system revisited. In: Abcha, N., Pelinovsky, E., Mutabazi, I. (eds.) Nonlinear Waves and Pattern Dynamics, pp. 3–43. Springer, Cham (2018). https://doi.org/10.1007/978-3-319-78193-8_1

Dutykh, D., Katsaounis, T., Mitsotakis, D.: Finite volume schemes for dispersive wave propagation and runup. J. Comput. Phys. **230**, 3035–3061 (2011). https://doi.org/10.1016/j.jcp.2011.01.003

Fenton, J.D.: The long wave equations. Alternative Hydraulics Paper 1, http://johndfenton.com/Papers/01-The-long-wave-equations.pdf (2010)

Glimsdal, S., Pedersen, G.K., Harbitz, C.B., Løvholt, F.: Dispersion of tsunamis: does it really matter? Nat. Hazards Earth. Syst. Sci. **13**, 1507–1526 (2013). https://doi.org/10.5194/nhess-13-1507-2013

Harten, A., Osher, S.: Uniformly high-order accurate nonscillatory schemes. SIAM J. Numer. Anal. **24**, 279–309 (1987). https://doi.org/10.1137/0724022

Horrillo, J., Kowalik, Z., Shigihara, Y.: Wave dispersion study in the Indian Ocean-tsunami of December 26, 2004. Marine Geodesy **29**(3), 149–166 (2006). https://doi.org/10.1080/01490410600939140

Korteweg, D.J., de Vries, G.: On the change of form of long waves advancing in a rectangular channel, and on a new type of long stationary waves. Phil. Mag. **39**, 422–443 (1895). https://doi.org/10.1080/14786449508620739

Kundu, P.K. Fluid Mechanics. Academic Press, San Diego (1990)
Løvholt, F., Pedersen, G., Bazin, S., Kühn, D., Bredesen, R.E., Harbitz, C.: Stochastic analysis of tsunami runup due to heterogeneous coseismic slip and dispersion. J. Geophys. Res.–Oceans **117**, C03047 (2012). https://doi.org/10.1029/2011jc007616
Madsen, P.A., Banijamali, B., Schäffer, H.A., Sørensen, O.R.: Boussinesq type equations with high accuracy in dispersion and nonlinearity. In: Edge, B. (ed.) Coastal Engineering 25th Conference 1996, vol. 1, pp. 95–108. ASCE, Hørsholm, Denmark (1997). https://doi.org/10.1061/9780784402429.008
Pelinovsky, E.N., Mazova, R.K.: Exact analytical solutions of nonlinear problems of tsunami wave run-up on slopes with different profiles. Natural Hazards **6**(3), 227–249 (1992). https://doi.org/10.1007/BF00129510
Peregrine, D.H.: Long waves on a beach. J. Fluid Mech. **27**, 815–827 (1967). https://doi.org/10.1017/s0022112067002605
[Lord] Rayleigh: On waves. Phil. Mag. **1**, 257–279 (1876)
Russell, J.S.: Report on waves. In: Report of the Fourteenth Meeting of the British Association for the Advancement of Science, pp. 311–390 (1844)
de Saint-Venant, A.J.C.B.: Théorie du mouvement non permanent des eaux, avec applications aux crues des rivières et à l'introduction des marées dans leur lit. C. R. Acad. Sci. Paris **73**, 147–154 (1871)
Stokes, G.G.: Report on recent researches in hydrodynamics. Rep. Br. Assoc. Adv. Sci. 1–20 (1846)
Yan, Z., Zhang, H.: New explicit and exact travelling wave solutions for a system of variant Boussinesq equations in mathematical physics. Phys. Lett. A **252**(6), 291–296 (1999). https://doi.org/10.1016/S0375-9601(98)00956-6

Chapter 4
Nonlinear Models of Finite Amplitude Interfacial Waves in Shallow Two-Layer Fluid

Oxana Kurkina, Andrey Kurkin, Efim Pelinovsky, Yury Stepanyants, and Tatiana Talipova

Abstract We present an example of systematic use of asymptotic methods for the description of long waves at the interface between two fluids of different densities. The governing equations for weakly nonlinear long interfacial waves are consistently derived for a general case of nonpotential flow and taking into account surface tension between two layers. Particular attention is given to the situations when some terms in the resulting fifth order KdV-type evolution equation become small, vanish, or change their sign.

4.1 Introduction

The dynamics of internal waves (IWs) is studied traditionally within the framework of simplified models. In many occasions the stratified medium in which IWs may propagate consists of several clearly distinguishable layers separated by thin sheets

O. Kurkina (✉) · A. Kurkin
Nizhny Novgorod State Technical University n.a. R.E. Alekseev, Nizhny Novgorod, Russia

E. Pelinovsky
Federal Research Center Institute of Applied Physics of the Russian Academy of Sciences (IAP RAS), Nizhny Novgorod, Russia

National Research University–Higher School of Economics, Nizhny Novgorod, Russia

University of Southern Queensland, Toowoomba, QLD, Australia
e-mail: pelinovsky@hydro.appl.sci-nnov.ru

Y. Stepanyants
University of Southern Queensland, Toowoomba, QLD, Australia
e-mail: Yury.Stepanyants@usq.edu.au

T. Talipova
Nizhny Novgorod State Technical University n.a. R.E. Alekseev, Nizhny Novgorod, Russia

Federal Research Center Institute of Applied Physics of the Russian Academy of Sciences (IAP RAS), Nizhny Novgorod, Russia

A. Berezovski, T. Soomere (eds.), *Applied Wave Mathematics II*, Mathematics of Planet Earth 6, https://doi.org/10.1007/978-3-030-29951-4_4

in which the properties of the medium rapidly vary. In such occasions even a mostly smoothly stratified medium can be adequately considered as a layered medium with a different density in each layer. In many cases it is quite sufficient to consider a two- or three-layer model to study the peculiarities of IWs (Helfrich and Melville, 2006).

The most popular and well studied environment of this kind is the two-layer model. It has minimum free parameters and is still applicable for the analysis of laboratory experiments as well as for the interpretation of field data (Ostrovsky and Stepanyants, 2005; Apel et al., 2007). As an example of a mathematical description of a classic water wave problem within the framework of this model, we consider first in detail the linear problem (Leibovich and Seebass, 1974).

The Korteweg–de Vries (KdV) equation is widely used to describe long nonlinear IWs (Ablowitz and Segur, 1981; Apel et al., 2007). Equally popular are its various generalisations that contain terms responsible for dissipation, inhomogeneity of the medium, shear flows, rotation of the fluid as a whole, etc. (Grimshaw et al., 1998, 2002a, 2010a; Apel et al., 2007; Holloway et al., 1999).

However, in many cases, the derivation of the basic model equation is not fully consistent. The reason is that initially it is assumed that the fluid motion is potential, and the KdV equation (derived more or less consistently) is complemented manually with essentially nonpotential terms describing the effects mentioned above. The derivation of the basic equation itself is often presented in a complicated form.

Therefore it seems reasonable to describe here in detail an example of the relatively simple and consistent derivation of the generalised fifth order KdV equation for IWs in a two-layer fluid, taking into account the surface tension between the layers (liquids in the layers can be immiscible, in general). The derivation is not constrained by the assumption of potential motion in the layers. Therefore, it is equally applicable both to the potential (irrotational) and nonpotential flows. In the latter case, such effects as the viscosity, inhomogeneity (including the vertical shear of background flow), rotation of the fluid as a whole, etc., can be included without any difficulty or loss of generality.

The derived equation is analysed in order to identify special particular cases, interesting both from the viewpoint of their mathematical structure and features of their solutions, and from the viewpoint of practical applications. We have selected the cases when the resulting equation is applicable to the description of IWs both in the ocean and in the laboratory experiments, as well as to physicochemical industrial processes that involve IWs in thin layers of immiscible liquids such as kerosene and water, or oil and water, or other combinations of liquids.

4.2 Linear Theory of Internal Waves in Layered Fluids

Following the traditional approach (see, e.g., Miropol'sky, 2001), we consider first the linearised system of equations of motion of an ideal incompressible layered fluid in the vertical xOz plane, where x is the horizontal coordinate and z is the vertical coordinate directed upwards, u and w are the components of velocity in the

x- and z-directions, respectively, t is time, g is the acceleration due to gravity. Let us represent the density ρ and pressure p as

$$\rho = \overline{\rho}(z) + \tilde{\rho}(x, z, t), \quad p = \overline{p}(z) + \tilde{p}(x, z, t), \tag{4.1}$$

where $\overline{\rho}(z)$ and $\overline{p}(z)$ are the density and the mean pressure of the layered fluid at rest, which are related by the hydrostatic equation:

$$\overline{p}_z = -g\overline{\rho}. \tag{4.2}$$

The linearised equations of motion are:

$$\overline{\rho} u_t = -\tilde{p}_x, \tag{4.3}$$

$$\overline{\rho} w_t = -\tilde{p}_z - g\tilde{\rho}, \tag{4.4}$$

where subscripts denote partial derivatives. The continuity equation

$$\rho_t + (\rho u)_x + (\rho w)_z = 0 \tag{4.5}$$

with the condition of incompressibility

$$\rho_t + u\rho_x + w\rho_z = 0 \tag{4.6}$$

reduces to

$$u_x + w_z = 0. \tag{4.7}$$

The linearised condition of incompressibility (4.6) takes the form:

$$\tilde{\rho}_t + w\overline{\rho}_z = 0. \tag{4.8}$$

The set of Eqs. (4.3), (4.4), (4.7), (4.8) should be supplemented by boundary conditions. For internal waves we will use the "rigid lid approximation" on the surface and assume that the bottom is impermeable. In other words, there is no vertical motion both at the surface $z = z_s$ and at the bottom $z = z_b$:

$$w(z = z_s) = 0, \tag{4.9}$$

$$w(z = z_b) = 0. \tag{4.10}$$

Equation (4.7) allows the introduction of the stream function ψ, which is defined as follows:

$$u = \psi_z, \quad w = -\psi_x. \tag{4.11}$$

The solution of the linearised set of Eqs. (4.3), (4.4), (4.7), and (4.8) with the boundary conditions (4.9), (4.10) can be sought in the form of a harmonic wave propagating in the x-direction:

$$\psi = f(z)e^{-i(kx-\omega t)}, \tag{4.12}$$

where $f(z)$ is an integrable function that describes the vertical structure of the fluid motion generated by the propagation of this wave, k is the wavenumber, and ω is the wave frequency.

Then using (4.3), (4.4), (4.7), and (4.11) and eliminating p' and ρ', we obtain the equation to determine $f(z)$:

$$(\overline{\rho} f')' - k^2\left(\overline{\rho} + \frac{g\overline{\rho}'}{\omega^2}\right) f = 0, \tag{4.13}$$

where the prime denotes the full derivative with respect to z.

In accordance with (4.9), (4.10), and with the definition (4.11), the boundary conditions for (4.13) can be written as:

$$f(z_b) = 0, \quad f(z_s) = 0. \tag{4.14}$$

We consider here the case where the fluid density in a layered fluid is described by a piecewise constant function with discontinuities of the first kind at certain points $\{z_k\}$. The derivative of unperturbed density can be expressed in terms of generalised functions:

$$\overline{\rho}' = \left[\overline{\rho}'(z)\right] + \sum_k [\overline{\rho}]_{z_k}\, \delta(z - z_k), \tag{4.15}$$

where $[\overline{\rho}]_{z_k} = \overline{\rho}(z_k + 0) - \overline{\rho}(z_k - 0) \equiv \Delta\rho_k$ is the density jump at the point $z = z_k$ and $\left[\overline{\rho}'(z)\right]$ is a piecewise continuous function.

Let us assume that the density function $\overline{\rho} = \overline{\rho}(z)$ has a discontinuity at $z = z_0$ where its value changes from $\overline{\rho}_1$ to $\overline{\rho}_2 \leq \overline{\rho}_1$. As usual, we assume that the layer of heavier fluid is below the layer of lighter fluid. The boundary condition at the interface $z = z_0$ between the layers can be obtained by integrating (4.13) over the ε-neighbourhood of the discontinuity:

$$\int\limits_{z_0-\varepsilon}^{z_0+\varepsilon} (\overline{\rho} f')'\, \mathrm{d}z - k^2 \int\limits_{z_0-\varepsilon}^{z_0+\varepsilon} \overline{\rho} f\, \mathrm{d}z - \frac{k^2 g}{\omega^2} \int\limits_{z_0-\varepsilon}^{z_0+\varepsilon} \overline{\rho}' f\, \mathrm{d}z = 0. \tag{4.16}$$

Using the mean value theorem, it is easy to show that $\int\limits_{z_0-\varepsilon}^{z_0+\varepsilon} \overline{\rho} f \mathrm{d}z = 0$ when $\varepsilon \to 0$, because $\overline{\rho}(z) > 0$ is a decreasing bounded function, and f is an integrable function.

Substituting expression (4.15) into (4.16) and applying the rule of integrating the delta function

$$\int_{-\infty}^{+\infty} f(z)\delta(z-z_0)\,\mathrm{d}z = \int_{z_0-\varepsilon}^{z_0+\varepsilon} f(z)\delta(z-z_0)\,\mathrm{d}z \equiv f(z_0), \tag{4.17}$$

we obtain the following boundary condition:

$$\overline{\rho}_b f'(z_0+0) - \overline{\rho}_a f'(z_0-0) + \frac{g}{\omega^2}(\overline{\rho}_a - \overline{\rho}_b) f(z_0) = 0. \tag{4.18}$$

Even though functions $\overline{\rho} f'$ and $\overline{\rho}'$ change abruptly at $z = z_0$, function $f(z)$ is continuous at this point:

$$\lim_{z\to z_0-0} f = \lim_{z\to z_0+0} f = f(z_0). \tag{4.19}$$

The solution of Eqs. (4.13), (4.14), (4.18), (4.19) allows us to determine function $f(z)$, as well as the relationship between the frequency ω and the wave number k of the harmonic wave, and thus, find the dispersion relation.

Once $f(z)$ is found, u, w, and $\tilde{\rho}$ can be obtained using the real parts of the expressions:

$$u = \mathrm{Re}\left[f'(z)e^{i(kx-\omega t)}\right],\quad w = \mathrm{Re}\left[-ikf(z)e^{i(kx-\omega t)}\right],\quad \tilde{\rho} = \mathrm{Re}\left(-\frac{i}{\omega} w \frac{\mathrm{d}\overline{\rho}}{\mathrm{d}z}\right). \tag{4.20}$$

For (4.13) with boundary conditions (4.14), the Sturm–Bôcher theorem guarantees the existence of eigenvalues ω^2. From this theorem it follows that each of the eigenfunctions f_n, $n = 0, 1, 2, \ldots$ (solutions of the system Eqs. (4.13), (4.14) corresponding to the parameter $\lambda = \lambda_n$) has $n+2$ zeros in the interval $z_s \le z \le z_b$. Here $\lambda = \omega^{-2}$ ($\lambda = \lambda_0, \lambda_1, \ldots$ changes in the range from Λ_1 to $\Lambda_2 > \Lambda_1$), and n is the number of the particular eigenfunction.

Thus, the vertical structure of the lowest mode (that has the largest frequency and phase velocity) is described by the function f_0. This function has no zeros in the bulk of the fluid but vanishes at the bottom and surface. Each subsequent (higher) mode will have a smaller frequency and lower phase velocity and one more zero within the bulk of the fluid. For the N-layer fluid bounded by the solid horizontal bottom and surface, there exist $N-1$ internal wave modes (Yih, 1960; Leibovich and Seebass, 1974).

4.3 Linear Theory of Monochromatic Wave Propagation in a Two-Layer Fluid

Let us consider the propagation of a harmonic wave in the simplest model of layered fluids, a two-layer fluid. The lower layer ($z_b \le z \le z_0$) of the fluid has the density $\rho_1 = \text{const}$, and the upper layer ($z_0 \le z \le z_s$) has the density $\rho_2 = \text{const}$, where $\rho_1 > \rho_2$ (Fig. 4.1).

In this case (4.13) with boundary conditions (4.14) take the form

$$f_i'' - k^2 f_i = 0, \quad i = 1, 2, \qquad f_1(z_b) = 0, \qquad f_2(z_s) = 0 \tag{4.21}$$

and conditions (4.15) and (4.16) can be written as

$$f_1(z_0) = f_2(z_0), \quad \rho_2 f_2' - \rho_1 f_1' - \frac{g}{\omega^2}(\rho_2 - \rho_1) f_1(z_0) = 0, \tag{4.22}$$

where

$$f = f_1(z), \quad z_b \le z \le z_0, \qquad f = f_2(z), \quad z_0 \le z \le z_s. \tag{4.23}$$

Substituting the solution of (4.21) in terms of hyperbolic functions

$$f_1 = A \sinh k(z - z_b), \quad f_2 = B \sinh k(z - z_s), \tag{4.24}$$

into (4.22), and denoting the thickness of the lower layer $z_0 - z_b = h_1$, and the thickness of the upper layer $z_s - z_0 - h_2$, one can find the dispersion relation of the

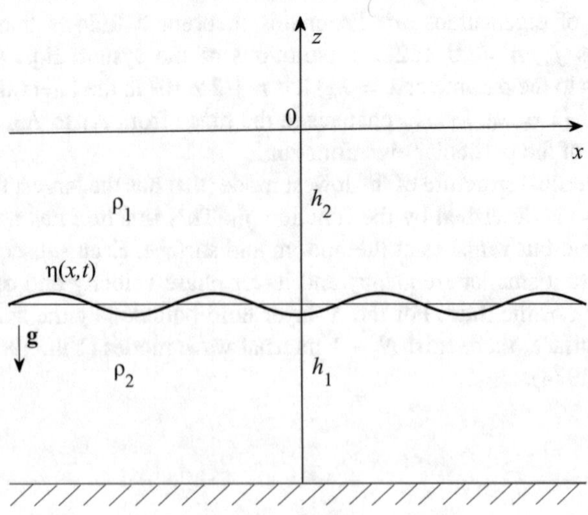

Fig. 4.1 Interfacial waves in a two-layer fluid.

internal wave:

$$\omega^2 = \frac{kg(\rho_1 - \rho_2)}{\rho_1 \coth kh_1 + \rho_2 \coth kh_2} \,. \tag{4.25}$$

Equation (4.25) highlights the fact that only one wave mode exists in the considered two-layer stratification. Two branches of relation (4.25), $\omega_{1,2} = \pm\sqrt{\omega^2(k)}$, correspond to waves propagating in opposite directions.

In practical applications it is almost always the case that the Boussinesq approximation can be used:

$$\rho_2 = \rho, \quad \rho_1 = \rho + \Delta\rho, \quad \frac{\Delta\rho}{\rho} \ll 1, \tag{4.26}$$

where $\Delta\rho \ll 1$, but $\Delta\rho g = \text{const}$. The dispersion relation (4.25) in the Boussinesq approximation takes the form:

$$\omega^2 = \frac{kg\Delta\rho}{\rho\,(\coth kh_1 + \coth kh_2)} \,. \tag{4.27}$$

Waves are called long if their wavelength substantially exceeds the water depth. For long internal waves (in the shallow water approximation), the relation $kH = k(h_1 + h_2) \ll 1$ is valid, where H is the total fluid depth, $H = h_1 + h_2 = z_s - z_b$. At small k, corresponding to long waves, function $\omega = \omega(k)$ can be expanded into the Taylor series in the neighbourhood of $k = 0$ with the necessary accuracy up to $O[(kH_2)^m]$:

$$\omega = \omega(0) + \sum_{n=1}^{m} \left.\frac{\mathrm{d}^n\omega}{\mathrm{d}k^n}\right|_{k=0} \frac{1}{n!} k^n + o(k^m), \quad \omega(0) = 0 \,. \tag{4.28}$$

If in expansion (4.28) the terms of the order $o(kH)$ can be neglected (for long waves), then setting $m = 1$, we obtain the linear version of the dispersion relation valid for very long waves:

$$\omega = \left.\frac{\mathrm{d}\omega}{\mathrm{d}k}\right|_{k=0} k \,. \tag{4.29}$$

The coefficient of long wave dispersion (the Boussinesq dispersion) can be determined by considering the next approximation of shallow water theory, where the first nonzero correction to (4.29) is also taken into account. The relevant procedure leads to the expression:

$$\omega = \left.\frac{\mathrm{d}\omega}{\mathrm{d}k}\right|_{k=0} k + \left.\frac{\mathrm{d}^3\omega}{\mathrm{d}k^3}\right|_{k=0} k^3 = ck - \beta k^3, \tag{4.30}$$

where c is the phase speed of linear long interfacial waves, given by the following expression:

$$c = \sqrt{gHl(1-l)\frac{\Delta\rho}{\rho}}, \tag{4.31}$$

and β is the dispersion coefficient:

$$\beta = \frac{cl(1-l)H^2}{6}. \tag{4.32}$$

We have introduced the notation $l = h_1/H$. As can be seen from expressions (4.31) and (4.32), for a fixed value of wave number, the depth of the liquid H and the density jump $\Delta\rho$ at the interface, the phase velocity of the wave has a (global) maximum at $l = 1/2$. An increase in the difference $\Delta\rho$ in density values of the lower and upper layers also leads to the increase in phase velocity for the same values of H, l and k. Since for every point (k_0, ω_0) of the dispersion curve the value and sign of phase speed c is determined by the tangent of the angle of inclination of the secant, drawn from the origin to the point (k_0, ω_0), and the group velocity is equal to the tangent angle at (k_0, ω_0), then according to (4.27) the two-layer fluid under consideration has a normal dispersion (Ostrovsky and Potapov, 2002).

The vertical structure of the stream function of a two-layer fluid in the long wave approximation without dispersion is determined by (4.22)–(4.24) with the accuracy of a constant multiple and has the form:

$$\begin{aligned} f(z) &= \tilde{C}(z - z_b)k, \qquad z_b \le z < z_0, \\ f(z) &= -\frac{h_1}{h_2}\tilde{C}(z - z_s)k, \quad z_0 \le z \le z_s, \end{aligned} \tag{4.33}$$

where $\tilde{C}$ is an arbitrary constant. From the physical point of view, it is convenient to choose this constant so that

$$f_{\max}(z^* = h_1) = 1, \tag{4.34}$$

which yields

$$\tilde{C} = \frac{1}{kh_1}. \tag{4.35}$$

The so-called mode function $f(z)$, determined by (4.33) and (4.35), represents a piecewise linear (triangular) function with a maximum at the interface of the layers and zeros at the bottom and surface of the fluid.

The framework of the linear theory is, strictly speaking, only applicable for studies of infinitely small amplitude waves, for which the effects of nonlinearity can be neglected. A natural generalisation of this theory, called the weakly nonlinear approach, allows the study of waves of small but finite amplitude. A specific balance of nonlinear effects and dispersion in many environments, including the layered fluids considered in this chapter, leads to the existence of long solitary waves and solitons. Their appearance, dynamics and propagation can be described in terms of stationary localised solutions of nonlinear evolution equations of the KdV family with coefficients depending on the parameters of the medium. The coefficients of such equations generally depend on the parameters of the medium. Even though the existence of this type of waves has been discovered long time ago, there are many unexplored issues related to the conditions of the existence, properties, appearance and dynamics of solitary internal waves and wave packets.

The derivation of nonlinear evolution equations for waves in a layered fluid from the governing equations has been carried out in the literature for about 50 years mainly using asymptotic methods. The derivation of nonlinear models is also possible through the systematic perturbation of the Hamiltonian representing the governing system (Craig and Groves, 2000; Craig et al., 2004). Along with the use of the technique of generalised functions mentioned in the previous sections on linear waves, an alternative method can be used. This method reduces to the consideration of physical fields and formulation of equations for each layer separately, matching solutions obtained for each layer at the moving and unknown a priori interfaces between the layers. Technically this method looks a bit simpler, despite the large dimension of the set of partial differential equations which must be solved. Below we will demonstrate the application of such an approach to the derivation of weakly nonlinear evolution equations describing long waves in a two-layer fluid.

4.4 Fully Nonlinear Equations of Motion

Let us consider propagation of nonlinear internal waves at the interface between two ideal fluids of finite depth. Let us denote the displacement of the interface from its horizontal position by $\eta(x,t)$, the horizontal components of fluid velocity by $u_{1,2}$, and the vertical components by $w_{1,2}$, where the subscripts 1 and 2 refer to the lower and upper layers, respectively.

We present the system of hydrodynamic equations for each of the layers. For $z_b < z < z_0 + \eta$ (in the lower layer) it reads:

$$\frac{\partial u_1}{\partial x} + \frac{\partial w_1}{\partial z} = 0,$$

$$\rho_1 \left(\frac{\partial u_1}{\partial t} + u_1 \frac{\partial u_1}{\partial x} + w_1 \frac{\partial u_1}{\partial z} \right) + \frac{\partial p_1}{\partial x} = 0, \tag{4.36}$$

$$\rho_1 \left(\frac{\partial w_1}{\partial t} + u_1 \frac{\partial w_1}{\partial x} + w_1 \frac{\partial w_1}{\partial z} + g \right) + \frac{\partial p_1}{\partial z} = 0.$$

For $z_0 + \eta < z < z_s$ (in the upper layer) it reads:

$$\frac{\partial u_2}{\partial x} + \frac{\partial w_2}{\partial z} = 0,$$

$$\rho_2 \left(\frac{\partial u_2}{\partial t} + u_2 \frac{\partial u_2}{\partial x} + w_2 \frac{\partial u_1}{\partial z} \right) + \frac{\partial p_2}{\partial x} = 0, \tag{4.37}$$

$$\rho_2 \left(\frac{\partial w_2}{\partial t} + u_2 \frac{\partial w_2}{\partial x} + w_2 \frac{\partial w_2}{\partial z} + g \right) + \frac{\partial p_2}{\partial z} = 0.$$

The boundary conditions at the interface between the layers $z = z_0 + \eta(x, t)$ are:

$$\frac{\partial \eta}{\partial t} + u_1 \frac{\partial \eta}{\partial x} = w_1,$$

$$\frac{\partial \eta}{\partial t} + u_2 \frac{\partial \eta}{\partial x} = w_2, \tag{4.38}$$

$$p_2 - p_1 = \sigma \eta_{xx} \left[1 + (\eta_x)^2 \right]^{-3/2},$$

where σ is the coefficient of surface tension between fluid layers with different physical properties.

The boundary conditions for the vertical velocity component at the impermeable bottom and surface are:

$$w_1(z = z_b) = 0, \quad w_2(z = z_s) = 0. \tag{4.39}$$

They express the above assertions that (i) the vertical velocity vanishes at the bottom and (ii) the water surface does not move in the vertical direction. Here we used the traditional "rigid lid" approximation at the surface, which filters out the surface mode.

Equations (4.36)–(4.37) with boundary conditions (4.38) at the interface between the layers and (4.39) at the bottom and surface are the basic equations in the Eulerian formulation.

4.5 Asymptotic Expansion

To proceed with the asymptotic procedure, it is necessary to identify small parameters characterising the problem. This is commonly done based on the geometry of the environment and the scale of phenomena of interest.

We focus on long waves, that is, waves whose horizontal scale (wavelength) L far exceeds the total depth of the fluid H. Differently from the perfectly linear framework, the characteristic amplitude of disturbances A (as well as the wave height) is assumed to be finite but small compared to the water depth H. This difference in the magnitudes makes it possible to introduce certain small parameters and to proceed in so-called weakly nonlinear framework. Specifically, we introduce the following small parameters: $\varepsilon = A/H$ that characterises the nonlinearity and $\mu = H^2/L^2$ that characterises the dispersion of waves. It is traditionally assumed that these parameters have the same order of magnitude: $\varepsilon \sim \mu$.

The boundary conditions (4.38) are formulated at the interface $\eta(x,t)$ of the layers. The location of this interface is not known beforehand and thus is a function to be determined. Assuming the smallness of all perturbations in the system, all unknown functions that depend on the vertical coordinate in these boundary conditions can be expanded into a Taylor series in terms of small deviations from the undisturbed horizontal level:

$$\phi(x, z = z_0 + \eta, t) = \sum_{n=0}^{\infty} \frac{\eta^n}{n!} \frac{\partial \phi}{\partial z^n}\bigg|_{z=z_0}. \tag{4.40}$$

The dynamic condition (the third equation in (4.38), below referred to as $(4.38)_3$ is written in a fully nonlinear form. However, further calculations show that in many occasions (and with the accuracy accepted in this study for the derivation of the final equations) the linear approximation of this condition is sufficient to take into account the effect of surface tension.. This follows from the first correction term in the expansion of the curvature of the interface

$$\eta_{xx}\left[1 + (\eta_x)^2\right]^{-3/2} \approx \eta_{xx}\left[1 - 3(\eta_x)^2/2\right].$$

This term is cubic in amplitude and fourth order of smallness in dispersion and thus very small for small ε and $\mu = H^2/L^2$. Therefore, it is acceptable to ignore this term within the limits of the required accuracy. In order to avoid unnecessary complications of calculations, we further restrict ourselves to the traditional representation of the curvature of the surface as $\partial^2\eta/\partial x^2$.

The next stage of the asymptotic procedure consists of expanding the unknown functions into a series in terms of the small parameter ε:

$$\begin{aligned}
\eta &= \varepsilon(\eta_0 + \varepsilon\eta_1 + \varepsilon^2\eta_2 + \ldots),\\
u_1 &= \varepsilon(u_{10} + \varepsilon u_{11} + \varepsilon^2 u_{12} + \ldots),\\
u_2 &= \varepsilon(u_{20} + \varepsilon u_{21} + \varepsilon^2 u_{22} + \ldots),\\
w_1 &= \varepsilon^{3/2}(w_{10} + \varepsilon w_{11} + \varepsilon^2 w_{12} + \ldots),\\
w_2 &= \varepsilon^{3/2}(w_{20} + \varepsilon w_{21} + \varepsilon^2 w_{22} + \ldots),\\
p_1 &= p_{10} - \rho_1 g z + \varepsilon(p_{11} + \varepsilon p_{12} + \varepsilon^2 p_{13} + \ldots),\\
p_2 &= p_{20} - \rho_2 g z + \varepsilon(p_{21} + \varepsilon p_{22} + \varepsilon^2 p_{23} + \ldots).
\end{aligned} \tag{4.41}$$

Next, we change the variables (x, t) to the "slow" time and "slow" coordinate (Engelbrecht et al., 1988). Let us denote by c the phase velocity of long linear waves (which is yet to be determined too) and introduce "slow" variables:

$$\xi = \varepsilon^{1/2}(x - ct), \quad \tau = \varepsilon^{3/2}t. \tag{4.42}$$

The derivatives in space and time coordinates can be written in terms of slow variables as follows:

$$\frac{\partial}{\partial x} = \varepsilon^{1/2}\frac{\partial}{\partial \xi}, \quad \frac{\partial}{\partial t} = -\varepsilon^{1/2}c\frac{\partial}{\partial \xi} + \varepsilon^{3/2}\frac{\partial}{\partial \tau}.$$

Substituting the series (4.41) into (4.36)–(4.37) and the series (4.40) for u_1, u_2, w_1, w_2, p_1, and p_2 into boundary conditions (4.38) and using the boundary conditions (4.39), as well as the slow variables (4.42) makes it possible to solve successively the equations in each order in ε. The equations obtained from $(4.36)_1$, $(4.37)_1$, (4.38), (4.39) are divided by $\varepsilon^{3/2}$, and from $(4.36)_2$ and $(4.37)_2$—by $\varepsilon^{1/2}$.

One of the advantages of such a consistent approach is that at each stage the explicit expressions for the corresponding corrections (higher order terms) in the series of quantities representing all physical fields (velocity components and pressure) in each layer are connected with the terms of the asymptotic series for the displacement of the interface for which the evolution equation is derived. After solving this equation for the interface location analytically or numerically, one can easily describe the full dynamics of a two-layer fluid throughout the entire fluid column, including all physical fields induced by internal waves, not just the dynamics of the interface.

This allows the solving of further problems associated with the dynamic effects, to carry out the comparison with the results of laboratory experiments, full nonlinear modelling, etc. The characteristics of nonlinear flows, especially in the presence of solitary waves, are very different from laminar flow (for example, resistance force, substance transfer, etc.), which can lead to various interesting effects. Based on the derived model equation and its solutions, a method for flow control can be proposed. This is an example of a possible practical application of the problem considered here.

Let us analyse the equations obtained in each order of the parameter ε. For simplicity, we set $z_0 = 0$.

The Zeroth Order From $(4.36)_1$ and $(4.37)_1$ we obtain the terms on the order of ε^0:

$$\frac{\partial u_{10}}{\partial \xi} + \frac{\partial w_{10}}{\partial z} = 0, \quad \frac{\partial u_{20}}{\partial \xi} + \frac{\partial w_{20}}{\partial z} = 0. \tag{4.43}$$

These linear equations allow us to seek solutions in the form of separated variables:

$$u_{10}(\xi, z, \tau) = A_1^u(\xi, \tau)U_1(z), \quad u_{20}(\xi, z, \tau) = A_2^u(\xi, \tau)U_2(z),$$
$$w_{10}(\xi, z, \tau) = A_1^w(\xi, \tau)W_1(z), \quad w_{20}(\xi, z, \tau) = A_2^w(\xi, \tau)W_2(z). \tag{4.44}$$

Equations (4.43) with the expressions (4.44) lead to the relations:

$$\frac{A_1^w}{\partial A_1^u/\partial \xi} = -\frac{U_1}{dW_1/dz} = c_1 = \text{const}, \quad \frac{A_2^w}{\partial A_2^u/\partial \xi} = -\frac{U_2}{dW_2/dz} = c_2 = \text{const}, \tag{4.45}$$

where c_1 and c_2 are arbitrary constants.

Let us discuss the dimension of the introduced functions. Traditionally, when separating variables in wave problems, the functions that determine the structure of the solution in the waveguide coordinate (in our case, on the vertical coordinate z), are assumed to be dimensionless, and the functions depending on the coordinate of wave propagation have the dimension of the original quantity. Following this principle, we will treat functions $U_{1,2}$ and $W_{1,2}$ as nondimensional, and $A_{1,2}^u$, $A_{1,2}^w$ as having the dimension of velocity [m·s^{-1}]. Under this condition, the constants c_1 and c_2 have the dimensions of length [m]. It is convenient to define c_1 and c_2 as follows:

$$c_1 = cT = \text{const}, \quad c_2 = cT = \text{const},$$

where c is the previously introduced phase velocity of linear long waves (to be determined further), and T is the characteristic time scale of the wave process (for example, wave period).

From the boundary conditions (4.39) at the bottom and surface we obtain:

$$W_1(z = z_b) = 0, \quad W_2(z = z_s) = 0. \tag{4.46}$$

Equations $(4.36)_2$, $(4.36)_3$, $(4.37)_2$, and $(4.37)_3$ are satisfied identically at the zero order ε^0. The boundary conditions (4.38) in this approximation lead to the following equations, respectively:

$$c\frac{\partial \eta_0}{\partial \xi} + A_1^w W_1\big|_{z=0} = 0, \quad c\frac{\partial \eta_0}{\partial \xi} + A_2^w W_2\big|_{z=0} = 0,$$

$$[p_{20} - p_{10} - (\rho_2 - \rho_1)gz]\big|_{z=0} \equiv [p_{20} - p_{10}]\big|_{z=0} = 0.$$

By taking into account relations (4.45), we obtain:

$$T\frac{\partial A_{1,2}^u}{\partial \xi} W_{1,2}\big|_{z=0} = -\frac{\partial \eta_0}{\partial \xi}, \quad p_{20} = p_{10}. \tag{4.47}$$

Since functions $W_{1,2}$ are determined up to a constant multiple, the following normalisation condition can be imposed:

$$W_{1,2}(z = 0) = 1. \tag{4.48}$$

Then relations (4.47) yield:

$$A^u_{1,2} = -\eta_0/T\,.$$

The First Order At the next approximation (terms on the order of ε^1) $(4.36)_1$ and $(4.37)_1$ lead to the following equations:

$$\frac{\partial u_{11}}{\partial \xi} + \frac{\partial w_{11}}{\partial z} = 0\,, \quad \frac{\partial u_{21}}{\partial \xi} + \frac{\partial w_{21}}{\partial z} = 0\,. \tag{4.49}$$

On the same order in ε from $(4.36)_2$ and $(4.37)_2$ we obtain:

$$-\rho_1 c^2 \frac{\partial \eta_0}{\partial \xi}\frac{dW_1}{dz} + \frac{\partial p_{11}}{\partial \xi} = 0\,, \quad -\rho_2 c^2 \frac{\partial \eta_0}{\partial \xi}\frac{dW_2}{dz} + \frac{\partial p_{21}}{\partial \xi} = 0\,. \tag{4.50}$$

Equations $(4.36)_3$ and $(4.37)_3$ give:

$$\frac{\partial p_{11}}{\partial z} = 0\,, \quad \frac{\partial p_{21}}{\partial z} = 0\,. \tag{4.51}$$

Solving (4.50) for pressures yields:

$$p_{11} = \rho_1 c^2 \eta_0 \frac{\mathrm{d}W_1}{\mathrm{d}z} + \tilde{p}_{11}\,, \quad p_{21} = \rho_2 c^2 \eta_0 \frac{\mathrm{d}W_2}{\mathrm{d}z} + \tilde{p}_{21}\,,$$

where $\tilde{p}_{11}$, $\tilde{p}_{21}$ are integration constants. From (4.51) then it follows that:

$$\frac{\mathrm{d}^2 W_1}{\mathrm{d}z^2} = 0\,, \quad \frac{\mathrm{d}^2 W_2}{\mathrm{d}z^2} = 0\,.$$

Therefore, $W_{1,2} = a_{1,2}z + b_{1,2}$ are linear functions of z. From normalisation conditions (4.48) we obtain $b_{1,2} = 1$ and from boundary conditions at the bottom and surface (4.46) we find $a_1 = 1/h_1$, $a_2 = -1/h_2$. Thus, we have:

$$W_1 = \frac{z}{h_1} + 1\,, \quad W_2 = -\frac{z}{h_2} + 1\,.$$

Taking into account the obtained relations, we can reduce (4.38) at the first order ε^1 to the following equations:

$$w_{11}|_{z=0} = \frac{\partial \eta_0}{\partial \tau} + \frac{2c}{h_1}\eta_0 \frac{\partial \eta_0}{\partial \xi} - c\frac{\partial \eta_1}{\partial \xi}\,, \tag{4.52}$$

$$w_{21}|_{z=0} = \frac{\partial \eta_0}{\partial \tau} + \frac{2c}{h_2}\eta_0 \frac{\partial \eta_0}{\partial \xi} - c\frac{\partial \eta_1}{\partial \xi}\,, \tag{4.53}$$

$$\eta_0\left(-\frac{\rho_1 c^2}{h_1}+\rho_1 g-\frac{\rho_2 c^2}{h_2}-\rho_2 g\right)-\tilde{p}_{11}+\tilde{p}_{21}=0. \tag{4.54}$$

Boundary conditions at the external boundaries (4.39) in this approximation at ε^1 are:

$$w_{11}(z=-h_1)=0,\quad w_{21}(z=h_2)=0. \tag{4.55}$$

Then from (4.54) it follows that

$$\tilde{p}_{21}=\tilde{p}_{11}.$$

The (phase) speed of infinitely long waves can be determined now in the linear approximation as:

$$c^2=\frac{g(\rho_1-\rho_2)h_1h_2}{\rho_1 h_2+\rho_2 h_1}.$$

This expression reduces to (4.31) in the Boussinesq approximation.

Now we can express all the terms of the lowest nonzero order of ε in (4.41) expressed through the one unknown function η_0—the displacement of fluid interface in the lowest approximation:

$$u_{10}=\frac{c}{h_1}\eta_0,\quad u_{20}=\frac{c}{h_2}\eta_0,$$

$$w_{10}=-c\frac{\partial\eta_0}{\partial\xi}\left(\frac{z}{h_1}+1\right),\quad w_{20}=-c\frac{\partial\eta_0}{\partial\xi}\left(-\frac{z}{h_2}+1\right),$$

$$p_{11}=\frac{\rho_1 c^2}{h_1}\eta_0+\tilde{p}_{11},\quad p_{21}=-\frac{\rho_2 c^2}{h_2}\eta_0+\tilde{p}_{11}.$$

The Second Order At ε^2 (4.36)$_1$ and (4.37)$_1$ lead to the following equations:

$$\frac{\partial u_{12}}{\partial\xi}+\frac{\partial w_{12}}{\partial z}=0,\quad \frac{\partial u_{22}}{\partial\xi}+\frac{\partial w_{22}}{\partial z}=0. \tag{4.56}$$

From Eqs. (4.36)$_2$ and (4.37)$_2$ we obtain at ε^2:

$$\rho_1 c\left(\frac{1}{h_1}\frac{\partial\eta_0}{\partial\tau}+\frac{c}{h_2^1}\eta_0\frac{\partial\eta_0}{\partial\xi}-\frac{\partial u_{11}}{\partial\xi}\right)+\frac{\partial p_{12}}{\partial\xi}=0, \tag{4.57}$$

$$\rho_2 c\left(\frac{1}{h_2}\frac{\partial\eta_0}{\partial\tau}+\frac{c}{h_2^2}\eta_0\frac{\partial\eta_0}{\partial\xi}-\frac{\partial u_{21}}{\partial\xi}\right)+\frac{\partial p_{22}}{\partial\xi}=0. \tag{4.58}$$

Equations (4.36)$_3$ and (4.37)$_3$ give:

$$\rho_1 c^2 \frac{\partial^2 \eta_0}{\partial \xi^2}\left(\frac{z}{h_1}+1\right)+\frac{\partial p_{12}}{\partial z}=0, \quad \rho_2 c^2 \frac{\partial^2 \eta_0}{\partial \xi^2}\left(-\frac{z}{h_2}+1\right)+\frac{\partial p_{22}}{\partial z}=0. \tag{4.59}$$

Solving these equations for the pressure in the layers, we get:

$$p_{12}=-\rho_1 c^2 \frac{\partial^2 \eta_0}{\partial \xi^2} z\left(\frac{z}{2h_1}+1\right)+\tilde{p}_{12}(\xi, \tau),$$

$$p_{22}=-\rho_2 c^2 \frac{\partial^2 \eta_0}{\partial \xi^2} z\left(\frac{z}{2h_2}+1\right)+\tilde{p}_{22}(\xi, \tau).$$

Equations (4.38) at ε^2 have the form:

$$w_{12}\big|_{z=0}=\left[\frac{\partial \eta_1}{\partial \tau}+\frac{c}{h_1}\left(\eta_0 \frac{\partial \eta_1}{\partial \xi}+\frac{\partial \eta_0}{\partial \xi}\eta\right)-\eta_0 \frac{\partial w_{11}}{\partial z}+u_{11}\frac{\partial \eta_0}{\partial \xi}-c\frac{\partial \eta_2}{\partial \xi}\right]\Bigg|_{z=0}, \tag{4.60}$$

$$w_{22}\big|_{z=0}=\left[\frac{\partial \eta_1}{\partial \tau}-\frac{c}{h_2}\left(\eta_0 \frac{\partial \eta_1}{\partial \xi}+\frac{\partial \eta_0}{\partial \xi}\eta\right)-\eta_0 \frac{\partial w_{21}}{\partial z}+u_{21}\frac{\partial \eta_0}{\partial \xi}-c\frac{\partial \eta_2}{\partial \xi}\right]\Bigg|_{z=0}, \tag{4.61}$$

$$\eta_1 g(\rho_1-\rho_2)-\sigma \frac{\partial^2 \eta_0}{\partial \xi^2}-\tilde{p}_{12}+\tilde{p}_{22}=0. \tag{4.62}$$

Boundary conditions at the bottom and surface (4.39) are:

$$w_{12}(z=-h_1)=0, \quad w_{22}(z=h_2)=0. \tag{4.63}$$

From (4.62) we have:

$$\tilde{p}_{22}=\tilde{p}_{12}-\eta_1 g(\rho_1-\rho_2)+\sigma \frac{\partial^2 \eta_0}{\partial \xi^2}.$$

Eliminating u_{11} and u_{21} from Eqs. (4.49), (4.57), and (4.58), we find:

$$w_{11}=\frac{c}{2}\frac{\partial^3 \eta_0}{\partial \xi^3}\left(\frac{z^3}{3h_1}+z^2\right)-\frac{c}{h_1^2}\eta_0 \frac{\partial \eta_0}{\partial \xi}z-\frac{1}{h_1}\frac{\partial \eta_0}{\partial \tau}z-\frac{1}{\rho_1 c}\frac{\partial \tilde{p}_{12}}{\partial \xi}z+\tilde{w}_{11}(\xi, \tau), \tag{4.64}$$

$$w_{21} = \frac{c}{2}\frac{\partial^3\eta_0}{\partial\xi^3}\left(-\frac{z^3}{3h_2}+z^2-\frac{2\sigma}{c^2\rho+2}\right)-\frac{c}{h_2^2}\eta_0\frac{\partial\eta_0}{\partial\xi}z+\frac{1}{h_2}\frac{\partial\eta_0}{\partial\tau}z-$$

$$-\frac{1}{\rho_2 c}\frac{\partial\tilde{p}_{12}}{\partial\xi}z+\frac{g(\rho_1-\rho_2)}{\rho_2 c}\frac{\partial\eta_1}{\partial\xi}\tilde{w}_{21}(\xi,\tau). \tag{4.65}$$

Next we use boundary conditions at the surface and bottom (4.55):

$$\frac{ch_1^2}{3}\frac{\partial^3\eta_0}{\partial\xi^3}+\frac{c}{h_1}\eta_0\frac{\partial\eta_0}{\partial\xi}+\frac{\partial\eta_0}{\partial\tau}+\frac{h_1}{\rho_1 c}\frac{\partial p_{12}}{\partial\xi}+w_{11}(\xi,\tau)=0, \tag{4.66}$$

$$\left(\frac{ch_2^2}{3}-\frac{\sigma h_2}{\rho_2 c}\right)\frac{\partial^3\eta_0}{\partial\xi^3}-\frac{c}{h_2}\eta_0\frac{\partial\eta_0}{\partial\xi}+\frac{\partial\eta_0}{\partial\tau}-\frac{h_2}{\rho_2 c}\frac{\partial\tilde{p}_{12}}{\partial\xi}+$$

$$+\frac{g(\rho_1-\rho_2)h_2}{\rho_2 c}\frac{\partial\eta_1}{\partial\xi}+\tilde{w}_{21}(\xi,\tau)=0, \tag{4.67}$$

and conditions (4.52)–(4.54) at the interface of the layers, from which we get:

$$\tilde{w}_{11}=\frac{2c}{h_1}\eta_0\frac{\partial\eta_0}{\partial\xi}+\frac{\partial\eta_0}{\partial\tau}-c\frac{\partial\eta_1}{\partial\xi}, \tag{4.68}$$

$$\tilde{w}_{21}=-\frac{2c}{h_2}\eta_0\frac{\partial\eta_0}{\partial\xi}+\frac{\partial\eta_0}{\partial\tau}-c\frac{\partial\eta_1}{\partial\xi}. \tag{4.69}$$

Equations (4.66) and (4.67) with the help of (4.68) and (4.69) can be converted to the following equations:

$$\frac{ch_1^2}{3}\frac{\partial^3\eta_0}{\partial\xi^3}+\frac{3c}{h_1}\eta_0\frac{\partial\eta_0}{\partial\xi}+2\frac{\partial\eta_0}{\partial\tau}-c\frac{\partial\eta_1}{\partial\xi}+\frac{h_1}{\rho_1 c}\frac{\partial\tilde{p}_{12}}{\partial\xi}=0, \tag{4.70}$$

$$\left(\frac{ch_2^2}{3}-\frac{\sigma h_2}{\rho_2 c}\right)\frac{\partial^3\eta_0}{\partial\xi^3}-\frac{3c}{h_2}\eta_0\frac{\partial\eta_0}{\partial\xi}+2\frac{\partial\eta_0}{\partial\xi}-\frac{h_2}{\rho_2 c}\frac{\partial\tilde{p}_{12}}{\partial\xi}+\frac{g(\rho_1-\rho_2)h_2}{\rho_2 c}\frac{\partial\eta_1}{\partial\xi}=0. \tag{4.71}$$

From these equations $\partial\tilde{p}_{12}/\partial\xi$ can be eliminated and the following KdV equation for the interface perturbation can be obtained:

$$\frac{\partial\eta_0}{\partial\tau}+\alpha\eta_0\frac{\partial\eta_0}{\partial\xi}+\beta\frac{\partial^3\eta_0}{\partial\xi^3}=0. \tag{4.72}$$

The coefficients at the nonlinear and dispersion terms (often called nonlinear and dispersion coefficients) of this equation are:

$$\alpha = \frac{3c}{2h_1h_2}\frac{\rho_1 h_2^2 - \rho_2 h_1^2}{\rho_1 h_2 + \rho_2 h_1}, \quad \beta = \frac{ch_1h_2}{6}\frac{\rho_1 h_1 + \rho_2 h_2 - 3\sigma}{\rho_1 h_2 + \rho_2 h_1}. \tag{4.73}$$

The parameter β, obtained from the asymptotic procedure in the Boussinesq approximation and in the absence of surface tension coincides with the dispersion coefficient (4.32) (the coefficient at the linear dispersive term in the relevant equation) obtained from the expansion of dispersion relation (4.25) in the series of small wavenumbers k.

Continuing further the procedure of derivation of a wave equation in the higher orders in small parameter ε, we collect all the first correction terms in the series (4.41), expressed in terms of functions η_0 and η_1. The vertical velocities are found from (4.64), (4.65), (4.68), (4.69) and any of (4.70), (4.71):

$$w_{11} = \left[\frac{c}{6h_1}z^3 + \frac{c}{2}z^2 + \left(\frac{\rho_1 h_2 + 2\rho_2 h_1}{\rho_1 h_1 h_2}\beta - \frac{\rho_2 c^2 h_2 - 3\sigma}{3c\rho_1}\right)z - \beta\right]\frac{\partial^3 \eta_0}{\partial \xi^3} +$$
$$+\left[\left(\frac{\rho_1 h_2 + 2\rho_2 h_1}{\rho_1 h_1 h_2}\alpha - c\frac{\rho_1 h_2^2 - 3\rho_2 h_1^2}{\rho_2 h_1^2 h_2^2}\right)z - \alpha + \frac{2c}{h_1}\right]\eta_0\frac{\partial \eta_0}{\partial \xi} - c\left(\frac{z}{h_1} + 1\right)\frac{\partial \eta_1}{\partial \xi},$$

$$w_{21} = \left[-\frac{c}{6h_2}z^3 + \frac{c}{2}z^2 + \left(\frac{\beta}{h_2} - \frac{ch_2}{3}\right)z - \beta\right]\frac{\partial^3 \eta_0}{\partial \xi^3} +$$
$$+\left[\left(\frac{\alpha}{h_2} + \frac{2c}{h_2^2}\right)z - \alpha - \frac{2c}{h_2}\right]\eta_0\frac{\partial \eta_0}{\partial \xi} - c\left(1 - \frac{z}{h_2}\right)\frac{\partial \eta_1}{\partial \xi}.$$

From (4.50) we find the components of the horizontal velocity:

$$u_{11} = \left(-\frac{c}{2h_1}z^2 - cz - \frac{\rho_1 h_2 + 2\rho_2 h_1}{\rho_1 h_1 h_2}\beta + \frac{\rho_2 c^2 h_2 - 3\sigma}{3c\rho_1}\right)\frac{\partial^2 \eta_0}{\partial \xi^2} +$$
$$+\left(-\frac{\rho_1 h_2 + 2\rho_2 h_1}{2\rho_1 h_1 h_2}\alpha + c\frac{\rho_1 h_2^2 - 3\rho_2 h_1^2}{2\rho_1 h_1^2 h_2^2}\right)\eta_0^2 + \frac{c}{h_1}\eta_1,$$

$$u_{21} = \left(\frac{c}{2h_1}z^2 - cz - \frac{\beta}{h_2} + \frac{ch_2}{3}\right)\frac{\partial^2 \eta_0}{\partial \xi^2} - \left(\frac{\alpha}{2h_2} + \frac{c}{h_2^2}\right)\eta_0^2 - \frac{c}{h_2}\eta_1.$$

The pressure components can be expressed from (4.59) with the help of any of (4.68), (4.69):

$$p_{12} = \left(-\frac{\rho_1 c^2}{2h_1} z^2 - \rho_1 c^2 z - \frac{2\rho_2 c\beta}{h_2} + \frac{\rho_2 c^2 h_2}{3} - \sigma \right) \frac{\partial^2 \eta_0}{\partial \xi^2} - $$

$$- \frac{\rho_2 c}{h_2} \left(\frac{3c}{2h_2} + \alpha \right) \eta_0^2 + \frac{\rho_1 c^2}{h_1} \eta_1 ,$$

$$p_{22} = \left(\frac{\rho_1 c^2}{2h_1} z^2 - \rho_2 c^2 z - \frac{2\rho_2 c\beta}{h_2} + \frac{\rho_2 c^2 h_2}{3} \right) \frac{\partial^2 \eta_0}{\partial \xi^2} - \frac{\rho_2 c}{h_2} \left(\frac{3c}{2h_2} + \alpha \right) \eta_0^2 - \frac{\rho_2 c^2}{h_2} \eta_1 .$$

The resulting expressions can be used to construct the equation of the next approximation.

The Third Order All equations in the third order in ε are rather complicated. It is recommended that the program of symbolic computations (e.g., Maple) should be used to obtain them. Here we only describe briefly the main steps necessary to derive the nonlinear evolution equation in the next order on a small parameter.

To start with, the derived Eqs. $(4.36)_3$ and $(4.37)_3$ at ε^3 are integrated in z. Doing so makes it possible to express the pressure components $p_{13}(\xi, z, \tau)$ and $p_{23}(\xi, z, \tau)$. These expressions contain arbitrary additive functions $\tilde{p}_{13}(\xi, \tau)$ and $\tilde{p}_{23}(\xi, \tau)$ that do not depend on z. The connection between the functions $\tilde{p}_{13}(\xi, \tau)$ and $\tilde{p}_{23}(\xi, \tau)$ can be found from boundary condition (4.38) at ε^3. Further, (4.56) is used to eliminate horizontal velocities u_{12} and u_{22} from equations derived from Eqs. $(4.36)_2$ and $(4.37)_2$ at ε^3.

The expressions for the vertical velocity components w_{12} and w_{22} also contain arbitrary additive functions $\tilde{w}_{13}(\xi, \tau)$ and $\tilde{w}_{23}(\xi, \tau)$. These functions can be specified using boundary conditions (4.60), (4.61). Finally, using boundary conditions (4.63) we eliminate $\tilde{p}_{13}(\xi, \tau)$. The resulting equation is as follows:

$$\frac{\partial \eta_1}{\partial \tau} + \alpha \left(\eta_0 \frac{\partial \eta_1}{\partial \xi} + \eta_1 \frac{\partial \eta_0}{\partial \xi} \right) + \beta \frac{\partial^3 \eta_1}{\partial \xi^3} + \alpha_1 \eta_0^2 \frac{\partial \eta_0}{\partial \xi} +$$

$$+ \gamma_1 \eta_0 \frac{\partial^3 \eta_0}{\partial \xi^3} + \gamma_2 \frac{\partial \eta_0}{\partial \xi} \frac{\partial^2 \eta_0}{\partial \xi^2} + \beta_1 \frac{\partial^5 \eta_0}{\partial \xi^5} = 0. \qquad (4.74)$$

Here α and β are given by (4.73), and the coefficients of the higher order terms are:

$$\alpha_1 = -\frac{3c(h_1^4\rho_2^2 + 8\rho_1\rho_2 h_1^3 h_2 + 14\rho_1\rho_2 h_1^2 h_2^2 + 8\rho_1\rho_2 h_1 h_2^3 + h_2^4\rho_1^2)}{8h_1^2 h_2^2(\rho_2 h_1 + \rho_1 h_2)^2},$$

$$\gamma_1 = -\frac{7(h_1^3 - h_2^3)\rho_1\rho_2 c^2 + 5h_1^2 h_2 \rho_2^2 c^2 + 2h_1 h_2(h_1 - h_2)\rho_1\rho_2 c^2 - 5h_1 h_2^2 \rho_1^2 c^2 - \sigma_1}{12c(\rho_2 h_1 + \rho_1 h_2)^2},$$

$$\gamma_2 = -\frac{1}{24c(\rho_2 h_1 + \rho_1 h_2)}\times$$

$$\times\left[31(h_1^3 - h_2^3)\rho_1\rho_2 c^2 + 23h_1^2 h_2 \rho_2^2 c^2 + 8h_1 h_2(h_1 - h_2)\rho_1\rho_2 c^2 - 23h_1 h_2^2 \rho_1^2 c^2 + \sigma_2\right],$$

$$\beta_1 = \frac{ch_1h_2}{90}\frac{\rho_1\rho_2(h_1^4 + h_2^4 + 15h_1^2h_2^2/2) + (19h_1h_2/4)(\rho_1^2h_1^2 + \rho_2^2h_2^2) - \sigma_3}{(\rho_2 h^1 + \rho_1 h_2)^2},$$

where

$$\sigma_1 = 3\sigma(h_1^2\rho_2 - h_2^2\rho_1), \quad \sigma_2 = 15\sigma(h_1^2\rho_2 - h_2^2\rho_1),$$

$$\sigma_3 = \frac{15\sigma h_1 h_2}{2c^2}\left(\rho_1 h_1 + \rho_2 h_2 + \frac{3\sigma}{2c^2}\right).$$

We introduce the notation for the partial sum of the series of interfacial displacement $\zeta = \eta_0 + \varepsilon\eta_1 = \eta + O(\varepsilon^2)$. It describes the perturbation of the interface up to the terms of the order of ε inclusive. Combining (4.72) and (4.74), we obtain the higher order KdV equation, which is valid in the same approximation in ε:

$$\frac{\partial\zeta}{\partial\tau} + \alpha\zeta\frac{\partial\zeta}{\partial\xi} + \beta\frac{\partial^3\zeta}{\partial\xi^3} + \varepsilon\left(\alpha_1\zeta^2\frac{\partial\zeta}{\partial\xi} + \gamma_2\frac{\partial\zeta}{\partial\xi}\frac{\partial^2\zeta}{\partial\xi^2} + \beta_1\frac{\partial^5\zeta}{\partial\xi^5}\right) = 0. \qquad (4.75)$$

This is the appropriate equation for IWs at the interface of two layers, generally speaking, immiscible fluids. We emphasise that in the process of its derivation we never used the condition of potentiality of fluid motion. Therefore, such a derivation can be successfully used to obtain equations of a viscous and/or rotating fluid. An equivalent fifth order KdV equation was derived in (Lee and Bearssley, 1974; Koop and Butler, 1981) using the assumption of potential flow in each of the layers, but without taking into account the surface tension between the layers.

The expansion procedure for fluids with any finite number of layers is technically very similar to the presented procedure. It is, however, much more tedious even for a three-layer fluid. The reason is that the set of unknown functions increases. Accordingly, the system of partial differential equations becomes more complicated. The handling of such systems requires a special caution with the calculations.

The asymptotic procedure, if necessary, can be extended to the higher orders in the small parameter. Doing so is required only in specific singular cases when (e.g., due to particular combinations of environmental parameters) some coefficients of Eq. (4.75) vanish. Below we will consider such special cases.

4.6 Analysis of Particular Cases

For the sake of simplicity we introduce the dimensionless parameters a, b and s:

$$a = \rho_2/\rho_1 < 1, \quad b = h_2/h_1, \quad s = 2\sigma\left[(\rho_1+\rho_2)gH^2\right]^{-1}$$

and express the coefficients of Eq. (4.75) through these parameters:

$$\frac{c^2}{gH} = \frac{b(1-a)}{(1+b)(a+b)}, \qquad \alpha^* = \frac{\alpha H}{c} = \frac{3}{2}\frac{(1+b)(b^2-a)}{b(a+b)},$$

$$\beta^* = \frac{\beta}{cH^2} = \frac{1}{12}\frac{2b(1+ab)(1-a)-3(1+a)(a+b)(1+b)^2 s}{(1+b)^2(a+b)(1-b)},$$

$$\alpha_1^* = \frac{\alpha_1 H^2}{c} = \frac{3}{8}\frac{(a^2+8ab+14ab^2+8ab^3+b^4)(1+b)^2}{b^2(a+b)^2}, \tag{4.76}$$

$$\gamma_1^* = \frac{\gamma_1}{cH} = -\frac{5a^2b+2ab-7ab^3+7a-2b^2a-5b^2}{12(1+b)(a+b)^2} + \frac{(1+b)(1+a)(a-b^2)s}{8b(1-a)(a+b)},$$

$$\gamma_2^* = \frac{\gamma_2}{cH} = -\frac{23a^2b+8ab-31ab^3+31a-8b^2a-23b^2}{24(1+b)(a+b)^2} - \frac{5(1+b)(1+a)(a-b^2)s}{16b(1-a)(a+b)},$$

$$\beta_1^* = \frac{\beta_1}{cH^4} = -\frac{(1+a)^2s^2}{32(1-a)^2} - \frac{(1+a)(ab+1)bs}{24(1-a)(1+b)^2(a+b)} + \frac{(4ab^4+19b+30b^2a+19a^2b^3+4a)b}{360(a+b)^2(1+b)^4}.$$

The dimensionless parameter s responsible for the surface tension contribution is inverse proportional to the square of fluid depth H. Therefore, its effect will be significant only in the case of very thin liquid layers (of the order of a few cm or less). Nevertheless, an equation containing s can be of interest from the point of view of its practical applications to some technological processes with fluid motion in thin layers. If we set $s = 0$ in (4.76), then our approach leads to the notation that is equivalent to the results obtained in (Koop and Butler, 1981).

Graphs of normalised coefficients $\alpha^*, \beta^*, \alpha_1^*, \beta_1^*, \gamma_1^*$ and γ_2^* are shown in Fig. 4.2. As one can see from these graphs, the effect of surface tension can be important only in a fairly narrow range of parameters. One can also observe that some coefficients can vanish and change their signs.

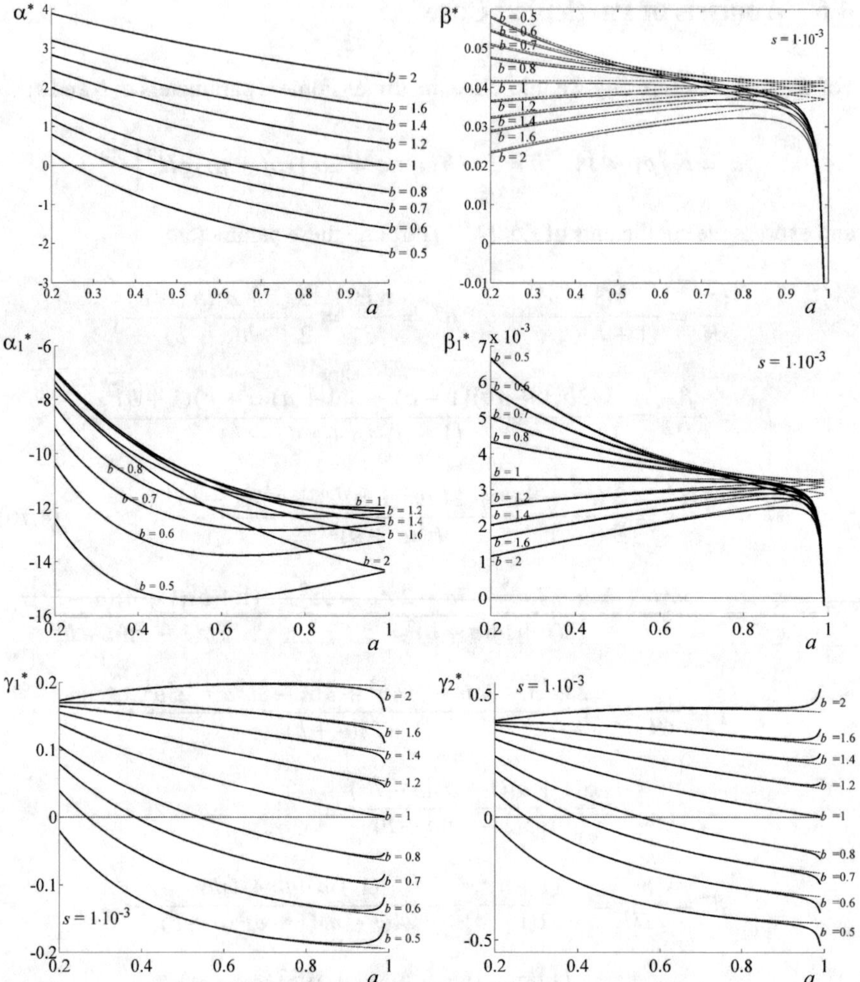

Fig. 4.2 Normalised coefficients of Eq. (4.75) as functions of the parameter $a = \rho_2/\rho_1$. Dashed lines show the corresponding dependencies when $\sigma = 0$ ($s = 0$).

In particular, the coefficient α at the quadratic nonlinear term of (4.75) (also called coefficient of quadratic nonlinearity) vanishes when $a_{cr} = b^2$. This feature and its consequences are well known (see, for example, (Djordjevic and Redekopp, 1978)). The dispersion coefficient β in (4.75) becomes zero when

$$s_{cr} = \frac{2b(1-a)(1+ab)}{3(1+b)^2(a+b)(1+a)}.$$

If both these coefficients vanish simultaneously, the expressions for the remaining coefficients of Eq. (4.74) take the form:

$$\alpha_1^* = -\frac{3(1+b)^2}{b}, \quad \beta_1^* = \frac{b^4 - b^3 + b^3 - b + 1}{90(1+b)^4}, \quad \gamma_1^* = \frac{-1+b}{6(1+b)}, \quad \gamma_2^* = \frac{-1+b}{3(1+b)},$$

and the basic Eq. (4.75) in the vicinity of such critical points noticeably simplifies. Of particular interest is the situation when the coefficients of quadratic nonlinearity α and third order dispersion β are of the same order, and the coefficients γ_1 and γ_2 are negligible. This can happen when $b \cong 1$, i.e., when the thicknesses of the layers are almost equal. In the vicinity of such a "double" critical point, (4.75) reduces to the Gardner–Kawahara equation:

$$\frac{\partial \hat{\zeta}}{\partial \tau} + \alpha \hat{\zeta} \frac{\partial \hat{\zeta}}{\partial \xi} + \beta \frac{\partial^3 \hat{\zeta}}{\partial \xi^3} + \alpha_1 \hat{\zeta}^2 \frac{\partial \hat{\zeta}}{\partial \xi} + \beta_1 \frac{\partial^5 \hat{\zeta}}{\partial \xi^5} = 0, \tag{4.77}$$

where $\hat{\zeta} = \varepsilon \zeta$. The dynamics of solitary waves within the framework of this equation has been studied in (Kurkina et al., 2015b).

If $\beta_1 = 0$, then (4.77) reduces to the well-known completely integrable Gardner equation (Ablowitz and Segur, 1981; Slyunyaev and Pelinovsky, 1999; Slyunyaev, 2001; Grimshaw et al., 2002b, 2010b). The coefficient at the cubic nonlinear term α_1 can have either sign depending on the fluid stratification (Grimshaw et al., 2002a; Kurkina et al., 2011, 2015a). If it is negative (which usually occurs in a two-layer fluid), soliton solutions to (4.77) have a limited amplitude (Apel et al., 2007; Ostrovsky, 2015). When the amplitude of a soliton approaches the limit, the soliton infinitely broadens.

The Gardner equation has been a subject of study by many authors (see, e.g., (Apel et al., 2007; Pelinovsky et al., 2007; Ostrovsky, 2015) and references therein). Within the framework of this equation there are strongly nonlinear effects which cannot be described by the classic KdV equation (Grimshaw et al., 1999, 2002b, 2010b; Slyunyaev and Pelinovsky, 1999; Slyunyaev, 2001). In particular, there are solitons propagating on the top of a very wide almost limiting amplitude soliton (the "table-top soliton"), and the formation of two soliton trains of different polarity at the initial stage of disintegration of initial pulse type perturbation with the amplitude exceeding the limiting value (in such case the table-top soliton plays a role of a pedestal).

The existence of table-top solitons and the possibility of propagation of other types of solitons through it were confirmed within the framework of more general nonlinear models (Helfrich and Melville, 2006). If the parameter $\alpha_1 > 0$, the Gardner equation (4.77) with $\beta_1 = 0$ permits solutions in the form of nonstationary solitary waves, breathers (see, e.g., (Ostrovsky, 2015) and references therein). Such situations with positive α_1 can occur in real oceanic conditions (Grimshaw et al., 2010a; Kurkina et al., 2015b).

In another extreme case, when $\alpha_1 = 0$, but $\beta_1 \neq 0$, (4.77) reduces to the Kawahara equation, which is not completely integrable, but which has a wide family of soliton solutions, including solitons with oscillating asymptotics (Kawahara, 1972; Kawahara and Takaoka, 1988).

Unlike the KdV equation, the higher order model (4.75) is not a Hamiltonian equation in general and does not preserve energy. However, in the particular cases when it reduces to completely integrable models, it clearly becomes Hamiltonian. Besides, there is one more particular case of $\alpha_1 = 0$, and $\gamma_2 = 2\gamma_1$ when (4.75) becomes Hamiltonian but remains nonintegrable.

With the help of a near-identity transformation, it can be mapped approximately into one of a number of integrable Hamiltonian equations. In particular, the appropriately chosen near-identity transformations allows the reduction of (4.75) to the classical KdV equation. With the help of such transformation there were obtained particular solutions for the higher order KdV equation (4.75) from the known solutions of the KdV equation (e.g., the two-soliton solution extending the relevant KdV solution (Marchant and Smyth, 1996; Kraenkel, 1998; Marchant, 1999) and the undular bore solution (Marchant and Smyth, 2006)).

More details can be found in (Khusnutdinova et al., 2018) where it was shown also that (4.75) has a variety of exact solitary wave solutions. Some of them represent the embedded solitons which can coexist with the linear waves (they are embedded into the continuous spectrum of linear waves). Others represent regular solitons. Stability of embedded solitons and their interaction with the regular solitons were studied numerically in (Khusnutdinova et al., 2018).

4.7 Conclusion

We have considered here the dynamics of weakly nonlinear long interfacial waves in a two-layer fluid under the action of gravity and capillary effects. The governing Euler equations were reduced to the family of Korteweg–de Vries-type equations. This was done with the help of an asymptotic procedure with two small parameters: the ratio of wave amplitude to water depth and ratio of water depth to wavelength.

In contrast to previous studies, we did not use the assumption of potential flow. Therefore, our analysis can be easily extended for the inclusion of fluid rotation, shear flows, and viscosity effects. In the first order of perturbation theory, the famous Korteweg–de Vries equation has been derived. All coefficients of the KdV equation were obtained in the explicit form. It was shown that some coefficients in certain conditions can change sign. When the coefficients become very small of vanish it is necessary to take into account higher order terms in the asymptotic procedure with the rescaling of small parameters.

In the next order, the extended fifth order KdV equation has been derived. All coefficients of this equation were obtained for a two-layer fluid and analysed then in details. Special attention has been given to the Gardner and Gardner–Kawahara equations which appear as particular cases of general (extended) fifth order KdV equation (4.75). This equation can be reduced to a completely integrable Hamiltonian equation with the help of a near identity transformation.

Equation (4.75) in its general form has a variety of exact solitary type solutions. Some of them represent embedded solitons whereas others represent regular solitons. The interesting issues for the further study are the stability of embedded solitons, details of their interactions with the regular solitons or other perturbations, as well as their dynamics in viscous inhomogeneous fluids.

We have also presented explicit expressions for various characteristics of a wave field (the horizontal and vertical components of velocity field, nonhydrostatic correction to the pressure). These expressions are derived in the form of the series up to the second order in small parameters. These series can be used for different practical purposes such as the analysis of vorticity fields and possible hydrodynamic instabilities developing in the course of propagation of interfacial waves, analysis of Lagrangian trajectories of fluid particles, transport phenomena due to the nonlinear drift of fluid particles, dynamical effects on submerged bodies, etc.

Acknowledgements This study was financially supported by the state task program in the sphere of scientific activity of the Ministry of Science and Higher Education of the Russian Federation (projects No 5.4568.2017/6.7 and 5.1246.2017/4.6), grant of the President of the Russian Federation (NSh-2685.2018.5), Program "Nonlinear Dynamics" of the Russian Academy of Sciences and grants of the Russian Foundation for Basic Research (RFBR No 18-02-00042, 19-05-00161).

References

Ablowitz, M.J., Segur, H.: Solitons and the Inverse Scattering Transform. SIAM Studies in Applied and Numerical Mathematics, vol. 4, Philadelphia (1981). https://doi.org/10.1137/1.9781611970883

Apel, J., Ostrovsky, L.A., Stepanyants, Y.A., Lynch, J.F.: Internal solitons in the ocean and their effect on underwater sound. J. Acoust. Soc. Am. **121**(2), 695–722 (2007). https://doi.org/10.1121/1.2395914

Craig, W., Groves, M.D.: Normal forms for wave motion in fluid interfaces. Wave Motion **31**, 21–41 (2000). https://doi.org/10.1016/s0165-2125(99)00022-0

Craig W., Guyenne P., Kalisch H.: A new model for large amplitude long internal waves. C. R. Mech. **332**(7), 525–530 (2004). https://doi.org/10.1016/j.crme.2004.02.026

Djordjevic, V.D., Redekopp, L.G.: The fission and disintegration of internal solitary waves moving over two-dimensional topography. J. Phys. Oceanogr. **8**(6), 1016–1024 (1978). https://doi.org/10.1175/1520-0485(1978)008<1016:tfadoi>2.0.co;2

Engelbrecht, J., Pelinovsky, E.N., Fridman, V.E.: Nonlinear Evolution Equations. Longman Scientific & Technical, Harlow (1988)

Grimshaw, R.H.J., Ostrovsky, L.A., Shrira, V.I., Stepanyants, Yu.A.: Long nonlinear surface and internal gravity waves in a rotating ocean. Surveys Geophys. **19**(4), 289–338 (1998)

Grimshaw, R., Pelinovsky, E., Talipova, T.: Solitary wave transformation in a medium with sign-variable quadratic nonlinearity and cubic nonlinearity. Physica D **132**, 40–62 (1999). https://doi.org/10.1016/s0167-2789(99)00045-7

Grimshaw, R., Pelinovsky, E., Poloukhina, O.: Higher-order Korteweg–de Vries models for internal solitary waves in a stratified shear flow with a free surface. Nonlin. Processes Geophys. **9**, 221–235 (2002a). https://doi.org/10.5194/npg-9-221-2002

Grimshaw, R., Pelinovsky, D., Pelinovsky, E., Slunyaev, A.: Generation of large-amplitude solitons in the extended Korteweg–de Vries equation. Chaos **12**(4), 1070–1076 (2002b). https://doi.org/10.1063/1.1521391

Grimshaw, R., Talipova, T., Pelinovsky, E., Kurkina, O.: Internal solitary waves: propagation, deformation and disintegration. Nonlin. Processes Geophys. **17**, 633–649 (2010a). https://doi.org/10.5194/npg-17-633-2010

Grimshaw, R., Slunyaev, A., Pelinovsky, E.: Generation of solitons and breathers in the extended Korteweg–de Vries equation with positive cubic nonlinearity. Chaos **20**(1), 013102 (2010b). https://doi.org/10.1063/1.3279480

Helfrich, K.R., Melville, W.K.: Long nonlinear internal waves. Annu. Rev. Fluid Mech. **38**, 395–425 (2006). https://doi.org/10.1146/annurev.fluid.38.050304.092129

Holloway, P., Pelinovsky, E., Talipova, T.: A generalised Korteweg–de Vries model of internal tide transformation in the coastal zone. J. Geophys. Res.–Oceans **104**(18), 333–350 (1999). https://doi.org/10.1029/1999jc900144

Kawahara, T.: Oscillatory solitary waves in dispersive media. J. Phys. Soc. Japan **33**, 260–264 (1972). https://doi.org/10.1143/jpsj.33.260

Kawahara, T., Takaoka, M.: Chaotic motions in oscillatory soliton lattice. J. Phys. Soc. Japan **57**(11), 3714–3732 (1988). https://doi.org/10.1143/jpsj.57.3714

Khusnutdinova, K., Stepanyants, Y., Tranter, M.: The influence of the nonlinear dispersion on the shapes of solitary waves. Phys. Fluids **30**, 022104 (2018)

Koop, C., Butler, G.: An investigation of internal solitary waves in a two-fluid system. J. Fluid Mech. **112**, 225–251 (1981). https://doi.org/10.1017/s0022112081000372

Kraenkel, R.A.: First-order perturbed Korteweg–de Vries solitons. Phys. Rev. E **57**(4), 4775–4777 (1998). https://doi.org/10.1103/physreve.57.4775

Kurkina, O.E., Kurkin, A.A., Soomere, T., Pelinovsky, E.N., Rouvinskaya, E.A.: Higher-order (2+4) Korteweg–de Vries-like equation for interfacial waves in a symmetric three-layer fluid. Phys. Fluids **23**(11), 116602-1 (2011). https://doi.org/10.1063/1.3657816

Kurkina, O., Kurkin, A., Rouvinskaya, E., Soomere, T.: Propagation regimes of interfacial solitary waves in a three-layer fluid. Nonlin. Processes Geophys. **22**, 117–132 (2015a). https://doi.org/10.5194/npg-22-117-2015

Kurkina, O., Singh, N., Stepanyants, Y.: Structure of internal solitary waves in two-layer fluid at near-critical situation. Commun. Nonlin. Sci. Numer. Modeling **22**, 1235–1242 (2015b). https://doi.org/10.1016/j.cnsns.2014.09.018

Lee, C.Y., Beardsley, R.C.: The generation of long nonlinear internal waves in a weakly stratified shear flow. J. Geophys. Res.–Oceans **79**(3), 453–462 (1974). https://doi.org/10.1029/jc079i003p00453

Leibovich, S., Seebass, A.R. (eds.): Nonlinear Waves. Cornell University Press, London (1974)

Marchant, T.R.: Asymptotic solitons of the extended Korteweg–de Vries equation. Phys. Rev. E **59**(3), 3745–3748 (1999). https://doi.org/10.1103/PhysRevE.59.3745

Marchant, T.R., Smyth, N.F.: Soliton interaction for the extended Korteweg–de Vries equation. IMA J. Appl. Math. **56**(2), 157–176 (1996). https://doi.org/10.1093/imamat/56.2.157

Marchant, T.R., Smyth, N.F.: An undular bore solution for the higher-order Korteweg–de Vries equation. J. Phys. A Math. Gener. **39**(37), L563–L569 (2006), https://doi.org/10.1088/0305-4470/39/37/L02

Miropol'sky, Yu.Z.: Dynamics of Internal Gravity Waves in the Ocean. Springer, Heidelberg (2001). https://doi.org/10.1007/978-94-017-1325-2

Ostrovsky, L.A., Pelinovsky, E.N., Shrira, V.I., Stepanyants Y.A.: Beyond the KdV: Post-explosion development. Chaos **25**(9), 097620 (2015). https://doi.org/10.1063/1.4927448

Ostrovsky, L.A., Potapov, A.I.: Modulated Waves: Theory and Applications. Johns Hopkins University Press, Baltimore (2002)

Ostrovsky, L.A., Stepanyants, Y.A.: Internal solitons in laboratory experiments: Comparison with theoretical models. Chaos **15**(3) 037111 (2005). https://doi.org/10.1063/1.2107087

Pelinovsky, E., Polukhina, O., Slunyaev, A., Talipova, T.: Internal solitary waves, In: Grimshaw, R.H.J. (ed.) Solitary Waves in Fluids, pp. 85–110. WIT Press, Southampton (2007). https://doi.org/10.2495/978-1-84564-157-3/04

Slyunyaev, A.V., Pelinovsky, E.N.: Dynamics of large-amplitude solitons. J. Exp. Theor. Phys. **89**(1), 173–181 (1999). https://doi.org/10.1134/1.558966

Slyunyaev, A.V.: Dynamics of localized waves with large amplitude in a weakly dispersive medium with a quadratic and positive cubic nonlinearity. J. Exp. Theor. Phys. **92**(3), 529–534 (2001). https://doi.org/10.1134/1.1364750

Yih, C.S.: Gravity waves in a stratified fluid. J. Fluid Mech. **8**(4), 481–508 (1960). https://doi.org/10.1017/s002211206000075x

Chapter 5
Emergence of Solitonic Structures in Hierarchical Korteweg–de Vries Systems

Andrus Salupere, Martin Lints, and Lauri Ilison

Abstract We explore numerically different types of solutions of a hierarchical Korteweg–de Vries equation (Giovine and Oliveri, Meccanica **30**(4), 341–357 (1995)) that describes *inter alia* wave propagation in microstructured (dilatant granular) materials. This equation contains three material parameters that collectively determine the type of solutions. The simulations focus on the effect the microstructure has on the field of motion driven by harmonic, cnoidal and sech^2-type initial waves. The simulations employ a Fourier transform based pseudospectral method and have been performed for a wide range of material parameters. The results are interpreted in terms of the properties of Korteweg–de Vries solitons. The analysis reveals considerable evolution and transformations of the field of motions during the propagation of signals of various kind due to the effect of the microstructure. A large part of the numerically tracked solutions have properties that match the core properties of solitons. On many occasions the emerging waves in the system propagate at a constant speed, keep their shape and interact with other similar entities elastically. The number of emerging solitons markedly depends on the ratio of dispersion parameters for micro- and macrostructure, and on the shape of the initial wave.

5.1 Introduction

Modelling nonlinear wave propagation in continuous media with microstructure is one of the gradually increasing challenges in contemporary mechanics and material science (see, e.g., (Maugin, 1999; Erofeev, 2003; Engelbrecht et al., 2005; Fish et al., 2012) and references therein). Microstructured materials are characterised by the existence of intrinsic internal spatial scales. Such usually concealed features of the matter, like the size of a grain or a crystallite, the periodicity of certain internal

A. Salupere (✉) · M. Lints · L. Ilison
Department of Cybernetics, School of Science, Tallinn University of Technology, Tallinn, Estonia
e-mail: salupere@ioc.ee; martin.lints@ioc.ee

A. Berezovski, T. Soomere (eds.), *Applied Wave Mathematics II*, Mathematics of Planet Earth 6, https://doi.org/10.1007/978-3-030-29951-4_5

lattices, the distance between the microcracks, etc. may induce the scale dependence in the equations that govern the properties of the material, and wave propagation in such materials.

Historically, two main approaches have been developed to describe the impact of microstructure on properties of various materials. Followers of the discrete approach (see, e.g., (Maugin, 1999)) focus first of all on the modelling of dynamics and consequences of elastic behaviour of crystal lattices in the matter. The continuum approach (Mindlin, 1964; Eringen, 1966, 1999) addresses the emerging features of microstructured materials in terms of higher order partial differential equations whose spatial derivatives eventually reflect various nonlinear effects stemming from the presence of structures at different scales. One of the basic ideas of Eringen (1966, 1999) is that macro- and microstructure of the material continua should be separated and corresponding conservation laws should be formulated separately for both structures. An alternative approach is to take the microstructural quantities into account in one set of conservation laws (Maugin, 1993a).

Another, quite recently introduced view regarding the description of the properties of microstructured materials relies on the concept of internal variables. The basic idea is to model microstructures as internal fields (Berezovski et al., 2011).

All these approaches have developed rapidly. While Porubov and Osokina (2019) and De Domenico et al. (2018) work with lattice models, Mindlin-type models are applied in (Pastrone, 2010; Pastrone and Engelbrecht, 2016; Majorana and Tracinà, 2019). Multiscale modelling is employed in (Settimi et al., 2019) to elucidate the influence of microcracks on the propagating waves in elastic microcracked bars. A hierarchical model to describe the microstructure and strength of lath martensite is introduced in (Galindo-Nava and del Castillo, 2015). A recent review of predictive nonlinear theories for multiscale modelling of heterogeneous materials is given in (Matouš et al., 2017).

5.1.1 Continuum Description of Granular Materials

The continuum description of granular materials was introduced in early studies by Goodman and Cowin (1972) with several additions by Bedford and Drumheller (1983). The most important scale factor in such descriptions is the average diameter of grains. This quantity is naturally related to the wavelength of the excitation of propagating waves in environments that support wave propagation. The dependence of the properties of the material on the specific scale involves both dispersive and nonlinear effects. A fundamental consequence is that solitary waves and solitons can exist in such a media if these two effects are specifically balanced.

In this chapter the focus is on the simulation of the emergence of solitonic structures and solitary wave interactions in dilatant granular materials. The basic idea of the model is to consider suspension of grains (particles) in a compressible fluid (the rest of the material). The fluid density is assumed to be small compared to the particle density, and rotation of particles is neglected. If the grains are incompressible, the one-dimensional (1D) equations of their motion can be reduced to the celebrated Korteweg–de Vries (KdV) equation (Korteweg and de Vries, 1895)

$$\psi(u) = u_t + uu_x + \alpha u_{xxx} = 0, \tag{5.1}$$

where $\psi(u)$ is sometimes called the KdV operator, α characterises the relative magnitude of dispersive effects (represented by the term αu_{xxx}) compared to the intensity of nonlinear interactions (represented by the term uu_x) and subscripts denote partial derivatives with respect to spatial coordinate x and time t.

However, if grains are compressible, one needs a more complicated (higher order) equation to follow the dynamics and fate of disturbances of the system. We shall employ a hierarchical Korteweg–de Vries (HKdV) equation

$$u_t + uu_x + \alpha_1 u_{xxx} + \beta\,(u_t + uu_x + \alpha_2 u_{xxx})_{xx} = 0 \tag{5.2}$$

that governs 1D wave propagation in dilatant granular materials (Giovine and Oliveri, 1995). Here α_1, α_2 are macro- and microlevel dispersion parameters, respectively, and β is a parameter involving the ratio of the grain size and the wavelength. The parameter β can be negative (if particles have high kinetic energy) as well as positive (if the kinetic energy of particles is small) (Giovine and Oliveri, 1995).

Equation (5.2) involves two KdV operators, $\psi_1(u)$ and $[\psi_2(u)]_{xx}$. The first three terms $\psi_1(u)$ describe motions in the macrostructure. The second KdV operator ($[\psi_2(u)]_{xx}$, terms in the brackets) accounts for motions in the microstructure. The parameter β controls the magnitude of influence of the microstructure. Therefore, (5.2) is hierarchical in the Whitham's sense (Whitham, 1974).

If $|\beta| \to 0$, the influence of microstructure can be neglected and the wave "feels" only the macrostructure. *Vice versa*, if $1/|\beta| \to 0$, only the influence of the microstructure is "felt" by the wave. The limiting case $\beta = 0$ results in the standard KdV equation with classic soliton solutions.

In this model the rotation of particles is neglected (Giovine and Oliveri, 1995). An extended continuum theory has been proposed for granular media where particle rotation is taken into account as well (Giovine, 2008). Daraio and Nesterenko have derived an equation for wave propagation in 1D chains of spherical beads (see (Daraio et al., 2005) and references therein). A hierarchical multiscale framework is proposed to model the mechanical behaviour of granular media (Guo and Zhao, 2014). This framework employs a rigorous hierarchical coupling between the finite element method and the discrete element method.

5.1.2 Higher Order KdV Like Equations

Equation (5.2) includes a fifth order dispersive term and complicated nonlinear terms and therefore can be considered as a higher order KdV-type equation. Due to the very rich physical background, equations that are somehow related to the KdV equation (including modified, generalised, higher order, etc. KdV equations and KdV hierarchies) have been intensely studied. The present book contains an overview of the derivation and use of one equation of this kind, an extended Gardner's equation, for internal gravity waves on the interface of a two-layer fluid (Kurkina et al., 2019).

The influence of higher order nonlinear terms on the shape and fate of generic wave solutions and solitary waves for mechanical systems governed by an extended KdV equation (a generalisation of the 5th order KdV equation) is addressed in (Porubov, 2003; Porubov et al., 2005). A variety of methods (tanh, sin–cos, tanh–coth, sinh–cosh and rational tanh–coth methods) have been applied for analytical study of various higher order KdV equations (Wazwaz, 2007, 2008). This approach resulted in exact solutions with distinct physical structure such as solitons, compactons, periodic and solitary patterns.

A modified KdV equation turned out to be a convenient tool for modelling the behaviour of undular bores and also allows for analytical undular bore solutions of the initial value problem (Marchant, 2008). Approaches based on the KdV hierarchy are also promising for solving several problems in nonlinear optics (Horsley, 2016). Solutions of the Ostrovsky equation (another higher order modification of the KdV equation) are applied to address the initial value problem for the Boussinesq–Klein–Gordon equation (Khusnutdinova and Tranter, 2018).

A hierarchy of generalised KdV equations is proposed to model wave propagation in generalised continua (Christov, 2015). This approach revealed the existence of peaked, compact travelling wave solutions ("peakompactons") in such systems. Similar equations are also employed to study capillary gravity waves in deep water (Kalisch, 2007). One of the derived systems hosts so-called approximate waves. Numerically found solutions of this type are compared to the Benjamin equation which arises in the special case of one-way propagation. As a spectral approximation of the KdV equation converges exponentially fast to the true solution if the Fourier basis is used and if the solution is analytic in a fixed strip on the real axis (Bjørkavåg and Kalisch, 2007), such exercises evidently reflect reality fairly well.

5.1.3 Solitonic Structures

The history of solitons in science goes back to the middle of the 19th century when John Scott Russell discovered waves what nowadays are known as solitons in natural experiments (Russell, 1844). After a half century, Korteweg and de Vries published their seminal paper (Korteweg and de Vries, 1895) where they derived a nonlinear equation that is today known as the KdV equation and highlighted its sech^2-type solitary wave and periodic cnoidal wave solutions. The contemporary soliton concept was introduced by Zabusky and Kruskal (1965). An excellent short review on history of solitons can be found in (Maugin, 2011).

It is often said that classic solitons can exist due to a certain balance between dispersion and nonlinearity. These phenomena are represented by terms u_{xxx} and uu_x, respectively, in the KdV equation. In a multitude of wave systems, the phase speed (of wave crests) and group speed (of wave energy) depend on the length (period) of the particular wave component. Consequently, each wave component travels with its own speed. This feature causes dispersion of wave packets, equivalently, a gradual lengthening of initially localised disturbances that contain many wave components. Nonlinear effects of certain kinds may counteract this process and produce compact signals.

The classic definition is that a soliton is a solitary solution of an evolution equation (solitary wave) that propagates at a constant speed and shape and interacts with other similar entities elastically. In other words, solitons restore their speed and shape after interactions (Drazin and Johnson, 1989). The classic equations that have soliton solutions are integrable and possess an infinite number of conservation laws. However, the property of integrability is a purely mathematical concept that is only conditionally applicable to real motions. Therefore, the classic definition is obviously too limiting for a proper description of the variety of physical phenomena that are mostly governed by nonintegrable equations and in which energy loss (or wave radiation) is an intrinsic part of the game. Solutions to such equations still often exhibit properties that are characteristic to solitons. These developments have led to the situation where the "border line" between solitons and long living solitary solutions of nonintegrable equations is vanishing and it is safe to say that a precise definition of solitons does not exist anymore.

The analysis of nonintegrable equations is a challenge because the powerful analytical methods for integrable systems do not work in these cases. This situation calls for widespread use of numerical methods to find general solutions for these problems as well as to track the fate of single solutions under arbitrary initial conditions. This approach has also a long history in the field of integrable systems. For example, the KdV equation was already numerically integrated under harmonic initial conditions by Zabusky and Kruskal (1965).

One of the most intriguing problems in this context is whether soliton-like or solitonic solutions can appear in nonintegrable systems, and under which conditions it may happen. If the solitonic character of a solution is detected, then new questions arise: how many solitons may emerge, is the solution periodic in time, does recurrence take place, do soliton trajectories form regular patterns, is it possible to use the data measured and gathered from the relevant experiments for solving inverse problems for the determination of material properties, etc. (see, e.g., (Salupere et al., 1994, 1996, 2003b,a; Peterson and van Groesen, 2000; Soomere, 2004; Engelbrecht and Salupere, 2005; Ilison and Salupere, 2005; Janno and Engelbrecht, 2005; Christov, 2009, 2012; Engelbrecht et al., 2011; Salupere, 2009)).

This chapter provides a concentrated overview of numerical studies into the properties and fate of various solutions to the HKdV equation. We start from an insight into the formulation of the problem and the basic components of the numerical method. The underlying numerical experiments were carried out with various initial conditions and wave shapes, and focused on the formation of solitonic solutions. The evolution of harmonic initial conditions was tracked in (Ilison and Salupere, 2003; Salupere et al., 2005). Sech^2-type initial conditions were employed in (Ilison et al., 2007; Salupere et al., 2008; Ilison, 2009; Ilison and Salupere, 2009; Salupere and

Ilison, 2010) and cnoidal initial conditions in (Lints et al., 2013; Salupere et al., 2014). The analyses highlight the role of different values of material parameters α_1, α_2, and β on the formation of different appearances and properties of solitonic solutions of the HKdV equation (5.2). This extensive material and the variety of findings from different numerical experiments make it possible to draw several conclusions about the nature of solutions to this equation that apparently are universal for different initial conditions.

5.2 Numerical Technique and Initial Conditions

In order to simulate the formation and propagation of different solitonic structures within the framework of the HKdV equation (5.2), the equation is integrated numerically under three types of initial conditions (IC): classic sinusoidal (harmonic waves), localised disturbances in the form of one or two sech^2-type pulses, and periodic cnoidal waves.

For numerical integration the pseudospectral method (PsM) based on discrete Fourier transformation (DFT) (Fornberg, 1998) was applied. A more detailed description of the numerical technique can be found in (Ilison et al., 2007; Ilison and Salupere, 2009; Salupere, 2009; Salupere et al., 2014). Numerical calculations were carried out with SciPy (Jones et al., 2007) software package using the FFTW library (Frigo and Johnson, 2005) for the DFT and the F2PY (Peterson, 2009) generated Python interface to the ODEPACK Fortran code (Hindmarsh, 1983) for the ODE solver.

5.2.1 Harmonic Initial Conditions

To analyse how harmonic waves are modified when they enter an environment where further wave propagation is governed by the HKdV equation, this equation is integrated using sinusoidal initial condition

$$u(x,0) = \sin x\,, \quad 0 \le x < 2\pi\,, \tag{5.3}$$

and periodic boundary conditions

$$u(x + 2n\pi, t) = u(x,t)\,, \quad n = 0, \pm 1, \pm 2, \ldots . \tag{5.4}$$

The main goals of these experiments were to (i) analyse and characterise the time-space behaviour of the solutions in three-dimensional space of material parameters α_1, α_2, and β, (ii) identify possible long living solution types, and (iii) highlight whether and under which values of the material parameters the solitonic solutions may emerge.

5.2.2 Localised Initial Conditions

As the HKdV equation (5.2) can be written as a linear superposition of two KdV operators $\psi_i = u_t + uu_x + \alpha_i u_{xxx}$, $i = 1, 2$ in the form

$$\psi_1 + \beta(\psi_2)_{xx} = 0, \tag{5.5}$$

it is natural to assume that, at least for small values of β, its solution mimics the classic soliton solution of the KdV equation. Following this line of thinking, we chose a single initial solitary wave

$$u(x, 0) = A \operatorname{sech}^2 \frac{x - m\pi}{\delta}, \quad \delta = \sqrt{\frac{12\alpha_1}{A}}, \tag{5.6}$$

as the initial condition for the HKdV equation (Ilison et al., 2007; Ilison and Salupere, 2009). This expression presents an analytical solution with a height of A to the KdV equation $\psi_1 = 0$ (Zabusky and Kruskal, 1965). The quantity δ is often called the width of the KdV soliton and the notations of height and amplitude coincide for KdV solitons.

The simulations revealed that four types of numerical solutions of the HKdV equation may be generated depending on the particular values of material parameters α_1, α_2, and β: (i) single (KdV) soliton, (ii) soliton ensemble (with weak tails), (iii) solitary wave with a tail, and (iv) solitary wave with a tail and certain additional wave packets. A wave or other component of the system is associated with a tail when it is not stable, usually continuously losing its energy and amplitude, or propagating rapidly away from the major energy containing region. The tail in case (ii) is sometimes so weak that it is practically indistinguishable from solitary wave profiles (that is, its presence is not visible as an additional local extremum of the numerical solution) as well as from spectral quantities.

For the purposes of the presentation in this chapter we run similar simulations using the two-wave initial condition

$$u(x, 0) = A_1 \operatorname{sech}^2 \frac{x - 16\pi}{\delta_1} + A_2 \operatorname{sech}^2 \frac{x - 48\pi}{\delta_2}. \tag{5.7}$$

This pulse consists of two sech^2-type localised solitary waves with amplitudes A_1 and A_2. Even though each component of (5.7) is an analytical solution to the KdV equation (equivalently, it satisfies the first KdV operator in Eq. (5.2) (Zabusky and Kruskal, 1965)), their sum is not. The solitary waves are shifted by 16π and 48π, respectively, with respect to the origin $x = 0$. The composite wave (5.7) is applied in order to simulate interactions of solitary waves that eventually emerge from different localised pulses. As above, the quantities

$$\delta_1 = \sqrt{\frac{12\alpha_1}{A_1}} \quad \text{and} \quad \delta_2 = \sqrt{\frac{12\alpha_1}{A_2}} \tag{5.8}$$

represent the widths of the pulses in the context of KdV solitons. Boundary conditions are periodic in space with the length of the period of 64π:

$$u(x + 64k\pi, t) = u(x, t), \quad k = \pm 1, \pm 2, \pm 3, \dots . \tag{5.9}$$

Similarly to the above, numerical experiments with initial conditions (5.7) made it possible to track interactions between (i) two single (KdV) solitons, (ii) solitons from two different ensembles, (iii) two solitary waves with tails, (iv) two solitary waves with tails and additional wave packets.

These four particular cases mimic the solution types employed above. The main goal was to follow the time-space behaviour of solutions to understand whether the interactions of the waves have solitonic nature.

5.2.3 *Cnoidal Initial Conditions*

The appearance and nature of solutions of the HKdV equation obviously depend on the material parameters α_1, α_2, and β and on the initial conditions. All initial conditions discussed above, both harmonic and sech^2-shaped functions (Ilison and Salupere, 2003; Salupere et al., 2005; Ilison, 2009) are limiting cases of cnoidal waves. Both these solutions were derived and the term 'cnoidal wave' (in analogy with a sinusoidal wave) was proposed by Korteweg and de Vries in their seminal paper (Korteweg and de Vries, 1895). The profiles of waves from this class follow the elliptic cosine cn squared, that is, their shape matches the function cn^2. Like sech^2-shaped solitary waves, cnoidal waves form a class of valid analytic solutions to the KdV equation.

Let us consider a cnoidal wave that is an analytical solution for the equation $\psi_1 = 0$. If $\alpha_1 = \alpha_2$ or $\beta = 0$, the HKdV equation (5.2) reduces to the KdV equation $\psi_1 = 0$. Therefore, such a cnoidal wave would propagate without any distortion for unlimited time. This feature suggests that the alteration of an initially cnoidal wave during propagation in the HKdV system highlights the effects of interaction between micro- and macrostructure in a manner that can be more intuitively understood compared to processes that occur with the harmonic wave or sech^2-type solitary pulses.

An additional advantage of the use of cnoidal waves compared to the sech^2 pulses is their periodic nature and existence of infinitely smooth derivatives (Korn and Korn, 2000). This means that their propagation will not generate any oscillations due to the Gibbs phenomenon when using the Fourier transform to solve (5.2) by pseudospectral methods.

A cnoidal wave solution for the KdV equation is derived as follows (Bhatnagar, 1979). The starting point is the assumption that the KdV equation $\psi_1 = 0$ has uniform wave train solutions in the form $u(x,t) = u(\xi)$, $\xi = x - ct$. For this kind of solution the KdV equation reduces to

$$-cu_\xi + uu_\xi + \alpha_1 u_{\xi\xi\xi} = 0. \tag{5.10}$$

Integrating (5.10) gives a cubic equation

$$-\frac{1}{2}cu^2 + \frac{1}{6}u^3 + \frac{1}{2}\alpha_1 u_\xi^2 + D_1 u + D_2 = 0, \tag{5.11}$$

from which it is possible to arrive at the cnoidal wave solution

$$u(\xi) = A\,\mathrm{cn}^2\left(\sqrt{\frac{A-\eta}{12\alpha_1}}\,\xi, \sqrt{m}\right). \tag{5.12}$$

This infinite wave train has a period

$$P = 4\sqrt{\frac{3\alpha_1}{A-\eta}}\,K(\sqrt{m}), \tag{5.13}$$

where A is the wave height (also called the profile amplitude if the wave profile minimum is at zero), $m = \sqrt{A/(A-\eta)}$ is the (elliptic) parameter (square root of the elliptic modulus), η is one of the zeros of (5.11), and $K(\sqrt{m})$ is Legendre's complete elliptic integral of the first kind (Korn and Korn, 2000).

Equation (5.13) can be used to calculate the spatial period of the wave, knowing its shape parameter m or to find the elliptic parameter m for a given dispersion parameter α_1 and amplitude A. In this work, A, α_1 and P are set to find m and η. The initial wave (5.12) will then propagate in case of the KdV equation without any change, similarly to a soliton. If the propagation is governed by the HKdV equation, the wave will eventually change. This kind of initial condition is also flexible for studying the interaction of solitary waves for various propagation distances by changing its spatial period.

In the limit $m \to 1$, a cnoidal wave (5.12) transforms to the above discussed sech^2-type initial wave

$$u(x) = A\,\mathrm{sech}^2\sqrt{\frac{A}{12\alpha_1}}x, \tag{5.14}$$

with a maximum at the origin. For smaller m values (around ≤ 0.7), the cnoidal wave (5.12) transforms to a $\sin^2$-type wave that resembles a harmonic wave (Bhatnagar,

1979). Such changes in the initial wave shape owing to variations in the parameter m may modify the number of emerging solitons.

The shape and period of the cnoidal wave are linked by (5.13):

$$\frac{2K(\sqrt{m})}{P} = \sqrt{\frac{A-\eta}{12\alpha_1}} = \sqrt{\frac{A}{12m\alpha_1}}. \tag{5.15}$$

The HKdV equation, like the KdV equation, has a shape selectivity property. This quality means that the initial wave will generally evolve into a solution determined by the dispersion parameter(s) of the equation (Christov and Velarde, 1995; Kliakhandler et al., 2000; Porubov, 2003; Porubov et al., 2003). This process is sensitive with respect to the initial wave shape. Therefore, a similar number of solitons will emerge whenever various cnoidal wave shapes satisfy the condition (Salupere et al., 2014)

$$m \sim \frac{A}{\alpha_1} P^2, \tag{5.16}$$

provided the ratio α_1/α_2 (which determines the number of emerging solitons) is constant (Ilison, 2009). Additionally, for some values of m, α_1, and A it is possible to closely replicate the shape of the sech^2 wave by means of the pseudospectral method, while preserving the smoothness of the initial wave and suppressing the Gibbs phenomenon.

5.3 Results and Discussion

Results of numerical experiments will be presented and discussed in this section for all three types of initial conditions described in Sect. 5.2. Simulation with different types of initial conditions had slightly different goals in terms of the emergence of (solitary) wave structures. For this reason different ranges of material parameters are used in simulations of different kinds of initial conditions. The overarching aim was a better understanding of general characteristics, such as symmetry of solutions, and the character of dispersion.

It is easy to check that the solutions of the HKdV equation (5.2) have the following symmetry property with respect to the parameters α_1 and α_2:

$$u(x, t, \alpha_1, \alpha_2) = -u(-x, t, -\alpha_1, -\alpha_2). \tag{5.17}$$

The dispersion analysis is based on tracking the quantity

$$R = \alpha_1 - 2\alpha_2\beta k^2 + \alpha_2^2\beta k^4. \tag{5.18}$$

Dispersion is normal (the phase speed is higher than the group speed) if $R > 0$, and anomalous (the group speed is higher than the phase speed) if $R < 0$. The value $R = 0$ corresponds to the nondispersive case when all wave harmonics have an equal group speed.

The sign of R depends on material parameters α_1, α_2, and β as well as on the wavenumber k. If in a certain domain of parameters α_1, α_2, and β, the sign of R is defined (either $R > 0$ or $R < 0$ for any wavenumber k), then one has a pure normal or pure anomalous dispersion case in this domain. Otherwise, the character of dispersion depends on the wavenumber k. This can be called a mixed dispersion case (see (Ilison, 2009) for details).

5.3.1 Harmonic Initial Conditions

Numerical integration of the HKdV equation (5.2) was performed in the range of parameters $-0.4 \le \alpha_1 \le 0.4$, $-0.4 \le \alpha_2 \le 0.4$, $-0.0222 \le \beta \le 0.0222$. This domain was convenient in order to demonstrate the behaviour of typical solutions.

Solutions of the HKdV equation (5.2) excited by harmonic initial conditions can be divided into two main types. Firstly, a train of n interacting solitons emerges like in the case of the KdV equation (Fig. 5.1). This type is called soliton ensemble (SE) in this chapter. Secondly, a soliton ensemble and an ensemble of m near equal amplitude solitary waves (EAE) emerge simultaneously (Fig. 5.2).

Fig. 5.1 Emergence of an ensemble of positive solitons from a harmonic wave. Timeslice plot over two 2π periods in space, $\alpha_1 = 0.05$, $\alpha_2 = 0.03$, $\beta = 0.0111$, $0 < t < 20$. Reprinted from Salupere et al. (2005). Copyright © (2005) IMACS, with permission from Elsevier.

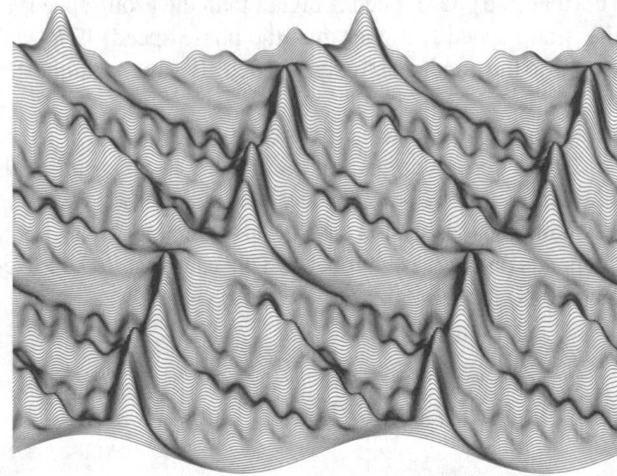

Fig. 5.2 Simultaneous emergence of a positive soliton ensemble (SE) and an ensemble of near equal amplitude solitary waves (EAE) from an initial harmonic wave. The EAE is suppressed. Timeslice plot over two 2π periods in space, $\alpha_1 = 0.05$, $\alpha_2 = 0.0525$, $\beta = 0.0111$, $0 < t < 20$. Reprinted from Salupere et al. (2005). Copyright © (2005) IMACS, with permission from Elsevier.

According to the symmetry property (5.17), the SE can be positive as well as negative. In Fig. 5.1 a positive SE is presented. Analysis of numerical results demonstrated that the behaviour of solitons in the ensemble is very similar to the behaviour of KdV solitons in case of harmonic initial conditions: solitons propagate at a constant speed and amplitude, they interact elastically and experience phase shift during interactions, the highest soliton is the fastest and propagates to the right, etc. (Zabusky and Kruskal, 1965; Salupere et al., 1996).

For the listed reasons this solution type was called a KdV ensemble in (Ilison and Salupere, 2003; Salupere et al., 2005). However this notion has caused misunderstanding in several discussions. To avoid ambiguity, we call this solution type a soliton ensemble here and below.

The symmetry property (5.17) requires that in case of negative SE the deepest soliton (the one that has the highest amplitude) is propagating to the left, and has a phase shift to the left during interactions.

The second solution type may involve an EAE, the appearance of which varies from a strongly suppressed one (Fig. 5.2), to a system with a comparable amplitude to the SE (Fig. 5.3). In some occasions the EAE dominates over the SE (Fig. 5.4). During such an amplification, solitary waves are stretched in the negative as well as in the positive direction (Salupere et al., 2005).

As solitary waves in the EAE propagate at nearly the same speed, they do not interact with each other. This means that waves from the EAE interact only with solitons from the SE. Through these interactions solitary waves from the EAE conserve their speed, shape, and amplitude. In other words, these interactions are

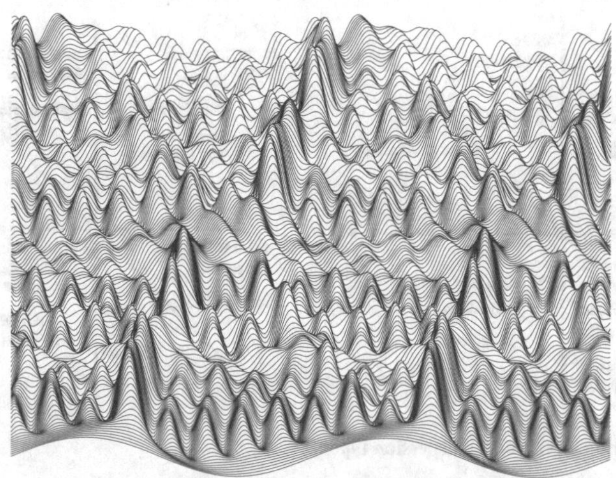

Fig. 5.3 Simultaneous emergence of a positive SE and an EAE from a harmonic wave. The EAE is dominating. Timeslice plot over two 2π periods in space, $\alpha_1 = 0.05$, $\alpha_2 = 0.067$, $\beta = 0.0111$, $0 < t < 20$.

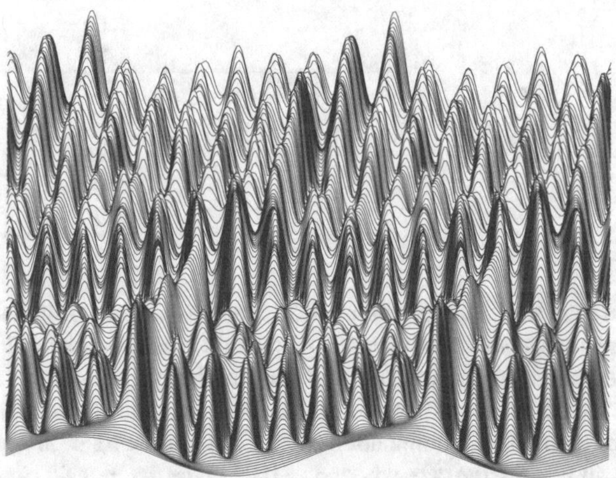

Fig. 5.4 Simultaneous emergence of a positive SE and an EAE from a harmonic wave. The EAE is dominating. Timeslice plot over two 2π periods in space, $\alpha_1 = 0.05$, $\alpha_2 = 0.07$, $\beta = 0.0111$, $0 < t < 20$. Reprinted from Salupere et al. (2005). Copyright © (2005) IMACS, with permission from Elsevier.

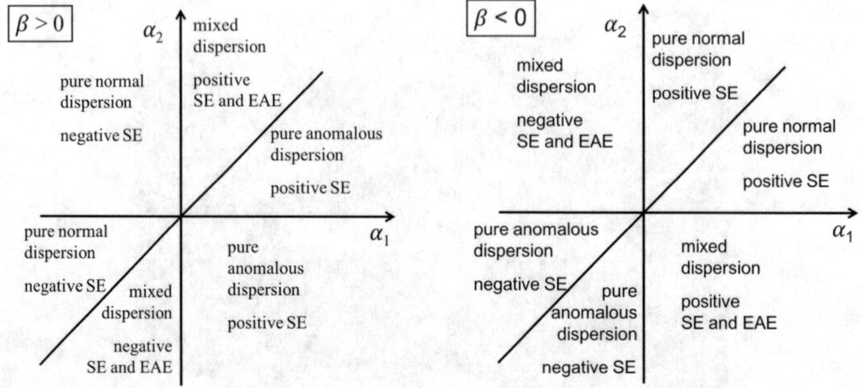

Fig. 5.5 Solution types versus dispersion types

Table 5.1 The number of solitary waves m in the dominating EA ensemble against the ratio α_2/α_1 for $\beta = 0.0111$ and $\beta = 0.0055$.

β	α_2/α_1	m
0.0111	1.11	9
	1.40	8
	1.80	7
	2.52	6
	3.47	5
0.0055	1.07	13
	1.27	12
	1.50	11
	1.81	10
	2.26	9

elastic and one can say that two solitonic structures have formed simultaneously. The relationships between solution types and dispersion types is visualised in Fig. 5.5.

For a fixed value of the parameter β the number m of solitary waves in the EAE is determined by the ratio α_2/α_1 (Table 5.1). The ratios presented in Table 5.1 correspond to the maximal domination of an EAE consisting of m solitary waves. For example, in Fig. 5.4 the case where $\beta = 0.0111, \alpha_2/\alpha_1 = 1.40$ can be seen, and therefore the number of solitary waves in the EAE (per one 2π space period) $m = 8$. If the ratio α_2/α_1 has values close to 1.40, one can still see an EAE of eight amplified solitary waves (Fig. 5.3). However, if the ratio α_2/α_1 continues to decrease (from 1.40 to 1.10), the amplification of the EAE weakens until an additional solitary wave appears in the EAE. After that the amplification starts again and at $\alpha_2/\alpha_1 = 1.11$ a dominating EAE occurs with $m = 9$.

The core conclusions from the numerical experiments are as follows. Harmonic initial conditions may lead to the simultaneous emergence of two solitonic structures, one of which is a soliton ensemble and the other consists of near equal amplitude solitary waves. Any of these structures can dominate over the other.

Table 5.2 The number of solitary waves m in the dominating EA ensemble against the parameter β for fixed values of α_1 and α_2.

α_1	α_2	β	m
0.03	0.09	0.0055	7
0.03	0.09	0.0111	4
0.03	0.09	0.0222	3
0.05	0.07	0.0055	9
0.05	0.07	0.0111	7
0.05	0.07	0.0222	5
0.09	0.19	0.0055	8
0.09	0.19	0.0111	5
0.09	0.19	0.0222	3

If the parameters α_1 and α_2 are fixed, a decrease in the parameter β leads to a higher maximum number of solitary waves in the EAE (Table 5.2). However, the particular case $\beta = 0$ results in a standard KdV soliton train. For a fixed value of β the number of waves m in the EAE is defined by the ratio α_2/α_1. Straight lines given by the ratio α_2/α_1 in Table 5.1 correspond to the dominating EAE-s of m solitary waves. An EAE can emerge in domains where (i) $\beta < 0$ and $\alpha_2\alpha_1 < 0$, and (ii) $\beta > 0$ and $\alpha_2\alpha_1 > 0$. The domination of the EAE over the SE can take place only in case (ii).

5.3.2 *Localised Initial Conditions*

The HKdV equation (5.2) is integrated numerically under sech^2-type initial conditions (5.7) and periodic boundary conditions (5.9) for $0 < \alpha_1 < 1, 0 < \alpha_2 < 1$ and $\beta = 111.11, 11.111, 1.111, 0.111, 0.0111$. The number of space grid points is $n = 4096$ and the length of the time interval $t_f = 100$.

This domain for material parameters α_1, α_2, and β was used in (Ilison et al., 2007; Ilison and Salupere, 2009) and serves as a basis for the definition of solution types in the case of a single sech^2-type initial condition (Sect. 5.2.2). Preliminary results of numerical simulations have been briefly described in Salupere and Ilison (2010). And additional sets of figures and data that support the analysis presented here can be found in an unpublished research report (Ilison and Salupere, 2008) that can be provided on request.

5.3.2.1 Interactions of Single Solitons

We start the analysis from the model case $\alpha_1 = \alpha_2$ when the HKdV equation (5.2) reduces to the KdV equation and the initial sech^2-pulse propagates at a constant speed and amplitude (Ilison et al., 2007; Ilison and Salupere, 2009). It is still of interest to simulate interactions between two initial pulses that have different amplitudes and

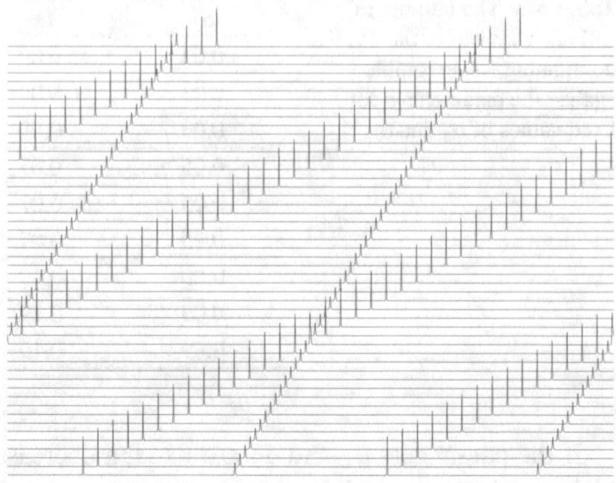

Fig. 5.6 Interactions of single solitons presented as the components of the initial double pulse (5.8) in the HKdV equation (5.2). Timeslice plot over two space periods for $\alpha_1 = \alpha_2 = 0.05$, $\beta = 0.111$, $0 \le t \le 100$, $A_1 = 15$, $A_2 = 5$.

therefore propagate at different speeds. In the case $\alpha_1 = \alpha_2 = 0.05$, $\beta = 0.111$, $A_1 = 15$, and $A_2 = 5$, the left-hand side solitary wave with amplitude $A_1 = 15$ propagates faster than the right-hand side one (with amplitude $A_2 = 5$). As a consequence, we can track an interaction of these two entities (see the corresponding timeslice plot in Fig. 5.6).

Not surprisingly, our analysis demonstrated that interactions between two emerging solitons are elastic as they restore their speeds and amplitudes after interaction. The interaction causes a phase shift. The higher amplitude soliton is shifted to the right and the lower amplitude soliton to the left. Between interactions both solitons propagate at a constant amplitude that is equal to the initial amplitude (Salupere and Ilison, 2010).

To verify the solitonic character of the solution, the behaviour of the solitons during interactions is compared to the behaviour of pure KdV solitons governed by the KdV equation. The phase shifts of the interacting components of a two-soliton solution of the KdV equation $u_t + uu_x + \alpha u_{xxx} = 0$ are:

$$\vartheta_1 = \Theta\Delta_1, \quad \vartheta_2 = -\Theta\Delta_2, \quad \Theta = \ln\frac{1+\sqrt{r}}{1-\sqrt{r}}, \quad \Delta_i = \sqrt{\frac{12\alpha}{A_i}}, \quad r = \frac{A_2}{A_1}. \tag{5.19}$$

Here ϑ_1, ϑ_2 are phase shifts of the higher and the lower soliton, respectively, $i = 1, 2$, and $A_1 > A_2$ are amplitudes of interacting solitons (Salupere et al., 2002).

In numerical experiments the velocities and phase shifts of interacting solitons are calculated from the "trajectories" of local maxima of the wave profiles. Note that three different types of trajectories can be distinguished (Salupere et al., 2002). One can conclude that between interactions, solitons restored their initial amplitudes A_i and speeds $c_i = A_i/3$, and numerically estimated phase shifts ϑ_i^{num} and phase shifts from (5.19) coincide. In other words, in case of $\alpha_1 = \alpha_2$ the components of the initial double pulse (5.8) behave exactly like KdV solitons.

5.3.2.2 Interactions of Soliton Ensembles

In the present subsection we examine the case if $\alpha_1 \neq \alpha_2$ when two different soliton ensembles with weak tails emerge from the initial double pulse (also called dual sech^2-type initial conditions) (5.7) (Ilison et al., 2007; Ilison and Salupere, 2009). The simulations of their evolution are performed for the following values of amplitudes and material parameters: $A_1 = 8$, $A_2 = 4$, $\alpha_1 = 1$, $\alpha_2 = 0.1$, and $\beta = 111.11$. The corresponding timeslice, pseudocolor and waveprofile maxima plots are presented in Figs. 5.7, 5.8, and 5.9, respectively.

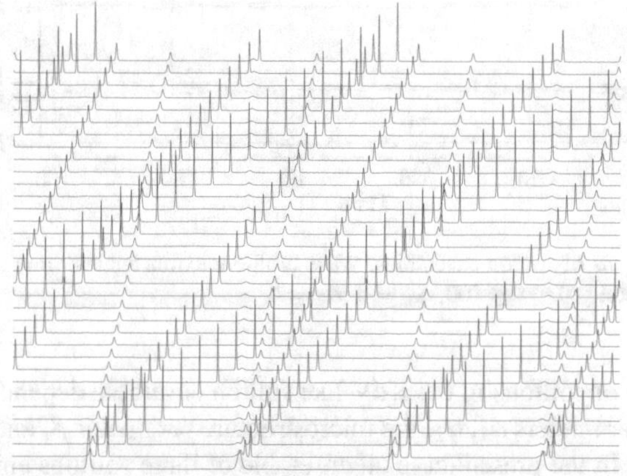

Fig. 5.7 Interactions of soliton ensembles. Timeslice plot over two space periods for $\alpha_1 = 1$, $\alpha_2 = 0.1$, $\beta = 111.11$, $0 \leq t \leq 100$, $A_1 = 8$, $A_2 = 4$.

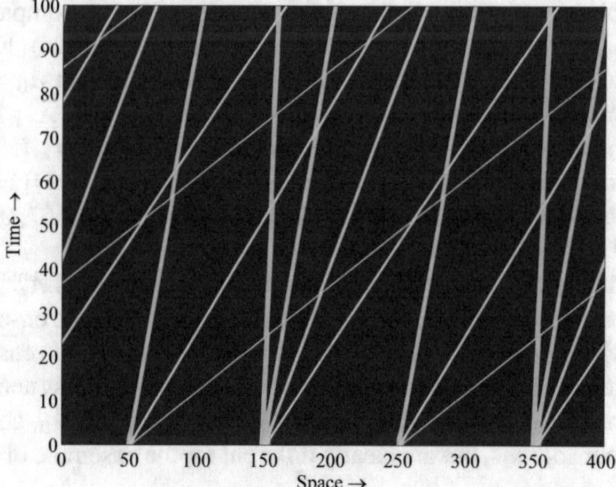

Fig. 5.8 Interactions of soliton ensembles. Pseudocolor plot over two space periods for $\alpha_1 = 1$, $\alpha_2 = 0.1$, $\beta = 111.11$, $0 \leq t \leq 100$, $A_1 = 8$, $A_2 = 4$.

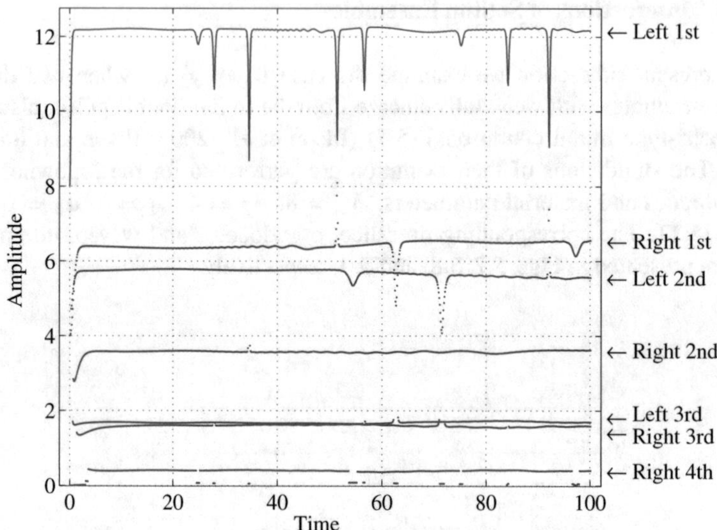

Fig. 5.9 Interactions of soliton ensembles. Wave profile maxima against time in case of $\alpha_1 = 1$, $\alpha_2 = 0.1$, $\beta = 111.11$, $0 \le t \le 100$, $A_1 = 8$, $A_2 = 4$.

The number of solitons in the KdV-like soliton ensemble depends on the values of dispersion parameters α_1, α_2, the microstructure parameter β, and the amplitude of the pulse A. In the present case an ensemble of three solitons emerges from the left-hand side initial pulse and an ensemble of four solitons from the right-hand side pulse.

To demonstrate the similarities and differences between the numerical solutions and the N-soliton solution of the KdV equation, let us compare the ratios of amplitudes of the emerging solitons. For the (normalised) KdV equation $u_t + uu_x + u_{xxx} = 0$, the initial pulse $u(x, 0) = A_0 \operatorname{sech}^2 x$ with $A_0 = 6N(N+1)$ results in a train of N solitons. In this train, the k-th soliton ($k = 1, \ldots, N$) propagates at the speed $c_k = 4(N + 1 - k)^2$ and its amplitude is $A_k = 3c_k$ (Newman et al., 1991). Therefore, the ratios $A_k/A_0 = 2(N + 1 - k)^2/[N(N + 1)]$ can be directly calculated. Consequently, for $N = 3$ we have $A_k/A_0 = [1.5, 0.667, 0.167]$, and for $N = 4$ this set is $A_k/A_0 = [1.6, 0.9, 0.4, 0.1]$.

The numerically evaluated soliton amplitudes A_k^{num} and ratios A_k^{num}/A_0 resulted in the values $A_k^{\text{num}}/A_0 \approx [1.52, 0.72, 0.21]$ for the left-hand side ensemble of three solitons and $A_k^{\text{num}}/A_0 \approx [1.6, 0.9, 0.4, 0.1]$ for the right-hand side ensemble of four solitons. Therefore, the ratios in question found for the KdV system A_k/A_0 and extracted from the simulations with the HKdV equation A_k^{num}/A_0, coincide for the ensemble of four solitons, but are clearly different for the ensemble of three solitons as $A_k^{\text{num}}/A_0 > A_k/A_0$.

Consequently, the ensemble of four solitons can be considered as a “pure” KdV soliton ensemble (without the tail) and the ensemble of three solitons could be interpreted as a KdV soliton ensemble with a (weak) tail. This difference highlights the importance of the microstructure on the propagation of pulses. The initial sech^2-pulses are (one-)soliton solutions of the KdV equation and thus only satisfy the equation that corresponds to the first KdV operator in (5.2). The formation of soliton trains evidently takes place due to the presence of the second KdV operator in (5.2) (Ilison and Salupere, 2009).

Solitons with different amplitudes propagate at different speeds and therefore interactions between emerging solitons generally take place in the HKdV system. Technically, it possible to trace here two types of interactions: (i) between solitons from different ensembles, and (ii) between solitons from the same ensemble. In both situations, during interactions, the solitons are phase shifted (Fig. 5.8) and amplitudes of higher solitons decrease (Fig. 5.9). However, after interactions the solitons almost restore their amplitudes (Fig. 5.9) and speeds (Fig. 5.8). All solitons also interact with the "tails" of the wave system. However, as the tails are weak, they do not influence the behaviour of solitons essentially. Their influence can be traced only as small modifications to wave profiles, where the tails can cause small oscillations in the location of maxima. The main outcome of the analysis is that that the observed interactions are nearly elastic and therefore the numerical solutions can be called solitonic.

5.3.2.3 Interactions of Solitons with Strong Tails

On some occasions, two solitons and strong tails emerge from the initial wave (5.7). This occurs, for example, for the amplitudes $A_1 = 15$ and $A_2 = 5$ of initial pulses and material parameters $\alpha_1 = 0.05, \alpha_2 = 0.07$, and $\beta = 111.11$. Due to the different initial amplitudes the emerged solitons propagate at different speeds and therefore interact with each other (Figs. 5.10, 5.11, 5.12).

Fig. 5.10 Interactions of solitons with strong tails. Wave profiles at $t = 0, 10, \ldots, 100$ over two space periods for $\alpha_1 = 0.03, \alpha_2 = 0.07, \beta = 111.11, A_1 = 15, A_2 = 5$.

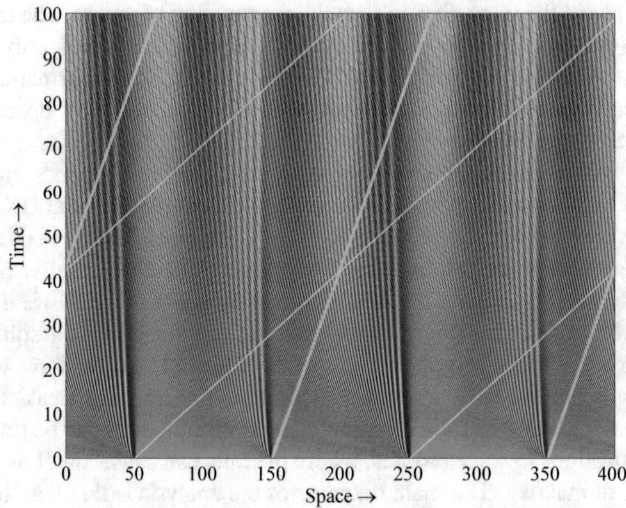

Fig. 5.11 Interactions of solitons with strong tails. Pseudocolor plot over two space periods for $\alpha_1 = 0.03$, $\alpha_2 = 0.07$, $\beta = 111.11$, $A_1 = 15$, $A_2 = 5$.

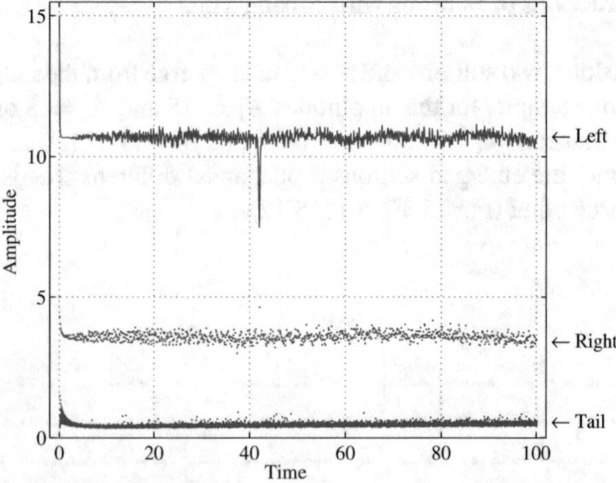

Fig. 5.12 Interactions of solitons with strong tails. Wave profile maxima against time in case of $\alpha_1 = 0.03$, $\alpha_2 = 0.07$, $\beta = 111.11$, $0 \leq t \leq 100$, $A_1 = 15$, $A_2 = 5$.

For this solution type a fairly strong tail emerges and influences the behaviour of emerged solitary waves essentially. As a result, the amplitudes of propagating solitary waves are lower than the amplitudes of the initial waves. Moreover, their amplitudes are not constant and exhibit substantial oscillations owing to the influence of the tails (Fig. 5.12).

The decrease in the left- and the right-hand side solitary wave amplitudes is proportional to the initial amplitudes. The propagating solitons are approximately 1.4 times lower than the initial waves (Fig. 5.12). Therefore, the shape of the initial wave is modified to be more appropriate to the actual solution of the equation. This phenomenon may be called a selection process. In other words, the selection means that during the propagation, the amplitude and the velocity of an initial solitary wave tend to the (finite) values predetermined by the equation coefficients (see (Christov and Velarde, 1995; Kliakhandler et al., 2000; Porubov, 2003; Porubov et al., 2003) for details). The interaction produces a phase shift in soliton trajectories. The higher solitary wave is shifted to the right and the lower solitary wave to the left. After the interaction, both waves almost restore their speeds and shapes. Therefore these solitary waves can be called solitons.

5.3.2.4 Interactions of Solitary Waves with Tails and Wave Packets

Solitary waves, tails, and wave packets may emerge simultaneously on some occasions. In such cases the envelope of the packet can propagate to the left or to the right and at much higher speed than that of the emerged solitary waves or the high frequency waves that form the packet (Ilison et al., 2007; Ilison and Salupere, 2009).

Here we present two examples. The first one corresponds to the set of parameters $\alpha_1 = 0.05, \alpha_2 = 0.11, \beta = 0.0111, 0 \leq t \leq 100, A_1 = 8, A_2 = 4$ (Figs. 5.13, 5.14). The second simulation was performed using the following parameters: $\alpha_1 = 0.09$, $\alpha_2 = 0.11, \beta = 0.0111, 0 \leq t \leq 100$, $A_1 = 12$, $A_2 = 2$ (Figs. 5.15, 5.16). All three components of the solution can be seen in timeslice plots (Figs. 5.13, 5.15).

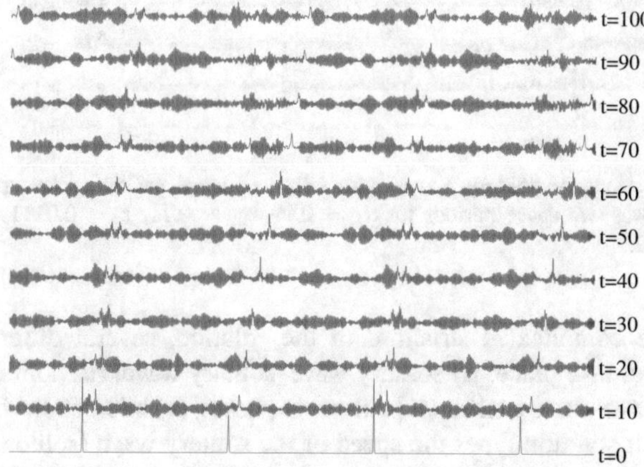

Fig. 5.13 Interactions of solitary waves with tails and wave packets. Wave profiles at $t = 0, 10, \ldots, 100$ over two space periods for $\alpha_1 = 0.05$, $\alpha_2 = 0.11$, $\beta = 0.0111$, $0 \leq t \leq 100$, $A_1 = 8$, $A_2 = 4$.

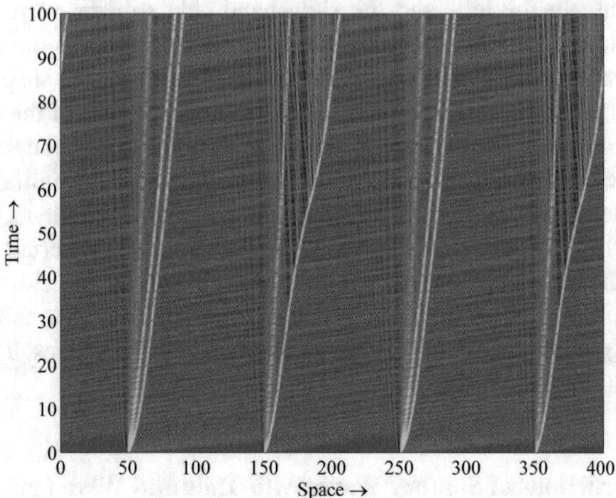

Fig. 5.14 Interactions of solitary waves with tails and wave packets. Pseudocolor plot over two space periods for $\alpha_1 = 0.05$, $\alpha_2 = 0.11$, $\beta = 0.0111$, $0 \leq t \leq 100$, $A_1 = 8$, $A_2 = 4$.

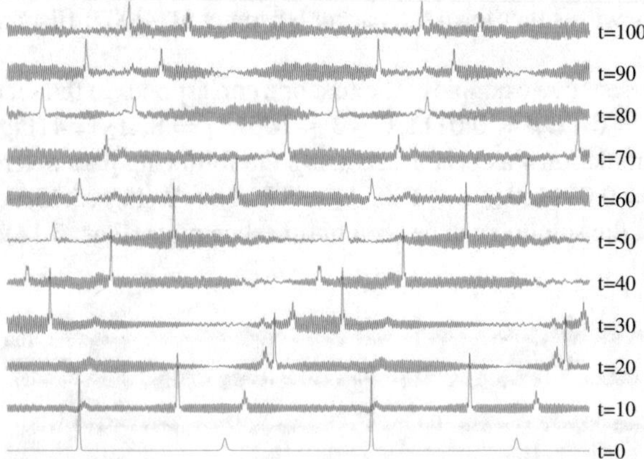

Fig. 5.15 Interactions of solitary waves with tails and wave packets. Wave profiles at $t = 0, 10, \ldots, 100$ over two space periods for $\alpha_1 = 0.09$, $\alpha_2 = 0.11$, $\beta = 0.0111$, $0 \leq t \leq 100$, $A_1 = 12$, $A_2 = 2$.

Due to the complicated structure of the solution, several different kinds of interactions can take place: (i) solitary wave–solitary wave, (ii) solitary wave–tail, (iii) solitary wave–wave packet, (iv) tail–wave packet, (v) interactions between wave packets. For all solution types the speed of the solitary wave (soliton) depends on the amplitude of the initial wave (Ilison et al., 2007; Ilison and Salupere, 2009). Not surprisingly, the higher is the initial wave, the higher is the emerged solitary wave and the higher is its speed.

Based on this knowledge it is natural to assume that similarly to the previous solution types, the interacting solitary waves emerge from different amplitude initial waves for this solution type. However, due to the emergence of different wave packets the situation here is more complicated. From a single initial pulse, one wave emerged besides the tail and wave packets. From a dual pulse, several solitary waves can emerge. However, interacting solitary waves were detected in a few cases only.

The influence of different wave packets on the behaviour of solitary waves can be so strong that their amplitudes decrease rapidly and it is practically impossible to distinguish between solitary waves and wave packets. Still, according to timeslice and pseudocolor plots (Figs. 5.13, 5.14), the emerged solitary waves are not completely suppressed. Essentially, a very strong selection (Christov and Velarde, 1995; Kliakhandler et al., 2000; Porubov, 2003; Porubov et al., 2003) takes place. The shapes of all solitary waves are altered to a certain critical amplitude level. This level can be several times lower than the amplitude of the initial wave.

In the first simulation the emerged solitary waves are selected nearly to the same amplitude level and they all propagate at nearly the same speed without interaction (Figs. 5.13, 5.14). The second simulation revealed that different solitary waves are selected to different amplitude levels and therefore interactions between solitary waves take place (Figs. 5.15, 5.16). Importantly, it is clear that these interactions are not elastic as the speeds and amplitudes of solitary waves are altered during interactions.

Notwithstanding these different interactions and the selection phenomenon, all three components of the solution persist over the whole integration time interval. In this sense the solution is stable. However, it is likely that in the current case the emerged solitary waves are not solitons, because either it is impossible to simulate their interactions or these interactions are not elastic.

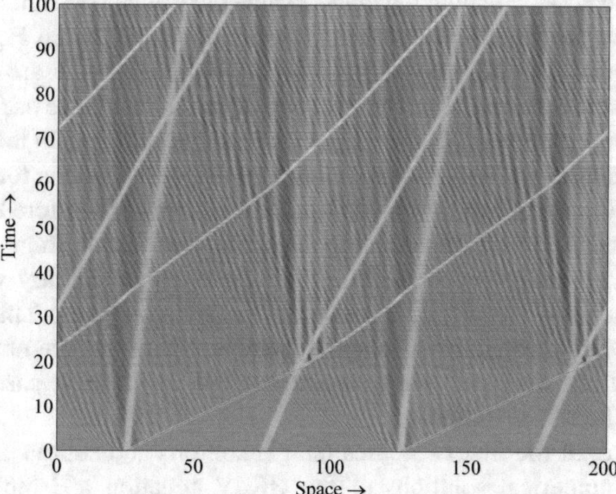

Fig. 5.16 Interactions of solitary waves with tails and wave packets. Pseudocolor plot over two space periods for $\alpha_1 = 0.09$, $\alpha_2 = 0.11$, $\beta = 0.0111$, $0 \le t \le 100$, $A_1 = 12$, $A_2 = 2$.

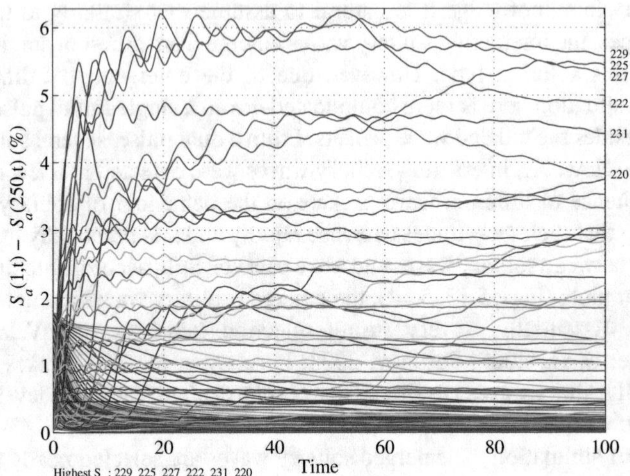

Fig. 5.17 Interactions of solitary waves with tails and wave packets. Time averaged spectral densities against time for $\alpha_1 = 0.05$, $\alpha_2 = 0.11$, $\beta = 0.0111$, $n = 4096$, $t_f = 100$, $A_1 = 8$, $A_2 = 4$.

The phenomenon of the formation of wave packets in terms of time averaged spectral densities (TANSD, see the Appendix) has been addressed in (Ilison and Salupere, 2009). In this framework, $S_{\mathrm{a}}(k, t_i)$ reflects the contribution of the k-th spectral density over the time interval $[0, t_i]$. The time evolution of these quantities are presented for the two examples of the fourth solution type given above (see the corresponding timeslice plots in Figs. 5.13 and 5.15).

Figures 5.17, 5.18 reveal that wave packets are formed by amplified higher order harmonics. The $S_{\mathrm{a}}(k, t)$ having the highest value determines the number of maxima (oscillations) in the given wave profile. One can conclude that in Fig. 5.17 several harmonics around $k = 229$ and in Fig. 5.18 for $140 \le k \le 143$ are amplified and therefore generate wave packets. The structure of the formed wave packets is slightly different from the case described in (Ilison and Salupere, 2009). In the present case, two sets of different wave packets emerge simultaneously. One is formed from the left hand and the other from the right hand initial pulse. Therefore the number of amplified harmonics can be significantly higher than in case of single initial pulses.

The presented analysis demonstrates that the emerged solitary waves interact elastically or nearly elastically on three occasions (solution types). This occurs when a single soliton is created, or a soliton ensemble is accompanied by a weak tail, or when a solitary wave coexists with a strong tail. In all these cases the solitary structures can be called solitons.

Furthermore, in the first two cases (single solitons and soliton ensembles) the behaviour of numerical solutions of the HKdV equation and the behaviour of analytical solutions of the KdV equation almost match each other. Therefore, it is to some extent justified to call these two solution types “single KdV soliton” and “KdV soliton ensemble”.

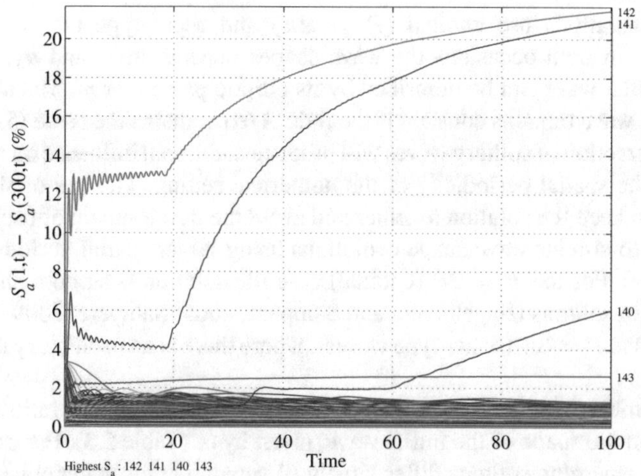

Fig. 5.18 Interactions of solitary waves with tails and wave packets. Time averaged spectral densities against time for $\alpha_1 = 0.09$, $\alpha_2 = 0.11$, $\beta = 0.0111$, $n = 4096$, $t_f = 100$, $A_1 = 12$, $A_2 = 2$.

The fourth solution type (i.e., solitary waves with tails and wave packets) behaves differently. The emerging solitary waves are altered by the wave packets essentially. In some cases interactions take place and on some other occasions they do not. Even if interactions take place, they are not elastic, and emerged waves thus cannot be called solitons.

5.3.3 *Cnoidal Initial Conditions*

The HKdV equation is integrated numerically under cnoidal initial conditions (5.12) and periodic boundary conditions with $P = 2\pi$ for $A = 1 \ldots 5$, $\alpha_1 = 0.3, 0.4$, $\alpha_2 = 0.0055 \ldots 0.042$ and $\beta = 11.111, 111.11$. For these values (5.13) gives the range of the elliptic parameter as $m = 0.5628 \ldots 0.9905$. The initial wave shape thus ranges from nearly sinusoidal ($m = 0.5628$) to a nearly sech^2-shaped signal (for $m = 0.9905$). For comparison, sech^2-shaped initial conditions are also used, with $P = 8\pi$.

The wave shape parameters are chosen to fit the macrostructure KdV operator of the HKdV equation, and therefore are related to the amplitude A and macrostructural dispersion parameter α_1. As mentioned above, if $\alpha_2 = \alpha_1$ or $\beta = 0$, the HKdV equation would yield simple periodic wave propagation without any change in shape. It is likely that in the general case $\alpha_1 \neq \alpha_2$ and $\beta \neq 0$ solitary waves will be generated and/or an oscillatory tail excited due to the effect of the microstructure (the impact of which is governed by β and α_2).

The simulations use cnoidal ($P = 2\pi$) and sech^2-type ($P = 8\pi$) initial conditions. In both occasions the wave shapes depend on A and α_1. The shape of the cnoidal wave can be described by its elliptic parameter m. The shape of the sech^2-type wave remains constant if the ratio A/α_1 remains the same (5.14).

The interaction of solitary waves will produce additional influence on the solution owing to the spatial periodicity of the numerical setting. The use of short spatial periods can keep the solution together and avoid the development of oscillatory tail compared to similar wave shape evolutions using larger spatial periods (Salupere et al., 2014). For the $P = 2\pi$ (cnoidal) case the solution is smooth and there are more hidden solitons (Engelbrecht and Salupere, 2005; Salupere, 2009, 2016) than for the choice $P = 8\pi$ (sech^2-type waves), where there is an oscillatory tail in many cases.

The number of emerging solitons primarily depends on the ratio α_1/α_2 and secondly on the shape of the initial wave, given by m (Table 5.3). For example, the number of emerging solitons differ largely (7 versus 11) for the cnoidal (Fig. 5.19) and sech^2-type (Fig. 5.20) initial waves with $\alpha_1/\alpha_2 = 72.7$, $A = 1$. This difference obviously is due to the different shapes of the initial conditions, as the parameters

Table 5.3 The main parameters of the simulations. The number of emerging HKdV solitons depends on the ratio α_1/α_2 and the shape of the initial wave (cnoidal wave parameter m). The shape of the sech^2-type wave is set using A and α_1. Here $\beta = 111.11$.

α_1/α_2	m	A	α_1	Cnoidal IC			sech^2 IC		
				Solitons	Visible	Hidden	**Solitons**	Visible	Hidden
72.7	0.98	5	0.4	**11**	8	3	**11**	11	Tail
72.7	0.95	4	0.4	**10**	8	2	**11**	10	1
72.7	0.91	3	0.4	**10**	8	2	**11**	10	1
72.7	0.80	2	0.4	**9**	7	2	**11**	10	1
72.7	0.56	1	0.4	**7**	6	1	**11**	9	2
14.3	0.98	5	0.4	**5**	4	1	**5**	5	0
14.3	0.95	4	0.4	**5**	4	1	**5**	5	0
14.3	0.91	3	0.4	**5**	4	1	**5**	5	0
14.3	0.80	2	0.4	**4**	4	0	**5**	5	0
14.3	0.56	1	0.4	**4**	3	1	**5**	5	0
11.8	0.99	5	0.3	**5**	4	1	**4**	4	Tail
11.8	0.98	4	0.3	**4**	4	0	**4**	4	Tail
11.8	0.95	3	0.3	**4**	4	0	**4**	4	Tail
11.8	0.88	2	0.3	**4**	3	1	**4**	4	0
11.8	0.67	1	0.3	**4**	3	1	**4**	4	0
7.1	0.99	5	0.3	**4**	3	1	**3**	3	Tail
7.1	0.98	4	0.3	**4**	3	1	**3**	3	Tail
7.1	0.95	3	0.3	**4**	3	1	**3**	3	Tail
7.1	0.88	2	0.3	**3**	3	0	**3**	3	Tail
7.1	0.67	1	0.3	**3**	2	1	**3**	3	0

Bold letters indicate the total values for different initial conditions for solitons

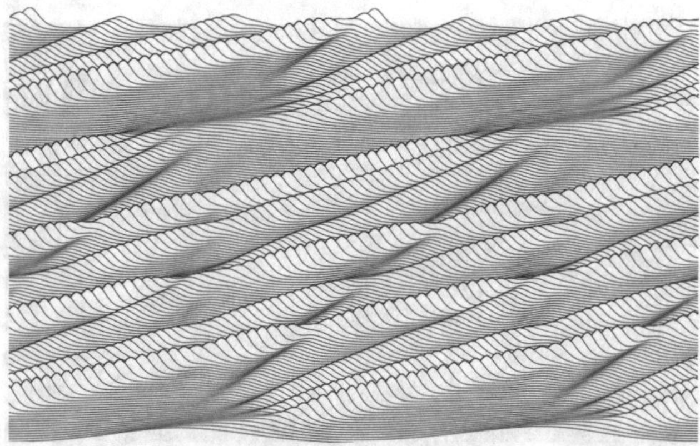

Fig. 5.19 Evolution of the cnoidal initial wave with $\alpha_1/\alpha_2 = 72.7$, $A = 1$, $m = 0.563$, $\alpha_1 = 0.4$. There are six visible solitons, and one hidden soliton.

Fig. 5.20 Evolution of the sech^2-type initial wave with $\alpha_1/\alpha_2 = 72.7$, $A = 1$, $\alpha_1 = 0.4$. There are nine visible and two hidden solitons.

α_i, β and A are identical. The cnoidal wave with $m = 0.5628$ is close to a harmonic wave and thus essentially different from the sech^2-shaped initial wave. The initial cnoidal wave generally produces far fewer solitons (including hidden solitons) than the sech^2-type initial condition.

To look deeper into this feature, let us compare results for the cnoidal (Fig. 5.21) and sech^2-shape initial waves (Fig. 5.22). As both simulations have identical material parameters $\alpha_1/\alpha_2 = 7.1, m = 0.980, A = 4$, the results should be similar. The shape of the cnoidal wave is much sharper than above because the parameter m is closer to 1, so its profile is comparable to sech^2-shape. Indeed, the number of visible solitons is three in both cases.

While the field of motions excited by a cnoidal wave has an additional hidden soliton, the one driven by the sech^2-shape signal has a weak oscillatory tail.

Fig. 5.21 Evolution of the cnoidal initial wave with $\alpha_1/\alpha_2 = 7.1$, $A = 4$, $m = 0.980$, $\alpha_1 = 0.3$. There are three visible solitons and one hidden soliton.

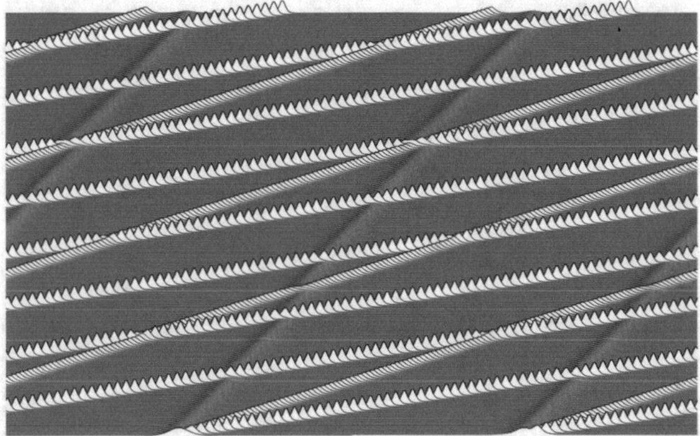

Fig. 5.22 Evolution of the sech^2-type initial wave with $\alpha_1/\alpha_2 = 7.1$, $A = 4$, $\alpha_1 = 0.3$. There are three solitons and a very weak oscillatory tail.

Therefore, when the initial waves are similar to each other (comparing sech^2 profile with a cnoidal profile) the number of emerging solitons becomes equal. This *inter alia* signals that the number of emerging solitons may be substantially affected by the shape of the initial wave.

The notion of hidden solitons denotes those which are missing in the soliton ensemble (train) that is formed at the very beginning of the integration time interval. In other words, they do not emerge before the first interactions between solitons take place. As the number of solitons increases, more solitons stay hidden. They can be detected via their influence on the higher solitons when these have no other interactions.

Such hidden structures are best detected from the fluctuations of the profiles of relatively large solitons near their maxima. The idea for their detection is that the presence of minor local minima or concavities in the amplitude of solitons is a characteristic feature of soliton interactions. If the amplitude curve of a higher soliton has such a minimum or a short concave section that cannot be associated with an interaction of any visible soliton, it is likely that such a feature indicates the existence of a hidden soliton. Hidden solitons can show up for very short time intervals after several interactions have taken place (Engelbrecht and Salupere, 2005; Salupere, 2009, 2016).

Let us give an example. Figure 5.23 shows the pseudocolor plot of cnoidal wave evolution in the HKdV system and Fig. 5.24 plots the maxima curves (time series of the locations of their maxima in space) of the detected solitons. The three highest solitons are already visible at the very beginning of the integration, but the fourth (the

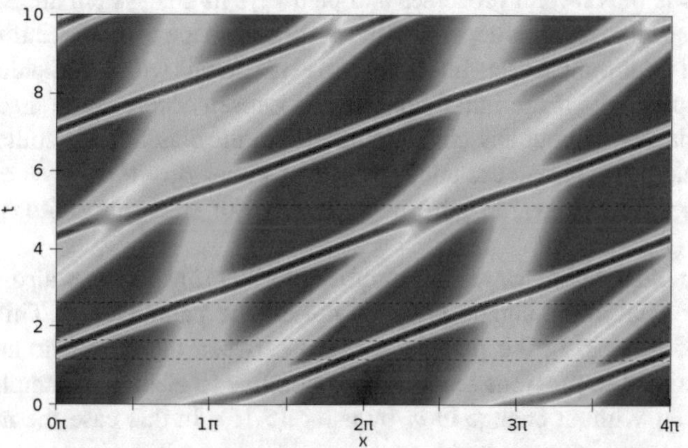

Fig. 5.23 Pseudocolor plot of the evolution of a cnoidal initial signal with $\alpha_1/\alpha_2 = 7.1$, $A = 4$, $m = 0.980$, $\alpha_1 = 0.3$.

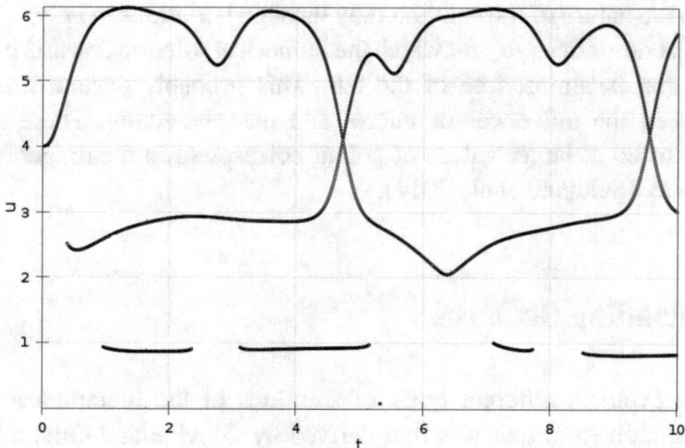

Fig. 5.24 Amplitude curves characterising the evolution of a cnoidal initial signal with $\alpha_1/\alpha_2 = 7.1$, $A = 4$, $m = 0.980$, $\alpha_1 = 0.3$.

hidden) soliton does not appear (as a waveprofile maximum) until several interactions have taken place. A single dot in the lowest part of Fig. 5.24 at $5 < t < 6$ is the only evidence of this hidden soliton in this type of graph.

The presence of such structures can be more reliably detected from concavities in the largest soliton amplitude in Fig. 5.24. In case of (almost) harmonic initial conditions (that is, for small parameter m of the initial cnoidal wave), the spectral amplitude maxima can point to the events of interaction between visible and hidden solitons. Unfortunately, the spectral data is not reliable for detecting hidden solitons from nonharmonic initial waves in the HKdV system. Hidden solitons exhibit stable propagation and thus are qualitatively different from the tail (Sect. 5.2.2).

The core finding from a comparison of numerical experiments with the HKdV equation driven by cnoidal input signals with those excited by a harmonic or localised input cases is that several processes and pathways in this system are universal and only some specific features depend on the particular shape of the excitation. There are still several important aspects inferred from experiments with the cnoidal excitation.

Not surprisingly, if the initial condition is chosen according to macrostructure, it will propagate without any change similarly to the classic KdV soliton when the effect of microstructure is excluded (by setting either $\beta = 0$ or $\alpha_1 = \alpha_2$). If this is not the case, the ratio α_1/α_2 governs the number of emerging solitons for a given value of β (Ilison, 2009).

For a cnoidal initial wave, the number of emerging solitons also depends on the shape of the initial signal, given by the elliptic parameter m. This allows the bridging of the gap in the parameter space between the harmonic and localised inputs discussed in previous sections. It is possible to change the amplitude A and dispersion α_1 without change in m by using (5.16). In this case the nature of the solution also remains unchanged.

The evolution of a harmonic initial wave in the KdV equation has been thoroughly studied, including counting hidden solitons (Salupere et al., 1994). For future simulations, the parameter space can be decreased by not including A or α_1 as they change the nature of the solution only through variations in m.

There exist ratios of α_1/α_2 for which the numerical solution consists of a solitonic wave train that is almost free of the tail. This probably occurs due to a good match between the influences of micro- and macrostructure. These ratios seem more pronounced at larger values of β that correspond to a stronger influence of microstructure (Salupere et al., 2014).

5.4 Concluding Remarks

This review explores different types of solutions of the hierarchical Korteweg–de Vries equation (5.2) that was first derived by Giovine and Oliveri (1995) and describes *inter alia* wave propagation in dilatant granular materials. The model originally considers suspension of grains in a compressible fluid, where fluid density is small compared to particle density and rotation of particles is neglected. This equation contains three material parameters that collectively determine the type of solutions. The numerical simulations have focused on the effect of the microstructure on the field of motions. They have been carried out using a discrete Fourier transformation based pseudospectral method.

The equation is closely linked to Korteweg–de Vries equation, to which it transforms to if grains are incompressible. This feature motivates the interpretation of the results first of all in terms of the appearance and classic properties of KdV solitons and other analytic solutions to the KdV equation. A generic process in the environment governed by the KdV equation and similar equations is the generation of solitary waves and solitonic structures from a wide variety of initial pulses.

For this reason, the initial waves vary from harmonic waves (that are not valid solutions to the KdV equation) to cnoidal waves and (as another limit case) to sech^2-type solitary waves (KdV solitons). The parameters of the incoming solitons have been chosen to accommodate the macrostructure operator ψ_1 of the equation under scrutiny. The analysis has revealed considerable evolution and transformations of the field of motions during the propagation of such signals due to the effect of the microstructure.

A particular goal of the simulations was to establish whether at least some of the numerical solutions of (5.2) have properties that match the core properties of solitons. In other words, we have carefully examined whether the emerging waves in the system propagate at a constant speed, keep their shape and interact with other similar entities elastically, i.e., restore their speed and shape after interactions.

The main findings and conclusions have been confirmed for a wide range of initial conditions and material parameters. For fixed values of the parameter β (that characterises the relative magnitude of the impact of microstructure) the number of emerging solitary waves is defined by the ratio α_2/α_1 of the dispersion parameters for micro- and macrostructure. On many occasions, the excited solitary waves do interact elastically or nearly elastically. This feature is predominant for three solutions types: (i) when a single soliton emerges, (ii) when a soliton ensemble is accompanied with a weak tail, and (iii) when a solitary wave interacts with a strong tail. The nature and number of emerging waves is governed not only by the material parameters but also by the shape of the initial wave.

In conclusion, it is safe to say that the complex HKdV model (5.2) exhibits a rich variety of solutions of different kind. Even though the basic appearance of this model suggests that at least some of its solutions are solitons, the presence of microstructure may easily destroy this property. Our simulations demonstrate that this model, albeit not integrable, keeps the fundamental solitonic character on most occasions but with some limitations.

Acknowledgements The authors are grateful for the possibility to collaborate with Jüri Engelbrecht during many long years and for many fruitful discussions. This research was supported by the Estonian Research Council via institutional block grant IUT33-24.

Appendix: Time Averaged Spectral Densities

The time averaged spectral densities (TANSD) serve as a tool to analyse formation of complex wave structures. We use the pseudospectral method (PsM) for numerical integration and therefore the discrete Fourier transform (DFT)

$$U(k,t) = \sum_{j=0}^{n-1} u\,(j\Delta x, t) \exp\left(-\frac{2\pi ijk}{n}\right) \tag{5.20}$$

is computed at every time step for the approximation of space derivatives. Here n is the number of space grid points, $\Delta x = 2\pi/n$ is the space step, i denotes imaginary unit, and $k = 0, \pm 1, \pm 2, \ldots, \pm\,(n/2-1)\,, -n/2$. If $U\,(k,t)$ is the DFT of the function $u(x,t)$, defined by (5.20), then spectral densities are:

$$S(k,t) = \frac{4U^2}{n^2}\,, \quad k = 1, \ldots, \frac{n}{2} - 1\,, \quad S(k,t) = \frac{2U^2}{n^2}\,, \quad k = \frac{n}{2}\,. \tag{5.21}$$

For each time instant t one can define normalised spectral densities

$$S_{\text{norm}}(k,t) = \frac{S(k,t)}{S_{\text{sum}}(t)} \times 100\%\,, \quad \text{where} \quad S_{\text{sum}}(t) = \sum_{k=1}^{n/2} S(k,t), \tag{5.22}$$

and time averaged normalised spectral densities (TANSD)

$$S_{\text{a}}(k,t) = \frac{\int_0^t S_{\text{norm}}(k,t)dt}{t}\,. \tag{5.23}$$

We have discrete values of spectral densities S and S_{norm} at discrete time moments t_j, i.e., we have $S(k,t_j)$ and $S_{\text{norm}}(k,t_j)$. Therefore, at $t = t_j$

$$S_{\text{a}}(k,t_j) = \frac{\sum_{m=1}^{j} S_{\text{norm}}(k,t_m)}{j}\,. \tag{5.24}$$

The time averaged spectral densities (5.24) reflect the contribution of the k-th spectral density (or amplitude) over the time interval $[0, t_j]$. The idea of applying TANSD comes from (Galgani et al., 1992) where “time average energies of single modes” are used in order to discuss the energy equipartition in systems of Fermi–Pasta–Ulam type. Compared with spectral densities, the analysis of TANSD gives a clearer understanding about the predomination of certain harmonics.

References

Bedford, A., Drumheller, D.: On volume fraction theories for discretized materials. Acta Mechanica **48**, 173–184 (1983). https://doi.org/10.1007/BF01170415

Berezovski, A., Engelbrecht, J., Maugin, G.A.: Generalized thermomechanics with dual internal variables. Arch. Appl. Mech. **81**, 229–240 (2011). https://doi.org/10.1007/s00419-010-0412-0

Bhatnagar, P.L.: Nonlinear Waves in One-Dimensional Dispersive Systems. Oxford University Press, Oxford (1979)

Bjørkavåg, M., Kalisch, H.: Exponential convergence of a spectral projection of the KdV equation. Phys. Lett. A **365**(4), 278–283 (2007). https://doi.org/10.1016/j.physleta.2006.12.085

Christov, C., Velarde, M.: Dissipative solitons. Phys. D **86**, 323–347 (1995). https://doi.org/10.1016/0167-2789(95)00111-G

Christov, I.: Internal solitary waves in the ocean: analysis using the periodic, inverse scattering transform. Math. Comput. Simulat. **80**, 192–201 (2009). https://doi.org/10.1016/j.matcom.2009.06.005

Christov, I.: Hidden solitons in the Zabusky–Kruskal experiment: Analysis using the periodic, inverse scattering transform. Math. Comput. Simulat. **82**(6), 1069–1078 (2012). https://doi.org/10.1016/j.matcom.2010.05.021

Christov, I.C.: On a hierarchy of nonlinearly dispersive generalized Korteweg–de Vries evolution equations. Proc. Estonian Acad. Sci. **64**(3), 212–218 (2015). https://doi.org/10.3176/proc.2015.3.02

Daraio, C., Nesterenko, V.F., Herbold, E.B., Jin, S.: Strongly nonlinear waves in a chain of Teflon beads. Phys. Rev. E **72**, 016603 (2005). https://doi.org/10.1103/PhysRevE.72.016603

De Domenico, D., Askes, H., Aifantis, E.C.: Capturing wave dispersion in heterogeneous and microstructured materials through a three-length-scale gradient elasticity formulation. J. Mech. Behav. Mater. **27**(5-6) (2018). https://doi.org/10.1515/jmbm-2018-2002

Drazin, P.G., Johnson, R.S.: Solitons: An Introduction. Cambridge University Press, Cambridge (1989). https://doi.org/10.1002/zamm.19900700817

Engelbrecht, J., Salupere, A.: On the problem of periodicity and hidden solitons for the KdV model. Chaos **15**, 015114 (2005). https://doi.org/10.1063/1.1858781

Engelbrecht, J., Berezovski, A., Pastrone, F., Braun, M.: Waves in microstructured materials and dispersion. Phil. Mag. **85**(33-35), 4127–4141 (2005). https://doi.org/10.1080/14786430500362769

Engelbrecht, J., Salupere, A., Tamm, K.: Waves in microstructured solids and the Boussinesq paradigm. Wave Motion **48**(8), 717–726 (2011). https://doi.org/10.1016/j.wavemoti.2011.04.001

Eringen, A.: Linear theory of micropolar elasticity. J. Math. Mech. **15**, 909–923 (1966). https://doi.org/10.1016/0020-7225(67)90004-3

Eringen, A.: Microcontinuum Field Theories I: Foundations and Solids. Springer, Berlin (1999)

Erofeev, V.I.: Wave Processes in Solids with Microstructure. World Scientific, Singapore (2003). https://doi.org/10.1142/5157

Fish, J., Filonova, V., Yuan, Z.: Reduced order computational continua. Comput. Methods Appl. Mech. Engrg. **221–222**, 104–116 (2012). https://doi.org/10.1016/j.cma.2012.02.010

Fornberg, B.: A Practical Guide to Pseudospectral Methods. Cambridge University Press, Cambridge (1998). https://doi.org/10.1017/CBO9780511626357

Frigo, M., Johnson, S.: The design and implementation of FFTW3. Proc. IEEE **93**(2), 216–231 (2005). https://doi.org/10.1109/JPROC.2004.840301

Galgani, L., Giorgilli, A., Martinoli, A., Vanzini, S.: On the problem of energy equipartition for large systems of the Fermi–Pasta–Ulam type: analytical and numerical estimates. Physica D **59**(4), 334–348 (1992). https://doi.org/10.1016/0167-2789(92)90074-W

Galindo-Nava, E., del Castillo, P.R.D.: A model for the microstructure behaviour and strength evolution in lath martensite. Acta Materialia **98**, 81–93 (2015). https://doi.org/10.1016/j.actamat.2015.07.018

Giovine, P.: An extended continuum theory for granular media. In: Capriz, G., Marino, P.M., Giovine, P. (eds.) Mathematical Models of Granular Matter. Lecture Notes in Mathematics, vol. 1937, pp. 167–192. Springer, Berlin (2008). https://doi.org/10.1007/978-3-540-78277-3_8

Giovine, P., Oliveri, F.: Dynamics and wave propagation in dilatant granular materials. Meccanica **30**(4), 341–357 (1995). https://doi.org/10.1007/BF00993418

Goodman, M.A., Cowin, S.C.: A continuum theory for granular materials. Arch. Rat. Mech. Anal. **44**(4), 249–266 (1972). https://doi.org/10.1007/BF00284326

Guo, N., Zhao, J.: A coupled FEM/DEM approach for hierarchical multiscale modelling of granular media. Int. J. Numer. Methods Eng. **99**(11), 789–818 (2014). https://doi.org/10.1002/nme.4702

Hindmarsh, A.C.: ODEPACK, a systematized collection of ODE solvers. In: Stepleman, R.S., et al. (eds.) Scientific Computing, pp. 55–64. North-Holland, Amsterdam (1983)

Horsley, S.A.R.: The KdV hierarchy in optics. J. Optics **18**(8), 085104 (2016). https://doi.org/10.1088/2040-8978/18/8/085104

Ilison, L.: Solitons and solitary waves in hierarchical Korteweg–de Vries type systems. PhD thesis, Tallinn University of Technology (2009)

Ilison, L., Salupere, A.: Solitons in hierarchical Korteweg–de Vries type systems. Proc. Estonian Acad. Sci. Phys. Math. **52**(1), 135–144 (2003)

Ilison, O., Salupere, A.: Propagation of sech^2-type solitary waves in higher-order KdV-type systems. Chaos Solitons Fractals **26**(2), 453–465 (2005). https://doi.org/10.1016/j.chaos.2004.12.045

Ilison, L., Salupere, A.: Interactions of Solitary Waves in Hierarchical KdV-Type System. Research Report Mech 291/08. Institute of Cybernetics at Tallinn University of Technology, Tallinn (2008)

Ilison, L., Salupere, A.: Propagation of sech^2-type solitary waves in hierarchical KdV-type systems. Math. Comput. Simulation. **79**, 3314–3327 (2009). https://doi.org/10.1016/j.matcom.2009.05.003

Ilison, L., Salupere, A., Peterson, P.: On the propagation of localised perturbations in media with microstructure. Proc. Estonian Acad. Sci. Phys. Math. **56**(2), 84–92 (2007)

Janno, J., Engelbrecht, J.: An inverse solitary wave problem related to microstructured materials. Inverse Problems **21**, 2019–2034 (2005). https://doi.org/10.1088/0266-5611/21/6/014

Jones, E., Oliphant, T., Peterson. P., et al: SciPy: Open source scientific tools for Python (2007). http://www.scipy.org

Kalisch, H.: Derivation and comparison of model equations for interfacial capillary-gravity waves in deep water. Math. Comput. Simulation **74**(2-3), 168–178 (2007)

Khusnutdinova, K., Tranter, M.: D'Alembert-type solution of the Cauchy problem for a Boussinesq-type equation with the Ostrovsky term. arXiv preprint arXiv:180808150 (2018). https://doi.org/10.1016/j.matcom.2006.10.008

Kliakhandler, I., Porubov, A., Velarde, M.: Localized finite-amplitude disturbances and selection of solitary waves. Phys. Rev. E **62**(4), 4959–4962 (2000). https://doi.org/10.1103/PhysRevE.62.4959

Korn, G.A., Korn, T.M.: Mathematical Handbook for Scientists and Engineers. Dover (2000). https://doi.org/10.1007/978-3-662-08549-3

Korteweg, D.J., de Vries, G.: On the change of form of long waves advancing in a rectangular canal, and on a new type of long stationary waves. Phil. Mag. **39**, 422–443 (1895). https://doi.org/10.1080/14786449508620739

Kurkina, O., Kurkin, A., Pelinovsky, E., Stepanyants, Y., Talipova, T.: Nonlinear models of finite-amplitude interfacial waves in shallow two-layer fluid. In: Berezovski, A., Soomere, T. (eds.) Applied Wave Mathematics II. Selected Topics in Solids, Fluids, and Mathematical Methods and Complexity, p. 61–88. Springer (2019) (this collection). https://doi.org/10.1007/978-3-030-29951-4_4

Lints, M., Salupere, A., Dos Santos, S.: Formation and detection of solitonic waves in dilatant granular materials: Potential application for nonlinear NDT. In: 7th International Workshop NDT in Progress: NDT of Lightweight Materials (2013) https://www.ndt.net

Majorana, A., Tracinà, R.: (2019) Exact and numerical solutions to a Mindlin microcontinuum model. arXiv preprint, arXiv:190102813 (2019)

Marchant, T.R.: Undular bores and the initial-boundary value problem for the modified Korteweg–de Vries equation. Wave Motion **45**(4), 540–555 (2008). https://doi.org/10.1016/j.wavemoti.2007.11.003

Matouš, K., Geers, M.G., Kouznetsova, V.G., Gillman, A.: A review of predictive nonlinear theories for multiscale modeling of heterogeneous materials. J. Comput. Phys. **330**, 192–220 (2017). https://doi.org/10.1016/j.jcp.2016.10.070

Maugin, G.A.: Material Inhomogeneities in Elasticity. Chapman & Hall, London (1993)

Maugin, G.A.: Nonlinear Waves in Elastic Crystals. Oxford University Press, Oxford (1999)

Maugin, G.A.: Solitons in elastic solids (1938–2010). Mech. Res. Commun. **38**(5), 341–349 (2011). https://doi.org/10.1016/j.mechrescom.2011.04.009

Mindlin, R.D.: Micro-structure in linear elasticity. Arch. Rat. Mech. Anal. **16**, 51–78 (1964). https://doi.org/10.1007/BF00248490

Newman, W.I., Campbell, D.K., Hyman, J.M.: Identifying coherent structures in nonlinear wave propagation. Chaos **1**(1), 77–94 (1991). https://doi.org/10.1063/1.165813

Pastrone, F.: Hierarchical structures in complex solids with microscales. Proc. Estonian Acad. Sci. **59**(2), 79–86 (2010). https://doi.org/10.3176/proc.2010.2.04

Pastrone, F., Engelbrecht, J.: Nonlinear waves and solitons in complex solids. Math. Mech. Solids **21**(1), 52–59 (2016). https://doi.org/10.1177/1081286515572245

Peterson, P.: F2PY: a tool for connecting Fortran and Python programs. Int. J. Comput. Sci. Engrng. **4**(4), 296–305 (2009). https://doi.org/10.1504/IJCSE.2009.029165

Peterson, P., van Groesen, E.: A direct and inverse problem for wave crests modelled by interactions of two solitons. Phys. D **141**(3-4), 316–332 (2000). https://doi.org/10.1016/S0167-2789(00)00037-3

Porubov, A.V.: Amplification of Nonlinear Strain Waves in Solids. World Scientific, Singapore (2003). https://doi.org/10.1142/5238

Porubov, A., Osokina, A.: Double dispersion equation for nonlinear waves in a graphene-type hexagonal lattice. Wave Motion **89**, 185–192 (2019). https://doi.org/10.1016/j.wavemoti.2019.03.013

Porubov, A.V., Gursky, V.V., Maugin, G.A.: Selection of localized nonlinear seismic waves. Proc. Estonian Acad. Sci. Phys. Math. **52**(1), 85–93 (2003)

Porubov, A., Maugin, G.A., Gursky, V., Krzhizhanovskaya, V.: On some localized waves described by the extended KdV equation. C. R. Mecanique **333**(7), 528–533 (2005). https://doi.org/10.1016/j.crme.2005.06.003

Russell, J.S.: Report on waves. Richard and John E Tailor (1844)

Salupere, A.: Pseudospectral method and discrete spectral analysis. In: Quak, E., Soomere, T. (eds.) Applied Wave Mathematics. Selected Topics in Solids, Fluids, and Mathematical Methods, pp. 301–333. Springer, Heidelberg (2009). https://doi.org/10.1007/978-3-642-00585-5_16

Salupere, A.: On hidden solitons in KdV related systems. Math. Comput. Simulat. **127**, 252–262 (2016). https://doi.org/10.1016/j.matcom.2014.04.012

Salupere, A., Ilison, L.: Numerical simulation of interaction of solitons and solitary waves in granular materials. In: Ganghoffer, J., Pastrone, F. (eds.) Mechanics of Microstructured Solids 2: Cellular Materials, Fibre Reinforced Solids and Soft Tissues, Lecture Notes in Applied and Computational Mechanics, vol. 50, pp. 21–28. Springer, Berlin (2010). https://doi.org/10.1007/978-3-642-05171-5_3

Salupere, A., Maugin, G.A., Engelbrecht, J.: Korteweg–de Vries soliton detection from a harmonic input. Phys. Lett. A **192**(1), 5–8 (1994). https://doi.org/10.1016/0375-9601(94)91006-5

Salupere, A., Maugin, G.A., Engelbrecht, J., Kalda, J.: On the KdV soliton formation and discrete spectral analysis. Wave Motion **23**, 49–66 (1996). https://doi.org/10.1016/0165-2125(95)00040-2

Salupere, A., Engelbrecht, J., Peterson, P.: Long-time behaviour of soliton ensembles. Part I—Emergence of ensembles. Chaos Solitons Fractals **14**, 1413–1424 (2002). https://doi.org/10.1016/S0960-0779(02)00069-3

Salupere, A., Peterson, P., Engelbrecht, J.: Long-time behaviour of soliton ensembles. Part II—periodical patterns of trajectories. Chaos Solitons Fractals **15**(1), 29–40 (2003a). https://doi.org/10.1016/S0960-0779(02)00070-X

Salupere, A., Engelbrecht, J., Peterson, P.: On the long-time behaviour of soliton ensembles. Math. Comput. Simulation **62**(1-2), 137–147 (2003b). https://doi.org/10.1016/S0378-4754(02)00178-7

Salupere, A., Engelbrecht, J., Ilison, O., Ilison, L.: On solitons in microstructured solids and granular materials. Math. Comput. Simulation **69**(5-6), 502–513 (2005). https://doi.org/10.1016/j.matcom.2005.03.015

Salupere, A., Ilison, L., Tamm, K.: On numerical simulation of propagation of solitons in microstructured media. In: Todorov, M. (ed.) Proceedings of the 34th Conference on Applications of Mathematics in Engineering and Economics, vol. 1067 of AIP Conference Proceedings, pp. 155–165. Melville, NY (2008). https://doi.org/10.1063/1.3030782

Salupere, A., Lints, M., Engelbrecht, J.: On solitons in media modelled by the hierarchical KdV equation. Arch. Appl. Mech. **84**(9), 1583–1593 (2014). https://doi.org/10.1007/s00419-014-0861-y

Settimi, V., Trovalusci, P., Rega, G.: Dynamical properties of a composite microcracked bar based on a generalized continuum formulation. Contin. Mech. Thermodyn. (2019). https://doi.org/10.1007/s00161-019-00761-7

Soomere, T.: Interaction of Kadomtsev–Petviashvili solitons with unequal amplitudes. Phys. Lett. A **332**(1-2), 74–81 (2004). https://doi.org/10.1016/j.physleta.2004.09.030

Wazwaz, A.M.: Analytic study for fifth-order KdV-type equations with arbitrary power nonlinearities. Commun. Nonlinear Sci. Numer. Simul. **12**(6), 904–909 (2007). https://doi.org/10.1016/j.cnsns.2005.10.001

Wazwaz, A.M.: New sets of solitary wave solutions to the KdV, mKdV, and the generalized KdV equations. Commun. Nonlinear Sci. Numer. Simul. **13**(2), 331–339 (2008). https://doi.org/10.1016/j.cnsns.2006.03.013

Whitham, G.: Linear and Nonlinear Waves. Wiley, New York (1974)

Zabusky, N.J., Kruskal, M.D.: Interaction of solitons in a collisionless plasma and the recurrence of initial states. Phys. Rev. Lett. **15**(6), 240–243 (1965). https://doi.org/10.1103/PhysRevLett.15.240

Chapter 6
Nonlinear Longitudinal Bulk Strain Waves in Layered Elastic Waveguides

Karima R. Khusnutdinova and Matthew R. Tranter

Dedicated to Jüri Engelbrecht on the occasion of his 80th birthday

Abstract We consider long longitudinal bulk strain waves in layered waveguides using Boussinesq-type equations. The equations are developed using lattice models, and this is viewed as an extension of the Fermi–Pasta–Ulam problem. We describe semianalytical approaches to the solution of scattering problems in delaminated waveguides, and to the construction of the solution of an initial value problem in the class of periodic functions, motivated by the scattering problems.

6.1 Introduction

Layered structures are frequently used in modern engineering constructions. The dynamical behaviour of layered structures depends not only on the properties of the bulk material, but also on the type of the bonding between the layers. For example, if layers have similar properties and the bonding between the layers is sufficiently soft, then the bulk strain soliton is replaced with a solitary wave radiating a copropagating oscillatory tail (Khusnutdinova et al., 2009). Experimental observations of both pure and radiating bulk strain solitons in layered bars have been discussed in (Dreiden et al., 2012).

The aim of this paper is to describe some efficient semianalytical approaches to the modelling of the scattering of pure and radiating solitary waves in delaminated areas of bonded layered structures developed in (Khusnutdinova and Samsonov, 2008; Khusnutdinova and Tranter, 2015, 2017), as well as some initial value problems motivated by these studies (Khusnutdinova and Moore, 2011; Khusnutdinova et al., 2014; Khusnutdinova and Tranter, 2019). In Sect. 6.2, the model equations are

K. R. Khusnutdinova (✉) · M. R. Tranter
Department of Mathematical Sciences, Loughborough University, Loughborough, UK
e-mail: K.Khusnutdinova@lboro.ac.uk; M.R.Tranter@lboro.ac.uk

A. Berezovski, T. Soomere (eds.), *Applied Wave Mathematics II*, Mathematics of Planet Earth 6, https://doi.org/10.1007/978-3-030-29951-4_6

discussed within the framework of lattice models which are presented as an extension of the famous Fermi–Pasta–Ulam (FPU) problem (also known as Fermi–Pasta–Ulam–Tsingou (FPUT) problem, see (Dauxois, 2008)). Section 6.3 is devoted to the scattering of pure and radiating solitary waves in bilayers with delamination. Section 6.4 is devoted to the discussion of the apparent zero mass contradiction between the solutions of the original problem formulation and its weakly nonlinear counterpart, which appears in periodic problems. We construct a solution of the initial value problem which bypasses this difficulty. We conclude in Sect. 6.5.

6.2 From Fermi–Pasta–Ulam Chain to the Model of a Layered Waveguide

The FPU problem is considered to be the starting point of modern nonlinear wave theory. In 1955, in Los Alamos, Fermi, Pasta and Ulam numerically studied the dynamics of an anharmonic chain of particles (Fermi et al., 1974) (see (Dauxois, 2008) for the discussion of the contribution by Mary Tsingou). The model consisted of identical equidistant (by a distance a) particles connected to their nearest neighbours by weakly nonlinear springs (Fig. 6.1).

Let u_n denote the displacement of the n-th particle from the equilibrium. Then, using Newton's second law one obtains:

$$m\ddot{u}_n = f(u_{n+1} - u_n) - f(u_n - u_{n-1}), \quad n = \overline{1, N},$$

where dots mean differentiation with respect to time t and $f(\Delta u)$ was given by $f(\Delta u) = k\Delta u + \alpha(\Delta u)^2$. The ends of the chain were fixed: $u_0 = u_{N+1} = 0$. Thus, the dynamics of the chain was described by the system

$$m\ddot{u}_n = k\left(u_{n+1} - 2u_n + u_{n-1}\right) + \alpha\left[(u_{n+1} - u_n)^2 - (u_n - u_{n-1})^2\right], \quad n = \overline{1, N},$$
$$u_0 = u_{N+1} = 0. \tag{6.1}$$

Typically, N was equal to 64.

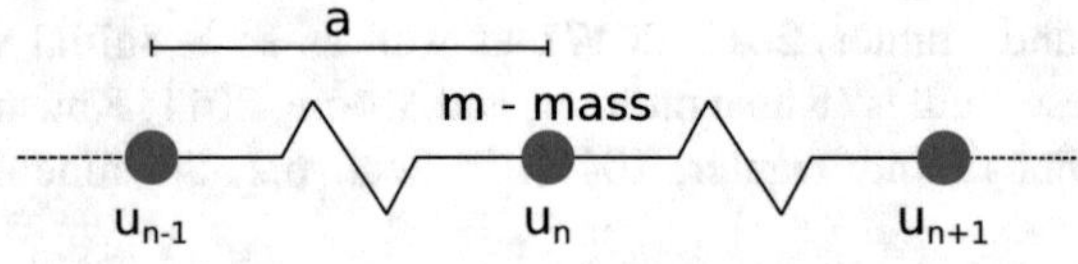

Fig. 6.1 A schematic of an FPU chain with particles of mass m and separation a, connected by weakly nonlinear springs.

The general solution of the linearised system ($\alpha = 0$) is given by a superposition of normal modes:

$$u_n^k(t) = A_k \sin\left(\frac{\pi k n}{N+1}\right) \cos\,(\omega_k t + \delta_k), \quad k = 1, \ldots, N,$$

$$\omega_k = 2\sqrt{\frac{k}{m}}\sin\left[\frac{\pi k}{2(N+1)}\right],$$

where A_k and δ_k are arbitrary constants. There is no energy transfer between the modes in the linear approximation. In the nonlinear chain ($\alpha \neq 0$), modes become coupled. It was expected that if all the initial energy was put into a single mode (or a few of the first modes), the nonlinear coupling would yield equal distribution of the energy among the normal modes. However, the numerical results were surprising: for example, if the energy was initially in the mode of lowest frequency, it returned almost entirely to that mode after interaction with a few other low frequency modes (FPU recurrence).

In 1965 this strange observation has motivated Zabusky and Kruskal to consider the FPU problem in the so-called continuum approximation (Zabusky and Kruskal, 1965). One assumes that

$$u_n(t) = u(x_n, t) = u(na, t), \quad u_{n\pm 1} = u(x_n \pm a, t),$$

and the displacement field u varies slowly justifying the Taylor expansion

$$u_{n\pm 1}(t) \approx u(x_n, t) \pm a u'(x_n, t) + \frac{1}{2}a^2 u''(x_n, t) \pm \pm \frac{1}{6}a^3 u'''(x_n, t) + \frac{1}{24}a^4 u''''(x_n, t) + \ldots, \tag{6.2}$$

where now primes denote differentiation with respect to spatial coordinate x. Substituting (6.2) into (6.1) and dropping the label n yields the equation

$$u_{tt} - c^2 u_{xx} = \varepsilon c^2 (u_x u_{xx} + \delta^2 u_{xxxx}), \tag{6.3}$$

where subscripts denote partial derivatives with respect to time t and spatial coordinate x, $c^2 = ka^2/m$, $\varepsilon = 2\alpha a/k$, and $\delta^2 = a^2/(12\varepsilon)$. Here, the leading order nonlinear and dispersive contributions are balanced at the same order of ε. This is the Boussinesq equation. It describes waves, which can propagate both to the right, and to the left (the two-way long wave equation).

A further reduction to the Korteweg–de Vries (KdV) equation was obtained by using an asymptotic multiple scales expansion of the solution of (6.3). We assume a solution of the form

$$u = f(\xi, T) + \varepsilon u^{(1)}(x, t) + \ldots, \quad \text{where} \quad \xi = x - ct, \quad T = \varepsilon t,$$

then (6.3) gives us

$$u_{tt}^{(1)} - c^2 u_{xx}^{(1)} = 2cf_{\xi T} + c^2 f_\xi f_{\xi\xi} + c^2\delta^2 f_{\xi\xi\xi\xi}\,.$$

The function $u^{(1)}$ will grow linearly in $\eta = x + ct$, unless

$$2cf_{\xi T} + c^2 f_\xi f_{\xi\xi} + c^2\delta^2 f_{\xi\xi\xi\xi} = 0\,.$$

By setting $q = \xi/6$, $\tau = cT/2$, this equation reduces to the canonical form of the KdV equation

$$q_\tau + 6qq_\xi + \delta^2 q_{\xi\xi\xi} = 0\,. \tag{6.4}$$

Zabusky and Kruskal numerically studied the dynamics of the KdV equation with sinusoidal initial conditions (for small δ^2, periodic boundary conditions), and discovered that the appearing solitons interact with each other elastically. They have called the emerging localised waves *solitons* because of the analogy with particles.

An extension of the FPU model in the form of the *dipole lattice model* (Fig. 6.2) was used in (Khusnutdinova et al., 2009) to obtain an equation describing nonlinear longitudinal bulk strain waves in a bar. It is convenient to view this system of coupled chains of particles as an anharmonic chain of oscillating dipoles $(P_n, \bar{P}_n)$ with 4 degrees of freedom: horizontal u_1^n and vertical u_2^n displacements of the geometrical centre of a dipole, in-plane rotation of the dipole axis (to an angle $\Delta\varphi^n$), and change of a distance $2u_4^n$ between the two poles of a dipole.

This is a symmetric version of the model proposed and considered in the linear approximation in connection with the dynamics of thin films in (Khusnutdinova, 1993). The model generalises the chain of dipoles (with the fixed distance between the poles) studied as a linear model in (Askar, 1985) (see also (Maugin, 1999)) and (Il'yushina, 1976), and as a nonlinear model in (Khusnutdinova, 1992).

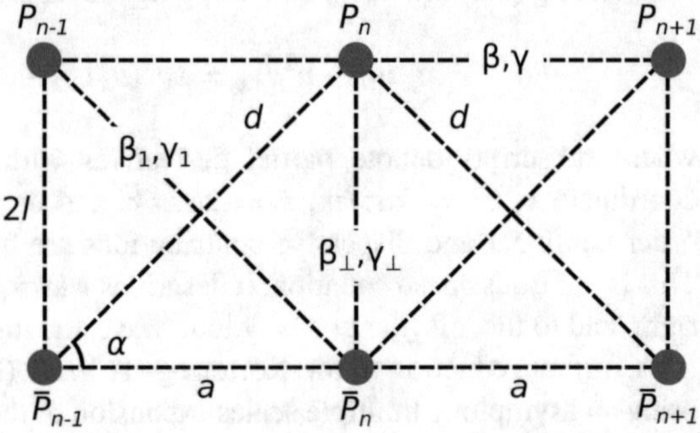

Fig. 6.2 The dipole lattice model.

The displacements of poles can be expressed in terms of the dipole coordinates as follows:

$$U_1^n = u_1^n - (l + u_4^n)\sin\Delta\varphi^n, \quad U_2^n = u_2^n + (l + u_4^n)\cos\Delta\varphi^n - l,$$
$$\overline{U_1^n} = u_1^n + (l + u_4^n)\sin\Delta\varphi^n, \quad \overline{U_2^n} = u_2^n - (l + u_4^n)\cos\Delta\varphi^n + l.$$

Assuming that rotations are small, so that $\Delta\varphi^n = u_3^n/l \ll 1$, we use truncated Taylor expansions in order to derive equations of motion up to quadratic terms:

$$U_1^n = u_1^n - u_3^n - \frac{u_3^n u_4^n}{l} + \ldots, \quad U_2^n = u_2^n + u_4^n - \frac{(u_3^n)^2}{2l} + \ldots, \text{ etc.}$$

Then, the kinetic energy of the n-th dipole has the form

$$T_n = \frac{M}{2}\left[(\dot{u}_1^n)^2 + (\dot{u}_2^n)^2 + (\dot{u}_4^n)^2 + \left(1 + \frac{u_4^n}{l}\right)^2 (\dot{u}_3^n)^2\right],$$

where $M = 2m$ is the dipole mass. The potential energy of the n-th dipole is defined by pairwise interactions between neighbouring particles:

$$\Phi_n = \Phi_{n,n+1} + \Phi_{\overline{n},\overline{n+1}} + \Phi_{n,\overline{n+1}} + \Phi_{\overline{n},n+1} + \Phi_{n-1,n} + \Phi_{\overline{n-1},\overline{n}} + \Phi_{n-1,\overline{n}} + \Phi_{\overline{n-1},n} + \Phi_{\perp}, \tag{6.5}$$

where overlines denote particles in the second ("bottom") row. Let the potential energy of interaction between any two neighbouring particles have the form

$$\Phi_*(\Delta r_*) = \frac{\widetilde{\beta}}{2}\Delta r_*^2 + \frac{\widetilde{\gamma}}{3}\Delta r_*^3 + \ldots, \tag{6.6}$$

where Δr_* is the change of a distance between the particles, and $(\widetilde{\beta}, \widetilde{\gamma})$ denotes one of three possible pairs of interaction constants (Fig. 6.2).

The change of a distance between a pole of the n-th dipole $\widetilde{P}_n$ and a pole of the $(n+1)$-th dipole $\widetilde{P}_{n+1}$ is given by

$$\Delta r_{\widetilde{n},\widetilde{n+1}} = \left[\left(\widetilde{U}_1^{n+1} - \widetilde{U}_1^n + r_0\cos\theta_0\right)^2 + \left(\widetilde{U}_2^{n+1} - \widetilde{U}_2^n + r_0\sin\theta_0\right)^2\right]^{1/2} - r_0, \tag{6.7}$$

where r_0 is the distance between the respective pair of poles in the equilibrium configuration. The change of a distance between the poles of the n-th dipole is given by $\Delta r_\perp = 2u_4^n$.

Let us introduce the differences $\Delta x = \widetilde{U}_1^{n+1} - \widetilde{U}_1^n$, $\Delta y = \widetilde{U}_2^{n+1} - \widetilde{U}_2^n$, and assume that these differences are small compared to r_0: $\Delta x/r_0 \ll 1$, $\Delta y/r_0 \ll 1$.

We can use the expansions

$$\Delta r_{\widetilde{n,n+1}} = \Delta x \cos\theta_0 + \Delta y \sin\theta_0 + \frac{1}{2r_0}(\Delta x \sin\theta_0 - \Delta y \cos\theta_0)^2 - \\ - \frac{1}{2r_0^2}(\Delta x \cos\theta_0 + \Delta y \sin\theta_0)(\Delta x \sin\theta_0 - \Delta y \cos\theta_0)^2 + \dots .$$

Finally, we can derive the Euler–Lagrange equations

$$\frac{\mathrm{d}}{\mathrm{d}t}\left(\frac{\partial T_n}{\partial \dot{u}_i^n}\right) - \frac{\partial T_n}{\partial u_i^n} + \frac{\partial \Phi_n}{\partial u_i^n} = 0, \quad i = \overline{1,4},$$

describing the nonlinear dynamics of the system, and consider the continuum approximation, as discussed before, for the FPU chain. The necessary calculations are much more tedious than that in the case of the FPU chain, but they can be performed using symbolic computations (Khusnutdinova et al., 2009).

As a result of these derivations, Khusnutdinova et al. (2009) showed that longitudinal waves asymptotically uncouple from other degrees of freedom, which become slaved to longitudinal waves. These waves are described by a Boussinesq-type equation that is asymptotically equivalent to a "doubly dispersive equation" (DDE) (Samsonov, 2001; Porubov, 2003). This equation has been earlier derived for the long longitudinal waves in a bar of rectangular cross section using the nonlinear elasticity approach (Khusnutdinova and Samsonov, 2008). In dimensional variables, the DDE for a bar of rectangular cross section $\sigma = 2a \times 2b$ has the form

$$f_{tt} - c^2 f_{xx} = \frac{\beta}{2\rho}(f^2)_{xx} + \frac{J\nu^2}{\sigma}(f_{tt} - c_1^2 f_{xx})_{xx}, \tag{6.8}$$

where

$$c = \sqrt{\frac{E}{\rho}}, \quad c_1 = \frac{c}{\sqrt{2(1+\nu)}}, \quad J = \frac{4}{3}ab(a^2 + b^2),$$
$$\beta = 3E + 2l(1-2\nu)^3 + 4m(1+\nu)^2(1-2\nu) + 6n\nu^2,$$

and ρ is the density, E is the Young modulus, ν is the Poisson ratio, while l, m, n are the Murnaghan moduli. Nondimensionalisation, Benjamin–Bona–Mahoney (BBM) type regularisation (Benjamin et al., 1972) of the dispersive terms, and scaling bring the equation to the form

$$f_{tt} - f_{xx} = \frac{1}{2}(f^2)_{xx} + f_{ttxx}. \tag{6.9}$$

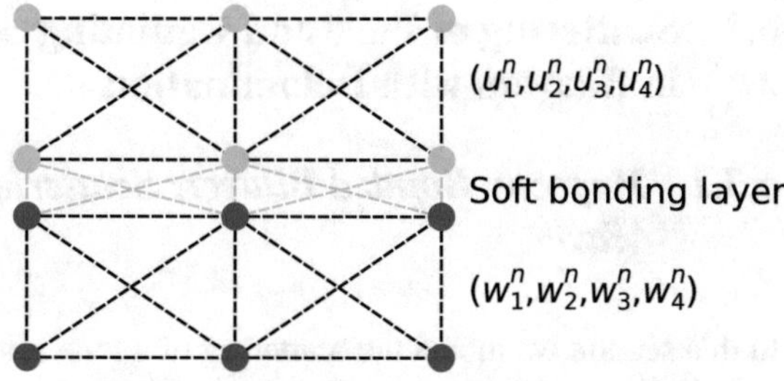

Fig. 6.3 The layered dipole lattice model.

A further extension in the form of the layered dipole lattice model (Fig. 6.3) was used in (Khusnutdinova et al., 2009) to study nonlinear waves in a bilayer with a sufficiently soft bonding layer (e.g., some types of adhesive bonding). It was shown that longitudinal waves can be modelled by a system of coupled regularised Boussinesq (cRB) equations (presented in the scaled, nondimensional form):

$$\begin{aligned} f_{tt} - f_{xx} &= \frac{1}{2}(f^2)_{xx} + f_{ttxx} - \delta(f-g), \\ g_{tt} - c^2 g_{xx} &= \frac{1}{2}\alpha(g^2)_{xx} + \beta g_{ttxx} + \gamma(f-g). \end{aligned} \tag{6.10}$$

Here, f and g denote the longitudinal strains in the layers, while the coefficients c, α, β, δ, γ are defined by the physical and geometrical parameters of the problem (Khusnutdinova et al., 2009).

In the symmetric case $c = \alpha = \beta = 1$ the system (6.10) admits the reduction $g = f$, where f satisfies the Boussinesq equation

$$f_{tt} - f_{xx} = \frac{1}{2}(f^2)_{xx} + f_{ttxx}, \tag{6.11}$$

which has particular solitary wave solutions:

$$f = 3(v^2-1)\,\mathrm{sech}^2\,\frac{\sqrt{v^2-1}(x-vt)}{2v}, \tag{6.12}$$

where v is the speed of the wave. In the cRB system of Eq. (6.10), when the characteristic speeds of the linear waves in the layers are close (i.e., c is close to 1), this pure solitary wave solution is replaced with a radiating solitary wave (Khusnutdinova et al., 2009; Khusnutdinova and Moore, 2011). This structure is a solitary wave radiating a copropagating one-sided oscillatory tail (see, e.g., (Benilov, 1993; Bona et al., 2008)). The radiating solitary waves emerge due to a resonance between a soliton and a harmonic wave, which can be deduced from the analysis of the linear dispersion relation of the problem (Khusnutdinova et al., 2009).

6.3 Scattering of Pure and Radiating Solitary Waves in Bilayers with Delamination

6.3.1 Perfectly Bonded Bilayer: Scattering of a Pure Solitary Wave

In this section we model the scattering of a long longitudinal strain solitary wave in a perfectly bonded bilayer with delamination at $x > 0$ (Fig. 6.4). The material of the layers is assumed to be the same (symmetric bar), while the material to the left and to the right of the $x = 0$ cross section can be different.

The problem is described by the set of scaled, nondimensional equations (Khusnutdinova and Samsonov, 2008)

$$
\begin{aligned}
u_{tt}^- - u_{xx}^- &= \varepsilon\left[-12u_x^- u_{xx}^- + 2u_{ttxx}^-\right], \quad x < 0,\\
u_{tt}^+ - c^2 u_{xx}^+ &= \varepsilon\left[-12\alpha u_x^+ u_{xx}^+ + 2\frac{\beta}{c^2}u_{ttxx}^+\right], \quad x > 0,
\end{aligned}
\tag{6.13}
$$

with associated continuity conditions

$$
u^-|_{x=0} = u^+|_{x=0}\,, \tag{6.14}
$$

$$
u_x^- + \varepsilon\left[-6\left(u_x^-\right)^2 + 2u_{ttx}^-\right]\Big|_{x=0} = c^2 u_x^+ + \varepsilon\left[-6\alpha\left(u_x^+\right)^2 + 2\frac{\beta}{c^2}u_{ttx}^+\right]\Big|_{x=0}, \tag{6.15}
$$

and appropriate initial and boundary conditions. Here, c, α, and β are constants defined by the geometrical and physical parameters of the structure, while ε is the small wave amplitude parameter. The functions $u^-(x,t)$ and $u^+(x,t)$ describe displacements in the bonded and delaminated areas of the structure, respectively. Condition (6.14) is continuity of longitudinal displacement, while condition (6.15) is the continuity of stress.

In what follows we assume that $\alpha = 1$ and $\beta = (n^2 + k^2)/[n^2\left(1 + k^2\right)]$, where n represents the number of layers in the structure and k is defined by the geometry of

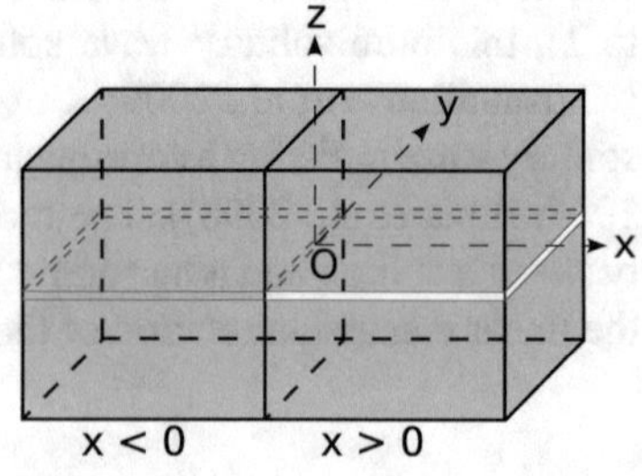

Fig. 6.4 An example of a bilayer with a perfect bond for $x < 0$ and complete delamination for $x > 0$.

the waveguide (Khusnutdinova and Samsonov, 2008). The cross-section $x = 0$ has width $2a$ and the height of each layer is b. In terms of these values $k = b/a$ and, as there are two layers in this example, $n = 2$.

Khusnutdinova and Tranter (2015) proposed direct and semianalytical numerical approaches to solving these two boundary value problems matched at $x = 0$. The semianalytical numerical approach was based on the weakly nonlinear solution constructed in (Khusnutdinova and Samsonov, 2008). The direct method was based on the use of finite difference techniques (Khusnutdinova and Tranter, 2015). This direct method was limited to only two sections in the structure at a time, so it could not be used for a short delamination region. The method was extended in (Tranter, 2019) so that it can solve for multiple sections in the bar and for multiple layers. Here we will outline the weakly nonlinear solution (Khusnutdinova and Samsonov, 2008; Khusnutdinova and Tranter, 2015) and present the leading order solution. To this end we look for a solution of the form

$$u^- = I(\xi_-, X) + R(\eta_-, X) + \varepsilon P(\xi_-, \eta_-, X) + O\left(\varepsilon^2\right),$$

$$u^+ = T(\xi_+, X) + \varepsilon Q(\xi_+, \eta_+, X) + O\left(\varepsilon^2\right),$$

where the characteristic variables are given by $\xi_- = x - t$, $\xi_+ = x - ct$, $\eta_- = x + t$, $\eta_+ = x + ct$, $X = \varepsilon x$. The leading order incident, reflected and transmitted waves are described by the functions $I(\xi_-, X)$, $R(\eta_-, X)$, $T(\xi_+, X)$, respectively. The functions $P(\xi_-, \eta_-, X)$ and $Q(\xi_+, \eta_+, X)$ describe the higher order corrections. Substituting this weakly nonlinear solution into (6.13) we find that the system is satisfied at leading order, while at $O(\varepsilon)$ we have

$$-2P_{\xi_-\eta_-} = \left(I_X - 3I_{\xi_-}^2 + I_{\xi_-\xi_-\xi_-}\right)_{\xi_-} + \left(R_X - 3R_{\eta_-}^2 + R_{\eta_-\eta_-\eta_-}\right)_{\eta_-} - \\ - 6\left(RI_{\xi_-} + IR_{\eta_-}\right)_{\xi_-\eta_-}. \tag{6.16}$$

To leading order the right-propagating incident wave $I = \int \tilde{I}\, d\xi_-$ is defined by the solution of the KdV equation

$$\tilde{I}_X - 6\tilde{I}\tilde{I}_{\xi_-} + \tilde{I}_{\xi_-\xi_-\xi_-} = 0. \tag{6.17}$$

Similarly the reflected wave $R = \int \tilde{R}\, d\eta_-$ satisfies the KdV equation

$$\tilde{R}_X - 6\tilde{R}\tilde{R}_{\eta_-} + \tilde{R}_{\eta_-\eta_-\eta_-} = 0. \tag{6.18}$$

Expressions for the higher order corrections can be found by substituting (6.17) and (6.18) into (6.16) and integrating with respect to both characteristic variables (Khusnutdinova and Samsonov, 2008).

Following the same steps for the second equation in (6.13), we find that the leading order transmitted wave $T = \int \tilde{T}\, d\xi_+$ is described by the equation

$$\tilde{T}_X - \frac{6}{c^2}\tilde{T}\tilde{T}_{\xi_+} + \frac{\beta}{c^2}\tilde{T}_{\xi_+\xi_+\xi_+} = 0. \tag{6.19}$$

To determine "initial conditions" for (6.18) and (6.19), we substitute the weakly nonlinear solution into continuity conditions (6.14) and (6.15) and retain terms at leading order. Firstly, we differentiate (6.14) with respect to time, giving

$$u_t^-\big|_{x=0} = u_t^+\big|_{x=0},$$

so to leading order we have

$$I_{\xi_-}\big|_{x=0} - R_{\eta_-}\big|_{x=0} = cT_{\xi_+}\big|_{x=0}, \tag{6.20}$$

and from (6.15) we have, to leading order,

$$I_{\xi_-}\big|_{x=0} + R_{\eta_-}\big|_{x=0} = c^2 T_{\xi_+}\big|_{x=0}. \tag{6.21}$$

Recalling that $\tilde{I} = I_{\xi_-}$ and similar relations for R and T, we solve the system (6.20), (6.21) for $\tilde{R}|_{x=0}$ and $\tilde{T}|_{x=0}$ in terms of $\tilde{I}|_{x=0}$ to obtain

$$\tilde{R}\big|_{x=0} = C_R\tilde{I}\big|_{x=0}, \quad \tilde{T}\big|_{x=0} = C_T\tilde{I}\big|_{x=0}, \tag{6.22}$$

where

$$C_R = \frac{c-1}{c+1}, \quad C_T = \frac{2}{c\,(c+1)}, \tag{6.23}$$

are the reflection and transmission coefficients, respectively. This shows that, if the materials are the same in the bonded and delaminated areas ($c = 1$), then $C_T = 1$, $C_R = 0$ and there is no leading order reflected wave.

A typical scenario is shown in Fig. 6.5 for the incident wave in the form of an exact solitary wave solution, with parameters $\alpha = 1, \beta = 5/8, c = 1, \varepsilon = 0.05$, with initial position $x = -200$ and initial speed $v = 1.03$. The weakly nonlinear solution (solution of (6.17), (6.18) and (6.19)) and the results of direct numerical simulations (solution of (6.13)–(6.15)) are presented for the strains $u_x^- = e^-$ and $u_x^+ = e^+$ at the initial moment of time and at $t = 1100$, respectively, when the wave is propagating in the delaminated section of the bar. The results are in good agreement, with a small phase shift for the lead soliton. Soliton fission occurs in the delaminated section of the bar, with two solitons generated from a single incident soliton (Khusnutdinova and Samsonov, 2008) (see also (Tappert and Zabusky, 1971; Pelinovsky, 1971) for the first studies of soliton fission in the context of water waves). As the material in both sections of the bar is the same, there is no leading order reflected wave.

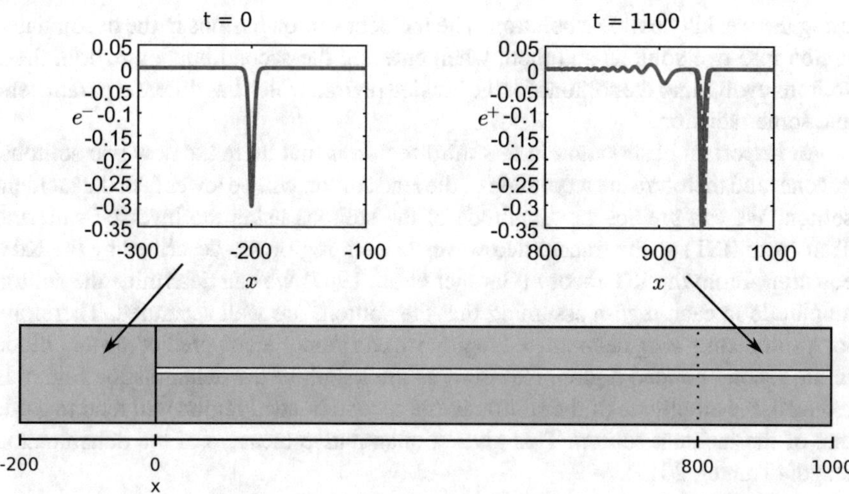

Fig. 6.5 The solution e^- and e^+ at the initial moment of time and at $t = 1100$, for the parameters $\alpha = 1$, $\beta = 5/8$, $c = 1$, $\varepsilon = 0.05$, with initial position $x = -200$ and initial speed $v = 1.03$, for direct numerical simulations (blue, solid line) and weakly nonlinear solution (red, dashed line).

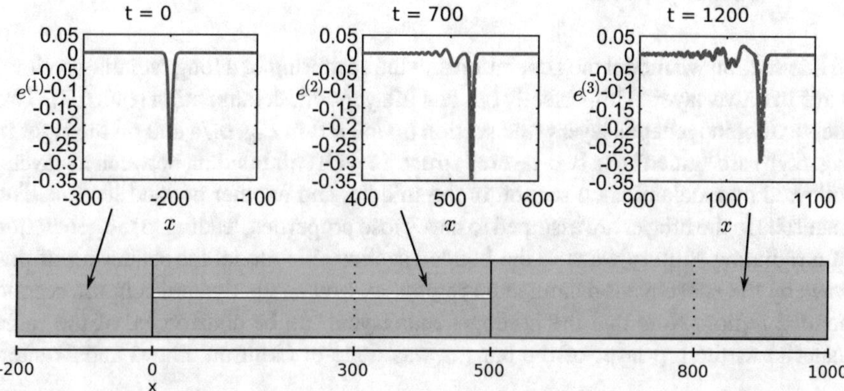

Fig. 6.6 The solutions $e^{(1)}$, $e^{(2)}$ and $e^{(3)}$ in a bilayer with three sections for the parameters $\alpha = 1$, $c = 1$, $\varepsilon = 0.05$ in all sections and $\beta = 1$ in the bonded sections with $\beta = 5/8$ in the delaminated region. The initial position is $x = -200$ and initial speed $v = 1.03$, for direct numerical simulations (blue, solid line) and weakly nonlinear solution (red, dashed line).

In physical applications it is often of interest to consider the case when the delamination is finite. In this paper we will present a single case for a bilayer with three sections, similarly to (Tranter, 2019). We take the same parameters as for Fig. 6.5 and present the results in Fig. 6.6, where we denote the strains in Sect. 6.1 as $e^{(1)}$, and so on for Sects. 6.2 and 6.3. We can see that again there is good agreement between the direct numerical simulations and the semianalytical solution constructed using the weakly nonlinear solution. The incident soliton fissions in the delaminated region into two solitons and then, when entering the second bonded region, these solitons evolve into the solitons for the bonded region, which has different parameters, and some radiation.

An important observation in this third region is that there are now two solitons, not one, and therefore the amplitude of the lead soliton will be lower than the incident soliton. We can predict the amplitude of the solitons using the Inverse Scattering Transform (IST) as the transmitted waves in each region are described by the KdV equation. From the IST theory (Gardner et al., 1967) we can determine the soliton amplitude in each region assuming that the solitons are well separated. Therefore, for a sufficiently long delaminated region we can theoretically predict the amplitude in the second bonded region. However, as the length of the delamination region is reduced, the amplitude of the soliton in the second bonded region will tend towards that of the incident soliton. This gives a quantitative measure of the delamination length (Tranter, 2019).

6.3.2 Imperfectly Bonded Bilayer: Scattering of a Radiating Solitary Wave

In this section we model the generation and the scattering of a long radiating solitary wave in a two-layered imperfectly bonded bilayer with delamination (Fig. 6.7). Two identical homogeneous layers (the section on the left in Fig. 6.7a and on the right in Fig. 6.7b) are "glued" to a two-layered structure with soft bonding between its layers, followed by a delaminated section in the middle, and another bonded section. The materials in the bilayer are assumed to have close properties, leading to the generation of a radiating solitary wave in the bonded section. We model the scattering of this wave by the subsequent delaminated region, as well as the dynamics in the second bonded region. Note that the homogeneous layers can be constructed of the same material as the top layer of the bar (as was done in (Khusnutdinova and Tranter,

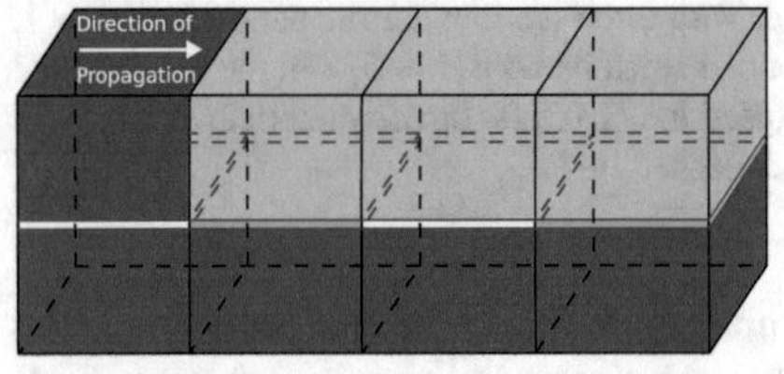

(a) Homogeneous section on the left.

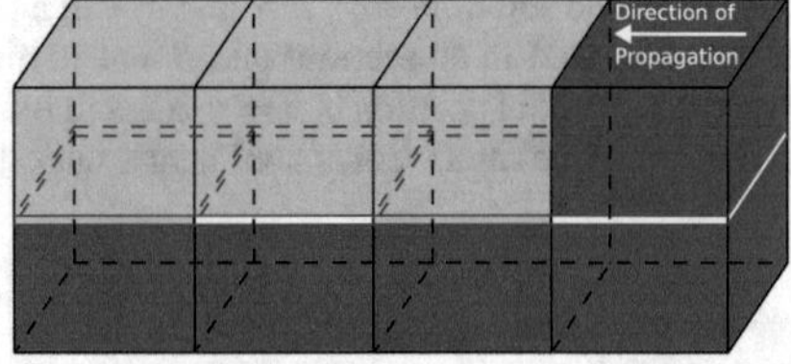

(b) Homogeneous section on the right.

Fig. 6.7 An example of a bilayer with a soft bond between the layers and a delaminated region between the bonded regions. A homogeneous section of the same material as the lower layer is attached to (a) the left-hand side of the structure, or (b) the right-hand side of the structure.

2017)) but using the same material as the lower layer will give a larger amplitude, as the characteristic speed is higher which leads to a higher transmission coefficient.

The problem is described by the following sets of scaled, nondimensional equations in the respective sections of the complex waveguide (Khusnutdinova et al., 2009; Khusnutdinova and Samsonov, 2008):

$$
\begin{aligned}
u_{tt}^{(1)} - c^2 u_{xx}^{(1)} &= 2\varepsilon\left(-6\alpha u_x^{(1)} u_{xx}^{(1)} + \beta u_{ttxx}^{(1)}\right), \\
w_{tt}^{(1)} - c^2 w_{xx}^{(1)} &= 2\varepsilon\left(-6\alpha w_x^{(1)} w_{xx}^{(1)} + \beta w_{ttxx}^{(1)}\right),
\end{aligned} \tag{6.24}
$$

for the section with two homogeneous layers,

$$
\begin{aligned}
u_{tt}^{(2,4)} - u_{xx}^{(2,4)} &= 2\varepsilon\left[-6u_x^{(2,4)} u_{xx}^{(2,4)} + u_{ttxx}^{(2,4)} - \delta\left(u^{(2,4)} - w^{(2,4)}\right)\right], \\
w_{tt}^{(2,4)} - c^2 w_{xx}^{(2,4)} &= 2\varepsilon\left[-6\alpha w_x^{(2,4)} w_{xx}^{(2,4)} + \beta w_{ttxx}^{(2,4)} + \gamma\left(u^{(2,4)} - w^{(2,4)}\right)\right],
\end{aligned} \tag{6.25}
$$

for the two bonded regions, and

$$
\begin{aligned}
u_{tt}^{(3)} - u_{xx}^{(3)} &= 2\varepsilon\left[-6u_x^{(3)} u_{xx}^{(3)} + u_{ttxx}^{(3)}\right], \\
w_{tt}^{(3)} - c^2 w_{xx}^{(3)} &= 2\varepsilon\left[-6\alpha w_x^{(3)} w_{xx}^{(3)} + \beta w_{ttxx}^{(3)}\right],
\end{aligned} \tag{6.26}
$$

for the delaminated region. Here the functions $u^{(i)}(x,t)$ and $w^{(i)}(x,t)$ describe longitudinal displacements in the "top" and "bottom" layers of the four sections of the waveguide, respectively. The values of the constants α, β, and c depend on the physical and geometrical properties of the waveguide, while the constants δ and γ depend on the properties of the soft bonding layer, and ε is a small amplitude parameter (Khusnutdinova et al., 2009; Khusnutdinova and Tranter, 2017).

These equations are complemented with continuity conditions for the longitudinal displacements and stresses at the interfaces between the sections, similarly to (6.14), (6.15), as well as the relevant initial and boundary conditions (Khusnutdinova and Tranter, 2015, 2017).

Again, the direct numerical modelling of this problem is difficult and expensive because one needs to solve several boundary value problems linked to each other via matching conditions at the boundaries. Therefore, we developed an alternative semianalytical approach based upon the use of several matched asymptotic multiple scales expansions and averaging with respect to the fast space variable at a constant value of one or another characteristic variable (Khusnutdinova and Tranter, 2015, 2017). We present an outline of the method here, omitting the matching at the interface between sections of the bilayer which can be found in (Khusnutdinova and Tranter, 2015, 2017).

To determine the weakly nonlinear solution to this system of equations, we firstly differentiate the governing equations with respect to x so that we can use a space averaging procedure, as the strain waves are localised. We denote $f^{(i)} = u_x^{(i)}$ and $g^{(i)} = w_x^{(i)}$ to obtain the equations 'in strains'

$$f_{tt}^{(1)} - c^2 f_{xx}^{(1)} = 2\varepsilon \left[-3\alpha \left(f^{(1)} \right)^2 + \beta f_{tt}^{(1)} \right]_{xx},$$

$$g_{tt}^{(1)} - c^2 g_{xx}^{(1)} = 2\varepsilon \left[-3\alpha \left(g^{(1)} \right)^2 + \beta g_{tt}^{(1)} \right]_{xx} \tag{6.27}$$

for $x < x_a$,

$$f_{tt}^{(2,4)} - f_{xx}^{(2,4)} = 2\varepsilon \left[-3 \left(f^{(2,4)} \right)^2 + f_{tt}^{(2,4)} \right]_{xx} - 2\varepsilon\delta \left(f^{(2,4)} - g^{(2,4)} \right),$$

$$g_{tt}^{(2,4)} - c^2 g_{xx}^{(2,4)} = 2\varepsilon \left[-3\alpha \left(g^{(2,4)} \right)^2 + \beta g_{tt}^{(2,4)} \right]_{xx} + 2\varepsilon\gamma \left(f^{(2,4)} - g^{(2,4)} \right) \tag{6.28}$$

for $x_a < x < x_b$ and $x > x_c$ (bonded regions), and

$$f_{tt}^{(3)} - f_{xx}^{(3)} = 2\varepsilon \left[-3 \left(f^{(3)} \right)^2 + f_{tt}^{(3)} \right]_{xx},$$

$$g_{tt}^{(3)} - c^2 g_{xx}^{(3)} = 2\varepsilon \left[-3\alpha \left(g^{(3)} \right)^2 + \beta g_{tt}^{(3)} \right]_{xx} \tag{6.29}$$

for $x_b < x < x_c$ (delaminated region).

We assume that all functions present in our expansions and their derivatives are bounded and sufficiently rapidly decaying at infinity. In the first region, the equation is identical in both homogeneous layers and therefore we assume the same incident wave in both, and consider asymptotic multiple scales expansions of the type

$$f^{(1)} = I(\nu, X) + R^{(1)}(\zeta, X) + \varepsilon P^{(1)}(\nu, \zeta, X) + O\left(\varepsilon^2\right),$$

$$g^{(1)} = I(\nu, X) + G^{(1)}(\zeta, X) + \varepsilon Q^{(1)}(\nu, \zeta, X) + O\left(\varepsilon^2\right),$$

where the characteristic variables are given by $\nu = x - ct$, $\zeta = x + ct$, and the slow variable $X = \varepsilon x$. Here, the functions I and $R^{(1)}$, $G^{(1)}$ represent the leading order incident and reflected waves, respectively, and $P^{(1)}$, $Q^{(1)}$ are the higher order corrections. Substituting the asymptotic expansion into the first equation in (6.27) and applying the averaging with respect to the fast space variable, we obtain the equations

$$I_X - 6\frac{\alpha}{c^2} I I_\nu + \beta I_{\nu\nu\nu} = 0, \tag{6.30}$$

$$R_X^{(1)} - 6\frac{\alpha}{c^2} R^{(1)} R_\zeta^{(1)} + \beta R_{\zeta\zeta\zeta}^{(1)} = 0. \tag{6.31}$$

Similarly, for the second layer we obtain

$$G_X^{(1)} - 6\frac{\alpha}{c^2} G^{(1)} G_\zeta^{(1)} + \beta G_{\zeta\zeta\zeta}^{(1)} = 0, \tag{6.32}$$

in addition to (6.30).

For the second section of the bar we expect radiating solitary waves to develop if the layers have close properties. We seek a weakly nonlinear solution of the form

$$f^{(2)} = T^{(2)}(\xi, X) + R^{(2)}(\eta, X) + \varepsilon P^{(2)}(\xi, \eta, X) + O\left(\varepsilon^2\right),$$

$$g^{(2)} = S^{(2)}(\xi, X) + G^{(2)}(\eta, X) + \varepsilon Q^{(2)}(\xi, \eta, X) + O\left(\varepsilon^2\right).$$

The characteristic variables are $\xi = x - t$, $\eta = x + t$ and X is the same as before, $T^{(2)}$ and $S^{(2)}$ represent the transmitted waves in the second section of the bar, where T is for the top layer and S is for the bottom layer. Similarly, $R^{(2)}$ and $G^{(2)}$ are the reflected waves, and the higher order corrections in this section are given by $P^{(2)}$ and $Q^{(2)}$, for the top and bottom layers, respectively. We substitute this weakly nonlinear solution into (6.28) and apply the averaging to obtain the system of equations

$$\left(T_X^{(2)} - 6T^{(2)} T_\xi^{(2)} + T_{\xi\xi\xi}^{(2)}\right)_\xi = \delta\left(T^{(2)} - S^{(2)}\right), \tag{6.33}$$

$$\left(S_X^{(2)} + \frac{c^2 - 1}{2\varepsilon} S_\xi^{(2)} - 6\alpha S^{(2)} S_\xi^{(2)} + \beta S_{\xi\xi\xi}^{(2)}\right)_\xi = \gamma\left(S^{(2)} - T^{(2)}\right). \tag{6.34}$$

This is a system of *coupled Ostrovsky equations* (note that the Ostrovsky equation was initially derived to describe long surface and internal waves in a rotating ocean (Ostrovsky, 1978; Grimshaw et al., 2007)), which appear naturally in the description of nonlinear waves in layered waveguides, both solid and fluid (Khusnutdinova and Moore, 2011; Alias et al., 2014).

Similarly, for the reflected waves in this region, we obtain

$$\left(R_X^{(2)} - 6R^{(2)} R_\eta^{(2)} + R_{\eta\eta\eta}^{(2)}\right)_\eta = \delta\left(R^{(2)} - G^{(2)}\right), \tag{6.35}$$

$$\left(G_X^{(2)} + \frac{c^2 - 1}{2\varepsilon} G_\eta^{(2)} - 6\alpha G^{(2)} G_\eta^{(2)} + \beta G_{\eta\eta\eta}^{(2)}\right)_\eta = \gamma\left(G^{(2)} - R^{(2)}\right). \tag{6.36}$$

Therefore, to leading order, the transmitted and reflected waves are described by two systems of coupled Ostrovsky equations.

We now consider the delaminated region and look for a weakly nonlinear solution to (6.29) of the form

$$f^{(3)} = T^{(3)}(\xi, X) + R^{(3)}(\eta, X) + \varepsilon P^{(2)}(\xi, \eta, X) + O\left(\varepsilon^2\right),$$
$$g^{(3)} = S^{(3)}(\nu, X) + G^{(3)}(\zeta, X) + \varepsilon Q^{(2)}(\nu, \zeta, X) + O\left(\varepsilon^2\right),$$

where the characteristic variables are the same as in previous sections. Substituting this into system (6.29) and averaging with respect to the fast space variable we obtain the equations

$$T_X^{(3)} - 6T^{(3)}T_\xi^{(3)} + T_{\xi\xi\xi}^{(3)} = 0, \tag{6.37}$$
$$S_X^{(3)} - 6\frac{\alpha}{c^2}S^{(3)}S_\nu^{(3)} + \beta S_{\nu\nu\nu}^{(3)} = 0, \tag{6.38}$$

describing transmitted waves, and the reflected waves are governed by the equations

$$R_X^{(3)} - 6R^{(3)}R_\eta^{(3)} + R_{\eta\eta\eta}^{(3)} = 0, \tag{6.39}$$
$$G_X^{(3)} - 6\frac{\alpha}{c^2}G^{(3)}G_\zeta^{(3)} + \beta G_{\zeta\zeta\zeta}^{(3)} = 0. \tag{6.40}$$

Finally, in the fourth region, we can make use of the same weakly nonlinear solution that was used in the second region and we obtain the same equations, but only for the functions describing transmitted waves in this region (no boundary to generate reflected waves). Therefore, with the weakly nonlinear solution

$$f^{(4)} = T^{(4)}(\xi, X) + \varepsilon P^{(4)}(\xi, \eta, X) + O\left(\varepsilon^2\right),$$
$$g^{(4)} = S^{(4)}(\xi, X) + \varepsilon Q^{(4)}(\xi, \eta, X) + O\left(\varepsilon^2\right),$$

the transmitted waves in this region are described by the coupled Ostrovsky equations

$$\left(T_X^{(4)} - 6T^{(4)}T_\xi^{(4)} + T_{\xi\xi\xi}^{(4)}\right)_\xi = \delta\left(T^{(4)} - S^{(4)}\right), \tag{6.41}$$

$$\left(S_X^{(4)} + \frac{c^2-1}{2\varepsilon}S_\xi^{(4)} - 6\alpha S^{(4)}S_\xi^{(4)} + \beta S_{\xi\xi\xi}^{(4)}\right)_\xi = \gamma\left(S^{(4)} - T^{(4)}\right). \tag{6.42}$$

In order to find "initial conditions" for the derived equations, we collect the expressions for the weakly nonlinear solutions and substitute them into the continuity conditions and retain terms at leading order, as was done for the perfectly bonded bilayer. See (Khusnutdinova and Tranter, 2015) for a detailed explanation of this procedure.

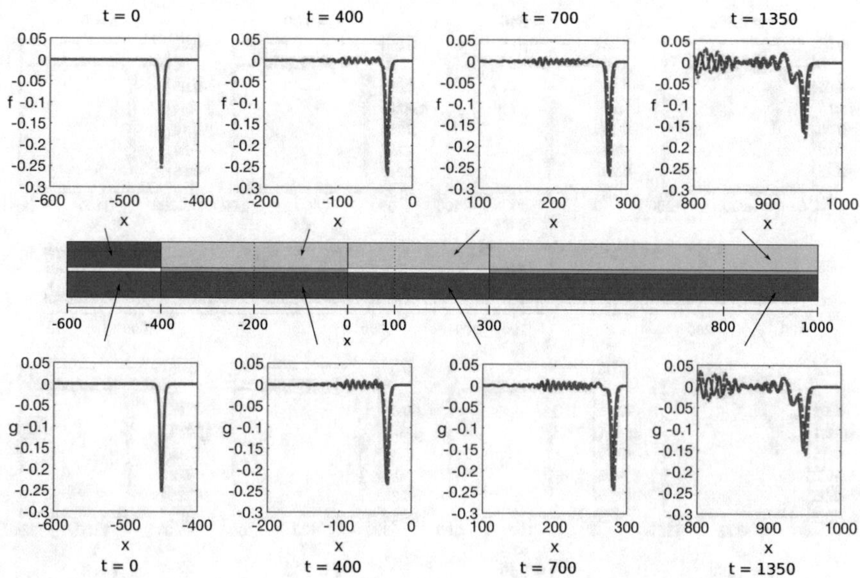

Fig. 6.8 The solutions f (top row) and g (bottom row) in each section of the bilayer, for the parameters $\alpha = \beta = 1.05$, $c = 1.025$, $\delta = \gamma = 0.8$ and $\varepsilon = 0.05$, with initial position $x = -450$ and initial speed $v = 1.05$, for direct numerical simulations (blue, solid line) and semianalytical method (red, dashed line). Two homogeneous layers, of the same material as the lower layer, are on the left and the waves propagate to the right.

A typical scenario is shown in Fig. 6.8 for the incident wave in the form of an exact solitary wave solution, with parameters $\alpha = \beta = 1.05$, $c = 1.025$, $\delta = \gamma = 0.8$ and $\varepsilon = 0.05$, with initial position $x = -450$ and initial speed $v = 1.05$. Two homogeneous layers, of the same material as the lower layer, are attached to the left of the bar. We see the generation of a radiating solitary wave in the bonded section of the bar, the separation of the solitary wave from its radiating tail in the delaminated section, and the re-coupling of the waves in the second bonded region.

A similar numerical experiment for the same parameters, with the homogeneous section on the right-hand side of the bilayer, is shown in Fig. 6.9. The results in this case are qualitatively similar, with a different length of radiating tail due to the change in the length of the relevant bonded section. We note that if the radiation wave packet in the second bonded region is closer to the leading wave when sending the waves from the right, then the delamination is closer to the left-hand side of the structure, and vice versa.

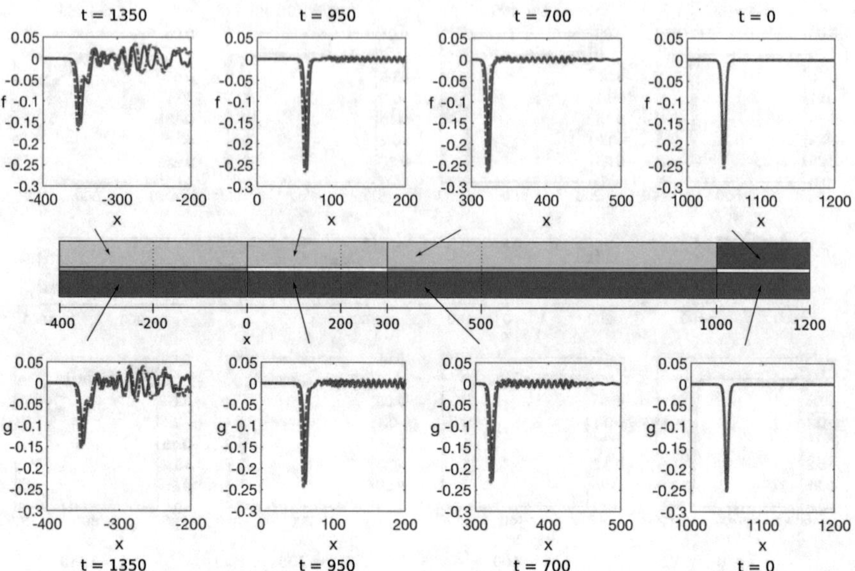

Fig. 6.9 The solutions f (top row) and g (bottom row) in each section of the bilayer, for the parameters $\alpha = \beta = 1.05$, $c = 1.025$, $\delta = \gamma = 0.8$ and $\varepsilon = 0.05$, with initial position $x = 1050$ and initial speed $v = 1.05$, for direct numerical simulations (blue, solid line) and semianalytical method (red, dashed line). Two homogeneous layers, of the same material as the lower layer, are on the right and the waves propagate to the left.

6.4 Initial Value Problem

When the characteristic speeds in the layers significantly differ, the dynamical behaviour is different (Khusnutdinova and Moore, 2011). To simplify the problem, let us assume that the material of the lower layer has much greater density (greater inertia), and consider the reduction $w = 0$. Then, the dynamics of the top layer is described (after appropriate scalings) by the *Boussinesq–Klein–Gordon* (BKG) equation (Khusnutdinova and Tranter, 2019)

$$u_{tt} - c^2 u_{xx} = \varepsilon \left[\frac{\alpha}{2} \left(u^2 \right)_{xx} + \beta u_{ttxx} - \gamma u \right].$$

This case is similar to a Toda lattice on an elastic substrate (Yagi and Kawahara, 2001). Such an equation has also arisen in the context of oceanic waves in a rotating ocean (Gerkema, 1996). While the accuracy of such single Boussinesq-type equations does not exceed the accuracy of the corresponding uni-directional models in the water wave context, they are valid two-directional models for various solid waveguides (see, e.g., (Samsonov, 2001; Porubov, 2003; Engelbrecht et al., 2017; Peets and Tamm, 2017; Garbuzov et al., 2019) and references therein).

The Ostrovsky equation (Ostrovsky, 1978), written in the general form

$$(\eta_t + \nu\eta\eta_x + \mu\eta_{xxx})_x = \lambda\eta$$

with some constant coefficients ν, μ and λ, implies that for any regular periodic solution on a finite interval (including the case of localised solutions on a large interval) the mean is zero for any $t \geq 0$:

$$\frac{1}{2L}\int_{-L}^{L} \eta \, \mathrm{d}x = 0.$$

However, the original BKG equation does not impose similar restrictions on its solutions or initial conditions (apparent *zero mass contradiction*).

This issue has been settled on the infinite line by considering a regularised Ostrovsky equation (Grimshaw, 1999). Similar regularisation was used for the Kadomtsev–Petviashvili equation (Ablowitz and Wang, 1997; Boiti et al., 1995), and the physical motivation has been discussed in (Grimshaw and Melville, 1989). It was shown that the mass rapidly adjusts within the *temporal boundary layer*. The main wave has zero mass, and the nonzero mass is transported to a large distance in the opposite direction to the propagation of this wave (Grimshaw and Helfrich, 2012).

In this section we outline the construction of the solution which bypasses this difficulty in the periodic case (Khusnutdinova et al., 2014; Khusnutdinova and Tranter, 2019). We consider the initial value problem on the interval $[-L, L]$:

$$u_{tt} - c^2 u_{xx} = \varepsilon \left[\frac{\alpha}{2}\left(u^2\right)_{xx} + \beta u_{ttxx} - \gamma u\right], \tag{6.43}$$

$$u|_{t=0} = F(x), \quad u_t|_{t=0} = V(x), \tag{6.44}$$

where F and V are sufficiently smooth $(2L)$-periodic functions, and both functions $F(x)$ and $V(x)$ may have nonzero mean values

$$F_0 = \frac{1}{2L}\int_{-L}^{L} F(x)\,\mathrm{d}x \quad \text{and} \quad V_0 = \frac{1}{2L}\int_{-L}^{L} V(x)\,\mathrm{d}x. \tag{6.45}$$

The mean value of u is calculated as

$$\langle u\rangle(t) = \frac{1}{2L}\int_{-L}^{L} u(x,t)\,\mathrm{d}x = F_0\cos\left(\sqrt{\varepsilon\gamma}t\right) + V_0\frac{\sin\left(\sqrt{\varepsilon\gamma}t\right)}{\sqrt{\varepsilon\gamma}}. \tag{6.46}$$

The initial value problem for the deviation from the oscillating mean value $\tilde{u} = u - \langle u\rangle(t)$ is given by

$$\tilde{u}_{tt} - c^2\tilde{u}_{xx} = \varepsilon\left[\alpha\left(F_0\cos\omega t + \frac{1}{\sqrt{\varepsilon}}\frac{V_0}{\sqrt{\gamma}}\sin\omega t\right)\tilde{u}_{xx} + \right.$$

$$\left. + \frac{\alpha}{2}\left(\tilde{u}^2\right)_{xx} + \beta\tilde{u}_{ttxx} - \gamma\tilde{u}\right], \tag{6.47}$$

and

$$\tilde{u}\big|_{t=0} = F(x) - F_0, \quad \tilde{u}_t\big|_{t=0} = V(x) - V_0, \tag{6.48}$$

where $\omega = \sqrt{\gamma\varepsilon}$.

We look for a weakly nonlinear solution of the form

$$\tilde{u}\,(x,t) = f^+\,(\xi_+, \tau, T) + f^-\,(\xi_-, \tau, T) + \sqrt{\varepsilon} P\,(\xi_-, \xi_+, \tau, T) + O\,(\varepsilon)\,, \tag{6.49}$$

where

$$\xi_\pm = x \pm ct, \quad \tau = \sqrt{\varepsilon} t, \quad T = \varepsilon t.$$

Here we aim to construct the solution up to and including $O\left(\sqrt{\varepsilon}\right)$ terms.

The first nontrivial equation appears at $O\left(\sqrt{\varepsilon}\right)$:

$$-4c^2 P_{\xi_-\xi_+} = 2cf^-_{\xi_-\tau} - 2cf^+_{\xi_+\tau} + \frac{\alpha V_0}{\sqrt{\gamma}} \sin\left(\sqrt{\gamma}\tau\right)\left(f^-_{\xi_-\xi_-} + f^+_{\xi_+\xi_+}\right). \tag{6.50}$$

Averaging with respect to x at constant ξ_- or ξ_+ yields the equations

$$2cf^-_{\xi_-\tau} + \frac{\alpha V_0}{\sqrt{\gamma}} \sin\left(\sqrt{\gamma}\tau\right) f^-_{\xi_-\xi_-} = 0 \tag{6.51}$$

and

$$2cf^+_{\xi_+\tau} - \frac{\alpha V_0}{\sqrt{\gamma}} \sin\left(\sqrt{\gamma}\tau\right) f^+_{\xi_+\xi_+} = 0. \tag{6.52}$$

Equations (6.51) and (6.52) are integrated using the method of characteristics, yielding

$$f^- = f^-\left(\xi_- + \frac{\alpha V_0}{2c\gamma}\cos\left(\sqrt{\gamma}\tau\right), T\right), \quad f^+ = f^+\left(\xi_+ - \frac{\alpha V_0}{2c\gamma}\cos\left(\sqrt{\gamma}\tau\right), T\right). \tag{6.53}$$

The appearance of the formulae (6.53) motivates the change of variables

$$\tilde{\xi}_- = \xi_- + \frac{\alpha V_0}{2c\gamma}\cos\left(\sqrt{\gamma}\tau\right), \quad \tilde{\xi}_+ = \xi_+ - \frac{\alpha V_0}{2c\gamma}\cos\left(\sqrt{\gamma}\tau\right), \tag{6.54}$$

instead of ξ_- and ξ_+. We can now rewrite the equation for P as $P_{\tilde{\xi}_-\tilde{\xi}_+} = 0$, which gives

$$P = g^-\left(\tilde{\xi}_-, \tau, T\right) + g^+\left(\tilde{\xi}_+, \tau, T\right). \tag{6.55}$$

At $O(\varepsilon)$, using the averaging, we obtain

$$g^{\pm}_{\tilde{\xi}_\pm} = -\frac{\alpha V_0}{4c^2\sqrt{\gamma}}\sin\left(\sqrt{\gamma}\tau\right) f^{\pm}_{\tilde{\xi}_\pm} \pm \frac{1}{2c} A^{\pm}\left(\tilde{\xi}_\pm, T\right)\tau \pm$$
$$\mp\left[\frac{\alpha^2 V_0^2}{16c^3\gamma}\left(\tau - \frac{\sin\left(2\sqrt{\gamma}\tau\right)}{2\sqrt{\gamma}}\right) - \frac{\alpha F_0}{2c\sqrt{\gamma}}\sin\left(\sqrt{\gamma}\tau\right)\right] f^{\pm}_{\tilde{\xi}_\pm\tilde{\xi}_\pm}, \tag{6.56}$$

where

$$A^{\pm}\left(\tilde{\xi}_\pm, T\right) = \left(\mp 2c f^{\pm}_T + \alpha f^{\pm} f^{\pm}_{\tilde{\xi}_\pm} + \beta c^2 f^{\pm}_{\tilde{\xi}_\pm\tilde{\xi}_\pm\tilde{\xi}_\pm}\right)_{\tilde{\xi}_\pm} - \gamma f^{\pm}. \tag{6.57}$$

Here, we omitted the homogeneous parts of the solutions for $g^{\pm}_{\tilde{\xi}_\pm}$. They can be shown to be equal to zero (Khusnutdinova and Tranter, 2019). To avoid secular terms we require

$$\left(\mp 2c f^{\pm}_T - \frac{\alpha^2 V_0^2}{8c^2\gamma} f^{\pm}_{\tilde{\xi}_\pm} + \alpha f^{\pm} f^{\pm}_{\tilde{\xi}_\pm} + \beta c^2 f^{\pm}_{\tilde{\xi}_\pm\tilde{\xi}_\pm\tilde{\xi}_\pm}\right)_{\tilde{\xi}_\pm} - \gamma f^{\pm} = 0. \tag{6.58}$$

Thus, we obtain two Ostrovsky equations for the left- and right-propagating waves. Equations (6.58) can be reduced to the standard form of the Ostrovsky equations by the change of variables

$$\hat{\xi}_\pm = \tilde{\xi}_\pm \mp \frac{\alpha^2 V_0^2}{16c^3\gamma} T.$$

Expansion (6.49) is also substituted into the initial conditions (6.48), which we satisfy at the respective orders of the small parameter.

The solution of the Cauchy problem (6.43), (6.44) for the original variable $u(x, t)$ up to and including $O\left(\sqrt{\varepsilon}\right)$ terms has the form

$$u(x,t) = V_0 \frac{\sin\left(\sqrt{\gamma}\tau\right)}{\sqrt{\varepsilon\gamma}} + F_0\cos\left(\sqrt{\gamma}\tau\right) +$$
$$+ f^- + f^+ + \sqrt{\varepsilon}\left[-\frac{\alpha V_0}{4c^2\sqrt{\gamma}}\sin\left(\sqrt{\gamma}\tau\right)\left(f^- + f^+\right) -\right.$$
$$\left. - \frac{\alpha}{2c\sqrt{\gamma}}\left(F_0\sin\left(\sqrt{\gamma}\tau\right) + \frac{\alpha V_0^2}{16c^2\gamma}\sin\left(2\sqrt{\gamma}\tau\right)\right)\left(f^-_{\tilde{\xi}_-} - f^+_{\tilde{\xi}_+}\right)\right] + O(\varepsilon), \tag{6.59}$$

where the functions $f^{\pm}\left(\tilde{\xi}_{\pm}, T\right)$ are solutions of the Ostrovsky equations (6.58), which should be solved subject to the initial conditions

$$f^{\pm}|_{T=0} = \frac{1}{2c}\left\{c\left[F\left(\tilde{\xi}_{\pm}\right) - F_0\right] \pm \int_{-L}^{\tilde{\xi}_{\pm}} (V(\sigma) - V_0)\, d\sigma\right\}. \tag{6.60}$$

Here $\tilde{\xi}_{\pm} = \xi_{\pm} \mp \frac{\alpha V_0}{2c\gamma}\cos\left(\sqrt{\gamma}\tau\right)$ (nonlinear characteristic variables when $V_0 \neq 0$).

To illustrate the validity of the constructed solution, we consider two examples. In both cases we will assume $c = \alpha = \beta = \gamma = 1$. Firstly let us consider a soliton on a raised pedestal, defined by the initial condition

$$u(x, 0) = A\,\text{sech}^2\frac{x}{\Lambda} + d_1, \tag{6.61}$$

$$u_t(x, 0) = \frac{2cA}{\Lambda}\,\text{sech}^2\frac{x}{\Lambda}\tanh\frac{x}{\Lambda} + d_2, \tag{6.62}$$

where d_1, d_2 are constants and we have

$$A = \frac{6ck^2}{\alpha}, \quad \Lambda = \frac{\sqrt{2c\beta}}{k}, \tag{6.63}$$

with $k = \sqrt{\alpha/3c}$. We take $d_1 = 5$, $d_2 = 0.5$, $\varepsilon = 0.001$ and present the results at $t = 1/\varepsilon$ in Fig. 6.10. We can see that there is a good agreement between the weakly nonlinear solution and the results of direct numerical simulations, with a small phase shift between the constructed solution and the direct numerical solution.

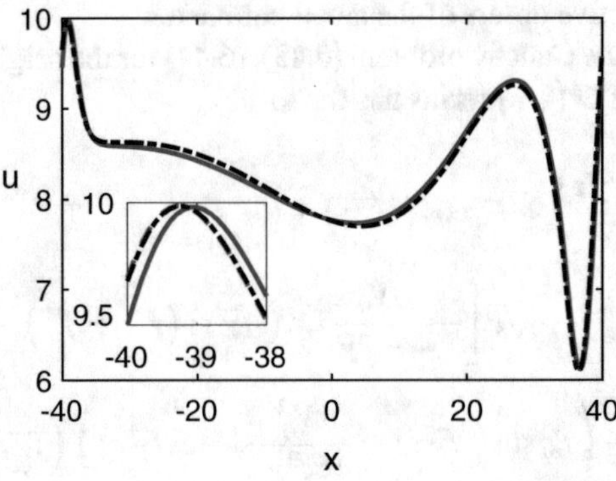

Fig. 6.10 A comparison of the numerical solution of the BKG equation (blue, solid line) and the constructed semianalytical solution including leading order (red, dashed line) and $O\left(\sqrt{\varepsilon}\right)$ (black, dash dotted line) terms, at $t = 1/\varepsilon$. Parameters are $L = 40$, $N = 800$, $k = 1/\sqrt{3}$, $c = \alpha = \beta = \gamma = 2$, $\varepsilon = 0.001$, $\Delta t = 0.01$, $\Delta T = \varepsilon\Delta t$, $d_1 = 5$, and $d_2 = 0.5$. There is a good agreement between the numerical solution and the constructed solution.

The second example we consider is for a cnoidal wave initial condition. The exact cnoidal wave solution of the equation

$$2cf_T^- + \alpha f^- f_{\xi_-}^- + \beta c^2 f_{\xi_-\xi_-\xi_-}^- = 0,$$

can be written in terms of the Jacobi elliptic function as follows (Johnson, 1997):

$$f^- = -\frac{6\beta c^3}{\alpha}\left\{ f_2 - (f_2 - f_3)\mathrm{cn}^2\left[\left(\xi^- + vT\right)\sqrt{\frac{f_1 - f_3}{2}} \,|\, m\right]\right\}, \qquad (6.64)$$

where

$$v = (f_1 + f_2 + f_3)\beta c^2, \quad m = \frac{f_2 - f_3}{f_1 - f_3}. \qquad (6.65)$$

This solution is parametrised by the constants $f_3 < f_2 < f_1$ such that the elliptic modulus $0 < m < 1$. The wave length can be calculated as

$$L = 2K(m)\sqrt{\frac{2}{f_1 - f_3}},$$

where $K(m)$ is the complete elliptic integral of the first kind. In our example we take $f_1 = 2 \times 10^{-5}$, $f_2 = 0$, $f_3 = -1/3$, giving $m \approx 0.999$. We take $c = \alpha = \beta = \gamma = 1$, $\varepsilon = 5 \times 10^{-4}$ and the same pedestal as for the soliton case, i.e., $d_1 = 5$ and $d_2 = 0.5$. The results are presented at $t = 1/\varepsilon$ in Fig. 6.11. We can see that again there is a good agreement between the weakly nonlinear solution and the results

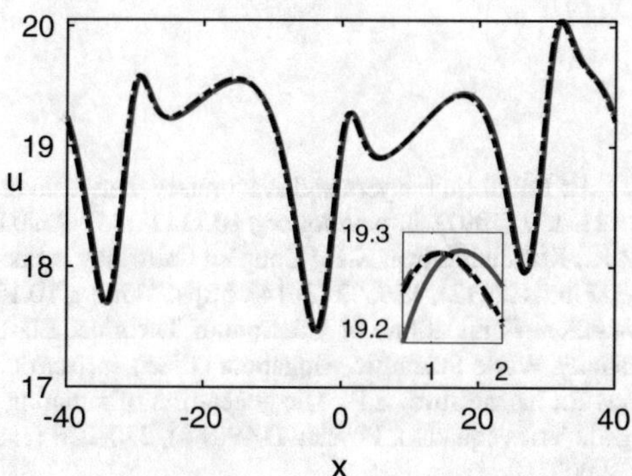

Fig. 6.11 A comparison of the numerical solution of the BKG equation (blue, solid line) and the constructed semianalytical solution including leading order (red, dashed line) and $O\left(\sqrt{\varepsilon}\right)$ (black, dash dotted line) terms, at $t = 1/\varepsilon$, for a cnoidal wave initial condition. Parameters are $L = 40$, $N = 800$, $c = \alpha = \beta = \gamma = 1$, $\varepsilon = 5 \times 10^{-4}$, $\Delta t = 0.01$, $\Delta T = \varepsilon \Delta t$, $d_1 = 5$, and $d_2 = 0.5$. There is a good agreement between the numerical solution and the constructed solution.

of direct numerical simulations, although the phase shift between the constructed solution and the direct numerical solution is larger in this case. The inclusion of higher order terms (at $O(\varepsilon)$) improved the solution a little but more terms from the expansion are required for better agreement.

6.5 Conclusions

In this chapter we discussed the scattering of long pure and radiating bulk strain solitary waves in a delaminated bilayer and a related initial value problem. The modelling was performed within the framework of the nondimensional Boussinesq-type equations. We highlighted key features of the behaviour of pure and radiating solitary waves in such delaminated bilayers, which could be used for introscopy of layered structures, in addition to traditional tools.

The fission of a single incident soliton into a group of solitons in the delaminated area of a perfectly bonded polymethyl methacrylate (PMMA) bilayer has been observed in (Dreiden et al., 2010). The generation of a radiating solitary wave and subsequent disappearance of the "ripples" in the delaminated area of a two-layered PMMA bar with the PCP (polychloroprene-rubber-based) adhesive has been observed in (Dreiden et al., 2012).

Our numerical modelling motivates further laboratory experimentation with other materials used in applications. We also discussed how a weakly nonlinear solution of the related initial value problem can be constructed, avoiding restrictions on the mass of the initial conditions, which opens the way to extending the semianalytical approaches to the scattering of solitary waves discussed in this paper to the scattering of periodic waves.

References

Ablowitz, M.J., Wang, X.P.: Initial time layers and Kadomtsev–Petviashvili-type equations. Stud. Appl. Math. **98**(2), 121–137 (1997). https://doi.org/10.1111/1467-9590.00043

Alias, A., Grimshaw, R.H., Khusnutdinova, K.R.: Coupled Ostrovsky equations for internal waves in a shear flow. Phys. Fluids **26**(12), 126603 (2014). https://doi.org/10.1063/1.4903279

Askar, A.: Lattice Dynamical Foundations of Continuum Theories: Elasticity, Piezoelectricity, Viscoelasticity, Plasticity. World Scientific, Singapore (1985). https://doi.org/10.1142/0192

Benilov, E.S., Grimshaw, R., Kuznetsova, E.P.: The generation of radiating waves in a singularly-perturbed Korteweg–de Vries equation. Physica D **69**(3-4), 270–278 (1993). https://doi.org/10.1016/0167-2789(93)90091-E

Benjamin, T.B., Bona, J.L., Mahony, J.J.: Model equations for long waves in nonlinear dispersive systems. Phil. Trans. Roy. Soc. Lond. A Math. Phys. Sci. **272**(1220), 47–78 (1972). https://doi.org/10.1098/rsta.1972.0032

Boiti, M., Pempinelli, F., Pogrebkov, A.: The KPI equation with unconstrained initial data. Acta Appl. Math. **39**(1-3), 175–192 (1995). https://doi.org/10.1007/978-94-011-0017-5_10

Bona, J.L., Dougalis, V.A., Mitsotakis, D.E.: Numerical solution of Boussinesq systems of KdV–KdV type: II. Evolution of radiating solitary waves. Nonlinearity **21**(12), 2825–2848 (2008). https://doi.org/10.1088/0951-7715/21/12/006

Dauxois, T.: Fermi, Pasta, Ulam, and a mysterious lady. Physics Today **6**, 55–57 (2008). https://doi.org/10.1063/1.2835154

Dreiden, G.V., Khusnutdinova, K.R., Samsonov, A.M., Semenova, I.V.: Splitting induced generation of soliton trains in layered waveguides. J. Appl. Phys. **107**(3), 034909 (2010). https://doi.org/10.1063/1.3294612

Dreiden, G.V., Khusnutdinova, K.R., Samsonov, A.M., Semenova, I.V.: Bulk strain solitary waves in bonded layered polymeric bars with delamination. J. Appl. Phys. **112**(6), 063516 (2012). https://doi.org/10.1063/1.4752713

Engelbrecht, J., Salupere, A., Tamm, K.: Waves in microstructured solids and the Boussinesq paradigm. Wave Motion **48**(8), 717–726 (2011). https://doi.org/10.1016/j.wavemoti.2011.04.001

Fermi, E., Pasta, J., Ulam, S.: Studies of Nonlinear Problems. Los Alamos Scientific Laboratory Report No. LA-1940, 1955. Lect. Appl. Math. **15**, 143–155 (1974)

Garbuzov, F.E., Khusnutdinova, K.R., Semenova, I.V.: On Boussinesq-type models for long longitudinal waves in elastic rods. Wave Motion **88**, 129–143 (2019). https://doi.org/10.1016/j.wavemoti.2019.02.004

Gardner, C.S., Greene, J.M., Kruskal, M.D., Miura, R.M.: Method for solving the Korteweg–deVries equation. Phys. Rev. Lett. **19**(19), 1095–1097 (1967). https://doi.org/10.1103/PhysRevLett.19.1095

Gerkema, T.: A unified model for the generation and fission of internal tides in a rotating ocean. J. Marine Res. **54**(3), 421–450 (1996). https://doi.org/10.1357/0022240963213574

Grimshaw, R.H.J.: Adjustment processes and radiating solitary waves in a regularised Ostrovsky equation. Eur. J. Mech. B/Fluids **18**(3), 535–543 (1999). https://doi.org/10.1016/s0997-7546(99)80048-x

Grimshaw, R., Helfrich, K.: The effect of rotation on internal solitary waves. IMA J. Appl. Math. **77**(3), 326–339 (2012). https://doi.org/10.1093/imamat/hxs024

Grimshaw, R., Melville, W.K.: On the derivation of the modified Kadomtsev–Petviashvili equation. Stud. Appl. Math. **80**(3), 183–202 (1989). https://doi.org/10.1002/sapm1989803183

Grimshaw, R.H.J., Ostrovsky, L.A., Shrira, V.I., Stepanyants, Y.A.: Long nonlinear surface and internal gravity waves in a rotating ocean. Surveys Geophys. **19**(4), 289–338 (1998). https://doi.org/10.1023/A:1006587919935

Il'yushina, E.A.: Towards formulation of elasticity theory of inhomogeneous solids with microstructure. Doctoral dissertation, PhD Thesis, Lomonosov Moscow State University (1976) (in Russian)

Johnson, R.S.: A Modern Introduction to the Mathematical Theory of Water Waves. Cambridge University Press (1997). https://doi.org/10.1017/cbo9780511624056

Khusnutdinova, K.R.: Nonlinear waves in a two-row system of particles. Vestnik Moskov. Univ. Ser. I Mat. Mekh. **2**, 71–76 (1992) (in Russian)

Khusnutdinova, K.R.: Wave dynamics of a medium constructed on the basis of a two-row system of particles. Deep Refinement of Hydrocarbon Material, **2**, 136–145 (1993) (in Russian)

Khusnutdinova, K.R., Moore, K.R.: Initial-value problem for coupled Boussinesq equations and a hierarchy of Ostrovsky equations. Wave Motion **48**(8), 738–752 (2011). https://doi.org/10.1016/j.wavemoti.2011.04.003

Khusnutdinova, K.R., Samsonov, A.M.: Fission of a longitudinal strain solitary wave in a delaminated bar. Phys. Rev. E **77**(6), 066603 (2008). https://doi.org/10.1103/physreve.77.066603

Khusnutdinova, K.R., Tranter, M.R.: Modelling of nonlinear wave scattering in a delaminated elastic bar. Proc. Roy. Soc. A Math. Phys. Eng. Sci. **471**(2183), 20150584 (2015). https://doi.org/10.1098/rspa.2015.0584

Khusnutdinova, K.R., Tranter, M.R.: On radiating solitary waves in bi-layers with delamination and coupled Ostrovsky equations. Chaos **27**(1), 013112 (2017). https://doi.org/10.1063/1.4973854

Khusnutdinova, K.R., Tranter, M.R.: D'Alembert-type solution of the Cauchy problem for the Boussinesq–Klein–Gordon equation. Stud. Appl. Math. **142**, 551–585 (2019). https://doi.org/10.1111/sapm.12263

Khusnutdinova, K.R., Samsonov, A.M., Zakharov, A.S.: Nonlinear layered lattice model and generalized solitary waves in imperfectly bonded structures. Phys. Rev. E **79**(5), 056606 (2009). https://doi.org/10.1103/physreve.79.056606

Khusnutdinova, K.R., Moore, K.R., Pelinovsky, D.E.: Validity of the weakly nonlinear solution of the Cauchy problem for the Boussinesq-type equation. Stud. Appl. Math. **133**(1), 52–83 (2014). https://doi.org/10.1111/sapm.12034

Maugin, G.A.: Nonlinear Waves in Elastic Crystals. Oxford University Press (1999)

Ostrovsky, L.A.: Nonlinear internal waves in a rotating ocean. Okeanologiya **18**, 119–125 (1978).

Peets, T., Tamm, K., Engelbrecht, J.: On the role of nonlinearities in the Boussinesq-type wave equations. Wave Motion **71**, 113–119 (2017). https://doi.org/10.1016/j.wavemoti.2016.04.003

Pelinovsky, E.N.: On the soliton evolution in inhomogeneous media. Appl. Mech. Techn. Phys. **6**, 80–85 (1971)

Porubov, A.V.: Amplification of Nonlinear Strain Waves in Solids. World Scientific (2003). https://doi.org/10.1142/5238

Samsonov, A.M.: Strain Solitons in Solids and how to construct them. Chapman and Hall/CRC, Boca Raton (2001).

Tappert, F.D., Zabusky, N.J.: Gradient-induced fission of solitons. Phys. Rev. Lett. **27**(26), 1774–1776 (1971). https://doi.org/10.1103/physrevlett.27.1774

Tranter, M.R.: Solitary wave propagation in elastic bars with multiple sections and layers. Wave Motion **86**, 21–31 (2019). https://doi.org/10.1016/j.wavemoti.2018.12.007

Yagi, D., Kawahara, T.: Strongly nonlinear envelope soliton in a lattice model for periodic structure. Wave Motion **34**(1), 97–107 (2001). https://doi.org/10.1016/s0165-2125(01)00062-2

Zabusky, N.J., Kruskal, M.D.: Interaction of "solitons" in a collisionless plasma and the recurrence of initial states. Phys. Rev. Lett. **15**(6), 240–243 (1965). https://doi.org/10.1103/physrevlett.15.240

Chapter 7
On Two-Dimensional Longitudinal Nonlinear Waves in Graphene Lattice

Alexey V. Porubov and Alena E. Osokina

Abstract A two-dimensional graphene lattice model with translational and angular interactions between the elements of two sublattices is considered. The nonlinearity is introduced via nonlinearly elastic translational forces between the masses in the lattice. The nonlinear strain behaviour is studied using the continuum limit of the lattice that results in the derivation of four coupled governing equations describing the dynamics of interacting sublattices. An asymptotic procedure is developed to account for weakly transversely perturbed nonlinear longitudinal plane waves giving rise to a two-dimensional nonlinear governing equation for longitudinal strains.

7.1 Introduction

The study of nonlinear processes in crystalline media is strongly connected with the development of the soliton theory (Zabusky and Deem, 1967; Zabusky and Kruskal, 1967; Askar, 1986; Andrianov et al., 2010) based on the continuum limit of the original discrete equations. Various nonlinear partial differential equations are obtained this way. Finding physically realistic solutions to such equations requires both the application of a wide spectrum of mathematical methods and physical intuition, which allows asymptotic simplification of the original complicated equations to the simpler model equations reflecting the essence of the phenomena in a straight and comprehensible manner.

In particular, it means the inclusion of nonlinear, dispersion, diffraction, and dissipation terms in the equations allowing to analyse their possible balances that giving rise to localised nonlinear waves. This combination distinguishes the works of Professor Jüri Engelbrecht, who made a significant contribution to the development

---

A. V. Porubov (✉)
Institute for Problems in Mechanical Engineering, Saint Petersburg, Russia

Peter the Great Saint Petersburg Polytechnic University (SPbPU), Saint Petersburg, Russia

A. E. Osokina
Peter the Great Saint Petersburg Polytechnic University (SPbPU), Saint Petersburg, Russia

A. Berezovski, T. Soomere (eds.), *Applied Wave Mathematics II*, Mathematics of Planet Earth 6, https://doi.org/10.1007/978-3-030-29951-4_7

of the theory of nonlinear waves in various media, including media with nonlinear elasticity (Engelbrecht, 1983, 1997), microstructured materials (Sertakov et al., 2014), soft tissues, nerve fibres (Engelbrecht et al., 2010), and rocks (Engelbrecht and Khamidullin, 1988).

The problem of the description of the behaviour of materials with microstructure consists in the development of an accurate model that is able to take into account the structural features of the medium. There are numerous papers on different lattice models (Born and Huang, 1954; Askar, 1986; Maugin, 1999; Ostoja-Starzewski, 2002; Askes and Metrikine, 2005; Manevich, 2005; Andrianov et al., 2010; Kevrekidis, 2010). Usually, most of them are dealing with linearly elastic translational interactions between the particles (Kirkwood, 1939; Born and Huang, 1954). Extensions usually concern the inclusion of nonlocal interactions (Kosevich and Savotchenko, 1999; Michelitsch et al., 2014) and additional degrees of freedom (Suiker et al., 2001; Potapov et al., 2009). In the latter case, the Cosserat model equations of motion were obtained from those of a lattice model. Various nonlinear partial differential equations are usually obtained as a result of the continuum consideration of the original crystalline discrete equations. The statement that takes into account additional degrees of freedom requires a two-dimensional (2D) consideration. Also, several physical phenomena cannot be modelled in the one-dimensional (1D) case, e.g., plane waves interaction.

Of special interest is a 2D hexagonal model since it is used for the modelling of the carbon nanotubes and graphene. The linearly elastic interactions in such lattice have been studied in (Peng et al., 2002; Zhang et al., 2002; Tovstik, 2012; Pedrielli et al., 2017). In addition to the previously mentioned lattice models, the separation into two sublattices is used in this paper. Besides translational interactions, angular interactions are taken into consideration.

The list of emerging developments in studies of nonlinear waves in crystal lattices is quite extensive due to the gradually increasing pool of experimental observation of such modes in a wide range of physical systems. Of particular interest are localised modes in models of the discrete nonlinear Schrödinger type and their applications in nonlinear optics and atomic physics (Brazhnyi and Konotop, 2004), or models of the Klein–Gordon type and their applications in mechanical and electrical systems (Koukouloyannis and Mackay, 2005), in the Fermi–Pasta–Ulam (FPU) type models arising in the examination of granular crystals with Hertzian (or modified Hertzian) interactions (Daraio et al., 2006; Khatri et al., 2009; Kevrekidis, 2010), and in materials engineering (Porter et al., 2015). It is known that nonlinear discrete equations usually cannot be solved analytically, and the continuum limit is needed for analytical consideration.

In this chapter, a 2D nonlinear graphene lattice is considered. The statement of the discrete problem is described in Sect. 7.2 as well as the derivation of the discrete nonlinear equations of motion. Section 7.3 deals with the continuum limit of the general equation for the case when transverse variations are weak. An asymptotic procedure is developed to account for weakly transversely perturbed nonlinear longitudinal plane waves giving rise to a 2D nonlinear governing equation for longitudinal strains. Then the influence of translational, angular and nonlinear stiffnesses on 2D longitudinal nonlinear wave localisation is studied. These results generalise our previous findings on the propagation of plane longitudinal waves (Porubov and Osokina, 2019).

7.2 Discrete Equations for Graphene Lattice

The main object of study in this chapter is a 2D hexagonal lattice (Fig. 7.1). The Cauchy–Born rule (Born and Huang, 1954; Zhang et al., 2002) assumes that the atoms in a material subjected to a homogeneous deformation move according to a single mapping from the undeformed to the deformed configuration. The Cauchy–Born rule that links the continuum model with interaction of particles requires the atomic structure of the materials to be centrosymmetric, because such a structure ensures the equilibrium of particles in a lattice. The hexagonal arrangement of the lattice elements does not meet this requirement. A lattice under "homogeneous deformation" at the cell level results in a deformation that may not be homogeneous inside the cell (Peng et al., 2002).

It is possible to modify the Cauchy–Born rule to describe hexagonal lattice structures by introducing a rigid body translation as an internal degree of freedom (DOF). A hexagonal lattice can be decomposed into two sublattices (marked by 1 and 2 in Fig. 7.1), each of which is centrosymmetric. Under a homogeneous deformation applied on the continuum level, each sublattice deforms according to

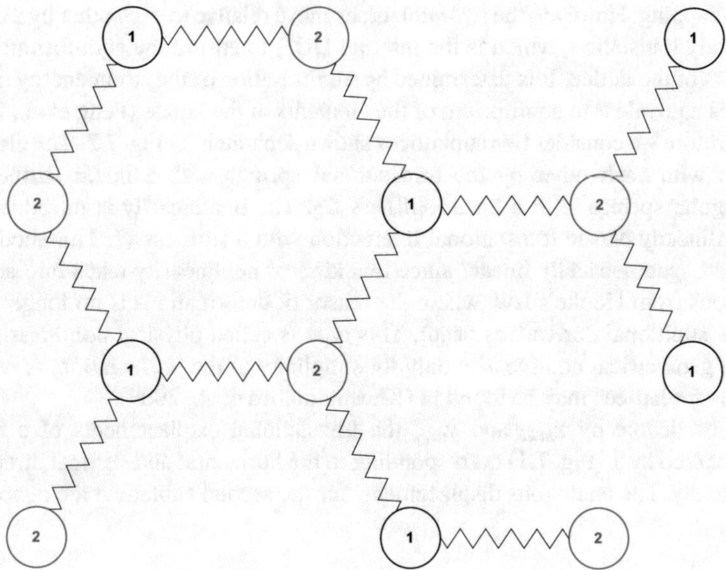

Fig. 7.1 Two-dimensional hexagonal lattice separated into two sublattices.

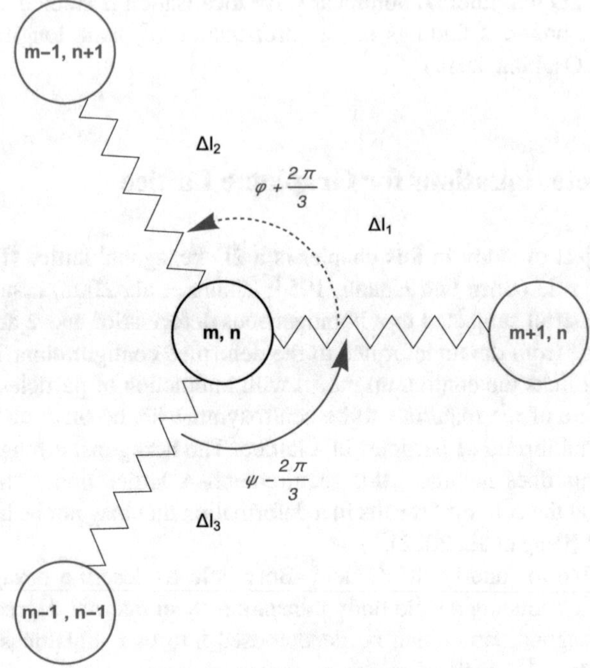

Fig. 7.2 The first sublattice 1

single mapping. However, the two sublattices move relative to each other by a certain rigid body translation, which is the internal DOF, to ensure the equilibrium of the elements of the lattice. It is determined by minimisation of the strain energy density which is equivalent to equilibrium of the elements of the lattice (Peng et al., 2002).

Therefore we consider two sublattices shown separately in Fig. 7.2. The elements interact with each other by the translational springs with a linear stiffness C_1 and angular springs with a linear stiffness C_2. The nonlinearity is introduced via the nonlinearly elastic translational interaction with a stiffness Q. This model can be called "geometrically linear" since this kind of nonlinearity takes into account deviations from Hooke's law, where the tensor of deformations is no longer linear and the additional derivatives occur. This type is called physical nonlinearity. We neglect geometrical nonlinearity only for simplicity of the presentation. A relevant example for lattices may be found in (Khusnutdinova et al., 2009).

Let us denote by $x_{m,n}$ and $y_{m,n}$ the translational displacements of a central mass marked by 1 (Fig. 7.1) corresponding to the horizontal and vertical directions, respectively. The analogous displacements for the second sublattice are denoted by

$X_{m,n}$ and $Y_{m,n}$. The potential energies for the sublattices are

$$\Pi_1 = C_1\left(\Delta l_1^2 + \Delta l_2^2 + \Delta l_3^2\right) + C_2 a^2\left(\phi^2 + \psi^2\right) + Q\left(\Delta l_1^3 + \Delta l_2^3 + \Delta l_3^3\right),$$
$$\Pi_2 = C_1\left(\Delta L_1^2 + \Delta L_2^2 + \Delta L_3^2\right) + C_2 a^2\left(\Phi^2 + \Psi^2\right) + Q\left(\Delta L_1^3 + \Delta L_2^3 + \Delta L_3^3\right), \tag{7.1}$$

where a is a distance between the lattice elements.

The translational elongations for the first sublattice (Fig. 7.2) are:

$$\Delta l_1 = X_{m+1,n} - x_{m,n},$$
$$\Delta l_2 = -\cos\frac{\pi}{3}(X_{m-1,n+1} - x_{m,n}) + \sin\frac{\pi}{3}(Y_{m-1,n+1} - y_{m,n}), \tag{7.2}$$
$$\Delta l_3 = -\cos\frac{\pi}{3}(X_{m,n} - x_{m-1,n-1}) - \sin\frac{\pi}{3}(Y_{m-1,n-1} - y_{m,n}).$$

Similarly, the translational elongations for the second sublattice (Fig. 7.3) are:

$$\Delta L_1 = X_{m,n} - x_{m-1,n},$$
$$\Delta L_2 = \cos\frac{\pi}{3}(x_{m+1,n+1} - X_{m,n}) + \sin\frac{\pi}{3}(y_{m+1,n+1} - Y_{m,n}), \tag{7.3}$$
$$\Delta L_3 = \cos\frac{\pi}{3}(x_{m+1,n-1} - X_{m,n}) - \sin\frac{\pi}{3}(y_{m+1,n-1} - Y_{m,n}).$$

The kinetic energy for the first sublattice is

$$K_1 = \frac{M}{2}\left(\dot{x}^2 + \dot{y}^2\right) + Ja^2\left(\dot{\varphi}^2 + \dot{\psi}^2\right), \tag{7.4}$$

where M is the mass of the particles in the lattice and J is the angular mass (moment of inertia).

The cosine formula is used to find the expressions for ϕ, ψ, Φ, and Ψ through the translational elongations. Let us introduce $\mathbf{r_{m+1}}$, $\mathbf{r_{m-1}}$ as the vectors going from the central particle with indices m, n to its neighbouring particles (Porubov and Osokina, 2019). Using the notations in Fig. 7.2 and the expression for a scalar product of two vectors, we obtain:

$$\cos\left(\frac{2\pi}{3} + \varphi\right) = \frac{\mathbf{r_{m+1}} \cdot \mathbf{r_{m-1}}}{|\,\mathbf{r_{m+1}}\,|\,|\,\mathbf{r_{m-1}}\,|}, \tag{7.5}$$

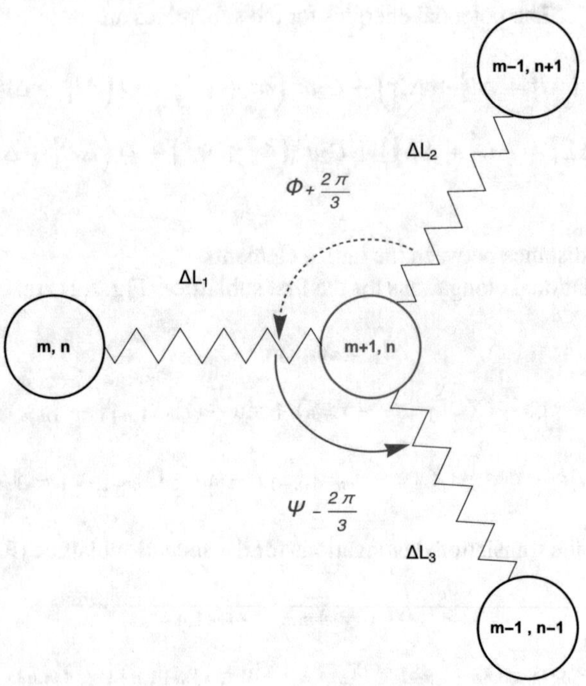

Fig. 7.3 The second sublattice 2

where the vectors $\mathbf{r_{m+1}}, \mathbf{r_{m-1}}$ are

$$\mathbf{r_{m+1}} = \mathbf{i}\left(l + X_{m+1,n} - x_{m,n}\right), \tag{7.6}$$

$$\mathbf{r_{n+1}} = -\mathbf{i}\left(l\cos\frac{\pi}{3} + X_{m-1,n+1} - x_{m,n}\right) + \mathbf{j}\left(l\sin\frac{\pi}{3} + Y_{m-1,n+1} - y_{m,n}\right) \tag{7.7}$$

and **i**, **j** are the unit vectors in the horizontal and vertical directions respectively. We consider only small variations in φ. This approximation allows us to expand the cosine function and the right-hand side in (7.5). Linearisation of the resulting equation leads to the approximate expression for φ:

$$\varphi = \frac{\sqrt{3}}{2l}\left(X_{m-1,n+1} - x_{m,n}\right) - \frac{1}{2l}\left(Y_{m-1,n+1} - y_{m,n}\right). \tag{7.8}$$

Similarly we obtain the approximate expression for ψ:

$$\psi = \frac{\sqrt{3}}{2l}\left(X_{m-1,n-1} - x_{m,n}\right) + \frac{1}{2l}\left(Y_{m-1,n-1} - y_{m,n}\right). \tag{7.9}$$

The expressions for the angular variations in the second sublattice are

$$\Phi = -\frac{\sqrt{3}}{2l}\left(x_{m+1,n+1} - X_{m,n}\right) - \frac{1}{2l}\left(y_{m+1,n+1} - Y_{m,n}\right), \tag{7.10}$$

$$\Psi = -\frac{\sqrt{3}}{2l}\left(x_{m+1,n-1} - X_{m,n}\right) + \frac{1}{2l}\left(y_{m+1,n-1} - Y_{m,n}\right). \tag{7.11}$$

The discrete equations are obtained using the Hamilton–Ostrogradsky variational principle

$$\frac{\mathrm{d}}{\mathrm{d}t}\frac{\partial K_1}{\partial \dot{x}_{m,n}} = -\frac{\partial \Pi_1}{\partial x_{m,n}}, \quad \frac{\mathrm{d}}{\mathrm{d}t}\frac{\partial K_1}{\partial \dot{y}_{m,n}} = -\frac{\partial \Pi_1}{\partial y_{m,n}}. \tag{7.12}$$

After substitution of the expressions for the kinetic and potential energy into (7.12), the equations for the first sublattice read:

$$\begin{aligned}
&(3J+M)\ddot{x}_{m,n} + \frac{\sqrt{3}}{2}J\left[\ddot{Y}_{m-1,n+1} - \ddot{Y}_{m-1,n-1} - \sqrt{3}(\ddot{X}_{m-1,n-1} - \ddot{X}_{m-1,n+1})\right] + \\
&+ \frac{C_1}{2}\left[6x_{m,n} - 4X_{m+1,n} - (X_{m-1,n-1} + X_{m-1,n+1}) + \sqrt{3}(Y_{m-1,n+1} - Y_{m-1,n-1})\right] + \\
&+ \frac{\sqrt{3}C_2}{2}\left[6x_{m,n} - 3(X_{m-1,n-1} + X_{m-1,n+1} + \sqrt{3}(Y_{m-1,n+1} - Y_{m-1,n-1})\right] + \\
&+ \frac{3Q}{8}\Big[(x_{m,n} - X_{m-1,n-1})^2 + (x_{m,n} - X_{m-1,n+1})^2 + 3(y_{m,n} - Y_{m-1,n-1})^2 + \\
&\qquad + 3(Y_{m-1,n+1} - y_{m,n})^2 + 2\sqrt{3}(x_{m,n} - X_{m-1,n-1})(y_{m,n} - Y_{m-1,n-1}) + \\
&\qquad + 2\sqrt{3}(x_{m,n} - X_{m-1,n+1})(Y_{m-1,n+1} - y_{m,n}) - 8(x_{m,n} - X_{m+1,n})^2\Big] = 0,
\end{aligned} \tag{7.13}$$

$$\begin{aligned}
&(J+M)\ddot{y}_{m,n} + \frac{J}{2}\left[\sqrt{3}(\ddot{X}_{m-1,n+1} - \ddot{X}_{m-1,n-1}) - (\ddot{Y}_{m-1,n-1} + \ddot{Y}_{m-1,n+1})\right] + \\
&+ \frac{C_1}{2}\left[6y_{m,n} + \sqrt{3}(X_{m-1,n+1} - X_{m-1,n-1}) - 3(Y_{m-1,n-1} + Y_{m-1,n+1})\right] + \\
&+ \frac{C_2}{2}\left[2y_{m,n} + \sqrt{3}(X_{m-1,n+1} - X_{m-1,n-1}) - (Y_{m-1,n-1} + Y_{m-1,n+1})\right] + \\
&+ \frac{3\sqrt{3}}{8}Q\Big[(x_{m,n} - X_{m-1,n-1})^2 - (x_{m,n} - X_{m-1,n+1})^2 + 3(y_{m,n} - Y_{m-1,n-1})^2 - \\
&\qquad - 3(Y_{m-1,n+1} - y_{m,n})^2 + 2\sqrt{3}(x_{m,n} - X_{m-1,n-1})(y_{m,n} - Y_{m-1,n-1}) - \\
&\qquad - 2\sqrt{3}(x_{m,n} - X_{m-1,n+1})(Y_{m-1,n+1} - y_{m,n})\Big] = 0.
\end{aligned} \tag{7.14}$$

Similarly, the kinetic energy for the second sublattice is

$$K_2 = \frac{M}{2}\left(\dot{X}^2 + \dot{Y}^2\right) + Ja^2\left(\dot{\Phi}^2 + \dot{\Psi}^2\right). \tag{7.15}$$

The Hamilton–Ostrogradsky variational principle for the second sublattice is

$$\frac{\mathrm{d}}{\mathrm{d}t}\frac{\partial K_2}{\partial \dot{X}_{m,n}} = -\frac{\partial \Pi_2}{\partial X_{m,n}}, \quad \frac{\mathrm{d}}{\mathrm{d}t}\frac{\partial K_2}{\partial \dot{Y}_{m,n}} = -\frac{\partial \Pi_2}{\partial Y_{m,n}}. \tag{7.16}$$

The substitution of the kinetic and potential energy into (7.16) yields the equations for the second sublattice:

$$\begin{aligned}
&(3J+M)\ddot{X}_{m,n} + \frac{\sqrt{3}}{2}J\left[\ddot{y}_{m+1,n-1} - \ddot{y}_{m+1,n+1} - \sqrt{3}(\ddot{x}_{m+1,n-1} - \ddot{x}_{m+1,n+1})\right] + \\
&+ \frac{C_1}{2}\left[6X_{m,n} - 4x_{m-1,n} - (x_{m+1,n-1} + x_{m+1,n+1}) + \sqrt{3}(y_{m+1,n-1} - y_{m+1,n+1})\right] + \\
&+ \frac{\sqrt{3}C_2}{2}\left[6X_{m,n} - 3(x_{m+1,n-1} + x_{m+1,n+1}) + \sqrt{3}(y_{m+1,n-1} - y_{m+1,n+1})\right] + \\
&+ \frac{3Q}{8}\Big[8(x_{m,n} - X_{m-1,n})^2 - (x_{m+1,n+1} - X_{m,n})^2 - (x_{m+1,n-1} - X_{m,n})^2 - \\
&\quad - 3(y_{m+1,n+1} - Y_{m,n})^2 - 3(Y_{m,n} - y_{m+1,n-1})^2 - \\
&\quad - 2\sqrt{3}(x_{m+1,n+1} - X_{m,n})(y_{m+1,n+1} - Y_{m,n}) - \\
&\quad - 2\sqrt{3}(x_{m+1,n-1} - X_{m,n})(Y_{m,n} - y_{m+1,n-1}\Big] = 0,
\end{aligned} \tag{7.17}$$

$$\begin{aligned}
&(J+M)\ddot{Y}_{m,n} + \frac{J}{2}\left[\sqrt{3}(\ddot{x}_{m+1,n-1} - \ddot{x}_{m+1,n+1}) - (\ddot{y}_{m+1,n-1} + \ddot{y}_{m+1,n+1})\right] + \\
&+ \frac{C_1}{2}\left[6Y_{m,n} + \sqrt{3}(x_{m+1,n-1} - x_{m+1,n+1}) - 3(y_{m+1,n-1} + y_{m+1,n+1})\right] + \\
&+ \frac{C_2}{2}\left[2Y_{m,n} + \sqrt{3}(x_{m+1,n-1} - x_{m+1,n+1}) - (y_{m+1,n-1} + y_{m+1,n+1})\right] + \\
&+ \frac{3\sqrt{3}}{8}Q\Big[(x_{m+1,n-1} - X_{m,n})^2 - (x_{m+1,n+1} - X_{m,n})^2 + 3(Y_{m,n} - y_{m+1,n-1})^2 - \\
&\quad - 3(y_{m+1,n+1} - Y_{m,n})^2 + 2\sqrt{3}(x_{m+1,n-1} - X_{m,n})(Y_{m,n} - y_{m+1,n-1}) - \\
&\quad - 2\sqrt{3}(x_{m+1,n+1} - X_{m,n})(y_{m+1,n+1} - Y_{m,n})\Big] = 0.
\end{aligned} \tag{7.18}$$

7.3 Continuum Limit for Weakly Transversely Perturbed Waves

7.3.1 Coupled Continuum Equations

In the following only weak transverse variations along the vertical axis are studied, as well as small vertical displacements. Also, nonlinear terms which act only in the horizontal direction are kept. The long wavelength continuum limit is used since the discrete nonlinear equations cannot be used for the analysis. Then the discrete variables $x_{m,n}$, $y_{m,n}$, $X_{m,n}$, and $Y_{m,n}$ are related to the continuum functions $u(x, y, t)$, $v(x, y, t)$, $U(x, y, t)$, and $V(x, y, t)$, respectively, while displacements of the neighbouring particles are expanded in the Taylor series, e.g.,

$$X_{m\pm1,n} = U \pm a\, U_x + \frac{1}{2} a^2\, U_{xx} + \dots .$$

We consider predominantly longitudinal waves propagating along the x-axis. On the one hand, we only keep the leading order nonlinear and dispersion terms in the equations of motions for u and U. On the other hand, only those terms that allow us to establish a connection between horizontal and vertical displacements are left in the equations of motion for the vertical displacements. Under these assumptions one obtains from (7.13)–(7.18):

$$\begin{aligned}
&(3J + M)u_{tt} + 3(C_1 + C_2)(u - U) - 3JU_{tt} - a(C_1 - 3C_2)U_x + 3JaU_{xtt} - \\
&- \frac{3a^2}{2}(C_1 + C_2)U_{xx} + \sqrt{3}a(C_1 + C_2)V_y - \sqrt{3}a^2(C_1 + C_2)V_{xy} - \frac{a^2}{2}(C_1 + 3C_2)U_{yy} - \\
&- \frac{3a^2}{2}JU_{xxtt} - \frac{a^3}{6}(C_1 - 3C_2)U_{xxx} - \frac{a^4}{8}(C_1 + C_2)U_{xxxx} - \frac{9Q}{4}(u - U)^2 + \\
&+ \frac{15Q}{2}a(u - U)U_x - \frac{9Q}{4}a^2(U_x)^2 = 0,
\end{aligned} \tag{7.19}$$

$$\begin{aligned}
&(3J + M)U_{tt} + 3(C_1 + C_2)(U - u) - 3Ju_{tt} + a(C_1 - 3C_2)u_x - 3Jau_{xtt} - \\
&- \frac{3a^2}{2}(C_1 + C_2)u_{xx} - \sqrt{3}a(C_1 + C_2)v_y - \sqrt{3}a^2(C_1 + C_2)v_{xy} - \frac{a^2}{2}(C_1 + 3C_2)u_{yy} - \\
&- \frac{3a^2}{2}Ju_{xxtt} + \frac{a^3}{6}(C_1 - 3C_2)u_{xxx} - \frac{a^4}{8}(C_1 + C_2)u_{xxxx} + \frac{9Q}{4}(u - U)^2 - \\
&- \frac{15Q}{2}a(u - U)u_x + \frac{9Q}{4}a^2(u_x)^2 = 0.
\end{aligned} \tag{7.20}$$

$$(3C_1 + C_2)(v - V) + \sqrt{3}a(C_1 + C_2)U_y + a(3C_1 + C_2)V_x - \sqrt{3}a^2(C_1 + C_2)U_{xy} = 0, \tag{7.21}$$

$$(3C_1 + C_2)(V - v) - \sqrt{3}a(C_1 + C_2)u_y - a(3C_1 + C_2)v_x - \sqrt{3}a^2(C_1 + C_2)u_{xy} = 0. \tag{7.22}$$

The displacements responsible for the motion of sublattices elements in the discrete equations are not physically reasonable in the continuum model. Indeed one cannot measure the displacements of the elements of sublattices in continua. The transformation of variables

$$U_1 = \frac{u + U}{2}, \quad U_2 = \frac{u - U}{2},$$

$$V_1 = \frac{v + V}{2}, \quad V_2 = \frac{v - V}{2}$$

allows us to describe the dynamics using the macrodisplacements U_1, V_1, and the variables u_1, v_1 accounting for internal variations or a microstructure. Then the measurable macrostrain may be introduced. These substitutions are introduced into equations of motion after obvious manipulations. Adding (7.19) and (7.20) yields:

$$\begin{aligned} &MU_{1,tt} - \sqrt{3}a(C_1 + C_2)V_{2,y} - \frac{a^2}{2}(C_1 + 3C_2)U_{1,yy} + a(C_1 - 3C_2)U_{2,x} - \\ &- 3aJU_{2,xtt} - \sqrt{3}a^2(C_1 + C_2)V_{1,xy} - \frac{3a^2}{2}(C_1 + C_2)U_{1,xx} - \\ &- \frac{3a^2}{2}JU_{1,xxtt} - \frac{a^3}{12}(C_1 - 3C_2)(U_{1,xxx} - U_{2,xxx}) - \frac{a^4}{16}(C_1 + C_2)(U_{1,xxxx}) - \\ &- Qa^2\left(\frac{15}{2}U_2U_{2,x} + \frac{9a}{4}U_{1,x}U_{2,x} + \frac{9a}{8}U_2U_{1,xx}\right) = 0. \end{aligned} \tag{7.23}$$

Subtraction of (7.19) from (7.20) results in

$$\begin{aligned} &(6J + M)U_{2,tt} + 6(C_1 + C_2)U_2 + \sqrt{3}a(C_1 + C_2)V_{1,y} + \frac{3a^2}{2}(C_1 + C_2)U_{2,xx} - \\ &- a(C_1 - 3C_2)U_{1,x} + 3aJ\,U_{1,xtt} + \sqrt{3}a^2(C_1 + C_2)V_{2,xy} - \\ &- \frac{a^3}{12}(C_1 - 3C_2)(U_{1,xxx} - U_{2,xxx}) - \frac{a^4}{16}(C_1 + C_2)U_{1,xxxx} - \\ &- Q\left[\frac{9}{2}U_2^2 + \frac{15}{2}aU_2U_{1,x} - \frac{9}{8}a^2\left(U_{1,x}^2 - U_{2,x}^2\right) + \frac{9}{8}a^2U_2\left(U_{1,xx} - U_{2,xx}\right)\right] = 0. \end{aligned} \tag{7.24}$$

Adding of (7.21) and (7.22) yields:

$$6(C_1 + C_2)U_2 + \sqrt{3}a(C_1 + C_2)V_{1,y} - a(C_1 - 3C_2)U_{1,x} = 0, \tag{7.25}$$

while subtraction of (7.21) from (7.22) results in

$$2(3C_1 + C_2)V_2 + \sqrt{3}a(C_1 + C_2)U_{1,y} + a(3C_1 + C_2)V_{1,x} = 0. \tag{7.26}$$

The coupled Eqs. (7.25), (7.26) are still difficult to analyse, and another asymptotic procedure is needed to decouple this system.

7.3.2 *Slaving Principle for Obtaining a Single Governing Equation*

Let us consider weakly nonlinear long elastic waves. This means that the strains are small, e.g., $U_{1,x}^2 << U_{1,x}$, and each derivative makes the term smaller, e.g., $U_{1,xxx} << U_{1,xx} << U_{1,x}$. Finally, we are interested in the cases when nonlinearity balances dispersion. This may only occur if the linear terms that contain higher order derivatives and nonlinear terms have the same magnitude. In this case the terms in the governing equations may be separated into different groups according to their order of magnitude. A systematic use of this procedure makes it possible to decouple (7.25), (7.26). This approach was previously used for the analysis of microstructured solids (Porubov and Pastrone, 2004) and can be called the slaving principle.

Assume that the solution U_2 to (7.24) has the form

$$U_2 = U_{21} + U_{22} + U_{23} + \ldots, \tag{7.27}$$

where $U_{23} << U_{22} << U_{21}$. Let us assume that U_{21} is of the same order as the other leading order term $U_{1,x}$ in (7.24). This assertion is reasonable for studying predominantly longitudinal waves (for which V_i are smaller than U_i) that propagate along the x axis and are slightly perturbed along the y axis. Leaving only these leading order terms, we obtain the equation for U_{21}:

$$a(3C_2 - C_1)U_{1,x} + 6(C_1 + C_2)U_{21} = 0.$$

Its solution is

$$U_{21} = \frac{a(C_1 - 3C_2)U_{1,x}}{6(C_1 + C_2)}. \tag{7.28}$$

Similarly, the next order terms are defined taking into account (7.28). The equation for U_{22} is

$$3a^2(C_1+C_2)U_{21,xx}+2(6J+M)U_{21,tt}+a\left[6JU_{1,xtt}-(C_1-3C_2)U_{21,x}\right]+$$

$$+15aQU_{1,x}U_{21}-9QU_{21}^2-\frac{1}{3}a^3(C_1-3C_2)U_{1,xxx}+(C_1+C_2)\left(2\sqrt{3}aV_{1,y}+12U_{22}\right)=0.$$

The solution of this equation is:

$$\begin{aligned}U_{22}=&\frac{a^2(C_1-3C_2)^2}{72(C_1+C_2)^2}U_{1,xx}-\frac{\sqrt{3}a}{6}V_{1,y}-\frac{a^3(C_1-3C_2)}{72(C_1+C_2)}U_{1,xx}-\\&-\frac{24aC_1J+aM(C_1-3C_2)}{36(C_1+C_2)^2}U_{1,xtt}-\frac{a^2Q(C_1-3C_2)(C_1+2C_2)}{12(C_1+C_2)^3}U_{1,x}^2.\end{aligned} \tag{7.29}$$

This procedure can be continued to obtain further terms in (7.27).

Equations (7.25), (7.26) are used to obtain the approximate relationships for V_1, V_2. The leading order part of (7.25) with (7.28) being taken into account is

$$-\frac{1}{3}a\left[\sqrt{3}a(7C_1+3C_2)U_{1,y}+6(3C_1+C_2)V_2\right]=0.$$

It leads to the solution for V_2:

$$V_2=-\frac{a(7C_1+3C_2)U_{1,y}}{2\sqrt{3}(3C_1+C_2)}. \tag{7.30}$$

Equation (7.26) together with (7.28) and (7.30) yields the following solution for $V_{1,x}$:

$$V_{1,x}=\frac{4C_1U_{1,yy}}{\sqrt{3}(3C_1+C_2)}.$$

Substitution of all obtained solutions into (7.23) yields a single governing equation for U_1:

$$U_{1,tt}-\alpha_1U_{1,xx}-\alpha_2U_{1,x}U_{1,xx}-\alpha_3U_{1,xxtt}-\alpha_4U_{1,xxxx}-\alpha_5U_{1,yy}=0, \tag{7.31}$$

where

$$\alpha_1=\frac{8a^2C_1(C_1+3C_2)}{3(C_1+C_2)},\quad \alpha_2=\frac{5a^3Q(C_1-3C_2)^2}{12(C_1+C_2)^2},\quad \alpha_3=\frac{4a^2C_1J}{C_1+C_2},$$

$$\alpha_4=\frac{a^4\left(7C_1^2+30C_1C_2-9C_2^2\right)}{36(C_1+C_2)},\quad \alpha_5=\frac{4a^2C_1(C_1+2C_2)}{3C_1+C_2}.$$

Therefore, an application of the slaving principle allows us to derive a single equation containing both nonlinear and dispersion terms for a study of the balance between nonlinearity and dispersion. The coefficients α_1, α_3, α_5 are always positive while α_4 can be of either sign depending on the angular stiffness C_2, and the sign of the coefficient α_2 at the nonlinear term entirely depends on the sign of the nonlinear stiffness Q.

7.4 Two-Dimensional Longitudinal Strain Waves

Equation (7.31) can be rewritten for longitudinal strains $W = U_{1,x}$ as follows:

$$W_{tt} - \alpha_1 W_{xx} - \alpha_2/2(W^2)_{xx} - \alpha_3 W_{xxtt} - \alpha_4 W_{xxxx} - \alpha_5 W_{yy} = 0. \qquad (7.32)$$

The equation is similar to that obtained in (Porubov et al., 2004) for longitudinal strain waves in a plate. It generalises the well-known Kadomtsev–Petviashvili equation (Kadomtsev and Petviashvili, 1970) by making it possible to consider waves propagating in both directions. The first two terms of Eq. (7.32) correspond to the linear wave equation. The third term accounts for the influence of nonlinearity. The next two terms describe dispersion of the waves while the remaining term is responsible for transverse diffraction.

Let us consider the linearised Eq. (7.32) by means of neglecting the term $\alpha_2/2(W^2)_{xx}$. Its harmonic solution in the conventional form $W = A\exp[i(kx - \omega t)]$ only exists if the dispersion relation

$$\omega = \sqrt{\frac{\alpha_1 - \alpha_4 k^2}{1 + \alpha_3 k^2}}$$

is satisfied. This relation imposes no restriction on the values of k at $\alpha_4 < 0$, while it bounds wave numbers at positive α_4 when $k < \sqrt{\alpha_1/\alpha_4}$.

Equation (7.32) is not integrable, however, it possesses particular analytical solutions. Of special interest are those that keep their shape and velocity of propagation. These solutions arise as a result of a balance between nonlinearity and dispersion. In particular, a localised bell-shaped plane travelling solitary wave solution is

$$W = \frac{3(V^2 - \alpha_1)}{\alpha_2}\,\mathrm{sech}^2\left[\sqrt{\frac{V^2 - \alpha_1}{4(\alpha_3 V^2 + \alpha_4)}}(x - Vt)\right].$$

The sign of the disturbance is defined by the sign of α_2 or nonlinear stiffness Q. When α_4 is positive, $V^2 - \alpha_1 > 0$, and the sign of the disturbance coincides with the sign of Q. For negative α_4, it is opposite to the sign of Q. This allows us to find out when a tensile or a compression longitudinal strain wave may propagate.

The plane wave solution does not vary in the transverse direction, however, it may be stable or unstable with respect to transverse perturbations. In our case the sign of α_5 is always positive and the conventional perturbation analysis reveals transverse stability of the plane solitary waves at positive α_4 (Porubov et al., 2004). In this case the wave cannot decay in the transverse direction. However, a resonant interaction of plane waves may result in a considerable increase in the wave height in the area of interaction (Porubov et al., 2005).

When a particular value of angular stiffness C_2 results in the negative sign of α_4, a transversely modulated plane solitary wave (Pelinovsky and Stepanyants, 1993) propagates as well as the 2D solitary wave or a lump (Porubov et al., 2005; Ablowitz and Segur, 1981).

7.5 Conclusions

The presented asymptotic technique allows us to reduce the coupled nonlinear continuum equations of the graphene lattice to a single nonlinear equation for longitudinal strain waves. It turns out that the angular stiffness in the original discrete model gives rise to various types of localisation phenomena of nonlinear waves. The described procedure can be extended to the study of weakly transversely perturbed shear waves. Previous studies of lattices of another structure revealed important variations in the governing equation from that of the longitudinal waves (Porubov and Berinskii, 2016; Porubov et al., 2018).

Another extension of the model may be reached by an inclusion of the geometrical nonlinearity in the discrete model (Khusnutdinova et al., 2009). Also, numerical simulations can be developed to study unsteady processes of the localisation of 2D nonlinear waves, as with the resonant interaction obtained in (Porubov et al., 2005).

Acknowledgements The work has been supported by the Russian Foundation for Basic Researches, grant No 17-01-00230-a.

References

Ablowitz, M.J., Segur, H.: Solitons and the Inverse Scattering Transform. SIAM Studies in Applied and Numerical Mathematics, vol. 4. SIAM, Philadelphia (1981). https://doi.org/10.1137/1.9781611970883

Askar, A.: Lattice Dynamical Foundations of Continuum Theories. Elasticity, Piezoelectricity, Viscoelasticity, Plasticity. World Scientific, Singapore (1985). https://doi.org/10.1142/0192

Andrianov, I.V., Awrejcewicz, J., Weichert, D.: Improved continuous models for discrete media. Math. Problems Engng. 986242 (2010). https://doi.org/10.1155/2010/986242

Askes, H., Metrikine, A.V.: Higher-order continua derived from discrete media: continualisation aspects and boundary conditions. Int. J. Solids Struct. **42**(1), 187–202 (2005). https://doi.org/10.1016/j.ijsolstr.2004.04.005

Born, M., Huang, K.: Dynamical Theory of Crystal Lattices. Clarendon Press, Oxford (1954)

Brazhnyi, V.A., Konotop, V.V.: Theory of nonlinear matter waves in optical lattices. Modern Phys. Lett. B, **18**(14), 627–651 (2004). https://doi.org/10.1142/S0217984904007190

Daraio, C., Nesterenko, V.F., Herbold, E.B., Jin, S.: Tunability of solitary wave properties in one-dimensional strongly nonlinear phononic crystals. Phys. Rev. E **73**(2), 026610 (2006). https://doi.org/10.1103/PhysRevE.73.026610

Engelbrecht, J.: Nonlinear Wave Processes of Deformation in Solids. Pitman Monographs and Surveys in Pure and Applied Mathematics, vol. 16. Pitman Advanced Publishing Program (1983)

Engelbrecht, J.: Nonlinear Wave Dynamics: Complexity and Simplicity. Kluwer Texts in the Mathematical Sciences, vol. 17. Springer, Dordrecht (1997). https://doi.org/10.1007/978-94-015-8891-1

Engelbrecht, J., Berezovski, A., Soomere, T.: Highlights in the research into complexity of nonlinear waves. Proc. Estonian Acad. Sci. **59**(2), 61–65 (2010). https://doi.org/10.3176/proc.2010.2.01

Engelbrecht, J., Khamidullin, Y.: On the possible amplification of nonlinear seismic waves. Phys. Earth Planet. Interiors **50**(1), 39–45 (1988). https://doi.org/10.1016/0031-9201(88)90089-1

Kadomtsev, B.B., Petviashvili, V.I.: On the stability of solitary waves in weakly dispersing media. Sov. Phys. Dokl. **15**(6), 539–541 (1970)

Kevrekidis, P.G.: Non-linear waves in lattices: past, present, future. IMA J. Appl. Math. **76**(3), 389–423 (2011). https://doi.org/10.1093/imamat/hxr015

Khatri, D., Daraio, C., Rizzo, P.: Coupling of highly nonlinear waves with linear elastic media. Proc. SPIE **7292**, Sensors and Smart Structures Technologies for Civil, Mechanical, and Aerospace Systems 2009, 72920P (2009). https://doi.org/10.1117/12.817574

Khusnutdinova, K.R., Samsonov, A.M., Zakharov, A.S.: Nonlinear layered lattice model and generalized solitary waves in imperfectly bonded structures. Phys. Rev. E **79**(5), 056606 (2009). https://doi.org/10.1103/PhysRevE.79.056606

Kirkwood, J. G.: The skeletal modes of vibration of long chain molecules. J. Chem. Phys. **7**(7), 506–509 (1939). https://doi.org/10.1063/1.1750479

Kosevich, A.M., Savotchenko, S.E.: Peculiarities of dynamics of one-dimensional discrete systems with interaction extending beyond nearest neighbors, and the role of higher dispersion in soliton dynamics. Low Temper. Phys. **25**(7), 550–557 (1999). https://doi.org/10.1063/1.593783

Koukouloyannis, V., MacKay, R.S.: Existence and stability of 3-site breathers in a triangular lattice. J. Phys. A Math. General **38**(5), 1021–1030 (2005). https://doi.org/10.1088/0305-4470/38/5/004

Manevich, A.I., Manevich, L.I.: The Mechanics of Nonlinear Systems with Internal Resonances. World Scientific, Singapore (2005). https://doi.org/10.1142/p368

Maugin, G. A.: Nonlinear Waves in Elastic Crystals. Oxford University Press, UK (1999)

Michelitsch, T.M., Collet, B., Wang, X.: Nonlocal constitutive laws generated by matrix functions: Lattice dynamics models and their continuum limits. Int. J. Engng. Sci. **80**, 106–123 (2014). https://doi.org/10.1016/j.ijengsci.2014.02.029

Ostoja-Starzewski, M.: Lattice models in micromechanics. Appl. Mech. Rev. **55**(1), 35–60 (2002). https://doi.org/10.1115/1.1432990

Pedrielli, A., Taioli, S., Garberoglio, G., Pugno, N.M.: Designing graphene based nanofoams with nonlinear auxetic and anisotropic mechanical properties under tension or compression. Carbon, **111**, 796–806 (2017). https://doi.org/10.1016/j.carbon.2016.10.034

Pelinovsky, D.E., Stepanyants, Y.A.: Self-focusing instability of plane solitons and chains of two-dimensional solitons in positive-dispersion media. Soviet J. Exp. Theor. Phys. **77**, 602–608 (1993)

Peng, Z., Yonggang, H., Geubelle, P.H., Kehchih, H.: On the continuum modeling of carbon nanotubes. Acta Mech. Sinica **18**(5), 528–536 (2002)

Porter, M.A., Kevrekidis, P.G., Daraio, C.: Granular crystals: Nonlinear dynamics meets materials engineering. Physics Today **68**(11), 44–50 (2015). https://doi.org/10.1063/pt.3.2981

Porubov, A.V., Berinskii, I.E.: Two-dimensional nonlinear shear waves in materials having hexagonal lattice structure. Math. Mech. Solids **21**(1), 94–103 (2016). https://doi.org/10.1177/1081286515577040

Porubov, A.V., Krivtsov, A.M., Osokina, A.E.: Two-dimensional waves in extended square lattice. Int. J. Non-Linear Mech. **99**, 281–287 (2018). https://doi.org/10.1016/j.ijnonlinmec.2017.12.008

Porubov, A.V., Osokina, A.E.: Double dispersion equation for nonlinear waves in a graphene-type hexagonal lattice. Wave Motion **89**, 185–192 (2019). https://doi.org/10.1016/j.wavemoti.2019.03.013

Porubov, A.V., Pastrone, F.: Non-linear bell-shaped and kink-shaped strain waves in microstructured solids. Int. J. Non-Linear Mech. **39**(8), 1289–1299 (2004). https://doi.org/10.1016/j.ijnonlinmec.2017.12.008

Porubov, A.V., Maugin, G.A., Mareev, V.V.: Localization of two-dimensional non-linear strain waves in a plate. Int. J. Non-Linear Mech. **39**(8), 1359–1370 (2004). https://doi.org/10.1016/j.ijnonlinmec.2003.12.002

Porubov, A.V., Tsuji, H., Lavrenov, I.V., Oikawa, M.: Formation of the rogue wave due to non-linear two-dimensional waves interaction. Wave Motion **42**(3), 202–210 (2005). https://doi.org/10.1016/j.wavemoti.2005.02.001

Potapov, A.I., Pavlov, I.S., Lisina, S.A.: Acoustic identification of nanocrystalline media. J. Sound Vibr. **322**(3), 564–580 (2009). https://doi.org/10.1016/j.jsv.2008.09.031

Sertakov, I., Engelbrecht, J., Janno, J.: Modelling 2D wave motion in microstructured solids. Mech. Res. Commun. **56**, 42–49 (2014). https://doi.org/10.1016/j.mechrescom.2013.11.007

Suiker, A.S.J., Metrikine, A.V., de Borst, R.: Comparison of wave propagation characteristics of the Cosserat continuum model and corresponding discrete lattice models. Int. J. Solids Struct. **38**(9), 1563–1583 (2001). https://doi.org/10.1016/S0020-7683(00)00104-9

Tovstik, P. E., Tovstik, T.P.: Static and dynamic analysis of two-dimensional graphite lattices. Mech. Solids **47**(5), 517–524 (2012). https://doi.org/10.3103/S0025654412050044

Zabusky, N.J., Deem, G.S.: Dynamics of nonlinear lattices I. Localized optical excitations, acoustic radiation, and strong nonlinear behavior. J. Comput. Phys. **2**(2), 126–153 (1967). https://doi.org/10.1016/0021-9991(67)90031-9

Zabusky, N.J., Kruskal, M.D.: Interaction of "solitons" in a collisionless plasma and the recurrence of initial states. Phys. Rev. Lett. **15**(6), 240–243 (1965). https://doi.org/10.1103/PhysRevLett.15.240

Zhang, P., Huang, Y., Geubelle, P.H., Klein, P.A., Hwang, K.C.: The elastic modulus of single-wall carbon nanotubes: a continuum analysis incorporating interatomic potentials. Int. J. Solids Struct. **39**(13-14), 3893–3906 (2002). https://doi.org/10.1016/S0020-7683(02)00186-5

Chapter 8
Shock Waves in Hyperbolic Systems of Nonequilibrium Thermodynamics

Tommaso Ruggeri and Shigeru Taniguchi

Abstract We present the state of the art of the mathematical theory of shock waves for hyperbolic systems. We start with a brief review of ideal shock waves discussing, in particular, the Riemann problem and the phase transition induced by shock waves in real gases. Then we consider dissipative systems and summarise the results concerning the behaviour of the shock thickness for increasing Mach number. In the last part, we present the framework of Rational Extended Thermodynamics theory of nonequilibrium rarefied gas and its theoretical predictions of shock waves in cases of both monatomic and polyatomic gases. Particular emphasis will be given to subshock formation and the related open problem.

8.1 Introduction

The shock wave is a mathematically and physically interesting phenomenon and has been a very important research topic for many years (Krehl, 2009). One of the main reasons is many examples of applications in various fields. The knowledge of shock waves is essential in aerodynamics for designing the optimal geometry of objects moving faster than the sound velocity, e. g., supersonic aircrafts or spacecrafts reentering the atmosphere of planets (Krehl, 2009). Shock waves are used to induce phase transitions or chemical reactions in material science. A typical example is the transformation of graphite to diamond induced by a shock wave (Morris, 1980). The supernova explosion is a typical example involving shock waves in astrophysics. The analysis of the supernova remnants gives us important information about its mechanism (Vink, 2012). There are numerous applications of shock waves in medical science. For example, the kidney stone in a human body can be destroyed by a weak

T. Ruggeri (✉)
University of Bologna, Bologna, Italy
e-mail: tommaso.ruggeri@unibo.it

S. Taniguchi
National Institute of Technology, Kitakyushu College, Kitakyushu, Japan
e-mail: taniguchi.shigeru@kct.ac.jp

A. Berezovski, T. Soomere (eds.), *Applied Wave Mathematics II*, Mathematics of Planet Earth 6, https://doi.org/10.1007/978-3-030-29951-4_8

shock wave generated by extracorporeal shock wave lithotripsy (ESWL) (Chaussy et al., 2007).

In Sect. 8.2, we start with a brief review of ideal shock waves in which the shock associated with hyperbolic systems of conservation laws is identified by its front, i.e., a moving surface that divides the space into two subspaces in which a continuous solution exists but where there is a jump of the field variables across the shock front. In particular, we discuss the celebrated Riemann problem that is mathematically solved only in the case of one space dimension.

Then we present some recent results on the phase transition induced by the shock wave in real gases. In Sect. 8.3, we consider dissipative systems of balance laws and we discuss the problem of shock thickness for increasing Mach number. In continuum thermomechanics there remains an open problem to compare theory with experiments.

In Sect. 8.4, we present the Rational Extended Thermodynamics (RET) theory for rarefied gas in nonequilibrium and we study the shock waves in this framework for both monatomic and polyatomic gases. In Sect. 8.5, we discuss the problems of subshock formation for any hyperbolic system of balance laws and comment on an open problem that exists related to this subject.

8.2 Ideal Shock Waves

It is well known that the systems of continuum theory are described by balance laws in integral form in which the variation in time of some global density field in the domain Ω must be balanced by the normal flux across the surface of the body Σ and by eventual production terms. With a regularity solution, any system of balance laws is assumed to be of the local form:

$$\frac{\partial \mathbf{u}}{\partial t} + \frac{\partial \mathbf{F}^i(\mathbf{u})}{\partial x_i} = \mathbf{f}(\mathbf{u}), \tag{8.1}$$

where (x_i, t) denotes space and time coordinates and $\mathbf{u}(x_i, t) \in \mathbb{R}^N$ is the unknown field vector of densities. The flux $\mathbf{F}^i$ $(i = 1, 2, 3)$ and the production $\mathbf{f}$ depend locally on the field variables. As usual, we adopt the summation convention, i.e., we take summation over repeated indices between 1 to 3. If $\mathbf{f} = 0$, the system (8.1) represents conservation laws.

An ideal shock wave is a moving surface that divides the space into two subspaces in which a continuous solution exists but where there is a jump of the field variables across the shock front. It is well known that a shock wave solution is a weak solution of (8.1) if and only if it satisfies the so-called *Rankine–Hugoniot (RH) conditions* across the shock front:

$$- s\, [\![\mathbf{u}]\!] + \left[\!\left[\mathbf{F}^i\right]\!\right] n_i = 0, \tag{8.2}$$

where $[\![g(\mathbf{u})]\!] = g(\mathbf{u}_1) - g(\mathbf{u}_0)$ for a generic function g and $(\mathbf{u}_1, \mathbf{u}_0)$ are, respectively, the values of $\mathbf{u}$ evaluated in the states behind (perturbed) and ahead (unperturbed) of the front surface. Here s denotes the velocity of the shock wave and $\mathbf{n} \equiv (n_i)$ is the unit normal to the front.

The set of all perturbed states $\mathbf{u}_1$ satisfying (8.2) for a given unperturbed state $\mathbf{u}_0$ is called the *Hugoniot locus* for the point $\mathbf{u}_0$ and is denoted as $\mathcal{H}(\mathbf{u}_0)$. If the unperturbed field $\mathbf{u}_0$ is known and we consider plane shocks with $\mathbf{n} =$ const, the RH relations furnish a system of N equations for the $N+1$ unknowns $\mathbf{u}_1$ and s. Thus *any one* among the $(N+1)-$tuple $(\mathbf{u}_1, s)$ may be chosen as the shock parameter. In the context of gas dynamics, a typical parameter is the unperturbed Mach number defined as:

$$M_0 = \frac{s - v_{0n}}{c_0}, \tag{8.3}$$

where v_{0n} and c_0 denote, respectively, the normal components of the velocity and the sound velocity in the unperturbed state. It is noticeable that the Mach number is defined with the difference of velocities and therefore it is possible to choose the reference frame moving with the shock front $s = 0$ without any loss of generality due to the Galilean invariance of the system.

8.2.1 Admissibility of Shock Waves

According to the theory of hyperbolic systems, not every solution of the RH relations corresponds to a physically meaningful shock wave. Thus, we need a criterion to select the perturbed states $\mathbf{u}_1$ that, together with $\mathbf{u}_0$, form *admissible* shocks.

Let λ and $\mathbf{r}$ be the eigenvalues and right eigenvectors of the following eigenvalue problem:

$$(\mathbf{A}_n - \lambda \mathbf{I})\mathbf{r} = 0, \tag{8.4}$$

where $\mathbf{A}^i = \nabla \mathbf{F}^i$, $\mathbf{A}_n = \mathbf{A}^i n_i$ and ∇ represents the gradient with respect of $\mathbf{u}$. The system is hyperbolic in time direction if all the *characteristic velocities* λ are real and the right eigenvectors are linearly independent. The system is called *genuinely nonlinear* if the following condition is satisfied:

$$\nabla \lambda \cdot \mathbf{r} \neq 0, \quad \forall \mathbf{u}. \tag{8.5}$$

Let us consider now one-dimensional space. When we deal with genuinely nonlinear waves, the selection rule is given by the *Lax condition* (Lax, 1973), according to which a shock wave is admissible if the shock velocity satisfies the inequality:

$$\lambda_0 < s < \lambda_1,$$

where $\lambda_0 \equiv \lambda(\mathbf{u}_0)$ and $\lambda_1 \equiv \lambda(\mathbf{u}_1)$ are the unperturbed and the perturbed characteristic velocities, respectively. The Lax condition turns out to be equivalent (at least for *weak shock waves*) to the condition of entropy growth across the shock (Dafermos, 2005).

When the genuine nonlinearity is lost in some hypersurface of the field variables, the Lax condition is substituted with the Liu condition (Liu, 1976, 1981) stating that a shock wave is admissible if

$$s \geqslant s_*, \ \forall s_* \in \{s_* : s_*(\mathbf{u}_* - \mathbf{u}_0) = \mathbf{F}(\mathbf{u}_*) - \mathbf{F}(\mathbf{u}_0), \ \mathbf{u}_* \in \mathcal{H}(\mathbf{u}_0) \text{ between } \mathbf{u}_0 \text{ and } \mathbf{u}_1\}.$$

This means that a shock is admissible if its speed, s, is not smaller than the speed of any other shock with the same unperturbed state $\mathbf{u}_0$ and with perturbed state $\mathbf{u}_*$ lying on the Hugoniot locus for $\mathbf{u}_0$ between $\mathbf{u}_0$ and $\mathbf{u}_1$. In this case, the entropy growth is not enough to give the admissibility and a superposition principle is needed (Liu and Ruggeri, 2003).

8.2.2 Riemann Problem for Conservation Laws

One of the main problems in hyperbolic systems is the Riemann problem. This problem was proposed by Riemann considering a gas that is initially separated into two regions by a thin diaphragm. The gases in the two regions are in different equilibrium thermodynamic states, respectively. The question raised by Riemann is what happens when the diaphragm is removed. In the literature, by extension of this problem, the Riemann problem deals with every solution of a system of conservation laws in one space dimension along the x axis when the initial data composed of two different constant states $(\mathbf{u}_L, \mathbf{u}_R)$ are connected with a jump at $x = 0$.

The Riemann problem for hyperbolic systems was completely solved by Lax (1973). It was shown that the solution of the Riemann problem for hyperbolic systems of conservation laws is a combination of the rarefaction waves, contact waves, and shock waves. Figure 8.1 shows a typical profile based on the system of Euler equations for an ideal gas.

A huge literature related to the Riemann problem exists, and in particular many numerical results have been obtained by using the Riemann solvers (see e.g., (Toro, 2009)).

Liu (1977a,b) noticed that Riemann initial data can be regarded as a rough approximation of continuous initial data containing sharp and rapid variations. Therefore he takes into account the presence of shock thickness, oscillations, noises, and continuous (although very steep) changes. The initial data connect continuously two different constant states $(\mathbf{u}_L, \mathbf{u}_R)$ for large $|x|$. This problem is called the *Riemann problem with structure*.

In the case of conservation laws of hyperbolic type, Liu (1977a,b) proved, roughly speaking, that the solution of this problem converges, for large time, to the solution of the associated Riemann problem with discontinuity of initial data $(\mathbf{u}_L, \mathbf{u}_R)$ at

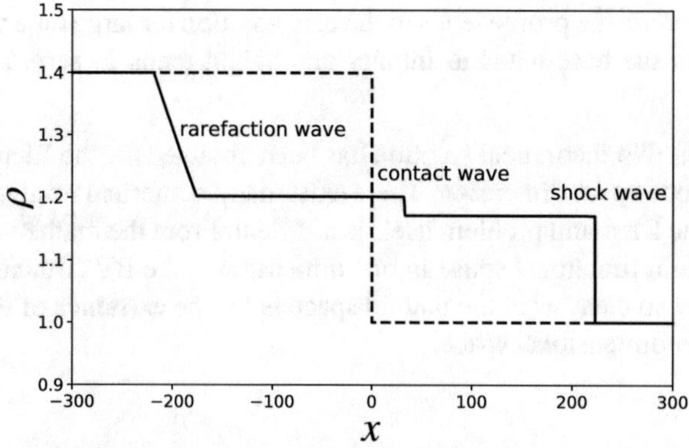

Fig. 8.1 Example of the numerical solution for the system of Euler equations of ideal gas. The mass density profile after some time (solid curve) and the initial profile (dotted line) are shown.

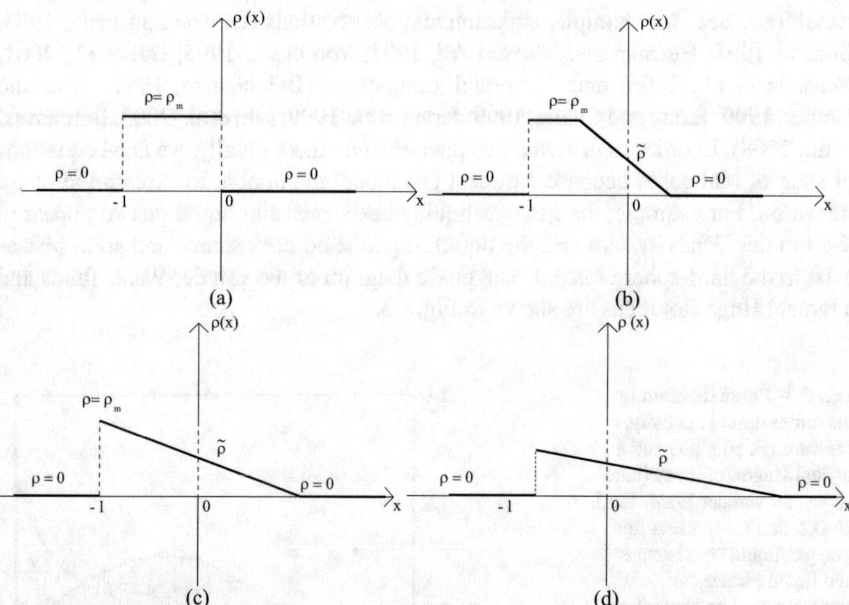

Fig. 8.2 Time evolution of the density in the traffic problem; (**a**) initial state, (**b**) formation of the rarefaction $\tilde{\rho}$, (**c**) the rarefaction reaches the last car at $t = t^*$ and (**d**) the state after t^*, the last car move out and the jump decays (see (Ruggeri, 2015)).

$x = 0$. This implies, in particular, if there is a perturbation of the same equilibrium state, after a long time the perturbation disappears and the equilibrium state is stable.

An elementary example of the Liu theorem is illustrated by the traffic model of Lighthill and Whitham (1955). Cars are initially at rest in front of the traffic light and start to go after the light become green. This is the special case of the Riemann problem with structure with initial data given by a square wave (Fig. 8.2a). According to the Liu theorem, the profile tends to the zero solution for large time with a triangle form in which the base tends to infinity and height tends to zero, preserving the initial area.

Open Problem No theoretical solution has been obtained for the Riemann problem in more than one space dimension. There exist many numerical results, however, the meaning of the Riemann problem itself is not clear. From the mathematical point of view, the natural functional space in one dimension is the BV (Bounded Variation) space. It is not so clear what the natural space is for the existence of the problem in more than one-dimensional space.

8.2.3 Shock Induced Phase Transition

It is well known that in some physical situations shock waves can produce phase transitions. See, for example, experimental observations (Hixson and Fritz, 1989; Boness, 1991; Herman and Elsayed-Ali, 1992; Yoo et al., 1993; Dai et al., 2001; Matsuda et al., 2006) and numerical simulations (Belonoshko, 1997; Woo and Greber, 1999; Jeong and Chang, 1999; Jeong et al., 1999; Luo et al., 2003; Berezovski et al., 2008). In order to describe this phenomenon theoretically, we need equations of state of real gases because the ideal gas model is not able to explain the phase transition. For example, the gas, gas/liquid coexistence and liquid phases appear in the van der Waals system and the liquid, liquid/solid coexistence and solid phases exist in the hard-sphere system. The phase diagram of the van der Waals fluids and a typical Hugoniot locus are shown in Fig. 8.3.

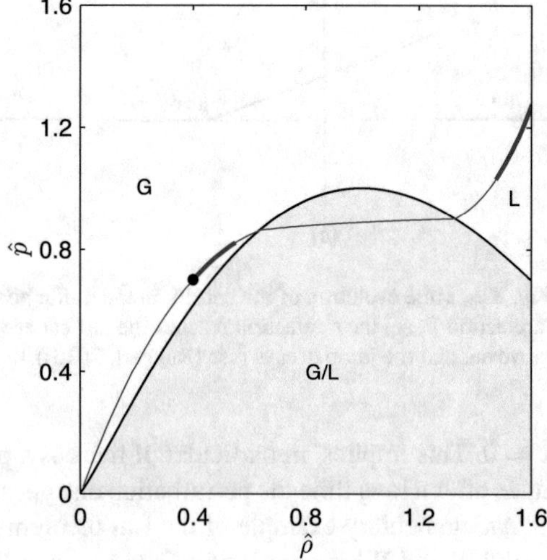

Fig. 8.3 Phase diagram in the dimensionless density, pressure $(\hat{\rho}, \hat{p})$ plane and a typical Hugoniot locus (blue curve) for van der Waals fluid. G, G/L and L represent the gas, gas/liquid coexistence and liquid phases, respectively. The black dot indicates the unperturbed state and the thick and thin parts of the Hugoniot locus, respectively, correspond to the admissible and inadmissible shock solutions (Zhao et al., 2011).

In the system of Euler equations, with the equations of state for real gases, the condition for a genuinely nonlinear system (8.5) is not always satisfied and therefore we need the Liu condition to check the admissibility (stability). By using the Liu condition, the stability of the shock waves inducing the phase transitions was studied and the following results were obtained.

(i) The possibility of the shock induced phase transition from liquid phase to solid phase in the hard-sphere system is given in (Zhao et al., 2008; Taniguchi et al., 2010; Zheng et al., 2010). The important role of the internal degrees of freedom of a molecule was also pointed out.

(ii) A complete classification of the admissibility of the shock induced gas $\to$ liquid phase transition in a van der Waals fluid was presented in (Zhao et al., 2011). Unusual shock waves called the *compressive upper shock* in which the perturbed density decreases across the Hugoniot locus $\mathbf{u}_0$ as the strength of a shock (the perturbed pressure) increases was also found (Taniguchi et al., 2010a).

(iii) The theoretical framework to analyse the shock waves in the three phases (gas, liquid and solid phases) in a unified way was also proposed by considering the system of hard spheres with attractive interactions (Taniguchi et al., 2011; Taniguchi and Sugiyama, 2012). By using this framework, the possibility of gas $\to$ solid phase transition was also discussed.

8.3 Dissipative Systems and Shock Structure

A shock wave is, in reality, not a discontinuous surface but has a structure with a sharp but continuous transition from an unperturbed state to a perturbed state. This effect is due to the presence of viscosity and heat conduction that have a dissipative character in contrast with the conservation laws of the Euler system. Typically both states are two different equilibrium states. In many gases, the thickness of a shock wave is of the order of the mean free path and this fact justifies the use of Euler fluid within the framework governed by the hyperbolic conservation laws and ideal shocks, as the thickness can be negligible in various applications.

Nevertheless, a very interesting problem is how the thickness of a shock wave changes when the unperturbed Mach number M_0 increases. The experiments of Alsmeter (1976) in monatomic gases show the surprising result that the thickness decreases until $M_0 \sim 3.1$ and then increases with the Mach number. This interesting phenomenon is a sort of benchmark for continuum theories because continuous theories have not been able to agree with experiments well until now.

The first tentative solution was with the parabolic theory of Thermodynamics of Irreversible Processes (TIP) which includes the Navier–Stokes–Fourier (NSF) theory. The famous tentative results are the ones of (Becker, 1922; Gilbarg and Paolucci, 1953), see, e.g., (Mott-Smith, 1951; Zoller, 1951; Foch, 1973). At the time of the work of Gilbarg and Paolucci (GB), the available experiments were to

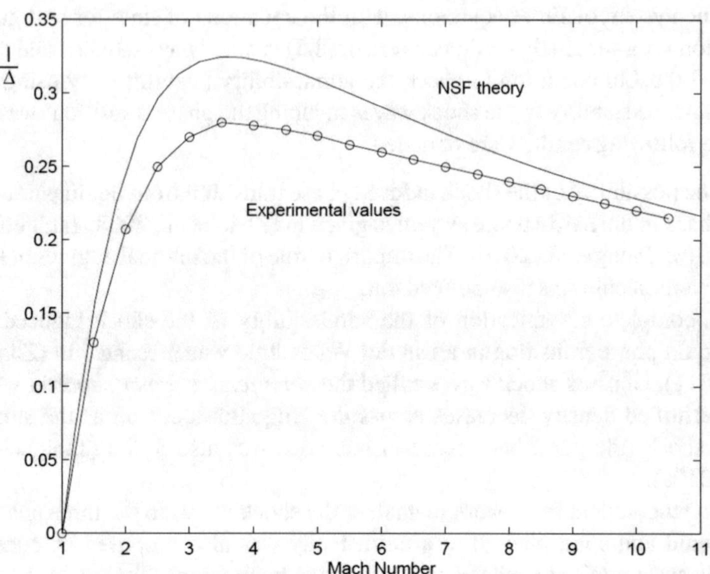

Fig. 8.4 The ratio l/Δ of the mean free path l and the shock thickness Δ as function of Mach number: experimental values for Ar (Alsmeter (1976) and theoretical results with the NSF model (Ruggeri, 1996).

not so high Mach numbers and we are in the branch where the thickness decreases. Ruggeri (1996) evaluates numerically the GB procedure also for high Mach numbers and the results of the NSF theory predict this change of behaviour of the thickness qualitatively, but the results are not satisfactory quantitatively except for very small Mach numbers (Fig. 8.4).

The shock wave structure can be obtained by parabolic and hyperbolic differential systems. In parabolic systems, the shock structure solution is usually continuous while in the hyperbolic system it is not always continuous. In Sect. 8.4, we shall discuss the tentative results based on the RET theory to explain the thickness behaviour within the framework of continuous mechanics with hyperbolic systems. For this aim, in the next subsection, we study the shock wave structure for a hyperbolic system of balance laws (8.1).

8.3.1 Shock Wave Structure

The shock wave structure is a regular solution of (8.1) depending on space and time through a single variable φ:

$$\mathbf{u} \equiv \mathbf{u}(\varphi), \quad \varphi = x_i n_i - st\,, \quad s = \text{const}, \quad \mathbf{n} \equiv (n_i) = \text{const}, \quad (\|\mathbf{n}\| = 1)\,, \tag{8.6}$$

with the boundary conditions at infinity

$$\lim_{\varphi\to\pm\infty} \mathbf{u}(\varphi) = \begin{cases} \mathbf{u}_0 \\ \mathbf{u}_1 \end{cases}, \quad \lim_{\varphi\to\pm\infty} \frac{d\mathbf{u}}{d\varphi} = 0, \tag{8.7}$$

i.e., a plane travelling wave solution connecting the two equilibrium constant states $\mathbf{u}_0$ and $\mathbf{u}_1$ satisfying $\mathbf{f}(\mathbf{u}_0) = \mathbf{f}(\mathbf{u}_1) = 0$. Substituting (8.6) into (8.1), we obtain the following system of ordinary differential equations (ODE):

$$(-s\mathbf{I} + \mathbf{A}_n)\frac{\mathrm{d}\mathbf{u}}{\mathrm{d}\varphi} = \mathbf{f}(\mathbf{u}) \tag{8.8}$$

with boundary conditions (8.7). As was noticed in (Ruggeri, 1993), when s approaches a characteristic eigenvalue λ, the solution may have a breakdown.

Concerning this breakdown of the continuous solution, there exists the following theorem proved by Boillat and Ruggeri (1998):

Theorem 1 (Subshock Formation) *Consider a system of N balance laws of which $M < N$ are conservation laws. Under the assumption that the system satisfies the entropy principle with convex entropy, a C^1 shock wave structure propagating with the velocity s greater than the maximum characteristic speed evaluated in the equilibrium state in front of the shock cannot exist, i.e., smooth solutions may exist only if $s \leq \lambda^{\max}(\mathbf{u}_0)$.*

The proof is based on the verification that smooth shock wave structure solutions must satisfy the entropy inequality and this inequality is violated when $s > \lambda^{\max}(\mathbf{u}_0)$. Therefore, when the shock velocity exceeds the maximum characteristic eigenvalue in the unperturbed state, a subshock arises necessarily.

The existence of the subshock formation with s slower than the maximum characteristic velocity in the unperturbed state will be discussed in Sect. 8.5.

8.3.2 Riemann Problem for Balance Laws

Even in the one-dimensional case, the theoretical solution of the Riemann problem for hyperbolic balance laws with production terms has not been obtained, in contrast to that of the conservation laws. In fact, there does not exist the rarefaction wave solution depending on x/t when there are the production terms. Nevertheless, there still exist many numerical results and a conjecture for large time behaviour.

As we have seen before, we consider a system of N balance laws with $M(< N)$ conservation laws. According to Boillat and Ruggeri (1997), we can define an *equilibrium subsystem* that is formed by the block of conservation laws evaluated at the equilibrium state.

By extending the conjecture of Liu (1993) given for a 2×2 system, Brini and Ruggeri (2006a,b); Mentrelli and Ruggeri (2006) proposed a conjecture on the large-time behaviour of the Riemann problems with and without structure for a system of balance laws. First, we study the Riemann problem for the equilibrium subsystem and obtain the usual result, i.e., a combination of ideal shocks, contact shocks and rarefaction waves.

According to this conjecture, the solutions of both Riemann problems with and without structure converge to the solutions for large time in which the ideal shock waves are replaced by the shock wave structures (with and without subshocks) of the full system while the rarefaction waves and contact shocks are the same as those of the equilibrium subsystem. The numerical tests support this conjecture for several physical systems of balance laws (Brini and Ruggeri, 2004, 2006a,b). In particular, it was numerically proved in a toy model proposed by Mentrelli and Ruggeri (2006).

The conjecture gives us an alternative method to obtain the shock wave structure with or without subshocks (discontinuity). In fact, according to the conjecture, if the initial Riemann data correspond to a shock family $\mathbb{S}$ of the equilibrium subsystem, the solution of the full system, for large time, converges to the corresponding shock wave structure. Therefore we can use the Riemann solver (see, e.g., (Toro, 2009)) or other numerical schemes for the system of balance laws (e.g., (Liotta et al., 2000)), for obtaining the shock structure profile with or without subshocks instead of solving the ODE (8.8). This strategy was adopted in studies of several shock phenomena and is particularly useful in the case of subshock formation (Arima et al., 2013; Conforto, 2017; Artale et al., 2018; Taniguchi and Ruggeri, 2018).

8.4 Rational Extended Thermodynamics and Shock Waves

It is well known that the theory of NSF is valid for phenomena where the assumption of local thermal equilibrium is well satisfied and therefore is no longer valid for describing phenomena beyond local equilibrium, like shock waves and ultrasonic waves (Herzfeld and Litovitz, 1959; Vincenti and Kruger, 1965; Bhatia, 1985; Zel'dovich and Raizer, 2002).

The Rational Extended Thermodynamics (RET) theory of gases (Müller and Ruggeri, 1998; Ruggeri and Sugiyama, 2015), in particular, has been established as a theory to explore highly nonequilibrium phenomena as mentioned above. It utilises the hierarchy structure of the moment equations of the kinetic theory as its justification at mesoscopic level.

In the RET framework, we adopt dissipative fluxes, such as viscous stress and heat flux, as independent variables in addition to the usual hydrodynamic variables, and assume a system of balance equations with local type constitutive equations. The closure is obtained by using universal principle for constitutive equations, in particular, the entropy principle or, at the kinetic level, by using the so-called maximum entropy principle for the truncated moment system associated with the Boltzmann equations.

The first attempt at using the RET theory was done in the case of monatomic gas with the 13 moments (Liu and Müller, 1983). The main result was that the closure based on the RET theory gives the same equations obtained by the kinetic considerations of Grad (1949). The shock structure was studied by Weiss (1995) and in a chapter written by him in the book (Müller and Ruggeri, 1998).

Independently of the number of moments, as the systems of RET theory are hyperbolic, the validity of the RET theory, as far as shock waves are concerned, it is up to the Mach number where the shock velocity s reaches the maximum characteristic velocity evaluated in equilibrium. Beyond this limitation, from the theorem by Boillat & Ruggeri (Theorem 1), a subshock emerges, and the model is no longer valid.

We need to increase the number of truncation N taking more fields into the model. In fact, according to a theorem stated in (Boillat and Ruggeri, 1997), the maximum characteristic velocity does not decrease with the number of moments. Thanks to the lower bound given in (Boillat and Ruggeri, 1997a), it tends to infinity for an infinite number of moments.

We need more moments in order to let the theory be valid for larger Mach numbers. In the limit of infinite Mach number, we substantially deal with the Boltzmann equation itself to predict smooth shock structure. Therefore, in principle, the RET theory is a continuum theory capable of describing the shock structure. In monatomic gas it needs many moments and therefore is not so useful. In this case, good agreement with experiments can be achieved directly by using the Boltzmann equation, and in fact the results presented by Bird (1970) are very good.

Instead, the RET theory gives very important results concerning shock structure even with few independent fields in the case of polyatomic gas as we will see in the next section.

8.4.1 Extended Thermodynamics for Polyatomic Gases and Shock Structure

The RET theory of rarefied polyatomic gases was developed in (Arima et al., 2011, 2013; Ruggeri and Sugiyama, 2015). This theory adopts the balance equations with binary hierarchy structure, where the dynamic pressure (nonequilibrium pressure) is properly taken into account. Note that the dynamic pressure vanishes in rarefied monatomic gases. The number of independent fields is now 14 (ET_{14}): the mass density ρ, velocity $\mathbf{v}$, temperature T, dynamic pressure Π, shear stress $\boldsymbol{\sigma}$, and heat flux $\mathbf{q}$. The first hierarchy consists of the balance equations for mass density, momentum density and momentum flux, and the second hierarchy consists of the balance equations for energy density and energy flux.

The binary hierarchy was justified and also derived from molecular RET theory (Pavić et al., 2013) by using the generalised kinetic theory where the distribution function depends also on an extra variable that represents the molecular internal degrees of freedom such as rotation and vibration (Borgnakke and Larsen, 1975; Bourgat et al., 1994).

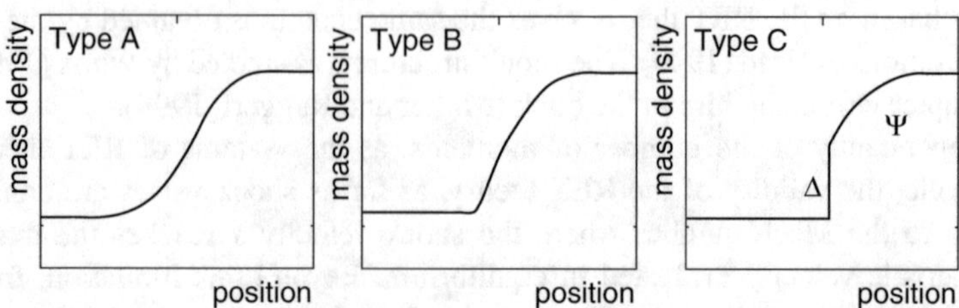

Fig. 8.5 Typical mass density profiles of three types of the shock wave structure in a rarefied polyatomic gas (Taniguchi et al., 2014): nearly symmetric profile (Type A), asymmetric profile (Type B) and the profile composed of thin and thick layers (Type C).

The theory of rarefied polyatomic gases with any number of fields was established in (Arima et al., 2014). The convergence to the singular limit of a monatomic gas when the molecular degrees of freedom $D \to 3$ was proved for ET_{14} in (Arima et al., 2013) and for the case with any number of fields in (Arima et al., 2016). A comprehensive literature for this subject can be found in Ruggeri and Sugiyama (2015).

In polyatomic gases, the shock wave structure may have quite different features from that in rarefied monatomic gases due to the existence of the slowly relaxing internal modes (rotational or vibrational modes) (Vincenti and Kruger, 1965; Zel'dovich and Raizer, 2002). A typical example is a rarefied carbon dioxide gas in which the vibrational modes are partially excited in some temperature range, including room temperature. The important features of the shock wave structure in such polyatomic gases are summarised as follows. Firstly, the thickness of a shock wave becomes several orders larger than the mean free path. Secondly, there exist three types of structures of the shock wave (Fig. 8.5): nearly symmetric profile (Type A), asymmetric profile (Type B) and the profile composed of thin and thick layers (Type C) (Smiley et al., 1952; Griffith and Bleakney, 1954; Smiley and Winkler, 1954; Griffith et al., 1956; Griffith and Kenny, 1957; Johanessen et al., 1962). As the unperturbed Mach number increases from one, the profile changes from Type A to Type B and to Type C.

Here, the successful results of the shock wave structure by the RET theory for polyatomic gases are summarised. Let us consider that a plane shock wave propagates along the x-axis. In this case the system of field equations is (Arima et al., 2011; Ruggeri and Sugiyama, 2015):

$$\begin{aligned}
&\frac{\partial \rho}{\partial t} + \frac{\partial}{\partial x}(\rho v) = 0, \\
&\frac{\partial \rho v}{\partial t} + \frac{\partial}{\partial x}\left(p + \Pi - \sigma + \rho v^2\right) = 0, \\
&\frac{\partial}{\partial t}(2\rho\varepsilon + \rho v^2) + \frac{\partial}{\partial x}\left[2\rho\varepsilon v + 2(p + \Pi - \sigma)v + \rho v^3 + 2q\right] = 0, \\
&\frac{\partial}{\partial t}\left[3(p + \Pi) + \rho v^2\right] + \frac{\partial}{\partial x}\left[(5p + 5\Pi - 2\sigma)v + \rho v^3 + \frac{5}{1 + \hat{c}_v} q\right] = -\frac{3\Pi}{\tau_\Pi},
\end{aligned}$$

$$
\begin{aligned}
&\frac{\partial}{\partial t}(p+\Pi-\sigma+\rho v^2)+\frac{\partial}{\partial x}\left\{3(p+\Pi-\sigma)v+\rho v^3+\frac{3}{1+\hat{c}_v}q\right\}=\frac{\sigma}{\tau_\sigma}-\frac{\Pi}{\tau_\Pi}\,,\\
&\frac{\partial}{\partial t}\left[2\rho\varepsilon v+2(p+\Pi-\sigma)v+\rho v^3+2q\right]+\\
&+\frac{\partial}{\partial x}\left[2\rho\varepsilon v^2+5(p+\Pi-\sigma)v^2+\rho v^4+2\left(\varepsilon+\frac{k_B}{m}T\right)p+\right.\\
&\qquad\left.+2\left(\varepsilon+2\frac{k_B}{m}T\right)(\Pi-\sigma)+\frac{10+4\hat{c}_v}{1+\hat{c}_v}qv\right]=-2\left[\frac{q}{\tau_q}+\left(\frac{\Pi}{\tau_\Pi}-\frac{\sigma}{\tau_\sigma}\right)v\right],
\end{aligned}
\tag{8.9}
$$

where ε, k_B, m, $\hat{c}_v$, τ_Π, τ_σ and τ_q are, respectively, the specific internal energy, the Boltzmann constant, the mass of a molecule, the dimensionless specific heat defined by $\hat{c}_v = (m/k_B)c_v$ with c_v being the specific heat, the relaxation times for the dynamic pressure, the shear stress and the heat flux. Here the vectorial and tensorial fields are assumed to be expressed as

$$
\mathbf{v}\equiv\begin{pmatrix}v\\0\\0\end{pmatrix},\quad \boldsymbol{\sigma}\equiv\begin{pmatrix}\sigma & 0 & 0\\ 0 & -\frac{1}{2}\sigma & 0\\ 0 & 0 & -\frac{1}{2}\sigma\end{pmatrix},\quad \mathbf{q}\equiv\begin{pmatrix}q\\0\\0\end{pmatrix}. \tag{8.10}
$$

The specific internal energy ε and the (equilibrium) pressure p are expressed in terms of the mass density and the temperature by adopting the equations of state for a rarefied polyatomic gas in the cases of both polytropic and nonpolytropic gases. The system (8.9) includes the NSF theory as a special case. In fact, by using the Maxwellian iteration (Ikenberry and Truesdell, 1956), the system (8.9) can be reduced to the system of the NSF theory with the following relationship between the relaxation times and the shear viscosity μ, the bulk viscosity ν, and the heat conductivity κ (Arima et al., 2011; Ruggeri and Sugiyama, 2015):

$$
\mu = p\tau_\sigma\,,\qquad \nu=\left(\frac{2}{3}-\frac{1}{\hat{c}_v}\right)p\tau_\Pi\,,\qquad \kappa=\left(1+\hat{c}_v\right)\frac{k_B}{m}p\tau_q\,. \tag{8.11}
$$

In contrast to the previous theories based on the Navier–Stokes–Fourier equations (proposed by Gilbarg and Paolucci (1953)) and the Bethe and Teller theory (Bethe and Teller, 1953) on the basis of the Meixner theory (Meixner, 1943, 1952), ET$_{14}$ can explain the shock wave structure of Types A, B and C in a unified way (Taniguchi et al., 2014). Figure 8.6 shows the theoretical mass density profile of Type C predicted by ET$_{14}$ for a moderately strong shock wave at $M_0 = 1.47$ with experimental data and prediction by the NSF theory. We see that ET$_{14}$ agrees with experimental results quantitatively, in contrast to the NSF theory.

Through the analysis of the shock wave structure, for gas with the above described features, the relaxation time for the dynamic pressure becomes much larger than the other relaxation times and the dynamic pressure plays an essential role. Therefore

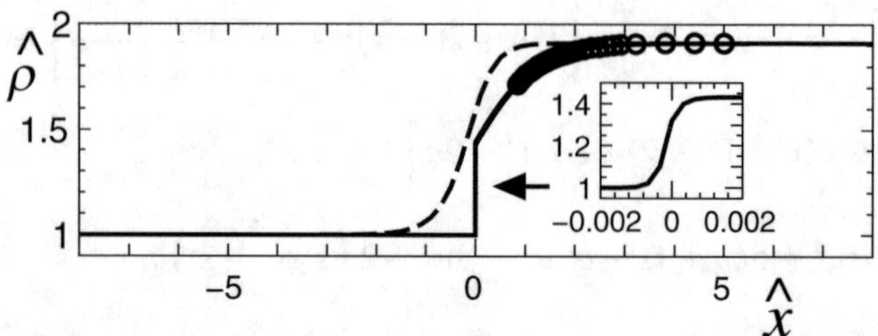

Fig. 8.6 Mass density profile of shock wave structure of Type C predicted by the RET theory (solid curve) for $M_0 = 1.47$ (Taniguchi et al., 2014). Corresponding experimental data in the thick layer (Johanessen et al., 1962) (circles) and the prediction by the NSF theory (dashed curve) are also shown.

the analysis was also undertaken on the basis of a simplified RET theory with only 6 independent fields (ET_6); ρ, $\mathbf{v}$, T and Π. In other words, in ET_6, we neglect the contributions due to shear stress and heat flux. The relevant system of equations is (Ruggeri and Sugiyama, 2015; Arima et al., 2012):

$$\begin{aligned}
&\frac{\partial \rho}{\partial t} + \frac{\partial}{\partial x}(\rho v) = 0, \\
&\frac{\partial \rho v}{\partial t} + \frac{\partial}{\partial x}\left[(p+\Pi) + \rho v^2\right] = 0, \\
&\frac{\partial}{\partial t}(2\rho\varepsilon + \rho v^2) + \frac{\partial}{\partial x}\left\{\left[2(p+\Pi) + 2\rho\varepsilon + \rho v^2\right] v\right\} = 0, \\
&\frac{\partial}{\partial t}\left[3(p+\Pi) - 2\rho\varepsilon\right] + \frac{\partial}{\partial x}\left\{\left[3(p+\Pi) - 2\rho\varepsilon\right] v\right\} = -\frac{3\Pi}{\tau_\Pi}.
\end{aligned} \tag{8.12}$$

The predictions by ET_{14} and ET_6 are almost the same except for a thin layer appearing in the profile of Type C, and the fine structure predicted by ET_{14} in the thin layer is replaced by a subshock within the limited resolution of ET_6 (Taniguchi et al., 2014). Therefore, we may use ET_6 in the case that the fine structure is not important.

Moreover, it was shown that the nonlinear version of ET_6 can be derived only from general principles without adopting the near equilibrium assumption (Arima et al., 2015). Therefore the nonlinear ET_6 is expected to be valid regardless of how far it is from local equilibrium, and therefore it is able to describe the shock wave structure for large Mach numbers (Taniguchi et al., 2016).

The successful results obtained from RET theories have also attracted the researchers working in the field of the kinetic theory, to confirm the predictions using the RET approach. Recently, if the bulk viscosity is large (the relaxation time for the dynamic pressure is large), the modified version of the ellipsoidal statistical (ES) model for polyatomic gases was shown to be able to explain the above discussed features and supports predictions by RET theories quantitatively (Kosuge et al., 2016; Kosuge and Aoki, 2018).

8.5 Subshock Formation with $s < \lambda^{\max}(\mathbf{u}_0)$

As was shown in Sect. 8.4, according to the numerical analysis of the case of rarefied monatomic gases, the RET theory predicts the subshock formation only when the shock velocity is greater than the maximum characteristic velocity in the unperturbed state. Based on this fact, it was conjectured that the subshock arises only with $s > \lambda^{\max}(\mathbf{u}_0)$. Furthermore, also in the case of polyatomic gases based on ET_{14}, we have the similar situation and this reinforces the conjecture. The singularity becomes regular except for the maximum characteristic velocity and ET_{14} predicts the subshock formation only when $s > \lambda^{\max}(\mathbf{u}_0)$.

In contrast to these results, the subshock formation with $s < \lambda^{\max}(\mathbf{u}_0)$ and the multiple subshocks were numerically observed in a binary mixtures of the Eulerian gas and the Grad systems of monatomic gases (Bisi et al., 2014; Conforto, 2017; Artale et al., 2018). However, the systems of binary mixtures have special properties such that the form of the balance equations for each component is the same as that of a single fluid and the coupling effect between the components comes only through the production terms.

Starting from a "toy" model proposed in (Mentrelli and Ruggeri, 2006), in order to understand these problematic results, the following 2×2 system (Taniguchi and Ruggeri, 2018) was considered:

$$\begin{aligned} \frac{\partial u}{\partial t} + \frac{\partial}{\partial x}\left(\frac{u^3}{3}\right) &= -\frac{u-v}{\tau}, \\ \frac{\partial v}{\partial t} + \frac{\partial}{\partial x}\left(\frac{v^5}{5}\right) &= -\frac{v-u}{\tau}, \end{aligned} \tag{8.13}$$

that can be rewritten as a combination of a conservation law and a balance law:

$$\begin{aligned} \frac{\partial}{\partial t}(u+v) + \frac{\partial}{\partial x}\left(\frac{u^3}{3} + \frac{v^5}{5}\right) &= 0, \\ \frac{\partial u}{\partial t} + \frac{\partial}{\partial x}\left(\frac{u^3}{3}\right) &= -\frac{u-v}{\tau}. \end{aligned} \tag{8.14}$$

Figure 8.7 shows the typical shock structures predicted by this model. As the velocity of the shock wave increases, the shock structure changes from continuous to the structure with a subshock ($s < \lambda^{\max}(\mathbf{u}_0)$) and then also to the structure with multiple subshocks ($s > \lambda^{\max}(\mathbf{u}_0)$). This is a clear counter example of the conjecture. It is noticeable that this simple toy model satisfies all important requirements of the system of the RET theory, namely, the entropy principle, convexity of the entropy, and dissipative character satisfying the Shizuta–Kawashima condition (Shizuta and Kawashima, 1985; Kawashima, 1987).

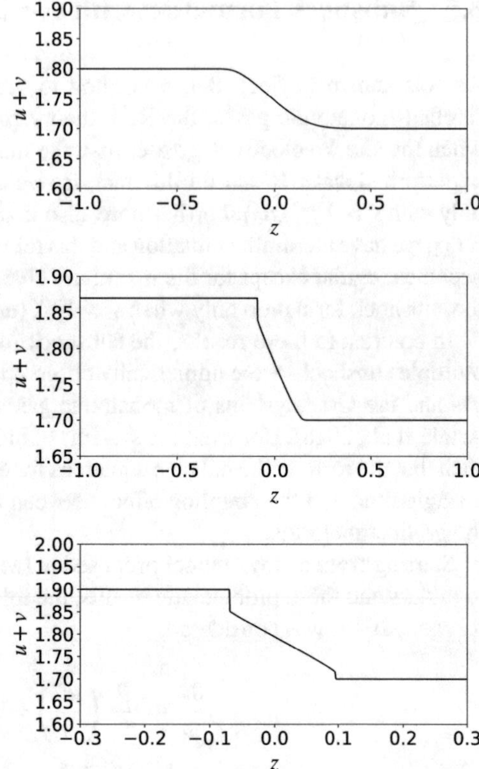

Fig. 8.7 Continuous shock structure with $s = 0.677$ (top), subshock formation with $s = 0.717$ ($< \lambda^{\max}(\mathbf{u}_0)$) (middle), and multiple subshock with $s = 0.735$ ($> \lambda^{\max}(\mathbf{u}_0)$) (bottom) (Taniguchi and Ruggeri, 2018). In this case, $\lambda^{\max}(\mathbf{u}_0) = 0.723$. $z \equiv x - st$.

Therefore, there is still an open and important problem to understand which property is hidden in the RET theory for which the subshock arises only for the shock velocity greater than the maximum characteristic velocity.

Acknowledgements This work was partially supported by National Group of Mathematical Physics GNFM-INdAM (TR) and by JSPS KAKENHI Grant Number JP16K17555 (ST).

References

Alsmeyer, H.: Density profiles in argon and nitrogen shock waves measured by the absorption of an electron beam. J. Fluid Mech. **74**, 497–513 (1976). https://doi.org/10.1017/s0022112076001912

Arima, T., Taniguchi, S., Ruggeri, T., Sugiyama, M.: Extended thermodynamics of dense gases. Continuum Mech. Thermodyn. **24**, 271–292 (2011). https://doi.org/10.1007/s00161-011-0213-x

Arima, T., Taniguchi, S., Ruggeri, T., Sugiyama, M.: Extended thermodynamics of real gases with dynamic pressure: An extension of Meixner's theory. Phys. Lett. A **376**, 2799–2803 (2012). https://doi.org/10.1016/j.physleta.2012.08.030

Arima, T., Taniguchi, S., Ruggeri, T., Sugiyama, M.: Monatomic rarefied gas as a singular limit of polyatomic gas in extended thermodynamics. Phys. Lett. A **377**, 2136–2140 (2013). https://doi.org/10.1016/j.physleta.2013.06.035

Arima, T., Mentrelli, A., Ruggeri, T.: Molecular extended thermodynamics of rarefied polyatomic gases and wave velocities for increasing number of moments. Ann. Phys. **345**, 111–140 (2014). https://doi.org/10.1016/j.aop.2014.03.011

Arima, T., Ruggeri, T., Sugiyama, M., Taniguchi, S.: Nonlinear extended thermodynamics of real gases with 6 fields, Int. J. Non-Linear Mech. **72**, 6–15 (2015). https://doi.org/10.1016/j.ijnonlinmec.2015.02.005

Arima, T., Ruggeri, T., Sugiyama, M., Taniguchi, S.: Monatomic gas as a singular limit of polyatomic gas in molecular extended thermodynamics with many moments. Ann. Phys. **372**, 83–109 (2016). (2016). https://doi.org/10.1016/j.aop.2016.04.015

Artale, V., Conforto, F., Martalò, G., Ricciardello, A.: Shock structure and multiple subshocks in Grad 10-moment binary mixtures of monatomic gases. Ricerche di Matematica (2018). https://doi.org/10.1007/s11587-018-0421-9(2018)

Belonoshko, A. B.: Atomistic simulation of shock wave-induced melting in argon. Science **275**, 955–957 (1997). https://doi.org/10.1126/science.275.5302.955

Becker, R.: Stoßwelle und Detonation. Z. f. Physik **8**, 321 (1922). https://doi.org/10.1007/bf01329605

Berezovski, A., Engelbrecht, J., Maugin, G.: Numerical Simulation of Waves and Fronts in Inhomogeneous Solids. World Scientific, Singapore (2008). https://doi.org/10.1142/6931

Bethe, H.A., Teller, E.: Deviations from Thermal Equilibrium in Shock Waves. (No. NP-4898; BRL-X-117). Engineering Research Institute, University of Michigan (1953). https://doi.org/10.2172/4420349

Bhatia, A.B.: Ultrasonic Absorption: An Introduction to the Theory of Sound Absorption and Dispersion in Gases, Liquids, and Solids. Dover, New York (1985)

Bird, G.A.: Aspects of the structure of strong shock waves. Phys. Fluids **13**, 1172 (1970). https://doi.org/10.1063/1.1693047

Bisi, M., Martalò, G., Spiga, G.: Shock wave structure of multi-temperature Euler equations from kinetic theory for a binary mixtures, Acta Appl. Math. **132**, 95–105 (2014). https://doi.org/10.1007/s10440-014-9939-3

Boillat, G., Ruggeri, T.: Hyperbolic principal subsystems: Entropy convexity and subcharacteristic conditions, Arch. Rational Mech. Anal. **137**, 305–320 (1997). https://doi.org/10.1007/s002050050030

Boillat, G., Ruggeri, T.: Moment equations in the kinetic theory of gases and wave velocities. Continuum Mech. Thermodyn. **9**(4), 205–212 (1997). https://doi.org/10.1007/s001610050066

Boillat, G., Ruggeri, T.: On the Shock Structure Problem for Hyperbolic System of Balance Laws and Convex Entropy. Continuum Mech. Thermodyn. **10**, 285–292 (1998). https://doi.org/10.1007/s001610050094

Boness, D.A.: Shock Wave Experiments and Electronic Band-Structure Calculations of Materials at High Temperature and Pressure, Ph.D. Thesis, University of Washington (1991)

Borgnakke, C., Larsen, P.S.: Statistical collision model for Monte Carlo simulation of polyatomic gas mixture, J. Comput. Phys. **18**, 405–420 (1975). https://doi.org/10.1016/0021-9991(75)90094-7

Bourgat, J.-F., Desvillettes, L., Le Tallec, P., Perthame, B.: Microreversible collisions for polyatomic gases and Boltzmann's theorem, Eur. J. Mech. B/Fluids **13**, 237–254 (1994)

Brini, F., Ruggeri, T.: The Riemann problem for a binary non-reacting mixture of Euler fluids. In: Monaco, R., Pennisi, S., Rionero, S., Ruggeri, T. (eds.): Proceedings WASCOM 2003, pp. 102–108. World Scientific, Singapore (2004). https://doi.org/10.1142/9789812702937_0013

Brini, F., Ruggeri, T.: On the Riemann problem in extended thermodynamics, In: Proceedings of the 10th International Conference on Hyperbolic Problems (HYP2004), Osaka, 13–17 Sept. 2004, vol. I, pp. 319–326. Yokohama Publisher Inc., Yokohama (2006a)

Brini, F., Ruggeri, T.: On the Riemann problem with structure in extended thermodynamics. Suppl. Rend. Circ. Mat. Palermo II **78**, 31–43 (2006b)

Chaussy, C.G., Eisenberger F., Forssmann, B.: Extracorporeal shockwave lithotripsy (ESWL): a chronology. J. Endourology, **21**, 1249–1253 (2007). https://doi.org/10.1089/end.2007.9880

Conforto, F., Mentrelli, A., Ruggeri, T.: Shock structure and multiple sub-shocks in binary mixtures of Eulerian fluids. Ricerche di Matematica **66**, 221–231 (2017). https://doi.org/10.1007/s11587-016-0299-3

Dafermos, C.: Conservation Laws in Continuum Physics, 2nd ed., Springer Verlag, Berlin (2005). https://doi.org/10.1007/3-540-29089-3

Dai, C. , Jin, X., Zhou, X., Liu, J., Hu, J.: Sound velocity variations and melting of vanadium under shock compression. J. Phys. D Appl. Phys. **34**, 3064–3070 (2001). https://doi.org/10.1088/0022-3727/34/20/310

Foch, J.D.: On higher order hydrodynamic theories of shock structure, Acta Physica Austriaca, suppl. **10**, 123–140 (1973). https://doi.org/10.1007/978-3-7091-8336-6_7

Gilbarg, D., Paolucci, D.: The structure of shock waves in the continuum theory of fluids. J. Rat. Mech. Anal. **2**, 617–642 (1953). https://doi.org/10.1512/iumj.1953.2.52031

Grad, H.: On the kinetic theory of rarefied gases. Comm. Pure Appl. Math. **2**(4) 331–407 (1949). https://doi.org/10.1002/cpa.3160020403

Griffith, W.C., Bleakney, W.: Shock waves in gases. Am. J. Phys. **22**, 597–612 (1954). https://doi.org/10.1119/1.1933855

Griffith, W.C., Kenny, A.: On fully-dispersed shock waves in carbon dioxide, J. Fluid Mech. **3**, 286–288 (1957). https://doi.org/10.1017/s0022112057000658

Griffith, W., Brickl, D., Blackman, V.: Structure of shock waves in polyatomic gases. Phys. Rev. **102**, 1209–1216 (1956). https://doi.org/10.1103/physrev.102.1209

Herman, J.W., Elsayed-Ali, H.E.: Superheating of Pb (111). Phys. Rev. Lett. **69**, 1228–1231 (1992). https://doi.org/10.1103/physrevlett.69.1228

Herzfeld, K.F., Litovitz, T.A.: Absorption and Dispersion of Ultrasonic Waves. Academic Press, New York (1959)

Hixson R.S., Fritz, J.N.: Acoustic velocities and phase transitions in molybdenum under strong shock compression. Phys. Rev. Lett. **62**, 637–640 (1989). https://doi.org/10.1103/physrevlett.62.637

Ikenberry, E., Truesdell, C.: On the pressure and the flux of energy in a gas according to Maxwell's kinetic theory. J. Rat. Mech. Anal. **5**, 1–54 (1956). https://doi.org/10.1512/iumj.1956.5.55001

Jeong, J. W., Lee I., and Chang, K. J.: Molecular-dynamics study of melting on the shock Hugoniot of Al. Phys. Rev. B **59**, 329–333 (1999). https://doi.org/10.1103/physrevb.59.329

Jeong, J.W., Chang, K.J.: Molecular-dynamics simulations for the shock Hugoniot meltings of Cu, Pd and Pt. J. Phys. Condens. Matter **11**, 3799–3806 (1999). https://doi.org/10.1088/0953-8984/11/19/302

Johannesen, N. H., Zienkiewicz, H.K., Blythe, P.A., Gerrard, J.H.: Experimental and theoretical analysis of vibrational relaxation regions in carbon dioxide. J. Fluid Mech. **13**, 213–224 (1962). https://doi.org/10.1017/s0022112062000634

Kawashima, S.: Large-time behaviour of solutions to hyperbolic-parabolic systems of conservation laws and applications. Proc. Roy. Soc. Edinburgh **106A**, 169–194 (1987). https://doi.org/10.1017/s0308210500018308

Kosuge, S., Aoki, K., Goto, T.: Shock wave structure in polyatomic gases: Numerical analysis using a model Boltzmann equation. AIP Conference Proceedings, vol. 1789, 180004 (2016). https://doi.org/10.1063/1.4967673

Kosuge, S., Aoki, K.: Shock-wave structure for a polyatomic gas with large bulk viscosity. Phys. Rev. Fluids **3**, 023401 (2018). https://doi.org/10.1103/physrevfluids.3.023401

Krehl, P.O.K.: History of Shock Waves, Explosions and Impact: A Chronological and Biographical Reference. Springer, Berlin (2009). https://doi.org/10.1007/978-3-540-30421-0

Lax, P.D.: Hyperbolic systems of conservation laws and the mathematical theory of shock waves. CBMS-NSF, Regional Conference Series in Applied Mathematics **11**, SIAM (1973). https://doi.org/10.1137/1.9781611970562

Lighthill, M.J., Whitham, G.B.: On kinematic waves: II. A theory of traffic on long crowded roads. Proc. Roy. Soc. A **229**, 317–345 (1955). https://doi.org/10.1098/rspa.1955.0089

Liotta, S.F., Romano, V., Russo, G.: Central scheme for balance laws of relaxation type. SIAM J. Numer. Anal. **38**, 1337–1356 (2000). https://doi.org/10.1137/s0036142999363061

Liu, T.-P.: The entropy condition and the admissibility of shocks. J. Math. Anal. Appl. **53**, 78–88 (1976). https://doi.org/10.1016/0022-247x(76)90146-3

Liu, T.-P.: Linear and nonlinear large-time behavior of solutions of general systems of hyperbolic conservation laws. Comm. Pure Appl. Math. **30**, 767–796 (1977). https://doi.org/10.1002/cpa.3160300605

Liu, T.-P.: Large-time behavior of solutions of initial and initial-boundary value problems of a general system of hyperbolic conservation laws. Commun. Math. Phys. **55**, 163–177 (1977). https://doi.org/10.1007/bf01626518

Liu, T.-P.: Admissible solutions of hyperbolic conservation laws. Mem. Am. Math. Soc. **240**, (1981). https://doi.org/10.1090/memo/0240

Liu, T.-P.: Nonlinear hyperbolic-dissipative partial differential equations. In: Ruggeri, T. (ed.) Recent Mathematical Methods in Nonlinear Wave Propagation. Lecture Notes in Mathematics, vol. 1640, pp. 103–136. Springer, Berlin (1996). https://doi.org/10.1007/bfb0093708

Liu, I-S., Müller, I.: Extended thermodynamics of classical and degenerate ideal gases. Arch. Rat. Mech. Anal. **83**, 285–332 (1983). https://doi.org/10.1007/bf00963838

Liu T.-P., Ruggeri, T: Entropy production and admissibility of shocks. Acta Mathematicae Applicatae Sinica, English Series **1**, 1–12 (2003). https://doi.org/10.1007/s10255-003-0074-6

Luo, S.N., Ahrens, T.J., Çağin, T., Strachan, A., Goddard III, W.A., Swift, D.C.: Maximum superheating and undercooling: Systematics, molecular dynamics simulations, and dynamic experiments. Phys. Rev. B **68**, 134206 (2003). https://doi.org/10.1103/physrevb.68.134206

Matsuda, A., Kondo K., Nakamura, K.G.: Nanosecond rapid freezing of liquid benzene under shock compression studied by time-resolved coherent anti-Stokes Raman spectroscopy. J. Chem. Phys. **124**, 054501 (2006). https://doi.org/10.1063/1.2165196

Meixner, J.: Absorption und Dispersion des Schalles in Gasen mit chemisch reagierenden und anregbaren Komponenten. I. Teil. Ann. Physik **43**, 470–487 (1943). https://doi.org/10.1002/andp.19434350608

Meixner, J.: Allgemeine Theorie der Schallabsorption in Gasen und Flussigkeiten unter Berucksichtigung der Transporterscheinungen. Acoustica **2**, 101–109 (1952)

Mentrelli, A., Ruggeri, T.: Asymptotic behavior of Riemann and Riemann with structure problems for a 2×2 hyperbolic dissipative system, Suppl. Rend. Circ. Mat. Palermo II **78**, 201–226 (2006)

Morris, D.G.: An investigation of the shock-induced transformation of graphite to diamond. J. Appl. Phys. **51**, 2059–2065 (1980). https://doi.org/10.1063/1.327873

Mott-Smith, H.W.: The solution of the Boltzmann equation for a shock wave, Phys. Rev. **82**, 885–892 (1951). https://doi.org/10.1103/physrev.82.885

Müller, I.; Ruggeri, T.: Rational Extended Thermodynamics. 2nd edn. Springer, New York (1998). https://doi.org/10.1007/978-1-4612-2210-1

Pavić, M., Ruggeri, T., Simić, S.: Maximum entropy principle for rarefied polyatomic gases. Physica A **392**, 1302–1317 (2013). https://doi.org/10.1016/j.physa.2012.12.006

Ruggeri, T.: Breakdown of shock-wave-structure solutions. Phys. Rev. E **47**, 4135–4140 (1993). https://doi.org/10.1103/physreve.47.4135

Ruggeri, T.: On the shock structure problem in non-equilibrium thermodynamics of gases. Transport Theory Stat. Phys. **25**, 567–574 (1996). https://doi.org/10.1080/00411459608220722

Ruggeri, T.: Principio di Entropia, Sistemi Simmetrici Iperbolici e Termodinamica Estesa. In: Bonfiglioli, A, Fioresi, R., Parmeggiani, A. (eds.) Topic in Mathematics, Bologna. Qauderni dell'Unione Matematica Italiana **55**, 137–150. Unione Matematica Italiana (Bologna) (2015)

Ruggeri, T., Sugiyama, M.: Rational Extended Thermodynamics beyond the Monatomic Gas. Springer, Heidelberg (2015). https://doi.org/10.1007/978-3-319-13341-6

Shizuta, Y., Kawashima, S.: Systems of equations of hyperbolic-parabolic type with applications to the discrete Boltzmann equation, Hokkaido Math. J. **14**, 249–275 (1985). https://doi.org/10.14492/hokmj/1381757663

Smiley, E.F., Winkler, E.H., Slawsky, Z.I.: Measurement of the vibrational relaxation effect in CO_2 by means of shock tube interferograms. J. Chem. Phys. **20**, 923–924 (1952). https://doi.org/10.1063/1.1700608

Smiley, E.F., Winkler, E.H.: Shock-tube measurements of vibrational relaxation. J. Chem. Phys. **22**, 2018–2022 (1954). https://doi.org/10.1063/1.1739984

Taniguchi, S., Ruggeri, T.: On the sub-shock formation in extended thermodynamics. Int. J. Non-Linear Mech. **99**, 69–78 (2018). https://doi.org/10.1016/j.ijnonlinmec.2017.10.024

Taniguchi, S., Sugiyama, M.: Shock-induced phase transitions in systems of hard spheres with attractive interactions. Acta Appl. Math. **122**, 473–483 (2012). https://doi.org/10.1007/s10440-012-9757-4

Taniguchi, S., Mentrelli, A., Zhao, N., Ruggeri, T., Sugiyama, M.: Shock-induced phase transition in systems of hard spheres with internal degrees of freedom. Phys. Rev. E **81**, 066307 (2010). https://doi.org/10.1103/PhysRevE.82.036324

Taniguchi, S., Mentrelli, A., Ruggeri, T., Sugiyama, M., Zhao, N.: Prediction and simulation of compressive shocks with lower perturbed density for increasing shock strength in real gases. Phys. Rev. E **82**, 036324 (2010). https://doi.org/10.1103/physreve.81.066307

Taniguchi, S., Zhao, N., Sugiyama, M.: Shock-induced phase transitions from gas phase to solid phase. J. Phys. Soc. Jpn. **80**, 083401 (2011). https://doi.org/10.1143/jpsj.80.083401

Taniguchi, S., Arima, T., Ruggeri, T., Sugiyama, M.: Effect of the dynamic pressure on the shock wave structure in a rarefied polyatomic gas. Phys. Fluids **26**, 016103 (2014). https://doi.org/10.1063/1.4861368

Taniguchi, S., Arima, T., Ruggeri, T., Sugiyama, M.: Thermodynamic theory of the shock wave structure in a rarefied polyatomic gas: Beyond the Bethe–Teller theory. Phys. Rev. E **89**, 013025 (2014). https://doi.org/10.1103/physreve.89.013025

Taniguchi, S., Arima, T., Ruggeri, T., Sugiyama, M.: Overshoot of the non-equilibrium temperature in the shock wave structure of a rarefied polyatomic gas subject to the dynamic pressure. Int. J. Non-Linear Mech. **79**, 66–75 (2016). https://doi.org/10.1016/j.ijnonlinmec.2015.11.003

Toro, E.: Riemann Solvers and Numerical Methods for Fluid Dynamics. Springer, Berlin (2009). https://doi.org/10.1007/b79761

Vincenti, W.G.; Kruger, C.H.Jr. Introduction to Physical Gas Dynamics. John Wiley and Sons, New York (1965)

Vink, J.: Supernova remnants: the X-ray perspective. Astron Astrophys Rev. **20**, 49 (2012). https://doi.org/10.1007/s00159-011-0049-1

Weiss, W.: Continuous shock structure in extended thermodynamics. Phys. Rev. E **52**, R5760–R5763 (1995). https://doi.org/10.1103/physreve.52.r5760

Woo M., Greber, I.: Molecular dynamics simulation of piston-driven shock wave in hard sphere gas. AIAA J. **37**, 215–221 (1999). https://doi.org/10.2514/2.692

Yoo, C.S., Holmes, N.C., Ross, M., Webb, D.J., Pike, C.: Shock temperatures and melting of iron at Earth core conditions. Phys. Rev. Lett. **70**, 3931–3934 (1993). https://doi.org/10.1103/physrevlett.70.3931

Zel'dovich, Ya.B., Raizer, Yu. P.: Physics of Shock Waves and High-Temperature Hydrodynamic Phenomena. Dover Publications, Mineola, New York (2002)

Zhao, N., Mentrelli, A., Ruggeri, T., Sugiyama, M.: Admissible shock waves and shock-induced phase transitions in a van der Waals fluids. Phys. Fluids **23**, 086101 (2011). https://doi.org/10.1063/1.3622772

Zhao, N., Sugiyama, M., Ruggeri, T.: Phase transition induced by a shock wave in hard-sphere and hard-disk systems. J. Chem. Phys. **129**, 054506 (2008). https://doi.org/10.1063/1.2936990

Zheng, Y., Zhao, N., Ruggeri, T., Sugiyama, M., Taniguchi, S.: Non-polytropic effect on shock-induced phase transitions in a hard-sphere system. Phys. Lett. A **374,** 3315–3318 (2010). https://doi.org/10.1016/j.physleta.2010.06.016

Zoller, K.: Zur Struktur des Verdichtungsstoßes. Zeitschr. f. Physik **130**, 1–38 (1951). https://doi.org/10.1007/bf01329729

Chapter 9
Linear and Nonlinear Theories for Thermoacoustic Waves in a Gas Filled Tube Subject to a Temperature Gradient

Nobumasa Sugimoto and Dai Shimizu

Abstract This article reviews briefly the linear and nonlinear theories of thermoacoustic waves in a gas filled tube subject to a temperature gradient. In the framework of fluid dynamics, asymptotic theories are developed by two parameters on the basis of a narrow tube approximation. One is a parameter measuring the order of nonlinearity, while the other is a parameter measuring diffusive effects by viscosity and heat conduction. Making use of these parameters as asymptotic ones, the full system of equations is reduced to compact and spatially one-dimensional (1D) equation(s). It is emphasised that the diffusive effects may be covered substantially by two cases where the layers are thin or thick enough in comparison with the tube radius.

9.1 Introduction

Over the last three decades, much interest has been attracted to thermoacoustic instability of a gas subject to a temperature gradient and ensuing emergence of self-excited oscillations in view of potential applicability to novel heat engines (Wheatley et al., 1983; Swift, 2017). The phenomena themselves are very interesting but physical mechanisms behind them are not easily understood. To explain the mechanisms, there are two approaches.

One is a thermodynamic view of a gas particle undergoing oscillations, i.e., a thermodynamic cycle of the particle in the vicinity of a wall in the presence of the temperature gradient (Wheatley et al., 1983; Swift, 2017). This view is intuitive and

N. Sugimoto (✉)
Department of Pure and Applied Physics, Kansai University, Osaka, Japan
e-mail: sugimoto@me.es.osaka-u.ac.jp

D. Shimizu
Department of Mechanical Engineering, Fukui University of Technology, Fukui, Japan
e-mail: shimizu@fukui-ut.ac.jp

A. Berezovski, T. Soomere (eds.), *Applied Wave Mathematics II*, Mathematics of Planet Earth 6, https://doi.org/10.1007/978-3-030-29951-4_9

consistent with Rayleigh's criterion of instability (Rayleigh, 1945; Howe, 1998), but it is valid only for infinitesimally small oscillations. As the amplitude becomes large, the cycle is not closed due to drift of the particle by acoustic streaming. Thus this Lagrangian view is not appropriate to quantify the phenomena. In contrast, the other is a fluid dynamical and Eulerian view, which enables a quantitative description. This approach was initially taken by Rott (1969, 1973) to derive marginal conditions of Taconis oscillations[1] discovered in the operation of liquid helium (Taconis et al., 1949).

In a quiescent gas subject to a steady temperature gradient, heat flows in the background and the gas is set in a state of thermal nonequilibrium. The heat flow is always exposed to various uncontrollable disturbances, and therefore not only the gas temperature but its density and pressure are fluctuating. If there exist eigenmodes of oscillations in a system containing the gas, it is possible that the disturbances would excite one of the modes to make the system unstable. However, because diffusive effects by viscosity and heat conduction are present in the gas, they counteract to suppress growth of the disturbances so that the quiescent state appears to be maintained.

As the temperature gradient becomes steeper and the magnitude of heat flow becomes greater, it may be expected that the diffusive effects can no longer suppress the instability but rather they promote it by taking energy from the heat flow. Usually the diffusive effects act to dissipate energy and damp disturbances. This occurs in closed systems. The present system is open, because the heat flow can supply energy infinitely to the system from outside of it. In this case, the dissipative effects play an adverse role, to take energy from the heat flow to grow disturbances. This may be understood by analogy with instability of a viscous fluid flow. Although Rott (1969, 1973) did not take such a view, this seems to be easily acceptable.

Most analytical work undertaken to date is mainly based on an equation derived by Rott (1969, 1973) for an infinitesimally small, harmonic disturbance. In fact, this equation, called the Rott equation, is regarded as being appropriate in the light of experimental results. It explains successfully marginal conditions of not only Taconis oscillations but also other thermoacoustic oscillations in tubes using a stack or a regenerator. However the Rott equation is limited to the linear harmonic disturbance and is incapable of describing a process leading to self-excited oscillations.

To analyse fully the phenomena from initial instability in linear regime to emergence of self-excited oscillations in nonlinear regime, the computational fluid dynamics (CFD) is promising. High fidelity simulations by the CFD are carried out by Scalo and his team (Scalo et al., 2015; Gupta et al., 2017). Even by the CFD,

[1]When a long thin tube with one end open is inserted into a dewar containing a liquid helium, and the open end approaches a liquid surface at several kelvin, a quiescent, gaseous helium in the tube begins to oscillate violently. Because the closed end is kept at a room temperature, there occurs a steep temperature gradient along the tube so that a heat flows toward the open end.

however, there are limitations. For example, a behaviour of the gas in tortuous flow passages within a regenerator is so complicated that it must be taken into account by an appropriate model. In addition, high fidelity simulations are not so easy to use because it needs expertise and resources of computers.

As the gap between the Rott equation and the CFD is very wide, the authors' team has developed linear and weakly nonlinear theories, aiming at going beyond the Rott equation to fill the gap. Distinction between the linear and nonlinear theories is made, of course, by the magnitude of a maximum pressure disturbance Δp relative to a reference pressure p_0 in a quiescent state. This is represented by ε $(= \Delta p/p_0)$. The diffusive effects appear significantly in the vicinity of a wall so they are measured typically by the thickness of the viscous diffusion layer relative to the tube radius R represented by δ $(= \sqrt{\nu/\omega}/R)$, ν and ω being, respectively, a kinematic viscosity and an angular frequency. Rather the thickness of the thermal diffusion layer $\sqrt{\kappa/\omega}$, κ being a thermal diffusivity, may be appropriate, but both thicknesses are comparable when the Prandtl number Pr $(= \nu/\kappa)$ is of order unity, though the thermal one is a little thicker for air $(Pr = 0.7)$.

Table 9.1 overviews the theories classified according to ε, δ and a type of disturbance. Target indicates which phenomena are expected to be quantified by the theories. Because the magnitude of disturbance ε in self-excited oscillations in reality is of order 10^{-3} to 10^{-1} at most, the whole process from initial instability to emergence of self-excited oscillations will be described by the weakly nonlinear theory for a small but finite value of ε $(0 \ll \varepsilon \ll 1)$. In this sense, the linear theory is valid only when the disturbance may be regarded as being infinitesimally small $(\varepsilon \to 0)$.

The diffusive effects are difficult to be taken for an arbitrary value of δ except in the linear theory. Thus they are treated separately for the case of thin layers $(\delta \ll 1)$ or that of thick layers $(\delta \gg 1)$ by asymptotic methods for small or large but finite magnitudes of δ. At any rate, there appears in between an intermediate range of a moderate value of δ. It turns out, however, that the two cases can cover it beyond their original ranges of validity so that the intermediate range is much narrower than imagined. This is demonstrated by good agreements of the marginal conditions derived by the asymptotic equations and those by the Rott equation (Sugimoto, 2019).

Section 9.2 describes the linear theory to derive the thermoacoustic wave equation in a general case of δ, from which the Rott equation in the frequency domain is reduced. Section 9.3 outlines the approximation of the thermoacoustic wave equation in the two cases, thin and thick diffusion layers. Section 9.4 extends the linear approximate theories to the weakly nonlinear regime by asymptotic methods systematically. Finally a summary of the theories is given with some discussions.

Table 9.1 Overview of the linear and nonlinear theories

	Magnitude of disturbance $\varepsilon = \Delta p / p_0$	Type of disturbance	Thickness of diffusion layers relative to tube radius $\delta = \sqrt{\nu/\omega}/R \sim \sqrt{\kappa/\omega}/R$			Target
			Thin ($\delta \ll 1$)	Moderate	Thick ($\delta \gg 1$)	
Linear theory	Infinitesimally small $\varepsilon \to 0$	Time-harmonic	Rott equation			Marginal condition Marginal oscillations
		Arbitrary	Thermoacoustic wave equation			Evolution from initial disturbance
			Linear boundary layer theory	Limit of thin and thick theory	Linear diffusion wave equation	
Weakly nonlinear theory	Small but finite $0 \ll \varepsilon \ll 1$	Arbitrary	Nonlinear boundary layer theory	Limit of thin and thick layer	Nonlinear diffusion wave equation	Evolution to self-excited oscillations, steady streaming
Fully nonlinear theory	Arbitrary	Arbitrary	CFD (Vortex shedding, flow separation, end effects, acoustic streaming, etc. associated with oscillatory flow and thermal fields)			Full simulation of phenomena

9.2 Linear Theory

Suppose a gas filled tube of radius R is subject to a nonuniform temperature T_w on the tube wall axially where no gravity is assumed (Fig. 9.1). Denoting axial and radial coordinates by x and r, respectively, T_w is assumed to depend on x only, so that T_w is axisymmetric. A gas temperature T_e in a quiescent state under a uniform pressure p_0 depends on x and also r in general.

As is usually the case, the wall temperature T_w changes along x so gradually that the following inequalities may hold:

$$\frac{R^2}{T_w}\left|\frac{\mathrm{d}^2T_w}{\mathrm{d}x^2}\right| \ll \frac{R}{T_w}\left|\frac{\mathrm{d}T_w}{\mathrm{d}x}\right| \ll 1. \tag{9.1}$$

In the following analysis, $R^2T_w^{-1}\mathrm{d}^2T_w/\mathrm{d}x^2$ and also $(RT_w^{-1}\mathrm{d}T_w/\mathrm{d}x)^2$ are neglected, and nonuniform effects of T_w are taken into account up to $\mathrm{d}T_w/\mathrm{d}x$. Then T_e may be set equal to T_w to the same approximation (Sugimoto, 2010).

The heat capacity of the tube wall is assumed to be large enough for the wall temperature to be unchanged even when the gas motion takes place. Since T_e is nonuniform axially, so is the density denoted by ρ_e. The gas is assumed to obey the law of ideal gas, so it holds that $\rho_e T_e = \rho_0 T_0$ by Charles' law, where the subscript 0 designates a value at a reference state. Note that the subscript e denotes the quantities in the quiescent state which are functions of x determined by $T_e(x)$.

With these settings, consider first a behaviour of infinitesimally small disturbance given to the quiescent state. On top of the linearisation, a narrow tube approximation is exploited where typical axial length of the disturbance L is much longer than R,

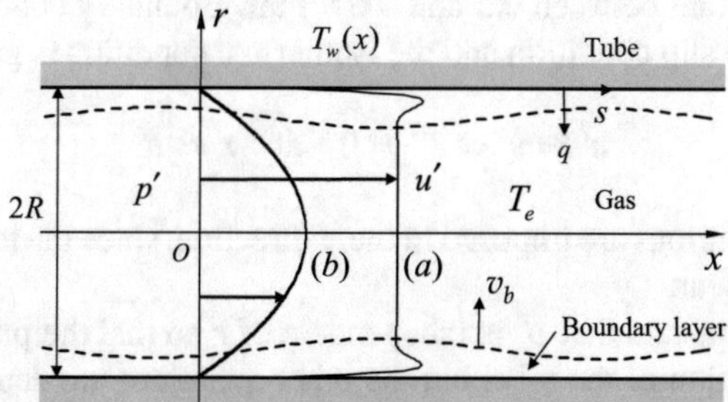

Fig. 9.1 Illustration of a circular tube of radius R filled with a thermoviscous gas and subject to a nonuniform temperature $T_w(x)$ along the tube wall, where x and r denote, respectively, the axial and radial coordinates, and T_e denotes the gas temperature in a quiescent state. The symbols p' and u' represent, respectively, the excess pressure over a uniform pressure p_0 and the axial velocity, while s and q represent, respectively, the shear stress acting on the gas at the wall surface and the heat flux into the gas through it. Profiles with (a) and (b) show qualitatively some instantaneous ones of u' over the cross section in the cases of thin and thick diffusion layers, respectively, and v_b denotes the radial velocity directed inward at the edge of a boundary layer in the case of thin diffusion layers.

i.e., $R/L \ll 1$. Thicknesses of viscous and thermal diffusion layers are arbitrary in comparison with the tube radius.

Denoting a disturbance by a prime $(\cdot)'$, equations of continuity, motion, energy, and equation of state for a thermoviscous gas are linearised around the quiescent state as follows:

$$\frac{\partial \rho'}{\partial t} + \frac{\partial}{\partial x}(\rho_e u') + \frac{1}{r}\frac{\partial}{\partial r}(r\rho_e v') = 0, \tag{9.2}$$

$$\rho_e \frac{\partial u'}{\partial t} = -\frac{\partial p'}{\partial x} + \frac{\mu_e}{r}\frac{\partial}{\partial r}\left(r\frac{\partial u'}{\partial r}\right), \tag{9.3}$$

$$0 = -\frac{\partial p'}{\partial r}, \tag{9.4}$$

$$\rho_e c_\mathrm{p}\left(\frac{\partial T'}{\partial t} + u'\frac{\mathrm{d}T_e}{\mathrm{d}x}\right) = \frac{\partial p'}{\partial t} + \frac{k_e}{r}\frac{\partial}{\partial r}\left(r\frac{\partial T'}{\partial r}\right), \tag{9.5}$$

$$\frac{p'}{p_0} = \frac{\rho'}{\rho_e} + \frac{T'}{T_e}, \tag{9.6}$$

where ρ, p, u, v, and T denote, respectively, density, pressure, axial velocity, radial velocity and temperature of the gas, t being time; c_p denotes a specific heat at constant pressure; and μ_e and k_e denote, respectively, the shear viscosity and heat conductivity at temperature T_e, which are assumed to obey a power law of T_e through

$$\frac{\mu_e}{\mu_0} = \frac{k_e}{k_0} = \left(\frac{T_e}{T_0}\right)^\beta, \tag{9.7}$$

where β is a constant between 0.5 and 0.6 for air. Boundary conditions at the wall surface require no slip condition and the isothermal condition is given by

$$u' = v' = T' = 0 \quad \text{at} \quad r = R. \tag{9.8}$$

No boundary conditions are imposed in the x direction, since the tube is so long that no end effects appear.

From (9.4) it appears that p' is independent of r so that the pressure is uniform over the cross section of the tube, but the other quantities are dependent on r. The uniformity of p' is an advantage for the analysis. Applying the Fourier transform with respect to t, all quantities may be expressed in terms of p'. In the context that follows, all quantities should be understood to be transformed. From (9.3), u' is solved in terms of p' by the use of the boundary condition. Substituting u' thus obtained into (9.5), T' is solved in terms of p'. Using (9.6), ρ' is expressed in terms of T' and therefore of p'. Substituting u' and ρ' thus obtained into (9.2), and solving for v', it follows from the boundary condition that a single equation for p' must be satisfied. This equation is finally transformed inversely to yield an equation for $p'(x, t)$.

The procedure mentioned is straightforward in principle but calculations are cumbersome. Rather it is illuminating to circumvent it by averaging each equation over the cross section of the tube with the boundary conditions. These equations are combined into the following form:

$$\frac{\partial^2 p'}{\partial t^2} - \frac{\partial}{\partial x}\left(a_e^2 \frac{\partial p'}{\partial x}\right) = \frac{2}{R}\left[\frac{a_e^2}{c_\mathrm{p} T_e}\frac{\partial q}{\partial t} - \frac{\partial}{\partial x}(a_e^2 s)\right], \tag{9.9}$$

where a_e stands for the local adiabatic sound speed at T_e defined by $\sqrt{\gamma p_0/\rho_e}$ or, equivalently, by $\sqrt{(\gamma-1)c_\mathrm{p} T_e}$, γ being the ratio of specific heat c_p at constant pressure to the one c_v at constant volume. The quantities s and q denote, respectively, the shear stress acting on the gas at the wall and heat flux flowing into the gas through the wall surface defined as follows:

$$s = \mu_e \frac{\partial u'}{\partial r}\bigg|_{r=R} \quad \text{and} \quad q = k_e \frac{\partial T'}{\partial r}\bigg|_{r=R}. \tag{9.10}$$

Equation (9.9) suggests that the heat flux and the wall friction act, respectively, as acoustic monopole and dipole (Howe, 1998). In the present context, s and q are expressed in terms of p' as

$$s = \sqrt{\nu_e}\mathcal{N}_\nu\left(\frac{\partial p'}{\partial x}\right) \tag{9.11}$$

and

$$\frac{\partial q}{\partial t} = c_\mathrm{p} T_e \sqrt{\nu_e}\left\{-\frac{\gamma-1}{\sqrt{Pr}}\mathcal{N}_\kappa\left(\frac{1}{a_e^2}\frac{\partial^2 p'}{\partial t^2}\right) + \right.$$

$$\left. + \frac{1}{T_e}\frac{\mathrm{d}T_e}{\mathrm{d}x}\left[\frac{1}{1-Pr}\mathcal{N}_\nu\left(\frac{\partial p'}{\partial x}\right) - \frac{1}{(1-Pr)\sqrt{Pr}}\mathcal{N}_\kappa\left(\frac{\partial p'}{\partial x}\right)\right]\right\}. \tag{9.12}$$

Here $\mathcal{N}_\nu$ denotes a functional defined by

$$\mathcal{N}_\nu\left(\frac{\partial p'}{\partial x}\right) = \int_{-\infty}^{t} \Theta\left[\frac{\nu_e(t-\tau)}{R^2}\right]\frac{\partial p'(x,\tau)}{\partial x}\mathrm{d}\tau. \tag{9.13}$$

The similar functional $\mathcal{N}_\kappa$ is defined by replacing ν_e in $\mathcal{N}_\nu$ with a thermal diffusivity κ_e $(= k_e/\rho_e c_\mathrm{p})$. The kernel function Θ is defined by

$$\Theta\left(\frac{\nu_e t}{R^2}\right) = \frac{1}{2\pi}\int_{-\infty}^{\infty}(-\mathrm{i}\omega)^{-\frac{1}{2}}\frac{I_1(1/\delta_e)}{I_0(1/\delta_e)}\mathrm{e}^{-\mathrm{i}\omega t}\,\mathrm{d}\omega, \tag{9.14}$$

with $\delta_e = (\mathrm{i}\nu_e/\omega)^{\frac{1}{2}}/R$. The symbols I_0 and I_1 denote, respectively, the modified Bessel functions of zeroth and first order. As the wall friction and heat flux are given in the form of the integrals from the remote past $\tau = -\infty$ to the present time $\tau = t$, the diffusive effects give rise to hereditary (memory) effects. Here it is important to notice that because $\mathcal{N}_\nu$ appears in (9.12) accompanied with the temperature gradient, the shear stress affects the heat flux, whereas the heat flux does not affect the shear stress. This holds only when the temperature gradient is present.

With the presented specification of s and q, (9.9) is expressed in the following form:

$$\frac{\partial^2 p'}{\partial t^2} - \frac{\partial}{\partial x}\left(a_e^2 \frac{\partial p'}{\partial x}\right) + \frac{2}{R}\frac{\partial}{\partial x}\left[a_e^2 \sqrt{\nu_e}\mathcal{N}_\nu\left(\frac{\partial p'}{\partial x}\right)\right] + \frac{2}{R}\frac{(\gamma-1)}{\sqrt{Pr}}\sqrt{\nu_e}\mathcal{N}_\kappa\left(\frac{\partial^2 p'}{\partial t^2}\right) -$$
$$-\frac{2}{R}\frac{a_e^2\sqrt{\nu_e}}{T_e}\frac{\mathrm{d}T_e}{\mathrm{d}x}\left[\frac{1}{1-Pr}\mathcal{N}_\nu\left(\frac{\partial p'}{\partial x}\right) - \frac{1}{(1-Pr)\sqrt{Pr}}\mathcal{N}_\kappa\left(\frac{\partial p'}{\partial x}\right)\right] = 0. \quad (9.15)$$

This is the final equation derived from (9.2)–(9.6). It is called the thermoacoustic wave equation (Sugimoto, 2010).

If a time harmonic disturbance is considered in the form of $p' = P(x)\mathrm{e}^{\mathrm{i}\omega t}$, where P and ω are, respectively, a complex pressure amplitude and an angular frequency, (9.15) becomes:

$$\frac{\mathrm{d}}{\mathrm{d}x}\left[(1-f_\nu)a_e^2\frac{\mathrm{d}P}{\mathrm{d}x}\right] + \left(\frac{f_\nu - f_\kappa}{1-\mathrm{Pr}}\right)\frac{a_e^2}{T_e}\frac{\mathrm{d}T_e}{\mathrm{d}x}\frac{\mathrm{d}P}{\mathrm{d}x} + \omega^2\left[1+(\gamma-1)f_\kappa\right]P = 0, \quad (9.16)$$

where f_ν and f_κ are defined, respectively, by

$$f_\nu = \frac{2I_1(\eta_e)}{\eta_e I_0(\eta_e)}, \quad f_\kappa = f\left(\sqrt{\mathrm{Pr}}\,\eta_e\right), \quad \text{with} \quad \eta_e = R\sqrt{\frac{\mathrm{i}\omega}{\nu_e}}. \quad (9.17)$$

The ordinary differential equation (9.16) of the second order for P is known as the Rott equation. As f_ν and f_κ result from $\mathcal{N}_\nu$ and $\mathcal{N}_\kappa$, respectively, they represent the viscous and thermal diffusion.

Finally, the equation of the acoustic energy can also be derived by taking the averages of (9.2)–(9.6) over the cross section of the tube. Denoting the average by a bar as

$$\overline{u'} = \frac{1}{\pi R^2}\int_0^R 2\pi r u'(x,r,t)\,\mathrm{d}r, \quad (9.18)$$

the equation becomes (Hyodo and Sugimoto, 2014)

$$\frac{\partial}{\partial t}\left(\frac{1}{2}\rho_e\overline{u'}^2 + \frac{1}{2}\frac{p'^2}{\rho_e a_e^2}\right) + \frac{\partial}{\partial x}\left(p'\overline{u'}\right) = \frac{2}{R}\left(\overline{u'}s + \frac{p'q}{\rho_e c_\mathrm{p} T_e}\right). \quad (9.19)$$

If the disturbance is periodic and the average is taken over its period, we have

$$\frac{\mathrm{d}I}{\mathrm{d}x} = \frac{2}{R}\left(\widetilde{\overline{u'}s} + \frac{\widetilde{p'q}}{\rho_e c_\mathrm{p} T_e}\right), \tag{9.20}$$

where the tilde means the mean over the period and I denotes the mean acoustic energy flux density (intensity) $p'\overline{u'}$ averaged over the cross section. Integrating (9.20) over a whole domain of x, the left-hand vanishes because of boundary or matching conditions. Hence the integral of the right-hand should vanish, which gives a condition for neutral oscillations to occur, i.e., a marginal condition of instability. This corresponds to Rayleigh's criterion of instability, though q here is the heat flux and not the heat release rate (Howe, 1998).

9.3 Approximation of the Thermoacoustic Wave Equation

As the thermoacoustic wave equation is the integro-differential equation, it is difficult to be solved. The kernel function in the integral on the right-hand side of (9.13) depends on the time interval $t - \tau$ between the present time t and a past time τ relative to the viscous diffusion time R^2/ν_e. If $\partial p'/\partial x$ is periodic in t with a period $2\pi/\omega$, then $\mathcal{N}_\nu$ is also periodic with the same period. Then the functional $\mathcal{N}_\nu$ may be approximated according to whether the period $2\pi/\omega$ is long or short in comparison with R^2/ν_e. This implies, equivalently, that the typical thickness of the diffusion layer $\sqrt{\nu_e/\omega}$ is thick or thin relative to R.

9.3.1 Case of Thin Diffusion Layers

When the layer is thin enough relative to R, i.e., $|\delta_e| \ll 1$, the thermoacoustic wave equation is approximated as

$$\frac{\partial^2 p'}{\partial t^2} - \frac{\partial}{\partial x}\left(a_e^2 \frac{\partial p'}{\partial x}\right) + \\ + \frac{2a_e^2\sqrt{\nu_e}}{R}\left[C\frac{\partial^{-\frac{1}{2}}}{\partial t^{-\frac{1}{2}}}\left(\frac{\partial^2 p'}{\partial x^2}\right) + \frac{(C + C_T)}{T_e}\frac{\mathrm{d}T_e}{\mathrm{d}x}\frac{\partial^{-\frac{1}{2}}}{\partial t^{-\frac{1}{2}}}\left(\frac{\partial p'}{\partial x}\right)\right] = 0, \tag{9.21}$$

where the fractional derivative of minus half order is defined by (Sugimoto, 1989)

$$\frac{\partial^{-\frac{1}{2}}}{\partial t^{-\frac{1}{2}}}\left(\frac{\partial p'}{\partial x}\right) \equiv \frac{1}{\sqrt{\pi}}\int_{-\infty}^{t}\frac{1}{\sqrt{t-\tau}}\frac{\partial p'}{\partial x}(x,\tau)\mathrm{d}\tau \tag{9.22}$$

and C and C_T are constants defined by

$$C = 1 + \frac{\gamma - 1}{\sqrt{Pr}}, \quad C_T = \frac{1}{2} + \frac{\beta}{2} + \frac{1}{\sqrt{Pr} + Pr}. \tag{9.23}$$

In this case, the diffusive effects are confined in thin layers, called a boundary layer[2] on the tube wall. The acoustic field may be divided into the boundary layer and a core region outside of it. In the core region, the diffusive effects are secondary. As is shown qualitatively by the profile (*a*) in Fig. 9.1, the axial velocity u' is almost uniform over a cross section of the core region, while v' is much smaller than u', and it increases linearly with r. At the edge of the boundary layer, v takes $-v_b$ defined by

$$v_b = \sqrt{\nu_e}\left[C\frac{\partial^{-\frac{1}{2}}}{\partial t^{-\frac{1}{2}}}\left(\frac{\partial u'}{\partial x}\right) + \frac{C_T}{T_e}\frac{\mathrm{d}T_e}{\mathrm{d}x}\frac{\partial^{-\frac{1}{2}} u'}{\partial t^{-\frac{1}{2}}}\right]. \tag{9.24}$$

The shear stress and the heat flux in this case are given, respectively, by

$$s = \sqrt{\nu_e}\frac{\partial^{-\frac{1}{2}}}{\partial t^{-\frac{1}{2}}}\left(\frac{\partial p'}{\partial x}\right) \tag{9.25}$$

and

$$q = \rho_e c_\mathrm{p} T_e \sqrt{\nu_e}\left[(C-1)\frac{\partial^{-\frac{1}{2}}}{\partial t^{-\frac{1}{2}}}\left(\frac{\partial u'}{\partial x}\right) + \left(C_T - \frac{1}{2} - \frac{\beta}{2}\right)\frac{1}{T_e}\frac{\mathrm{d}T_e}{\mathrm{d}x}\frac{\partial^{-\frac{1}{2}} u'}{\partial t^{-\frac{1}{2}}}\right]. \tag{9.26}$$

It is worth noting that $q/\rho_e c_\mathrm{p} T_e$ is of the same form as v_b. Although the corresponding coefficients are different ($C = 1.47$ and $C_T = 1.39$ for $\gamma = 1.4$, $Pr = 0.72$, and $\beta = 0.5$), it implies that v_b behaves almost in a similar way to the heat flux, i.e., the heat flux pushes (pulls) the edge of the boundary layer into (out of) the core region. Because $q/\rho_e c_\mathrm{p} T_e$ has the same dimension of speed as v_b, the product with p' means work on the gas in the core region and it inputs power into it, if a phasing between p' and q is favourable.

9.3.2 *Case of Thick Diffusion Layers*

When the layer is thick enough relative to R, i.e., $|\delta_e| \gg 1$, the acoustic field is dominated by diffusion. In this case, (9.15) is approximated to be the following

[2]By the boundary layer we mean both viscous and thermal diffusion layers together.

diffusion wave (advection) equation

$$\frac{\partial p'}{\partial t} - \frac{\partial}{\partial x}\left(\alpha_e \frac{\partial p'}{\partial x}\right) + \frac{\alpha_e}{T_e}\frac{\mathrm{d}T_e}{\mathrm{d}x}\frac{\partial p'}{\partial x} + \\ + \left[\frac{4}{3}\gamma - (\gamma - 1)Pr\right]\frac{\alpha_e}{a_e^2}\frac{\partial^2 p'}{\partial t^2} - \frac{1}{6}(1 + \beta + Pr)\frac{\alpha_e R^2}{\nu_e T_e}\frac{\mathrm{d}T_e}{\mathrm{d}x}\frac{\partial^2 p'}{\partial t \partial x} = 0, \tag{9.27}$$

where the diffusivity α_e is defined by

$$\alpha_e = \frac{a_e^2 R^2}{8\gamma \nu_e} = \frac{p_0 R^2}{8\mu_e}. \tag{9.28}$$

If no dependence of μ_e on T_e (i.e., $\beta = 0$) is assumed, α_e is a constant independent of x. However, if $\beta \neq 0$, α_e decreases as T_e increases.

The first three terms in (9.27) represent the lowest order relation and the fourth and fifth terms represent the second order correction. Since $\delta \gg 1$ (the tube radius is very narrow), the gas is thermally in a perfect contact with the wall and the gas temperature is equal to the local wall temperature to the lowest order, i.e., $T' = 0$. If no temperature gradient is present, the lowest order equation is simply a diffusion equation, and the disturbance is simply diffused, not propagated. In such a situation, the idea of sound speed loses its physical significance, though it might be thought that the isothermal sound speed would take over adiabatic sound speed. When the temperature gradient is present, however, the disturbance is not only diffused by the second term but also propagated (or advected) by the third term in the direction of the temperature gradient. This result is a new finding.

Since the tube is narrow, the viscous effects are primary and the axial velocity resembles that of Poiseuille flow in an incompressible fluid, as is shown by the profile (*b*) in Fig. 9.1. In fact, u' is given to the lowest order by a parabolic function as

$$u' = -\frac{1}{4\mu_e}\frac{\partial p'}{\partial x}\left(R^2 - r^2\right). \tag{9.29}$$

However, since the compressibility is present, v' is present and is given by

$$v' = \frac{1}{2p_0 R^2}\frac{\partial p'}{\partial t}(R^2 - r^2)r. \tag{9.30}$$

The higher order terms of T' are:

$$T' = \frac{1}{4k_e}\frac{\partial p'}{\partial t}\left(R^2 - r^2\right) + \frac{1}{64}\frac{Pr}{\rho_e \nu_e^2}\frac{\mathrm{d}T_e}{\mathrm{d}x}\frac{\partial p'}{\partial x}\left(3R^2 - r^2\right)\left(R^2 - r^2\right). \tag{9.31}$$

Using these relations, the shear stress and the heat flux on the tube wall are given, respectively, by

$$s = \frac{R}{2}\frac{\partial p'}{\partial x} \tag{9.32}$$

and

$$q = -\frac{R}{2}\left(\frac{\partial p'}{\partial t} + \frac{\gamma}{\gamma - 1}\frac{\alpha_e}{T_e}\frac{\mathrm{d}T_e}{\mathrm{d}x}\frac{\partial p'}{\partial x}\right). \tag{9.33}$$

The two terms on the right-hand side of (9.33) are comparable when $a_e/\omega L \sim \delta$. It is found from (9.32) that the wall friction $2\pi Rs$ balances with the pressure force $-\pi R^2 \partial p'/\partial x$ acting over the cross section. It is found from (9.33) that when the pressure increases temporarily at a certain point, the heat is released to the wall by the first term on the right-hand side of (9.33). The second term is expressed in terms of the shear stress. For a negative pressure gradient, that is, when $\partial p'/\partial x < 0$, the shear stress s is directed in the negative direction of x. If the temperature gradient is positive, then the heat flows into the gas, whereas if this gradient is negative, the heat flows out of the gas.

Using the approximate equations (9.21) and (9.27), marginal conditions of instability have been examined. For the Taconis oscillations, the marginal condition is available analytically by using (9.21) for a smooth temperature distribution in the form of a quadratic function (Sugimoto and Yoshida, 2007). For the Sondhauss tube, similarly, the marginal conditions are obtained analytically (Sugimoto and Takeuchi, 2009). In both cases, the marginal curves have two branches with respect to a minimum temperature ratio and appear to be qualitatively similar to those derived by the Rott equation with no smooth temperature distributions (Rott, 1973; Rott and Zouzoulas, 1976).

For a looped tube with a single stack inserted, the marginal conditions are derived analytically by using both (9.21) and (9.27) (Hyodo and Sugimoto, 2014), but no comparison is made with those derived by the Rott equation. The comparison is made in (Sugimoto, 2019) to verify the anticipation that the two cases cover substantially the whole domain of δ except for a narrow interval where the marginal temperature ratio takes a minimum.

9.4 Nonlinear Theory

The approximate equations derived from the linear thermoacoustic wave equation in the cases of thin and thick diffusion layers turn out to be valid up to $\delta \sim O(1)$ beyond their respective ranges of validity, i.e., $\delta \ll 1$ and $\delta \gg 1$. Consequently it is found that the intermediate range of a moderate value of δ is narrow and that it may substantially be covered by limits of the theories for thin and thick layers.

This encourages the extension of the approximate equations to the weakly nonlinear regime $0 \ll \varepsilon \ll 1$.

9.4.1 *Case of Thin Diffusion Layers*

As is the case with the linear theory, the acoustic field may be divided into the boundary layer and the core region outside of it. In the core region, nonlinear effects are now assumed to be primary rather than the diffusive ones. For self-excited oscillations in reality, ε is of order 10^{-1} at most, while the acoustic Reynolds number $a_0^2/\nu\omega$ is very large, e.g., 10^7 for 100 Hz, a_0 being a typical sound speed. The tube radius is assumed to be much smaller than a typical axial wavelength, but the starting point is not set in equations simplified by the narrow tube approximation but in the full equations. Assuming an almost plane wave, u' is almost uniform over the cross-section of the core region, while v' is much smaller than u'.

Averaging the full fluid dynamical equations over the cross section of the core region (not over the cross-section of the tube), the equations of continuity, axial motion and energy are reduced to 1D equations (Sugimoto and Shimizu, 2008a). Designating the averaged quantity by a attaching a bar $\overline{(\cdot)}_m$ with the subscript m, it follows that

$$\frac{\partial \bar{\rho}_m}{\partial t} + \frac{\partial}{\partial x}(\bar{\rho}_m \bar{u}_m) = \frac{2}{R}\bar{\rho}_m v_b, \tag{9.34}$$

$$\bar{\rho}_m \left(\frac{\partial \bar{u}_m}{\partial t} + \bar{u}_m \frac{\partial \bar{u}_m}{\partial x} \right) = -\frac{\partial \bar{p}_m}{\partial x}, \tag{9.35}$$

$$\bar{\rho}_m \bar{T}_m \left(\frac{\partial \bar{S}_m}{\partial t} + \bar{u}_m \frac{\partial \bar{S}_m}{\partial x} \right) = 0, \tag{9.36}$$

where v_b denotes the radial velocity at the edge of the boundary layer directed into the core region. The quantity $\bar{S}_m$ denotes the entropy, which is related to the pressure and the density through

$$\frac{\bar{\rho}_m}{\rho_e} = \left(\frac{\bar{p}_m}{p_0} \right)^{1/\gamma} \exp\left(-\frac{\bar{S}_m - S_e}{c_\mathrm{p}} \right), \tag{9.37}$$

where S_e is the entropy in the quiescent state at p_0 and T_e. Since the boundary layer is thin, it is described by the linear theory, so v_b is taken to be (9.24). With this relation, thus, the system of Eqs. (9.34)–(9.37) is closed so it can be solved when initial and boundary conditions are imposed.

In fact, the Taconis oscillations in a helium filled tube with one end open and the other closed (quarter wavelength tube) can be simulated by solving the system of equations. A smooth temperature distribution which increases from the open end

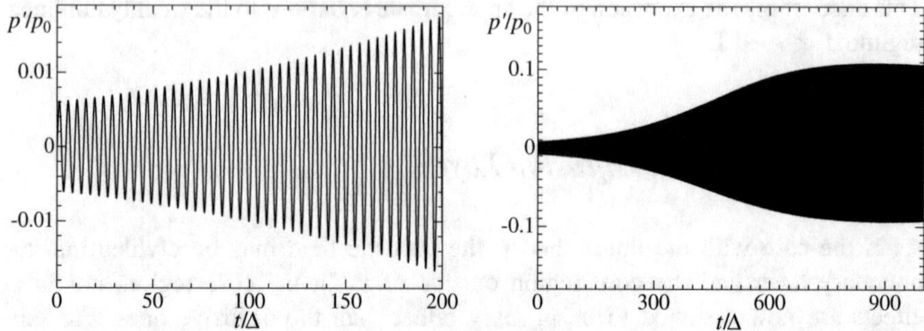

Fig. 9.2 Simulations of the Taconis oscillations in a helium filled quarter wavelength tube for the temperature ratio 50 at the hot (closed) end to the cold (open) end of the tube where a small disturbance is given initially at $t = 0$ and temporal profiles of p' at the closed end are shown over short and long times in the left and right figures, respectively, Δ being $4L/a_L$. Initial growth by instability (left) tends to be suppressed by nonlinearity to saturation of the amplitude (right). Reprinted from Sugimoto and Shimizu (2008a), with the permission of AIP Publishing.

monotonously to the closed end is imposed. The boundary condition at the open end is simply given by $p = p_0$ without taking account of radiation into free space.

Given a small disturbance initially, the implicit Crank–Nicolson finite difference scheme is used to solve for evolution of the disturbance. For the temperature ratio 50 at the hot (closed) end to the cold (open) end, Fig. 9.2 shows the temporal profiles of p' at the closed end over short and long times in left and right figures, respectively. Here $\Delta = 4L/a_L$, where L and a_L are, respectively, the tube length and the sound speed at the closed (hot) end. It is seen that the initial growth by instability (left) tends to be suppressed by nonlinearity to saturation of the amplitude (right). Thanks to the simulations, all field variables are available and related quantities such as fluxes are obtainable. For the details, see the papers (Sugimoto and Shimizu, 2008a; Sugimoto et al., 2008b; Shimizu and Sugimoto, 2010).

9.4.2 *Case of Thick Diffusion Layers*

Starting from the fluid dynamical equations and using $\delta = \sqrt{\nu/\omega}/R \gg 1$, the systematic asymptotic expansion with respect to ε and δ^{-1} is made. All field variables are evaluated beyond the relations described in Sect. 9.3.2 up to higher order terms. The lowest order relations are the same as those derived by the narrow tube approximation. To proceed to higher order, relations between the small parameters must be specified. In addition to δ, a parameter χ defined by $a/\omega L$ is introduced, a being a typical adiabatic sound speed. The analysis assumes that χ is large and comparable with δ (Sugimoto, 2016).

Specifying higher order terms step-by-step, the nonlinear diffusion wave (advection) equation is derived as follows:

$$\frac{\partial p'}{\partial t} - \frac{\partial}{\partial x}\left(\alpha_e \frac{\partial p'}{\partial x}\right) + \frac{\alpha_e}{T_e}\frac{\mathrm{d}T_e}{\mathrm{d}x}\frac{\partial p'}{\partial x} + \left[\frac{4}{3}\gamma - (\gamma - 1)Pr\right]\frac{\alpha_e}{a_e^2}\frac{\partial^2 p'}{\partial t^2} -$$
$$- \frac{1}{6}(1+\beta+Pr)\frac{\alpha_e R^2}{\nu_e T_e}\frac{\mathrm{d}T_e}{\mathrm{d}x}\frac{\partial^2 p'}{\partial t \partial x} - \frac{p'}{p_0}\frac{\partial p'}{\partial t} - \frac{\alpha_e}{p_0}\left(\frac{\partial p'}{\partial x}\right)^2 = 0. \quad (9.38)$$

This is the nonlinear version of (9.27). Unlike in the case of thin diffusion layers, nonlinear terms due to advection in the equation of motion are small. The nonlinear terms in (9.38) stem from the density change in the equation of continuity and they are free from the temperature gradient (Sugimoto, 2016).

The expressions for the shear stress and heat flux on the tube wall are lengthy and not reproduced here. If time periodic but unharmonic oscillations such as self-excited oscillations are concerned with a period τ, then they are expressed as

$$\frac{2}{R}\tilde{s} = \widetilde{\frac{\partial p'}{\partial x}} + \frac{R^2}{6\nu_e p_0}\left[2\widetilde{\frac{\partial p'}{\partial t}\frac{\partial p'}{\partial x}} + \frac{\alpha_e}{T_e}\frac{\mathrm{d}T_e}{\mathrm{d}x}\widetilde{\left(\frac{\partial p'}{\partial x}\right)^2}\right] \quad (9.39)$$

and

$$\frac{2}{R}\tilde{q} = -\left(\frac{\gamma}{\gamma - 1}\right)\frac{\alpha_e}{T_e}\frac{\mathrm{d}T_e}{\mathrm{d}x}\left(\widetilde{\frac{\partial p'}{\partial x}} + \widetilde{\frac{p'}{p_0}\frac{\partial p'}{\partial x}}\right), \quad (9.40)$$

where the tilde $\widetilde{(\cdot)}$ implies the mean over the period. Note that $\tilde{q}$ changes its sign, depending on the sign of the temperature gradient.

For the time periodic oscillations, (9.38) averaged over the period becomes

$$\frac{\partial}{\partial x}\left(\alpha_e \widetilde{\frac{\partial p'}{\partial x}}\right) + \frac{\alpha_e}{T_e}\frac{\mathrm{d}T_e}{\mathrm{d}x}\widetilde{\frac{\partial p'}{\partial x}} - \frac{\alpha_e}{p_0}\widetilde{\left(\frac{\partial p'}{\partial x}\right)^2} = 0. \quad (9.41)$$

The linear higher order terms vanish and (9.41) is expressed as

$$\frac{\alpha_e}{T_e}\widetilde{\left(\frac{\partial p'}{\partial x}\right)^2} = -\frac{\partial}{\partial x}\left(\frac{\alpha_e}{T_e}\widetilde{\frac{\partial p'}{\partial x}}\right)p_0. \quad (9.42)$$

This expression shows that the mean pressure gradient $\widetilde{\partial p'/\partial x}$ does not vanish due to the nonlinear terms so that the mean pressure varies along the tube. This relation also indicates that the mean of the pressure gradient squared is derived from the first order derivative of the mean pressure gradient.

Next, by evaluating the mean of (9.38) multiplied by $\alpha_e \partial p'/\partial x$, we have:

$$\frac{\alpha_e}{T_e^2}\widetilde{\frac{\partial p'}{\partial t}\frac{\partial p'}{\partial x}} = \frac{\partial}{\partial x}\left[\frac{\alpha_e^2}{2T_e^2}\widetilde{\left(\frac{\partial p'}{\partial x}\right)^2}\right], \tag{9.43}$$

to the lowest order. Using (9.42), (9.43) is derived from the second order derivative of the mean pressure gradient. Using these relations, $\tilde{s}$ is expressed in terms of the mean pressure gradient, while $\tilde{q}$ is expressed in terms of $\tilde{s}$ and the mean of $(p'/p_0)\partial p'/\partial x$. For both acoustic and thermoacoustic streamings, the means of the mass flux and energy flux are also expressed in terms of combinations of them (Sugimoto, 2016). In any event, p' must be obtained by solving (9.38).

Equation (9.38) is usually solved jointly with (9.34)–(9.37) for $\delta \ll 1$ for simulations in usual thermoacoustic devices. A simulation of thermoacoustic oscillations in a looped tube with a stack is now under way (Shimizu, 2018). For a shock wave to be captured, the thermoviscous effects neglected in (9.35) and (9.36) must be taken into account.

In this connection, it is noted that the approximate equations are spatially 1D and derived without taking account of an end of flow passage. When they are used, for example, in the vicinity of an end of the stack, appropriate matching conditions must be applied. As far as the linear theory is concerned, the continuity of mass flux and energy flux (which leads to the continuity of pressure) provide the matching conditions (Hyodo and Sugimoto, 2014). In the nonlinear case, treatment of the energy flux needs consideration.

9.5 Summary

The linear and nonlinear theories for thermoacoustic waves in a gas filled tube subject to a temperature gradient axially have been reviewed. As summarised in Table 9.1, they are classified according to the magnitude of disturbance and the thickness of viscous and thermal diffusion layers relative to the tube radius. It is surprising but useful to find that the diffusive effects may substantially be covered by the theories for thin and thick diffusion layers.

The assumptions and limitations of the theories are summarised. The effects of the temperature gradient are taken into account up to the first order derivative of T_w, so that the local gas temperature T_e in a quiescent state is uniform over the cross-section of the tube and is equal to the wall temperature T_w. The linear theory is based on the narrow tube approximation, by which the pressure is regarded as being uniform over the cross-section. This uniformity makes the analysis tractable. The weakly nonlinear theories start with the full equations and do not exploit the narrow tube approximation explicitly. As far as the lowest order relations are concerned, however, this approximation holds. In the higher order terms, nonuniformity of the pressure appears.

The assumption of a large heat capacity of the wall becomes questionable when the wall is thin and the amplitude of oscillations becomes large. In experiments, in fact, the wall temperature is affected considerably by oscillations. Therefore the heat conduction in the wall should be solved together with the gas oscillations where the heat fluxes are continuous across the wall surface. This is a hard task even in the linear theory (Swift, 1988; Sugimoto and Hyodo, 2012). When a combination of geometry, gas and solid materials satisfies a special condition, the thermoacoustic wave equation breaks down by nonuniformity, which may be called a resonance. Coupling between the gas and the solid needs to be clarified.

Acknowledgements The authors wish to thank Grants-in-Aid for Scientific Research (KAKENHI No. 26289036, No. 18H01375, and No. 18K03938) by the Japan Society for the Promotion of Science.

References

Gupta, P., Lodato, G., Scalo, C.: Spectral energy cascade in thermoacoustic shock waves. J. Fluid Mech. **831**, 358–393 (2017). https://doi.org/10.1017/jfm.2017.635

Howe, M.S.: Acoustics of Fluid–Structure Interactions. Cambridge University Press, Cambridge (1998). https://doi.org/10.1017/cbo9780511662898

Hyodo, H., Sugimoto, N.: Stability analysis for the onset of thermoacoustic oscillations in a gas-filled looped tube. J. Fluid Mech. **741**, 585–618 (2014). https://doi.org/10.1017/jfm.2013.621

[Lord] Rayleigh: The Theory of Sound, vol. II, pp. 230–234. Dover, New York (1945)

Rott, N.: Damped and thermally driven acoustic oscillations in wide and narrow tubes. Z. Angew. Math. Phys. **20**(2), 230–243 (1969). https://doi.org/10.1007/bf01595562

Rott, N.: Thermally driven acoustic oscillations. Part II: stability limit for helium. Z. Angew. Math. Phys. **24**(1), 54–72 (1973). https://doi.org/10.1007/bf01593998

Rott, N., Zouzoulas, G.: Thermally driven acoustic oscillations, part IV: tubes with variable cross-section. Z. Angew. Math. Phys. **27**(2), 197–224 (1976). https://doi.org/10.1007/bf01590805

Scalo, C., Lele, S.K., Hesselink, L.: Linear and nonlinear modelling of a theoretical travelling-wave thermoacoustic heat engine. J. Fluid Mech. **766**, 368–404 (2015). https://doi.org/10.1017/jfm.2014.745

Shimizu, D., Iwamatsu, T., Sugimoto, N.: Numerical simulations of thermoacoustic oscillations in a looped tube by asymptotic theories for thickness of diffusion layers. Proceedings of Meetings on Acoustics, **34**(1), 045025 (2018). https://doi.org/10.1121/2.0000888

Shimizu, D., Sugimoto, N.: Numerical study of thermoacoustic Taconis oscillations. J. Appl. Phys. **107**, 034910 (2010). https://doi.org/10.1063/1.3298465

Sugimoto, N.: 'Generalized' Burgers equations and fractional calculus. In: Jeffrey, A. (ed.) Nonlinear Wave Motion, pp. 162–179. Longman Scientific & Technical, Essex (1989)

Sugimoto, N.: Thermoacoustic-wave equations for gas in a channel and a tube subject to temperature gradient. J. Fluid Mech. **658**, 89–116 (2010). https://doi.org/10.1017/s0022112010001540

Sugimoto, N.: Nonlinear theory for thermoacoustic waves in a narrow channel and pore subject to a temperature gradient. J. Fluid Mech. **797**, 765–801 (2016). https://doi.org/10.1017/jfm.2016.295

Sugimoto, N.: Marginal conditions for the onset of thermoacoustic oscillations due to instability of heat flow. IMA J. Appl. Math. **84**(1), 118–144 (2019). https://doi.org/10.1093/imamat/hxy051

Sugimoto, N., Hyodo, H.: Effects of heat conduction in a wall on thermoacoustic-wave propagation. J. Fluid Mech. **697**, 60–91 (2012). https://doi.org/10.1017/jfm.2012.36

Sugimoto, N., Shimizu, D.: Boundary-layer theory for Taconis oscillations in a helium-filled tube. Phys. Fluids **20**(10), 104102 (2008a). https://doi.org/10.1063/1.2990763

Sugimoto, N., Shimizu, D., Kimura, Y.: Evaluation of mean energy fluxes in thermoacoustic oscillations of a gas in a tube. Phys. Fluids **20**(2), 024103 (2008b). https://doi.org/10.1063/1.2837176

Sugimoto, N., Takeuchi, T.: Marginal conditions for thermoacoustic oscillations in resonators. Proc. R. Soc. Lond. A **465**, 3531–3552 (2009). https://doi.org/10.1098/rspa.2009.0279

Sugimoto, N., Yoshida, M.: Marginal condition for the onset of thermoacoustic oscillations in a gas in a tube. Phys. Fluids **19**(7), 074101 (2007). https://doi.org/10.1063/1.2742422

Swift, G.W.: Thermoacoustic engines. J. Acoust. Soc. Am. **84**(4), 1145–1180 (1988). https://doi.org/10.1121/1.396617

Swift, G.W.: Thermoacoustics: A Unifying Perspective for Some Engines and Refrigerators. 2nd edn. ASA Press, Springer, Cham (2017)

Taconis, K.W., Beenakker, J.J.M., Nier, A.O.C., Aldrich, L.T.: Measurements concerning the vapour-liquid equilibrium of solutions of He^3 in He^4 below 2.19°K. Physica **15**(8-9), 733–739 (1949). https://doi.org/10.1016/0031-8914(49)90078-6

Wheatley, J., Hofler, T., Swift, G.W., Migliori, A.: An intrinsically irreversible thermoacoustic heat engine. J. Acoust. Soc. Am. **74**(1), 153–170 (1983). https://doi.org/10.1121/1.389624

Part III
Modelling and Mathematics

Chapter 10
Mathematics of Nerve Signals

Tanel Peets and Kert Tamm

Abstract Since the classical works of Hodgkin and Huxley (J. Physiol. **117**(4), 500–544 (1952)), it has become evident that the nerve function is a richer phenomenon than a set of electrical action potentials (AP) alone. The propagation of an AP is accompanied by mechanical and thermal effects. These include the pressure wave (PW) in axoplasm, the longitudinal wave (LW) in a biomembrane, the transverse displacement (TW) of a biomembrane and temperature changes (Θ). The whole nerve signal is, therefore, an ensemble of waves. The primary components (AP, LW, PW) are characterised by corresponding velocities. The secondary components (TW, Θ) are derived from the primary components and have no independent velocities of their own. In this chapter, the emphasis is on mathematical models rather than the physiological aspects. Based on models of single waves, a coupled model for the nerve signal is presented in the form of a nonlinear system of partial differential equations. The model equations are solved numerically by making use of the Fourier transform based pseudospectral method.

10.1 Introduction

Understanding signal propagation in nervous systems is important for the comprehension of functioning of living organisms as nerve signals control motion, behaviour and consciousness in many respects. The nervous system is made of neurons which are specialised cells that function as information processing units. They gather input signals through a branching structure known as a dendritic tree. If the inputs add up to a certain threshold limit, the electrical pulse is fired at the axon hillock and propagated along the nerve axon to the nerve terminals (synapses) (Nolte, 2015; Bresslof, 2014).

T. Peets (✉) · K. Tamm
Department of Cybernetics, School of Science, Tallinn University of Technology, Tallinn, Estonia
e-mail: tp@ioc.ee; kert@ioc.ee

A. Berezovski, T. Soomere (eds.), *Applied Wave Mathematics II*, Mathematics of Planet Earth 6, https://doi.org/10.1007/978-3-030-29951-4_10

Axons are built as cylindrical biomembranes that are made of a lipid bilayer, filled with intracellular fluid (axoplasm) and placed into a surrounding extracellular (interstitial) fluid. This structure is also called a nerve fibre. The lipid bilayer is a barrier for most water soluble molecules. It maintains an imbalance of electrically charged ions on both sides of the membrane. This gives rise to the membrane potential which is characteristic to all living cells. In excitable cells, such as nerve fibres, the ions can enter and leave the cell through specialised ion channels giving rise to a propagating electrical signal also known as action potential (AP). The contemporary understanding of axon physiology is described, for, example, in (Clay, 2005; Debanne et al., 2011).

The backbone of models describing the AP propagation is the celebrated Hodgkin–Huxley (HH) model (Hodgkin and Huxley, 1952), which is based on an electrical circuit analogy. This model has been extended by many authors by accounting for additional ion currents resulting in more detailed models (Courtemanche et al., 1998; Clay, 2005; Debanne et al., 2011). Simplified models that are useful for modelling purposes have been derived by Nagumo et al. (1962) and Engelbrecht (1981). The former is known as FitzHugh–Nagumo (FHN) model. It uses only one combined ion current instead of three. The latter (Engelbrecht, 1981) is a one-wave evolution equation.

Traditionally, the nerve function is explained only in terms of electrical signals propagating along the nerve fibres. Nowadays it is understood that the nerve signalling cannot be understood in terms of AP alone. Many experimental studies demonstrate that the relevant electrical signals are also accompanied by mechanical effects (deformation of the biomembrane, pressure in the axoplasm) and temperature changes (Abbott et al., 1958; Howarth et al., 1968; Terakawa, 1985; Tasaki, 1988, etc.). The whole signal is, therefore, an ensemble of primary and secondary components.

The primary components include the AP, the longitudinal wave (LW) in the biomembrane and the pressure wave (PW) in the axoplasm. Each of these components has its own propagation velocity. Secondary components of the signal are derived from the primary components and have no independent velocities. These components are the transverse displacement of a biomembrane (TW) and the temperature changes (Θ).

Kaufmann (1989) introduced the hypothesis that the electrical action potentials are inseparable from force, displacement, temperature, entropy and other membrane variables. Indeed, the AP generates also a PW in the axoplasm and a LW in the biomembrane. The LW means a local longitudinal compression wave which is coupled to the changes in the diameter of an axon (swelling) reflected by a TW. This process is characterised also by heat production and temperature (Θ) changes. A mathematical model describing the coupling of these signals has been proposed by Engelbrecht et al. (2018a,c).

The processes of signalling in nerve fibres are usually studied from the physiological viewpoint. However, mathematical modelling seems to increasingly gain attention because, beside the *in silico* simulations, modelling helps to reveal the concealed features of processes. As signalling is a dynamical process that can be described in terms of waves, attention is focused on the corresponding wave models.

In this chapter, some of the recent developments of wave modelling in nerve fibres are described following the preprint (Engelbrecht et al., 2019). We start from a description of the properties of single waves and continue with the analysis of their coupling in an ensemble of waves. Numerical details are given in the Appendix.

10.2 Mathematical Models Describing Wave Processes in Nerve Fibres

In this section, an overview of mathematical models describing individual components of the nerve signal is presented. The treatment here is somewhat simplified as most nerve fibres are covered by the myelin sheath which permits the ion change between intracellular and extracellular fluids only in certain nodes (Ranvier nodes). The giant squid axon used in many experiments is, however, unmyelinated (see, for example, (Bresslof, 2014)). Further on, we shall model the behaviour of unmyelinated axons. The emphasis is on the mathematical structure of the governing equations rather than on the extremely rich physiological aspects.

10.2.1 The Action Potential

Both the axoplasm and the interstitial fluid contain ions of sodium (Na^+) and potassium (K^+) as well as other ions (Hodgkin and Huxley, 1952; Scott, 1999). The relative concentration of ions create the transmembrane potential which at the equilibrium value is in the range of -50 to -100 mV. If a stimulus applied to a nerve is below a certain threshold value then the equilibrium value will be restored fast. If a stimulus is large enough (above the threshold), then an action potential will be formed and it starts to propagate along the nerve fibre. For a standard HH axon (Scott, 1999) with a diameter about 0.5 mm, the amplitude of an AP is about +50 mV and it has a typical overshoot of about 20 mV. The whole duration of a pulse is about 1 ms or in some cases even more. The propagation velocity depends on the axon diameter and is about 2 to 110 m$\cdot$ s^{-1} (Malmivuo and Plonsey, 1995). A typical shape of the AP is shown in Fig. 10.1.

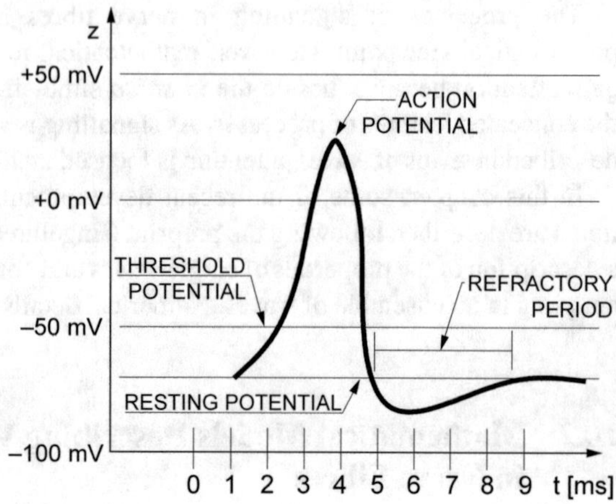

Fig. 10.1 A typical shape of an action potential (Engelbrecht, 2015b).

10.2.1.1 The Hodgkin–Huxley Model

The derivation of the HH model starts from the hyperbolic telegraph equations. In original notations they are as follows (Lieberstein, 1967):

$$\begin{aligned} \pi a^2 C_a \frac{\partial v}{\partial t} + \frac{\partial i_a}{\partial x} + 2\pi a I = 0, \\ \frac{L}{\pi a^2}\frac{\partial i_a}{\partial t} + \frac{\partial v}{\partial x} + \frac{R}{\pi a^2} i_a = 0, \end{aligned} \tag{10.1}$$

where v is the potential difference across the biomembrane, i_a is the axon current per unit length and I is ion current density. The system involves several constants: C_a is the axon self-capacitance per unit area per unit length, L is the axon specific self-inductance and R is the axon specific resistance. As usual, x and t are space coordinate and time, respectively and a is the radius of an axon.

It is possible to rewrite (10.1) as one second order partial differential equation (PDE):

$$\frac{\partial^2 v}{\partial x^2} - LC_a \frac{\partial^2 v}{\partial t^2} = RC_a \frac{\partial v}{\partial t} + \frac{2}{a} RI + \frac{2}{a} L \frac{\partial I}{\partial t}. \tag{10.2}$$

Since in electrophysiology, inductance L is usually assumed to be negligible, (10.2) can be rewritten as

$$\frac{\partial^2 v}{\partial x^2} = RC_a \frac{\partial v}{\partial t} + \frac{2}{a} RI. \tag{10.3}$$

This assumption has an important structural consequence. While (10.2) is a hyperbolic PDE, (10.3) is a parabolic PDE. The main difference is that in the case of a hyperbolic PDE, the disturbances travel at finite, and in case of a parabolic PDE, at infinite speeds. For this reason equations that describe travelling waves are usually hyperbolic PDEs. In case of (10.3), the finite speed of a travelling wave is ensured by the ion current $j = 4\pi a I$. Such equations are known as reaction-diffusion type equations. In the following we call them wave-like equations (Engelbrecht, 2015b; Billingham and King, 2001).

From the experiments, Hodgkin (1964) realised that the membrane current I can be divided into four components: the capacitive current I_C, ion currents of potassium I_K and sodium I_{Na}, and a leakage current I_L which consists of chloride and other ions. Denoting the maximum conductances of these ion components as g_{Na}, g_K, and g_L, the membrane current is

$$I = g_K n^4 (v - v_K) + g_{Na} m^3 h (v - v_{Na}) + g_L (v - v_L) + C_m \frac{\partial v}{\partial t}, \tag{10.4}$$

where C_m is the membrane capacitance per unit area and v_{Na}, v_K, v_L denote equilibrium potential of the corresponding ions. The phenomenological (hidden) variables n, m, and h govern the 'turning on' and 'turning off' of individual membrane conductances. Variable n governs the potassium conductance ('turning on') and the variables m, h govern the sodium conductance ('turning on' and 'turning off', respectively). The values of these variables are calculated from the following kinetic equations:

$$\begin{aligned} \frac{dn}{dt} &= \alpha_n(1-n) - \beta_n n, \\ \frac{dm}{dt} &= \alpha_m(1-m) - \beta_m m, \\ \frac{dh}{dt} &= \alpha_h(1-h) - \beta_h h. \end{aligned} \tag{10.5}$$

The expressions for the parameters in the kinetic equations have been determined experimentally (Hodgkin and Huxley, 1952):

$$\begin{aligned} \alpha_n &= \frac{0.01(10+V)}{\exp[(10+V)/10]-1}, & \beta_n &= 0.125\exp(V/80), \\ \alpha_m &= \frac{0.1(25+V)}{\exp[(25V)/10]-1}, & \beta_m &= 4\exp(V/18), \\ \alpha_h &= 0.07\exp(+V/20), & \beta_h &= \frac{1}{\exp[(30+V)/10]+1}. \end{aligned} \tag{10.6}$$

These coefficients are defined via milliseconds [ms] in units of [ms^{-1}] while potential V is in millivolts [mV].

The celebrated HH model has been tested by many experiments. Several modifications have been proposed to modify the ion current j (see, for example, (Courtemanche et al., 1998)). However, the HH model describes the AP as an electrical signal supported by ion currents and does not take into account other accompanying effects (Appali et al., 2010).

10.2.1.2 The Bonhoeffer–van der Pol Model

The HH model involves many variables and physical parameters which depend on the morphology of nerve fibres. This feature justifies numerous attempts to derive simplified models which could describe the process with a sufficient accuracy. For example, FitzHugh (1961) used the ideas of Bonhoeffer (1948) and van der Pol (1926) for deriving a model of the excitable oscillatory system. Here and below the notations from original papers are used. The starting point is the van der Pol equation

$$\ddot{x} + c(x^2 - 1)\dot{x} + x = 0, \tag{10.7}$$

where c is a positive constant and x denotes an oscillating quantity (amplitude). The dots denote differentiation with respect to time t. By using Liénard's transformation

$$y = \dot{x}/c + x^3/3 - x, \tag{10.8}$$

a system of first order equations is obtained:

$$\begin{aligned} \dot{x} &= c(y + x - x^3/3), \\ \dot{y} &= -x/c. \end{aligned} \tag{10.9}$$

The Bonhoeffer–van der Pol (BVP) model enlarges (10.9) into

$$\begin{aligned} \dot{x} &= c(y + x - x^3/3 + z), \\ \dot{y} &= -(x - a + by)/c, \end{aligned} \tag{10.10}$$

where

$$1 - 2b/3 < a < 1, \quad 0 < b < 1, \quad b < c^2. \tag{10.11}$$

Here a and b are constants and z is the stimulus intensity representing the membrane current.

The quantity x is related to the membrane potential and excitability and y to the accommodation and refractoriness. The phase plane analysis demonstrates the threshold and all-or-none phenomena together with the refractory part of the solution. Examples of numerical solutions using $a = 0.7$, $b = 0.8$, $c = 3$ and various values of z are described in (FitzHugh, 1961, Fig. 2).

The variables x, y, and z in the BVP model correspond to variables v and m, h and n, and I in the HH model, respectively. However, the HH model describes the propagating wave ((10.3) is a PDE) while the BVP model describes the standing (stationary) profile and (10.10) are ordinary differential equations (ODEs).

10.2.1.3 The FitzHugh–Nagumo Model

In original notations the FitzHugh–Nagumo (FHN) model is written like the BVP model (10.10) in the form (Nagumo et al., 1962):

$$
\begin{aligned}
h\frac{\partial^2 u}{\partial s^2} &= \frac{1}{c}\frac{\partial u}{\partial t} - w - \left(u - \frac{u^3}{3}\right), \\
c\frac{\partial w}{\partial t} &+ bw = a - u,
\end{aligned}
\tag{10.12}
$$

where u is the voltage and w is the recovery current. The constants h, c, b, a are positive and satisfy the following conditions:

$$1 > b > 0, \quad c^2 > b, \quad 1 > a > 1 - \frac{2}{3}b. \tag{10.13}$$

As noted by Nagumo et al. (1962), (10.12) serve as the distributed BVP model (10.10) which is an ODE. The possible governing equation

$$ch\frac{\partial^3 u}{\partial t \partial s^2} = \frac{\partial^2 u}{\partial t^2} - c\left(1 - u^2\right)\frac{\partial u}{\partial t} + u - a \tag{10.14}$$

can be derived from (10.12), where for simplicity $b = 0$.

The change of variables $x = s/\sqrt{ch}$, $z = 2a(a-u)/(a^2-1)$, $\mu = c(a^2-1)$, $\varepsilon = (a^2-1)/(4a^2)$ leads to the FHN equation (Nagumo et al., 1962):

$$\frac{\partial^3 z}{\partial t \partial x^2} = \frac{\partial^2 z}{\partial t^2} + \mu\left(1 - z + \varepsilon z^2\right)\frac{\partial z}{\partial t} + z, \tag{10.15}$$

where $\mu > 0$ and $3/16 > \varepsilon > 0$.

A modification of (10.12) changes the polynomial $u - 1/3u^3$ to its full form (Neu et al., 1997; Bountis et al., 2000). Then (10.12) can be represented in the form of two coupled equations (Engelbrecht et al., 2018b) using dimensionless variables z and j:

$$
\begin{aligned}
\frac{\partial z}{\partial t} &= z(z-a)(1-z) - j + D\frac{\partial^2 z}{\partial x^2}, \\
\frac{\partial j}{\partial t} &= \varepsilon(-j + bz),
\end{aligned}
\tag{10.16}
$$

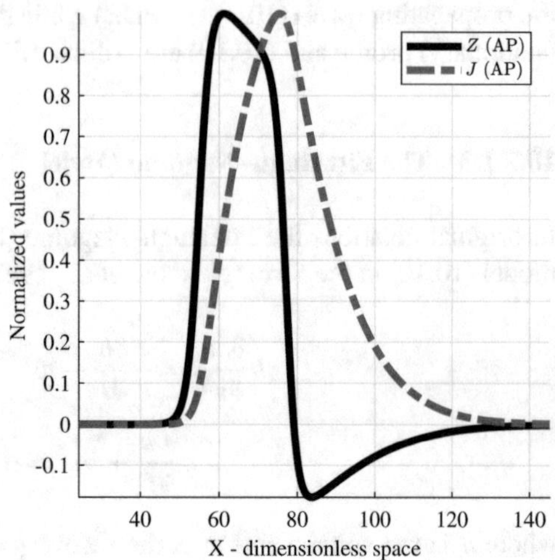

Fig. 10.2 Solution of the FHN equation (10.12). Dimensionless membrane voltage Z is plotted in solid black and dimensionless ion current J in dashed red. Numerical details and coefficients are discussed in the Appendix.

where D is a constant, ε is the timescale difference, and so-called activation parameters satisfy the conditions $0 < a < 1$ and $b > 0$.

The FHN model is simpler to analyse as it involves only one ion current instead of individual ion currents tied to the kinetic equations as in the HH model. Nowadays the FHN model is one of the cornerstones in the analysis of the nerve pulse and/or more generally, the dynamics of excitable media (Engelbrecht, 1991).

An example of numerical solutions of (10.12) indicates that the FHN model gives a correct shape of the AP (Fig. 10.2). The profiles depicted in (Fig. 10.2) propagate from right to left.

10.2.1.4 Evolution Equations

From wave mechanics, it is known that the classic wave equation describes two waves, one propagating to the right and another to the left. Under certain conditions, these waves can be separated. In this case, the result is an evolution equation which describes one wave, either propagating to the right or to the left. In mathematical terms, the leading derivative is then of the first order. The details of the derivation of such equations by the reductive perturbation method are described, e.g., in (Taniuti and Nishihara, 1983; Engelbrecht, 1983).

Models governing the propagation of the AP are as a rule derived from the telegraph equation where the inductance is neglected. The result is a parabolic equation which due to the existence of the ion current j leads to the propagating wave. The full telegraph equation involves also the inductance L and, consequently, leads to the finite velocity $c_0^2 = a/(2LC)$. The velocity of wave propagation is strongly influenced by the ion current and thus different from this value. A hyperbolic extension of the HH equation has been derived by Lieberstein (1967).

The existence of the finite velocity in the initial model permits to use the approaches that are applied for the derivation of evolution equations (Engelbrecht, 1983, 1991; Taniuti and Nishihara, 1983). Such derivations rely on the possibility of splitting the wave process into single waves. While two waves are governed by the conventional one-dimensional (1D) wave equations, the evolution equation governs the propagation of one wave only, either to the right or to the left. For example, the well-known Korteweg–de Vries and Burgers equations are widely used one-wave models.

The evolution equation for the nerve pulse has been derived by Engelbrecht (1981). The starting point of the derivation is the full telegraph equation (10.2) and a simplified description of the ion current as in case of the FHN model. The evolution equation in a moving frame $\xi = c_0 t - x$ is (Engelbrecht, 1981, 1991):

$$\frac{\partial^2 z}{\partial x \partial \xi} + f(z)\frac{\partial z}{\partial \xi} + g(z) = 0, \tag{10.17}$$

where

$$f(z) = \mu(b_0 + b_1 z + b_2 z^2), \quad g(z) = b_{00} z, \tag{10.18}$$

and μ, b_0, b_1, b_2, and b_{00} are constants.

Like the HH and FHN models, (10.17) is able to reflect the main properties of the action potential such as the existence of (i) a threshold for propagating the signal, (ii) a steady (stationary) pulse, and (iii) a refraction length. The moving frame includes the velocity c_0 determined from the telegraph equation, but the final velocity of a pulse is dictated by the ion current. The steady pulse is described by an ODE in terms of $\eta = x + \Lambda\xi$, where $\Lambda > 0$ reflects the deviation from the velocity c_0. This equation reads:

$$z'' + f(z)z' + \Lambda^{-1} g(z) = 0, \tag{10.19}$$

where $z' = dz/d\eta$.

Equation (10.19) is a Liénard type equation (Reissig et al., 1963) which governs the stationary pulse. Without loss of generality we can assume $b_{00} > 0$. The properties of (10.19) substantially depend on roots z_1, z_2 of $f(z) = 0$. If $z_1 = -z_2 \neq 0$, then (10.19) is the van der Pol equation. If $z_1 < 0$, $z_2 > 0$, then (10.19) describes the process in a lamp generator with a 'soft regime' (Andronov et al., 1959). Using Bendixson's theorem, it is possible to show that the van der Pol equation has a limit cycle. However, under the conditions of a lamp generator there is no limit cycle (Engelbrecht and Tobias, 1987).

Equation (10.19) can also be presented as a system

$$\begin{aligned} dz/d\eta &= y, \\ dy/d\eta &= -b_4 z - (b_0 + b_1 z + b_2 z^2)\mu, \end{aligned} \tag{10.20}$$

where $b_4 = b_{00}/\Lambda$.

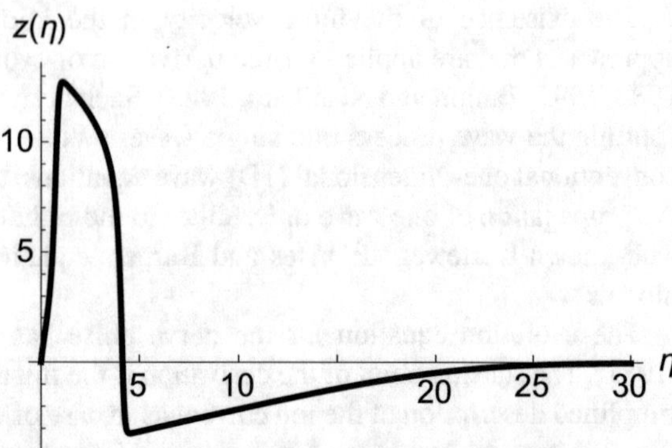

Fig. 10.3 An example solution of (10.20). Here $\mu = 3$, $b_{00} = 2.6$, $b_0 = 1$, $b_1 = -1$, and $b_2 = 0.1$.

It is easy to see that if $b_{00} \neq 0$, the origin (0, 0) is the only singular point of the system. If $b_0^2 - 4b_4 > 0$, this singular point is a node (Thompson, 1982). The phase trajectories are determined by the equation

$$dy/dz = -(b_0 + b_1 z + b_2 z^2) - b_4 z/y, \tag{10.21}$$

and the zero-isoclines by the equation

$$h(z) = -b_4 z/(b_0 + b_1 z + b_2 z^2). \tag{10.22}$$

Details of the solutions of (10.19) can be found in (Engelbrecht, 1991). An example of solutions of (10.20) (Fig. 10.3) demonstrates that the Liénard type equation is able to grasp the main shape of the AP.

10.2.2 Phenomena in Biomembranes

10.2.2.1 The Longitudinal Wave in a Biomembrane

A cylindrical biomembrane forms a wall of a nerve fibre separating intra- and extra-cellular fluids. It is made of a lipid bilayer and can be treated as a microstructured material. Such a biomembrane can be deformed. According to present understanding this feature could also be relevant for the nerve signal formation and propagation.

A mathematical model for a LW in a biomembrane is proposed by Heimburg and Jackson (2005). The model is based on the classic wave equation for the density change of the biomembrane $\Delta\rho_0 = u$ and on two assumptions.

The first assumption is that the compressibility of the biomembrane affects the velocity of sound c in the biomembrane. This impact can be, to a first approximation beyond the linear model, described as a quadratic function

$$c^2 = c_0^2 + pu + qu^2, \tag{10.23}$$

where c_0 is the is the velocity in the unperturbed state and p, q are constants determined from experiments.

The second assumption is that the propagation of sound in the biomembrane is dispersive (that is, the phase velocity is not constant but is a function of the wave number). Under these assumptions the following governing equation for the LW can be derived from the traditional 1D wave equation:

$$\frac{\partial^2 u}{\partial t^2} = \frac{\partial}{\partial x}\left[\left(c_0^2 + pu + qu^2\right)\frac{\partial u}{\partial x}\right] - h_1\frac{\partial^4 u}{\partial x^4}, \tag{10.24}$$

where the *ad hoc* dispersive term with constant h_1 has been added. This equation is called the Heimburg–Jackson (HJ) model in the following.

Equation (10.24) is a Boussinesq-type equation (Christov et al., 2007) that incorporates the following effects: (i) bidirectionality of waves, (ii) nonlinearity (of any order), and (iii) dispersion (of any order modelled by space and time derivatives of the fourth order at least).

The HJ model (10.24) contains only one dispersion term $h_1 u_{xxxx}$ which reflects the elastic properties of a biomembrane. However, it is known from theory and experiments of microstructured materials (e.g., (Porubov, 2003; Maurin and Spadoni, 2016a,b, etc.) that proper models should also consider inertial effects.

This shortcoming of the HJ model (10.24) was addressed by Engelbrecht et al. (2015) by means of inserting an additional dispersion term $h_2 u_{xxtt}$ (which is related to the inertial properties of the biomembrane) into this equation. The resulting equation reads:

$$\frac{\partial^2 u}{\partial t^2} = \frac{\partial}{\partial x}\left[\left(c_0^2 + pu + qu^2\right)\frac{\partial u}{\partial x}\right] - h_1\frac{\partial^4 u}{\partial x^4} + h_2\frac{\partial^4 u}{\partial x^2 \partial t^2}. \tag{10.25}$$

The additional dispersion term is also important from the viewpoint of revealing the concealed links between the onset and properties and propagation of signals. If the equation contains only one dispersion term $h_1 u_{xxxx}$, the phase velocity $c_{ph}^2 = c_0^2 + h_1 k^2$ tends to infinity as the wave number k increases (Engelbrecht et al., 2015; Peets and Tamm, 2015). Pulses, in general, contain a wide variety of harmonics. The potentially infinite velocity of high frequency components can cause problems in the analysis (Engelbrecht et al., 2015). Moreover, the presence of only the fourth order spatial derivatives can lead to instabilities in wave propagation (Maugin, 1999). If another dispersion term $h_2 u_{xxtt}$ is present, the phase velocity is bounded. The maximum velocity for high frequency harmonics $c_1^2 = h_1/h_2$ is defined by the ratio of the dispersion coefficients.

Heimburg and Jackson (2005, 2007) have shown that a solitary wave solution exists for (10.24). This solution has been analysed for a case of biomembrane when lipids are above melting transition ($p < 0, q > 0$). Perez-Camacho and Ruiz-Suarez (2017) have shown that negative amplitude solitary waves exist in lipid bilayers when $p > 0$, $q > 0$. Analytical and numerical studies of (10.24) and (10.25) have demonstrated the richness of the HJ model from the viewpoint of mathematical physics (Peets and Tamm, 2015; Tamm and Peets, 2015; Peets et al., 2016, 2019; Engelbrecht et al., 2017).

Equation (10.25) has two analytical solutions in dimensionless form (Peets et al., 2019):

$$V_1(\xi) = \frac{-6(1-c^2)}{P + P\sqrt{1-6(1-c^2)Q/P^2}\cosh\left[\xi\sqrt{(1-c^2)/(H_1-H_2c^2)}\right]}, \tag{10.26}$$

$$V_2(\xi) = \frac{-6(1-c^2)}{P - P\sqrt{1-6(1-c^2)Q/P^2}\cosh\left[\xi\sqrt{(1-c^2)/(H_1-H_2c^2)}\right]}, \tag{10.27}$$

where $V(\xi)$ is the amplitude, $\xi = X - cT$ is a moving frame with dimensionless space coordinate X and time T, c is a velocity, and P, Q, H_1, and H_2 are dimensionless coefficients. If $Q > 0$, only $V_1(\xi)$ represents a solitary wave. If $Q < 0$, (10.24) and (10.25) have two coexisting solutions called a soliton doublet (Peets et al., 2019).

An example solution (10.26) for the case of a biomembrane ($p < 0$, $q > 0$, Fig. 10.4) highlights that the inclusion of inertia allows the 'control' of the width of a solitary wave and results in a more localised solution.

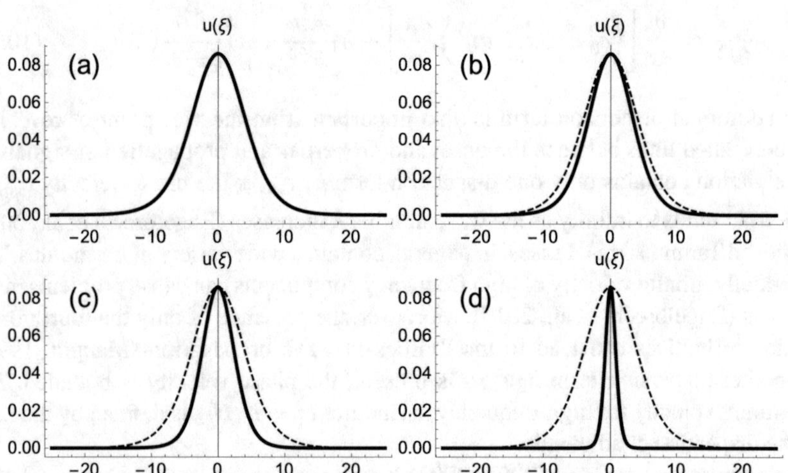

Fig. 10.4 The effect of the second dispersive term H_2U_{XXTT} on the width of a solitary wave. Here $P = -16$, $Q = 80$, $H_1 = 2$; (a) $H_2 = 0$, (b) $H_2 = 1$, (c) $H_2 = 2$ and (d) $H_2 = 3$. The solid lines represent solutions of (10.25) and the dashed lines depict solutions of (10.24) ($H_2 = 0$) (Engelbrecht et al., 2017).

10.2.2.2 The Transverse Displacement of the Biomembrane

It has been experimentally shown by a number of studies (Iwasa et al., 1980; Tasaki, 1988; Tasaki et al., 1989) that the propagation of an AP is accompanied by a swelling of a nerve fibre that travels synchronously with the AP. These early experiments (Gonzalez-Perez et al., 2016) have been repeated in more recent studies from the viewpoint of applications (Yang et al., 2018). A similar coupling of the electrical signal and mechanical pulses have been reported in excitable plant cells (Fillafer et al., 2017).

Equation (10.25) describes a LW in a biomembrane that is a primary component of a nerve signal. Engelbrecht et al. (2015) proposed that the transverse wave (TW) measured in experiments can be modelled in the same manner as in the theory of rods (Porubov, 2003):

$$w = -\nu a \frac{\partial u}{\partial x}, \tag{10.28}$$

where w is a transverse displacement and ν is the Poisson coefficient in the theory of rods (Porubov, 2003). Note that the TW has no independent velocity and is derived from the LW (equivalently, it travels synchronously with the LW). For this reason it is usually referred to as a secondary component of a signal (Fig. 10.5).

10.2.3 *The Pressure Wave in the Axoplasm*

The existence of a PW in the axoplasm has been experimentally demonstrated by Terakawa (1985). The axoplasm within a nerve fibre can be modelled like a viscous fluid (Gilbert, 1975). It means that a pressure wave (PW) could be described by

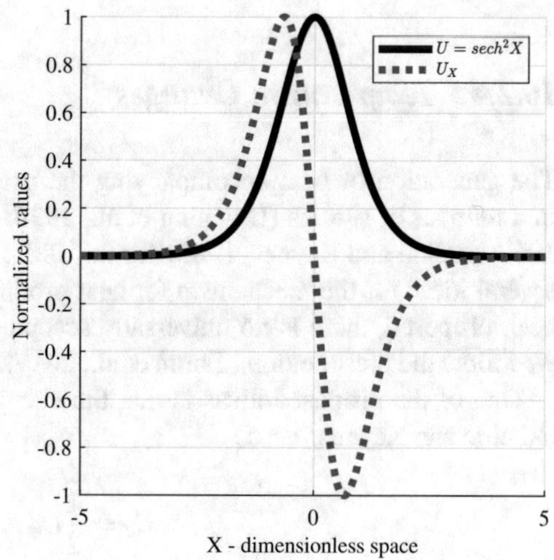

Fig. 10.5 A sketch of LW (solid black line) and TW (dotted blue line).

Navier–Stokes equations. In the 1D setting, such model in terms of longitudinal velocity v_x is (Tritton, 1988):

$$\rho\left(\frac{\partial v_x}{\partial t}+v_x\frac{\partial v_x}{\partial x}\right)+\frac{\partial p}{\partial x}-\mu_1\frac{\partial^2 v_x}{\partial x^2}=F, \tag{10.29}$$

where ρ is the density, p is the pressure, μ_1 is the viscosity parameter and F is the body force.

In the two-dimensional (2D) setting for waves in the fluid surrounded by a cylindrical tube, the governing equations are (Lin and Morgan, 1956):

$$\begin{aligned}\frac{\partial^2 p}{\partial t^2}&=c_f^2\left(\frac{\partial^2 p}{\partial x^2}+\frac{\partial^2 p}{\partial r^2}+\frac{1}{r}\frac{\partial p}{\partial r}\right),\\ \rho\frac{\partial^2 U_x}{\partial t^2}+\frac{\partial p}{\partial x}&=0,\\ \rho\frac{\partial^2 U_r}{\partial t^2}+\frac{\partial p}{\partial r}&=0,\end{aligned} \tag{10.30}$$

where (x, r) are cylindrical coordinates, U_x and U_r are the longitudinal and transverse displacements, respectively, and c_f is the velocity.

As the amplitude of the PW is very small (Terakawa, 1985), the simplest possibility could be to model PW by the 1D wave equation with an added viscous term

$$\frac{\partial^2 p}{\partial t^2}=c_f^2\frac{\partial^2 p}{\partial x^2}-\mu\frac{\partial p}{\partial t}, \tag{10.31}$$

where μ is the viscosity parameter.

10.2.4 Temperature Changes

The generation of heat accompanying the propagation of AP has been observed in a number of studies (Downing et al., 1926; Abbott et al., 1958; Howarth et al., 1968; Ritchie and Keynes, 1985; Tasaki, 1988; Tasaki and Byrne, 1992). Although several ideas for the mechanism for heat production and temperature changes have been proposed, there is no universally accepted mechanism in the literature. Here we follow the presentation (Tamm et al., 2019).

One of the proposed ideas is that the energy of the membrane capacitor E_c is (Ritchie and Keynes, 1985)

$$E_c=\frac{1}{2}C_m Z^2, \tag{10.32}$$

where C_m is the capacitance and Z is the amplitude of the AP (here and below dimensionless variables are used). As the heat energy $Q \approx E_c$, it can be deduced that Q should be proportional to Z^2. The standard Fourier law (the thermal conductivity equation) in its simplest 1D form is

$$Q = -k\Theta_X, \tag{10.33}$$

where Q is the heat flux, k is the thermal conductivity and Θ is the temperature. Note that heat energy is proportional to the negative temperature gradient. Combining (10.32) and (10.33), we obtain:

$$\Theta \propto \frac{C_m}{2k} \int Z^2 dX. \tag{10.34}$$

Experimental results by Abbott et al. (1958) demonstrate that the heat increase at the surface of the fibre is positive. It has also been argued that either $\Theta \propto Z$ or $\Theta_T \propto Z$, depending on the electrical properties of the system (Tasaki and Byrne, 1992). The potential Z of the AP or its square Z^2 might not be the only possible sources. For example, Heimburg and Jackson (2014) and Mussel and Schneider (2019) argued in favour of the idea that experimentally observed temperature changes might be the result of the propagating mechanical wave in the lipid bilayer. As there seems to be no clear consensus which quantities associated with the nerve pulse propagation are the sources of the thermal energy, we indicate here several possibilities for the coupling of the waves in the ensemble to reach the additional model equation for the temperature.

The aim is to cast these ideas into a mathematical form. As far as temperature is a function of space and time, we opt to use the classic heat equation. It is a parabolic PDE describing the distribution of heat (or the variation in temperature) in a given region over time. It is straightforward (Carslaw and Jaeger, 1959) to derive the heat equation in terms of temperature from the Fourier law by considering the conservation of energy

$$\Theta_T = \alpha\Theta_{XX}, \tag{10.35}$$

where α is the thermal diffusivity. In our case, a possible source term must be added to (10.35)

$$\Theta_T = \alpha\Theta_{XX} + F_3(Z, J, U). \tag{10.36}$$

The source term F_3 accounts for different assumptions proposed to describe the temperature generation and consumption. A mathematical foundation of heat production and temperature changes in nerve fibres is proposed by Tamm et al. (2019).

10.3 Modelling of an Ensemble of Waves

10.3.1 General Ideas

A natural generalisation of the description of the individual constituents of the nerve signal and models governing the primary (AP, PW, LW) and secondary (TW, Θ) components is to describe all these signals in a coupled network. Already Hodgkin (1964) stated that "In thinking about the physical basis of the AP perhaps the most important thing to do at the present moment is to consider whether there are any unexplained observations which have been neglected in an attempt to make the experiments fit into a tidy pattern." The same reasoning has been emphasised by Wilke (1912) and more recently by Bennett (2000) and Andersen et al. (2009).

A mathematical model for the nerve signal propagation in terms of coupled PDEs has been proposed in (Engelbrecht, 2015a; Engelbrecht et al., 2018c,a). Here we present the coupled model of nerve signal propagation following the ideas presented by Engelbrecht et al. (2018c): (i) the main component of the nerve signal is the AP which is a carrier of information and triggers all other processes, (ii) the mechanical waves in the axoplasm and in the surrounding biomembrane are generated due to the changes in the electrical signal (the AP or ion currents), (iii) the changes in the PW may also influence the waves in the biomembrane, (iv) the biomembrane can be deformed in the longitudinal and transverse direction, (v) the channels in the biomembrane can be opened and closed under the influence of electrical signals or by the mechanical inputs (Mueller and Tyler, 2014), (vi) the axoplasm in a fibre can be modelled as a viscous fluid where a PW is generated due to the electrical signal (Terakawa, 1985).

In order to unite these equations into one system involving the coupling effects, the crucial part must be added – the coupling forces. We focus here on the pulse type or bipolar type forcing. It is evident that the bipolar forcing to a wave equation leads to an energetically balanced solution (Engelbrecht et al., 2018a). As an example, the gradients of a membrane potential Z and ion current J (that is one of the coupling forces used in our simulations) are shown in Fig. 10.6.

In general terms, the wave ensemble in a nerve fibre is composed of primary components (AP, PW, LW). These waves when considered separately possess finite and possibly different velocities. In an ensemble, the velocities are synchronised as demonstrated in experiments. This feature needs further analysis (Engelbrecht et al., 2018a). The secondary (derived) components are the TW and temperature change Θ. These components have no independent velocities. They only emerge together with primary components in the proposed system. This means that the physical and physiological constraints must be accounted for in the governing equations of the primary components of the ensemble. The block scheme of signal generation with primary and secondary components of an ensemble of waves is shown in Fig. 10.7.

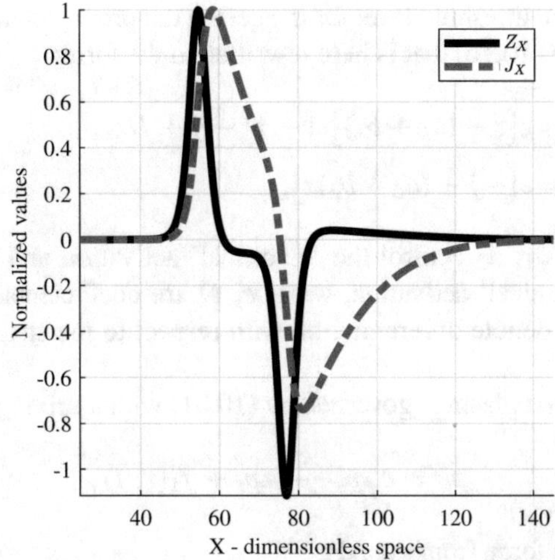

Fig. 10.6 An example of bipolar coupling forces – gradients of the membrane potential Z_X (solid black) and of the ion current J_X (dashed red). The AP calculated from the FHN is shown in Fig 10.2.

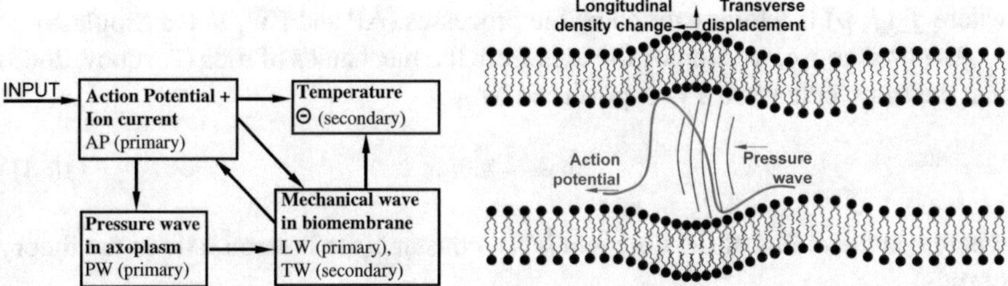

Fig. 10.7 The block diagram of the possible mathematical model for the ensemble of waves in the nerve fibre (left) and an artistic sketch of wave ensemble propagation in the axon (right).

10.3.2 Mathematical Model for an Ensemble of Waves

Mathematical modelling essentially means casting a real world problem into a mathematical formulation. This is a useful approach in gaining a better understanding of the underlying processes from a different viewpoint and for performing *in silico* experiments. A mathematical model is not a goal in itself and it needs to be validated by physical experiments. Such an approach is becoming standard, e.g., in systems biology (Noble, 2002; Vendelin et al., 2007).

Following the line of thinking of Engelbrecht et al. (2018a), we assume that the process is initiated by an electrical impulse $z(t)$ defined at certain initial instant $t = 0$ as follows:

$$z\,|_{t=0} = f(x), \tag{10.37}$$

where z exceeds the threshold level for triggering the propagation of the signal. The AP is described by (10.16) that is here rewritten in the form:

$$\begin{aligned} z_t &= z\big[z - (a_1 + b_1)\big](1 - z) - j + D z_{xx}\,, \\ j_t &= \varepsilon\big[-j + (a_2 + b_2) z\big], \end{aligned} \tag{10.38}$$

where parameters a_1, a_2 control the 'electrical' activation and parameters b_1, b_2 control the 'mechanical' activation, while ε, D are coefficients. Subscripts x and t here and further denote differentiation with respect to the spatial coordinate and time, respectively.

The PW in the axoplasm is governed by (10.31) with a driving force

$$p_{tt} = c_f^2 p_{xx} - \mu p_t + f_1(z, j)\,, \tag{10.39}$$

where $f_1(z, j)$ is a force from the AP.

The LW in the biomembrane is governed by (10.25)

$$u_{tt} = \left[\left(c_0^2 + pu + qu^2\right) u_x\right]_x - h_1 u_{xxxx} + h_2 u_{xxtt} + f_2(j, p)\,, \tag{10.40}$$

where $f_2(j, p)$ is a force exerted by the processes (AP and PW) in the axoplasm.

As noted above, we employ the idea from the mechanics of rods (Porubov, 2003) that the transverse wave TW is governed by

$$w = -kru_x\,, \tag{10.41}$$

where r is the radius of the fibre and k is a constant (the Poisson ratio in the theory of rods).

The crucial problem here is how to quantify the mechanism of coupling forces. The main hypothesis (Engelbrecht et al., 2018a,c) is as follows: the mechanical waves in the axoplasm and surrounding biomembrane are generated due to changes in electrical signals (the AP and ion current). The associated hypothesis is: the changes in the pressure wave PW may also influence the waves in a biomembrane. As changes are related to either space or time derivatives, it is assumed that the leading terms in coupling forces are in the form of z_x, z_t, j_x, j_t, p_x, and p_t.

The final equations for the primary components in a joint model of the ensemble behaviour in dimensionless variables and coordinates (Engelbrecht et al., 2018a,c) are:

$$Z_T = D_1 Z_{XX} - J + Z\big[Z - (A_1 + B_1) - Z^2 + (A_1 + B_1) Z\big], \tag{10.42}$$

$$J_T = \varepsilon_1\big[(A_2 + B_2)\, Z - J\big], \tag{10.43}$$

$$\begin{aligned} U_{TT} &= c^2 U_{XX} + NUU_{XX} + MU^2 U_{XX} + MU^2 U_{XX} + NU_X^2 + \\ &\quad + 2MUU_X^2 - H_1 U_{XXXX} + H_2 U_{XXTT} + F_2(P, J, Z)\,, \end{aligned} \tag{10.44}$$

$$P_{TT} = c_f^2 P_{XX} - \mu P_T + F_1(Z, J), \tag{10.45}$$

$$F_1 = \eta_1 Z_X + \eta_2 J_T + \eta_3 Z_T, \tag{10.46}$$

$$F_2 = \gamma_1 P_T + \gamma_2 J_T - \gamma_3 Z_T. \tag{10.47}$$

The secondary components (TW and temperature Θ) are calculated from the following equations:

$$W = \kappa U_X, \tag{10.48}$$

$$\Theta_T = \alpha \Theta_{XX} + F_3(Z, J, U), \tag{10.49}$$

where κ is a coefficient (in calculations $\kappa = 1$) and several versions of coupling forces F_3 are used in numerical solutions in the next section (Tamm et al., 2019):

$$F_3 = \tau_1 J^2, \quad F_3 = \tau_2 Z^2, \quad F_3 = \tau_3 Z_T + \tau_4 J_T. \tag{10.50}$$

Details of distinguishing the primary and secondary components are presented by Engelbrecht et al. (2018d). Here the capital letters Z, J, U, P, W, and Θ indicate dimensionless variables while X, T are dimensionless coordinates and other symbols denote parameters. The system of (10.42)–(10.46) is solved for

$$Z|_{T=0} = F_0(X), \tag{10.51}$$

where $F_0(X)$ is a pulse-like excitation.

10.3.3 Numerical Results

The model equations are solved using the pseudospectral method (PsM) (Engelbrecht et al., 2018c; Salupere, 2009), which is explained in the Appendix. The idea of the PsM is to write the PDEs in a form where all the time derivatives are on the left hand side (LHS) and all the spatial derivatives on the right-hand side (RHS) of the equations, and then apply properties of the Fourier transform for finding the spatial derivatives. This approach makes it possible to reduce the PDE into an ordinary differential equation (ODE) which can be solved by standard numerical methods.

We track numerically the fate of a localised initial pulse for Z in the middle of the space domain on the background of undisturbed (zero) initial state. This initial 'spark' is designed to exceed the threshold value for triggering a pulse and causes the AP to emerge and propagate to the left and right. All other waves are generated by the propagating AP via coupling terms added to the system as described above. We employ periodic boundary conditions as required for the PsM. We have used short enough time integration intervals to avoid interaction between counter-propagating waves at the boundaries. In the figures only left propagating waves are plotted.

The following parameter values have been used for the numerical simulations:

1. for the FHN Eqs. (10.42)–(10.43):
   $D_1 = 1,\ \varepsilon_1 = 0.018,\ A_1 = 0.2,\ A_2 = 0.2,\ B_1 = -0.05U,\ B_2 = -0.05U,$
2. for the improved Heimburg–Jackson model (10.44):
   $c^2 = 0.10,\ N = -0.05,\ M = 0.02,\ H_1 = 0.2,\ H_2 = 0.99,$
3. for the pressure (10.45):
   $c_f^2 = 0.09,\ \mu = 0.05,$
4. for the heat equation (10.49):
   $\alpha = 0.05,$
5. and the coupling coefficients are:
   $\gamma_1 = 0.008,\ \gamma_2 = 0.01,\ \gamma_3 = 3 \cdot 10^{-5};\ \eta_1 = 0.005,\ \eta_2 = 0.01,\ \eta_3 = 0.003,$
   $\tau_{1,2,3} = 5 \cdot 10^{-5},\ \tau_4 = 1 \cdot 10^{-3}.$

The parameters for the numerical scheme are $L = 64\pi$ (the length of the spatial domain), $T_f = 400$ (the end time for integration, figures are shown at $T = 400$), $n = 2048$ (the number of spatial grid points), the initial amplitude for Z is $Z_0 = 1.2$ and width of the initial sech^2 type pulse is $B_0 = 1.0$ (Engelbrecht et al., 2018c).

The primary components (AP, LW, PW) and Θ for the calculated ensemble of waves in dimensionless form are shown in Fig. 10.8 and the secondary component TW along with AP and LW in Fig. 10.9. The results of simulations are qualitatively similar to the experimental results (Terakawa, 1985; Yang et al., 2018). A detailed analysis of the ensemble of primary waves and secondary components can be found in (Engelbrecht et al., 2018a,c) where also other possible coupling forces are considered.

The results in Fig. 10.8 are obtained using a coupling term $F_3 = \tau_3 Z_T + \tau_4 J_T$ for Θ. The exact mechanism of heat production is not clear and different couplings can be used for the heat equation (10.49) resulting in different profiles (Tamm et al., 2019). Some of the temperature profiles that emerge under the proposed heat generation couplings are depicted in Fig. 10.10.

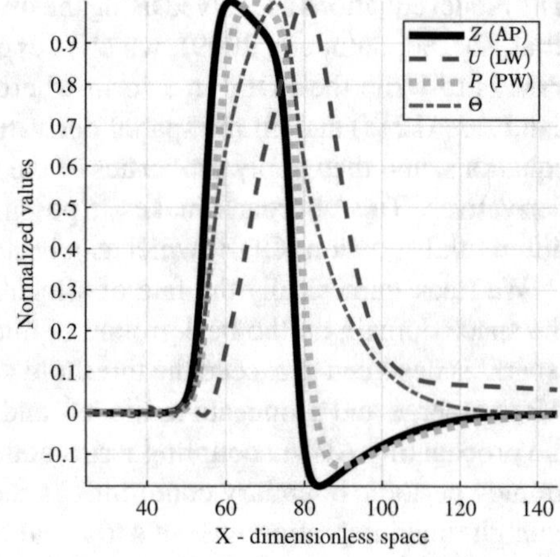

Fig. 10.8 An ensemble of waves representing the nerve signal. For Θ coupling term $F_3 = \tau_3 Z_T + \tau_4 J_T$ is used (see (10.50)). Numerical details are discussed in the Appendix.

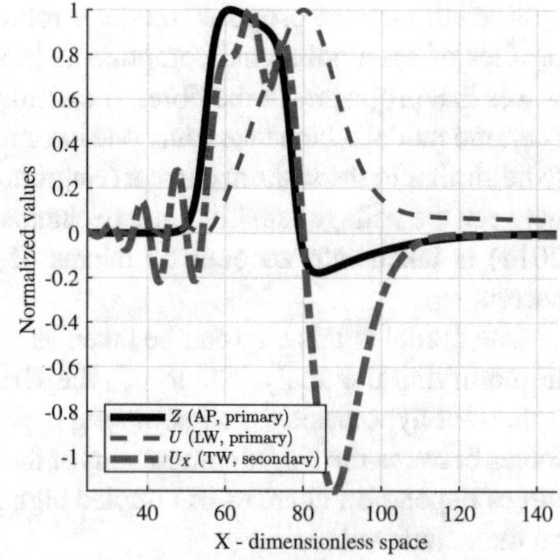

Fig. 10.9 The TW (dashed red) calculated from the LW (dashed blue) according to the idea described Sect. 10.2.2.2. The AP is shown in solid black. Numerical details are discussed in the Appendix.

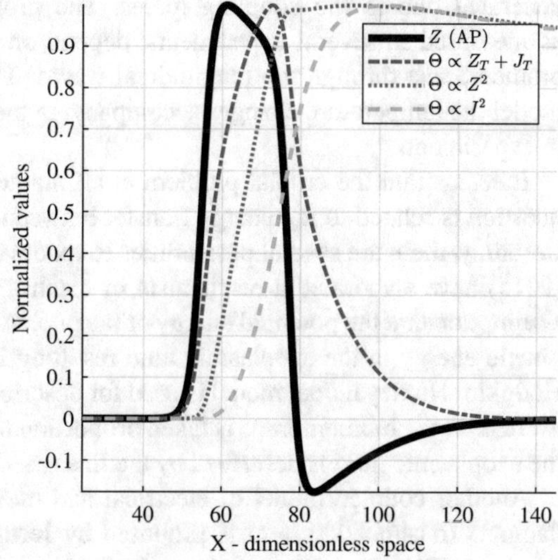

Fig. 10.10 The effect of different coupling terms on the solution of Θ. Numerical details are discussed in the Appendix.

10.4 Discussion

The material described above represents an attempt of straightforward modelling of complex processes in nerve fibres. Even though the presented model serves first of all as a proof of concept, it is fairly robust in the sense that it works properly under a wide range of conditions. However, after demonstrating that the basic equations of mathematical physics in their modified form replicate the backbone of nerve pulse dynamics, further studies could enlarge significantly the description of physiological effects. Here we have been working at the integrative level (Noble, 2002), trying to unite the different functional elements into a whole.

Even though the proposed model is robust, it describes qualitatively correctly the profiles of essential signal components in a nerve fibre in terms of an ensemble of waves that propagate in the fibre. The action potential is described by the FitzHugh–Nagumo model which takes only one ion current into account (the one that is assumed to be similar to the sodium current (Nagumo et al., 1962). A possibility to distinguish between the voltage sensitive and mechanosensitive ion channels (Mueller and Tyler, 2014) is taken into account by means of different parameters governing the ion current.

The model at this step can be taken as a proof of concept. The possible next step in modifying the analysis is to use the HH model instead of the FHN model. The important hypothesis for constructing the coupled model is related to the coupling forces between the single components of the full signal. It is assumed that the coupling forces depend on changes of coupled signals (pulses) i.e., on their derivatives, not on their static values.

By generating the action potential from an initial input, the other waves are generated due to the coupling forces. The properties of the transversal waves, as measured in several experiments, depend on the mechanical properties of the biomembrane through the longitudinal waves. There are several possible ways to model the temperature changes accompanying the action potential as demonstrated in experiments.

It seems that the crucial problem in all mathematical models of the process in question is related to the energy transfer between the components of the ensemble. Certainly, there are several possibilities to model the coupling. El Hady and Machta (2015) have elaborated a mechanism of electromechanical coupling based on the assumption that the potential energy of the process is stored in the biomembrane and kinetic energy in the axoplasmic fluid resulting in mechanical surface waves. The Hodgkin–Huxley model model is used for describing the action potential. The force exerted on the biomembrane is taken proportional to the square of the voltage while the axoplasmic fluid is described by the linearised Navier–Stokes equation.

Another coupled model of electrical and mechanical signals based on spring-dampers (dashpots) system is proposed by Jérusalem et al. (2014). Johnson and Winlow (2018) have discussed, on the basis of physiological effects, how to unite the Hodgkin–Huxley model model at the macroscopic level and the soliton model (Heimburg and Jackson, 2005) at the microscopic level. In their approach, the Na^+ ion current is the responsible electromechanical interaction.

As stressed by Coveney and Fowler (2005), coupled models across several processes could provide also the route for calculating many unknown parameters. That explains why systems biology pays so much attention to the modelling of biological complexity involving molecular and continuum approaches. The *in silico* experiments permit the coverage of a large range of possible physical parameters in order to find suitable sets verified by experiments *in vivo* or *in vitro*. This is also the case of nerve signals where the molecular structure of the biomembrane affects the process and the conjectured coupling forces need further studies and quantification. Although much is known about the fascinating process of nerve signalling, the full picture needs further concerted efforts of experimentalists and theoreticians. A recent detailed overview (Drukarch et al., 2018) reflects the contemporary insights in this field that could lead to "the formulation of a more extensive and complete conception of the nerve impulse".

In the modelling efforts described above, the path was from structures to dynamics, trying to grasp the leading effects. Such an approach is certainly robust and is used widely in systems biology (Kitano, 2002). Further modifications of the model should turn to details such as taking into account the various ionic currents, membrane channels fluctuations, general oxygen consumption, CO_2 output, the influence of carbohydrates and membrane proteins, etc together with elaborating the mechanisms of heat production and consumption. This will enhance the predictive power of the model.

Modelling at the interface of physiology and mathematics casts more light on the fascinating phenomenon of nerve pulse propagation. The structure of mathematical equations, especially of coupling forces helps to understand the causality of effects. Last but not least, as stressed by Kaufmann (1989), one cannot forget the physical background of biological processes.

Acknowledgements The authors are indebted to prof. Jüri Engelbrecht for his guidance and support over the years and for introducing the authors to the exciting field of mathematical physics.

This research was supported by the European Union through the European Regional Development fund (Estonian Programme TK 124) and by the Estonian Research Council (projects IUT 33-24, PUT 434).

Section 10.2.4 is derived in part from the article (Tamm et al., 2019) published in J. Non-Equilib. Thermodyn. (2019) aop: © De Gruyter, available online: https://doi.org/10.1515/jnet-2019-0012.

Appendix

The pseudospectral method (PsM) (Fornberg, 1988; Salupere, 2009) is used to solve the system of dimensionless model equations:

$$
\begin{aligned}
Z_T &= D Z_{XX} + Z\left[Z - C_1 - Z^2 + C_1 Z\right] - J, \quad J_T = \varepsilon\left[C_2 Z - J\right], \\
U_{TT} &= c^2 U_{XX} + N U U_{XX} + M U^2 U_{XX} + N U_X^2 + 2 M U U_X^2 - \\
&\quad - H_1 U_{XXXX} + H_2 U_{XXTT} + F_2(Z, J, P), \\
P_{TT} &= c_f^2 P_{XX} - \mu P_T + F_1(Z, J, U), \quad \Theta_T = \alpha \Theta_{XX} + F_3(Z, J, U).
\end{aligned}
\tag{10.52}
$$

Here Z is the action potential, J is the recovery current, $C_i = A_i + B_i$, A_i is the 'electrical' and B_i is the 'mechanical' activation coefficient, D, ε are coefficients, $U = \Delta\rho$ is the longitudinal density change in lipid layer, c is velocity of unperturbed state in lipid bilayer, M, N are the nonlinear coefficients, H_1, H_2 are the dispersion coefficients, $F_1 = \eta_1 Z_X + \eta_2 J_T + \eta_3 Z_T$ and γ_i are the coupling coefficients for the mechanical wave, η_i are the coupling coefficients for the pressure wave, P is the pressure, c_f is the characteristic velocity in the fluid, and $F_2 = \gamma_1 P_T + \gamma_2 J_T - \gamma_3 Z_T$ and μ is the (viscous) damping coefficient.

The notation Θ represents the (local) temperature and α is a coefficient characterising the temperature diffusion within the environment. 'Mechanical' activation coefficients in the action potential and ion current expressions are connected to the improved Heimburg–Jackson part of the model as $B_1 = -\beta_1 U$ and $B_2 = -\beta_2 U$ where β_1, β_2 are the mechanical coupling coefficients. In (10.52) either J_T or J_X are used as coupling forces and $F_3(Z, J, U)$ follows expressions (10.50).

Initial and Boundary Conditions

A sech^2 type localised initial condition is applied to Z in (10.52). We make use of the periodic boundary conditions for all the members of the model equations

$$
\begin{aligned}
&Z(X,0) = Z_0 \text{sech}^2 B_0 X\,, \qquad Z(X,T) = Z(X + 2Km\pi, T)\,,\\
&J(X,0) = 0, \quad J_T(X,0) = 0, \quad J(X,T) = J(X + 2Km\pi, T)\,,\\
&U(X,0) = 0, \quad U_T(X,0) = 0, \quad U(X,T) = U(X + 2Km\pi, T)\,,\\
&P(X,0) = 0, \quad P_T(X,0) = 0, \quad P(X,T) = P(X + 2Km\pi, T)\,,\\
&\Theta(X,0) = 0, \quad \Theta(X,T) = \Theta(X + 2Km\pi, T)\,,
\end{aligned}
\tag{10.53}
$$

where $m = 1, 2, \ldots, K$ is the total number 2π sections in the spatial period. The amplitude of the initial 'spark' is Z_0 and the width parameter is B_0. Such an initial condition is a narrow 'spark' in the middle of the considered space domain with the amplitude above the threshold resulting in the usual FHN action potential formation which then proceeds to propagate in the positive and negative directions of the 1D space domain under consideration.

For all other equations, we set the initial excitation to zero and make use of the same periodic boundary conditions. The solution representing the mechanical and pressure wave is generated over time as a result of coupling with the action potential and ion current parts in the model system. In the presented calculations wave interactions are not investigated and integration intervals in time are designed so that the tracked waves do not reach the boundaries. This means that the particular type of boundary conditions is unimportant; however, the pseudospectral method requires periodic boundary conditions.

We do not discuss here several potentially important aspects. For example, the action potentials (and ion currents tied to these) annihilate each other during the interaction (as expected) but the mechanical and pressure waves can keep on going through many interactions. This process is supported by the use of periodic boundary conditions and makes it possible to track numerous interactions within the modelled wave ensembles.

The Derivatives and Integration

For numerical solving of (10.52) we use the discrete Fourier transform (DFT) based PsM (Fornberg, 1988; Salupere, 2009). Variable Z can be represented in the Fourier space as

$$\widehat{Z}(k,T) = \mathrm{F}[Z] = \sum_{j=0}^{n-1} Z(j\Delta X, T)\exp\left(-\frac{2\pi \mathrm{i} jk}{n}\right), \tag{10.54}$$

where n is the number of space grid points ($n = 2^{12} = 4096$ in the presented simulations), $\Delta X = 2\pi/n$ is the space step, $k = 0, \pm 1, \pm 2, \ldots, \pm(n/2-1), -n/2$. Further, i is the imaginary unit, F denotes the DFT, and F^{-1} denotes the inverse DFT. The idea of the PsM is to approximate space derivatives by making use of the DFT:

$$\frac{\partial^m Z}{\partial X^m} = \mathrm{F}^{-1}\left[(\mathrm{i}k)^m \mathrm{F}(Z)\right], \tag{10.55}$$

reducing therefore the PDE to an ODE and making it possible to use standard ODE solvers for the integration with respect to time. The model (10.52) contains a mixed derivative and coupling force terms can be taken either as a space derivative (which can be found like in (10.54)) or time derivative (which is not suitable for a direct PsM application and needs to be handled separately).

For integration in time, (10.52) are rewritten as a system of first order ODE's after the modification to handle the mixed partial derivative term and a standard numerical integrator is applied. In the present paper ODEPACK FORTRAN code (Hindmarsh, 1983) ODE solver is used by making use of the F2PY (Peterson, 2005) generated Python interface. Handling the data and initialisation of the variables is done in Python by making use of the package SciPy (Jones et al., 2007).

Handling of Mixed Derivatives

Normally the PsM is intended for solving $u_t = \Phi(u, u_x, u_{2x}, \ldots, u_{mx})$ type equations. The presence of a mixed partial derivative $H_2 U_{XXTT}$ in (10.52) requires some modifications (Ilison et al., 2007; Ilison and Salupere, 2009; Salupere, 2009). This is done by rewriting in (10.52) the equation for U so that all partial derivatives with respect to time are at the left-hand side of the equation:

$$\begin{aligned} U_{TT} - H_2 U_{XXTT} = c^2 U_{XX} + NUU_{XX} + MU^2 U_{XX} + N\left(U_X\right)^2 + \\ + 2MU\left(U_X\right)^2 - H_1 U_{XXXX} + F_2(Z, J, P). \end{aligned} \tag{10.56}$$

This representation allows one to introduce a new variable $\Phi = U - H_2 U_{XX}$. Making use of properties of the DFT, one can express variable U and its spatial derivatives in terms of Φ:

$$U = \mathrm{F}^{-1}\left[\frac{\mathrm{F}(\Phi)}{1 + H_2 k^2}\right], \quad \frac{\partial^m U}{\partial X^m} = \mathrm{F}^{-1}\left[\frac{(\mathrm{i}k)^m \mathrm{F}(\Phi)}{1 + H_2 k^2}\right]. \tag{10.57}$$

Finally, in (10.52) the equation for U can be rewritten in terms of the variable Φ as

$$\begin{aligned}\Phi_{TT} &= c^2 U_{XX} + NUU_{XX} + MU^2 U_{XX} + N\,(U_X)^2 + \\ &+ 2MU\,(U_X)^2 - H_1 U_{XXXX} + F_2(Z, J, P),\end{aligned} \tag{10.58}$$

where all partial derivatives of U with respect to X are calculated in terms of Φ by using expression (10.57) and therefore one can apply the PsM for numerical integration of (10.58). Other Eqs. (10.52) are already written in the form which can be solved by the standard PsM.

The Time Derivatives P_T and J_T

The time derivatives P_T and J_T are found using different methods. For finding P_T it is enough to write the equation for P in (10.52) as two first order ODE's. This has to be done anyway as the integrator requires first order ODE's. It is possible to extract P_T from there directly:

$$\begin{aligned}P_T &= \bar{V}, \\ \bar{V}_T &= c_f^2 P_{XX} - \mu P_T + F_2(Z, J, U).\end{aligned} \tag{10.59}$$

For finding J_T a basic backward difference scheme is used:

$$J_T(n, T) = \frac{J(n, T) - J(n, (T - \Delta T))}{T - (T - \Delta T)} \approx \frac{\Delta J(n, T)}{\Delta T}, \tag{10.60}$$

where J is the ion current from (10.52), n is the spatial node number, T is the dimensionless time and dT is the integrator internal time step (which is variable). In the presented simulations the integrator is allowed to take up to 10^6 internal time steps between ΔT values to provide the desired numerical accuracy.

The Technical Details and Numerical Accuracy

As noted earlier, the calculations are carried out with the Python package SciPy (Jones et al., 2007) using the FFTW library (Frigo and Johnson, 2005) for the DFT and the F2PY (Peterson, 2005) generated Python interface to the ODEPACK FORTRAN code (Hindmarsh, 1983) for the ODE solver. The particular integrator used is the 'vode' with options set to nsteps $= 10^6$, rtol $= 10^{-11}$, atol $= 10^{-12}$ and $\Delta T = 2$.

Typically, the hyperbolic functions, like the hyperbolic secant $\text{sech}^2(X)$ in our initial conditions in (10.53), are defined around zero (Fig. 10.5). However, in the presented examples the spatial domain is taken from 0 to $2\pi K$. This means that the functions in (10.53) are actually shifted to the right (in direction of the positive axis of space) by πK. Thus, the shape typically defined around zero is actually in our case located in the middle of the spatial domain (or period). This is a matter of preference. In the present case the reason for doing so is the wish to have a more convenient mapping between the values of X and indexes. The numerical results would be the same if one would use a spatial period from $-\pi K$ to πK.

The 'discrete frequency function' k in (10.55) is typically formulated on the interval from $-\pi$ to π, however, we use a different spatial period than 2π and also shift our space from 0 to $2\pi K$ meaning that

$$k = \left[\frac{0}{K}, \frac{1}{K}, \frac{2}{K}, \ldots, \frac{n/2-1}{K}, \frac{n/2}{K}, -\frac{n/2}{K}, -\frac{n/2-1}{K}, \ldots, -\frac{n-1}{K}, -\frac{n}{K}\right], \tag{10.61}$$

where n is number of the spatial grid points uniformly distributed across our spatial period. The size of the Fourier spectrum is $n/2$. It is, in essence, the number of spectral harmonics used for approximating the periodic functions and their derivatives and K is the number of 2π sections in our space interval.

There are few different possibilities for handling the division by zero that occurs in (10.60) during the initialisation of the ODE solver and when the numerical iteration during the integration reaches the desired accuracy resulting in a zero length time step. For the initialisation of the numerical function an initial value of 1 is used for dT. This is just a technical nuance as during the initialisation the time derivative can be assumed zero without loss of generality anyway as there is no change in the value of $J(n, 0)$.

For handling the division by zero during the integration when ODE solver reaches the desired accuracy using values from two steps back from the present time for J and T is computationally the most efficient. Another straightforward alternative is to use a logical cycle inside the ODE solver for checking if dT would be zero but this is computationally inefficient. In the presented examples using a value two steps back in time for calculating J_T is used for all presented results involving J_T. The difference between the numerical solutions of the J_T with the scheme using a value one step back and additional check for division by zero and using two steps back in time scheme only if division by zero occurs is only approximately 10^{-6}. The gain from the elimination of this small difference is not worth the nearly twofold increase in the numerical integration time by using a much more complicated exact scheme.

The overall accuracy of the numerical solution is $\sim 10^{-7}$ for the fourth derivatives, $\sim 10^{-9}$ for the second derivatives and $\sim 10^{-11}$ for the time integrals. The accuracy of J_T is $\sim 10^{-6}$, which is adequate and roughly of the same order of magnitude as the accuracy of fourth spatial derivatives. The accuracy estimates are not based on the solving (10.52) with the presented parameters. They are based instead on using the same scheme with the same technical parameters for finding the derivatives of $\sin(x)$ and comparing these to an analytic solution. Even though in the PST the spectral filtering (suppression of the higher harmonics in the Fourier spectrum) is a common approach for increasing the stability of the scheme, in the presented simulations the filtering is not used. Instead, the highest harmonic (which tends to collect the truncation errors from the finite numerical accuracy of floating point numbers in the PST schemes) is monitored as a 'sanity check' of the scheme.

References

Abbott, B.C., Hill, A.V., Howarth, J.V.: The positive and negative heat production associated with a nerve impulse. Proc. R. Soc. B Biol. Sci. **148**(931), 149–187 (1958). https://doi.org/10.1098/rspb.1958.0012

Andersen, S.S.L., Jackson, A.D., Heimburg, T.: Towards a thermodynamic theory of nerve pulse propagation. Prog. Neurobiol. **88**(2), 104–13 (2009). https://doi.org/10.1016/j.pneurobio.2009.03.002

Andronov, A., Witt, A., Khaikin, S.: Theory of Oscillations. Phys. Math. Publ. Moscow (In Russian) (1959)

Appali, R., Petersen, S., Van Rienen, U.: A comparison of Hodgkin–Huxley and soliton neural theories. Adv. Radio Sci. **8**, 75–79 (2010). https://doi.org/10.5194/ars-8-75-2010

Bennett, M.V.L.: Electrical synapses, a personal perspective (or history). Brain Res. Rev. **32**(1), 16–28 (2000). https://doi.org/10.1016/S0165-0173(99)00065-X

Billingham, J., King, A.C.: Wave Motion. Cambridge Texts in Applied Mathematics. Cambridge University Press (2001). https://doi.org/10.1017/CBO9780511841033

Bonhoeffer, K.F.: Activation of passive iron as a model for the excitation of nerve. J. Gen. Physiol. **32**(1), 69–91 (1948). https://doi.org/10.1085/jgp.32.1.69

Bountis, T., Starmer, C.F., Bezerianos, A.: Stationary pulses and wave front formation in an excitable medium. Prog. Theor. Phys. Suppl. **139**, 12–33 (2000). https://doi.org/10.1143/PTPS.139.12

Bresslof, P.C.: Waves in Neural Media. Springer, New York (2014)

Carslaw, H., Jaeger, J.: Conduction of Heat in Solids, 2nd edn. Oxford Science Publications, Oxford (1959)

Christov, C.I., Maugin, G.A., Porubov, A.V.: On Boussinesq's paradigm in nonlinear wave propagation. C. R. Mec. **335**(9-10), 521–535 (2007). https://doi.org/10.1016/j.crme.2007.08.006

Clay, J.R.: Axonal excitability revisited. Prog. Biophys. Mol. Biol. **88**(1), 59–90 (2005). https://doi.org/10.1016/j.pbiomolbio.2003.12.004

Courtemanche, M., Ramirez, R.J., Nattel, S.: Ionic mechanisms underlying human atrial action potential properties: insights from a mathematical model. Am. J. Physiol. **275**(1), 301–321 (1998). https://doi.org/10.1152/ajpheart.1998.275.1.H301

Coveney, P.V., Fowler, P.W.: Modelling biological complexity: a physical scientist's perspective. J. R. Soc. Interface **2**(4), 267–280 (2005). https://doi.org/10.1098/rsif.2005.0045

Debanne, D., Campanac, E., Bialowas, A., Carlier, E., Alcaraz, G.: Axon physiology. Physiol. Rev. **91**(2), 555–602 (2011). doi10.1152/physrev.00048.2009

Downing, A.C., Gerard, R.W., Hill, A.V.: The heat production of nerve. Proc. R. Soc. B Biol. Sci. **100**(702), 223–251 (1926). https://doi.org/10.1098/rspb.1926.0044

Drukarch, B., Holland, H.A., Velichkov, M., Geurts, J.J., Voorn, P., Glas, G., de Regt, H.W.: Thinking about the nerve impulse: A critical analysis of the electricity-centered conception of nerve excitability. Prog. Neurobiol. **169**, 172–185 (2018). https://doi.org/10.1016/j.pneurobio.2018.06.009

El Hady, A., Machta, B.B.: Mechanical surface waves accompany action potential propagation. Nat. Commun. **6**, 6697 (2015). https://doi.org/10.1038/ncomms7697

Engelbrecht, J.: On theory of pulse transmission in a nerve fibre. Proc. R. Soc. Lond. **375**(1761), 195–209 (1981). https://doi.org/10.1098/rspa.1981.0047

Engelbrecht, J.: Nonlinear Wave Processes of Deformation in Solids. Pitman Monographs and Surveys in Pure and Applied Mathematics, vol. 16. Pitman Advanced Publishing Program (1983)

Engelbrecht, J.: An Introduction to Asymmetric Solitary Waves. Pitman Monographs and Surveys in Pure and Applied Mathematics, vol 56. Longman Scientific & Technical, Harlow (1996)

Engelbrecht, J.: Complexity in engineering and natural sciences. Proc. Estonian Acad. Sci. **64**(3), 249–255 (2015a). https://doi.org/10.3176/proc.2015.3.07

Engelbrecht, J.: Questions About Elastic Waves. Springer International Publishing, Cham (2015b). https://doi.org/10.1007/978-3-319-14791-8

Engelbrecht, J., Tamm, K., Peets, T.: On mathematical modelling of solitary pulses in cylindrical biomembranes. Biomech. Model. Mechanobiol. **14**(1), 159–167 (2015). https://doi.org/10.1007/s10237-014-0596-2

Engelbrecht, J., Tamm, K., Peets, T.: On solutions of a Boussinesq-type equation with displacement-dependent nonlinearities: the case of biomembranes. Phil. Mag. **97**(12), 967–987 (2017). https://doi.org/10.1080/14786435.2017.1283070

Engelbrecht, J., Peets, T., Tamm, K.: Electromechanical coupling of waves in nerve fibres. Biomech. Model. Mechanobiol. **17**(6), 1771–1783 (2018). https://doi.org/10.1007/s10237-018-1055-2

Engelbrecht, J., Peets, T., Tamm, K., Laasmaa, M., Vendelin, M.: On the complexity of signal propagation in nerve fibres. Proc. Estonian Acad. Sci. **67**(1), 28–38 (2018). https://doi.org/10.3176/proc.2017.4.28

Engelbrecht, J., Tamm, K., Peets, T.: Modeling of complex signals in nerve fibers. Med. Hypotheses **120**, 90–95 (2018). https://doi.org/10.1016/j.mehy.2018.08.021

Engelbrecht, J., Tamm, K., Peets, T.: Primary and secondary components of nerve signals. arXiv:1812.05335 [physics.bio-ph] (2018)

Engelbrecht, J., Tamm, K., Peets, T.: Mathematics of nerve signals. arXiv:1902.00011 [physics.bio-ph] (2019)

Engelbrecht, J., Tobias, T.: On a model stationary nonlinear wave in an active medium. Proc. R. Soc. Lond. A Math. Phys. Eng. Sci. **A411**(1840), 139–154 (1987). https://doi.org/10.1098/rspa.1987.0058

Fillafer, C., Mussel, M., Muchowski, J., Schneider, M.F.: On cell surface deformation during an action potential. arXiv:1703.04608 [physics.bio-ph] (2017)

FitzHugh, R.: Impulses and physiological states in theoretical models of nerve membrane. Biophys. J. **1**(6), 445–466 (1961). https://doi.org/10.1016/S0006-3495(61)86902-6

Fornberg, B.: A Practical Guide to Pseudospectral Methods. Cambridge University Press, Cambridge (1998). https://doi.org/10.1017/CBO9780511626357

Frigo, M., Johnson, S.: The design and implementation of FFTW 3. Proc. IEEE **93**(2), 216–231 (2005). https://doi.org/10.1109/jproc.2004.840301

Gilbert, D.S.: Axoplasm architecture and physical properties as seen in the Myxicola giant axon. J. Physiol. **253**, 257–301 (1975). https://doi.org/10.1113/jphysiol.1975.sp011190

Gonzalez-Perez, A., Mosgaard, L., Budvytyte, R., Villagran-Vargas, E., Jackson, A., Heimburg, T.: Solitary electromechanical pulses in lobster neurons. Biophys. Chem. **216**, 51–59 (2016). https://doi.org/10.1016/j.bpc.2016.06.005

Heimburg, T., Jackson, A.D.: On soliton propagation in biomembranes and nerves. Proc. Natl. Acad. Sci. USA **102**(28), 9790–5 (2005). https://doi.org/10.1073/pnas.0503823102

Heimburg, T., Jackson, A.: On the action potential as a propagating density pulse and the role of anesthetics. Biophys. Rev. Lett. **2**(1), 57–78 (2007). https://doi.org/10.1142/s179304800700043x

Heimburg, T., Jackson, A.D.: Thermodynamics of the nervous impulse. In: Nag, K. (ed.) Structure and Dynamics of Membranous Interfaces, pp. 317–339. John Wiley & Sons, Hoboken, NJ, USA (2014). https://doi.org/10.1002/9780470388495.ch12

Hindmarsh, A.: ODEPACK, A Systematized Collection of ODE Solvers, vol. 1. North-Holland, Amsterdam (1983)

Hodgkin, A.L.: The Conduction of the Nervous Impulse. Liverpool University Press (1964)

Hodgkin, A.L., Huxley, A.F.: A quantitative description of membrane current and its application to conduction and excitation in nerve. J. Physiol. **117**(4), 500–544 (1952). https://doi.org/10.1113/jphysiol.1952.sp004764

Howarth, J.V., Keynes, R.D., Ritchie, J.M.: The origin of the initial heat associated with a single impulse in mammalian non-myelinated nerve fibres. J. Physiol. **194**(3), 745–93 (1968). https://doi.org/10.1113/jphysiol.1968.sp008434

Ilison, L., Salupere, A.: Propagation of sech^2-type solitary waves in hierarchical KdV-type systems. Math. Comput. Simul. **79**(11), 3314–3327 (2009).

Ilison, L., Salupere, A., Peterson, P.: On the propagation of localized perturbations in media with microstructure. Proc. Estonian Acad. Sci. Phys. Math. **56**(2), 84–92 (2007)

Iwasa, K., Tasaki, I., Gibbons, R.: Swelling of nerve fibers associated with action potentials. Science **210**(4467), 338–339 (1980). https://doi.org/10.1126/science.7423196

Jérusalem, A., García-Grajales, J.A., Merchán-Pérez, A., Peña, J.M.: A computational model coupling mechanics and electrophysiology in spinal cord injury. Biomech. Model. Mechanobiol. **13**(4), 883–896 (2014). https://doi.org/10.1007/s10237-013-0543-7

Johnson, A.S., Winlow, W.: The soliton and the action potential – primary elements underlying sentience. Front. Physiol. **9** (2018). https://doi.org/10.3389/fphys.2018.00779

Jones, E., Oliphant, T., Peterson, P.: SciPy: open source scientific tools for Python (2007). http://www.scipy.org

Kaufmann, K.: Action Potentials and Electromechanical Coupling in the Macroscopic Chiral Phospholipid Bilayer. Caruaru, Brazil (1989)

Kitano, H.: Systems biology: a brief overview. Science **295**(5560), 1662–1664 (2002). https://doi.org/10.1126/science.1069492

Lieberstein, H.H.: On the Hodgkin–Huxley partial differential equation. Math. Biosci. **1**(1), 45–69 (1967). https://doi.org/10.1016/0025-5564(67)90026-0

Lin, T., Morgan, G.: Wave propagation through fluid contained in a cylindrical elastic shell. J. Acoust. Soc. Am. **28**(6), 1165–1176 (1956)

Malmivuo, J., Plonsey, R.: Bioelectromagnetism. Principles and Applications of Bioelectric and Biomagnetic Fields. Oxford University Press (1995). https://doi.org/10.1093/acprof:oso/9780195058239.001.0001

Maugin, G.A.: Nonlinear Waves in Elastic Crystals. Oxford University Press, Oxford (1999)

Maurin, F., Spadoni, A.: Wave propagation in periodic buckled beams. Part I: Analytical models and numerical simulations. Wave Motion **66**, 190–209 (2016). https://doi.org/10.1016/j.wavemoti.2016.05.008

Maurin, F., Spadoni, A.: Wave propagation in periodic buckled beams. Part II: Experiments. Wave Motion **66**, 210–219 (2016). https://doi.org/10.1016/j.wavemoti.2016.05.009

Mueller, J.K., Tyler, W.J.: A quantitative overview of biophysical forces impinging on neural function. Phys. Biol. **11**(5), 051001 (2014). https://doi.org/10.1088/1478-3975/11/5/051001

Mussel, M., Schneider, M.F.: It sounds like an action potential: unification of electrical, chemical and mechanical aspects of acoustic pulses in lipids. J. R. Soc. Interface **16**(151), 20180743 (2019). https://doi.org/10.1098/rsif.2018.0743

Nagumo, J., Arimoto, S., Yoshizawa, S.: An active pulse transmission line simulating nerve axon. Proc. IRE **50**(10), 2061–2070 (1962). https://doi.org/10.1109/JRPROC.1962.288235

Neu, J.C., Preissig, R., Krassowska, W.: Initiation of propagation in a one-dimensional excitable medium. Physica D **102**(3–4), 285–299 (1997). https://doi.org/10.1016/S0167-2789(96)00203-5

Noble, D.: Chair's introduction. In: Bock, G., Goode, J.A. (eds.) 'In Silico' Simulation of Biological Processes, pp. 1–3. John Wiley & Sons, Chichester (2002). https://doi.org/10.1002/0470857897.ch1

Nolte, D.D.: Introduction to Modern Dynamics: Chaos, Networks, Space and Time. Oxford University Press, Oxford (2015)

Peets, T., Tamm, K.: On mechanical aspects of nerve pulse propagation and the Boussinesq paradigm. Proc. Estonian Acad. Sci. **64**(3S), 331–337 (2015). https://doi.org/10.3176/proc.2015.3S.02

Peets, T., Tamm, K., Engelbrecht, J.: On the role of nonlinearities in the Boussinesq-type wave equations. Wave Motion **71**, 113–119 (2017). https://doi.org/10.1016/j.wavemoti.2016.04.003

Peets, T., Tamm, K., Simson, P., Engelbrecht, J.: On solutions of a Boussinesq-type equation with displacement-dependent nonlinearity: A soliton doublet. Wave Motion **85**, 10–17 (2019). https://doi.org/10.1016/j.wavemoti.2018.11.001

Perez-Camacho, M.I., Ruiz-Suarez, J.: Propagation of a thermo-mechanical perturbation on a lipid membrane. Soft Matter **13**, 6555–6561 (2017). https://doi.org/10.1039/C7SM00978J

Peterson, P.: F2PY: Fortran to Python interface generator. http://cens.ioc.ee/projects/f2py2e/ (2005)

van der Pol, B.: On "relaxation-oscillations". London, Edinburgh, Dublin Philos. Mag. J. Sci. **2**(11), 978–992 (1926). https://doi.org/10.1080/14786442608564127

Porubov, A.V.: Amplification of Nonlinear Strain Waves in Solids. World Scientific, Singapore (2003). https://doi.org/10.1142/5238

Reissig, R., Sansone, G., Conti, R.: Qualitative Theorie Nichtlinearer Differentialgleichungen. Edizioni Cremonese, Roma (1963)

Ritchie, J.M., Keynes, R.D.: The production and absorption of heat associated with electrical activity in nerve and electric organ. Q. Rev. Biophys. **18**(4), 451 (1985). https://doi.org/10.1017/S0033583500005382

Salupere, A.: The pseudospectral method and discrete spectral analysis. In: Quak, E., Soomere, T. (eds.) Applied Wave Mathematics. Selected Topics in Solids, Fluids and Mathematical Methods, pp. 301–334. Springer, Heidelberg (2009). https://doi.org/10.1007/978-3-642-00585-5_16

Scott, A.: Nonlinear Science. Emergence & Dynamics of Coherent Structures. Oxford University Press (1999)

Tamm, K., Peets, T.: On solitary waves in case of amplitude-dependent nonlinearity. Chaos Solitons Fractals **73**, 108–114 (2015). https://doi.org/10.1016/j.chaos.2015.01.013

Tamm, K., Engelbrecht, J., Peets, T.: Temperature changes accompanying signal propagation in axons. J. Non-Equilibr. Thermodyn. **44**(3), 277–284 (2019). https://doi.org/10.1515/jnet-2019-0012

Taniuti, T., Nishihara, K.: Nonlinear Waves. Pitman, Boston (1983)

Tasaki, I.: A macromolecular approach to excitation phenomena: mechanical and thermal changes in nerve during excitation. Physiol. Chem. Phys. Med. NMR **20**(4), 251–268 (1988)

Tasaki, I., Byrne, P.M.: Heat production associated with a propagated impulse in bullfrog myelinated nerve fibers. Jpn. J. Physiol. **42**(5), 805–813 (1992). https://doi.org/10.2170/jjphysiol.42.805

Tasaki, I., Kusano, K., Byrne, P.M.: Rapid mechanical and thermal changes in the garfish olfactory nerve associated with a propagated impulse. Biophys. J. **55**(6), 1033–1040 (1989). https://doi.org/10.1016/s0006-3495(89)82902-9

Terakawa, S.: Potential-dependent variations of the intracellular pressure in the intracellularly perfused squid giant axon. J. Physiol. **369**(1), 229–248 (1985). https://doi.org/10.1113/jphysiol.1985.sp015898

Thompson, J.: Instabilities and Catastrophes in Science and Engineering. Wiley, New York (1982)

Tritton, J.: Physical Fluid Dynamics. Oxford Sci. Publ. (1988)

Vendelin, M., Saks, V., Engelbrecht, J.: Principles of mathematical modeling and in silico studies of integrated cellular energetics. In: Saks, V. (ed.) Molecular System Bioenergetics: Energy for Life, pp. 407–433. Wiley, Weinheim (2007). https://doi.org/10.1002/9783527621095.ch12

Wilke, E.: On the problem of nerve excitation in the light of the theory of waves. Pflügers Arch. **144**, 35–38 (1912)

Yang, Y., Liu, X.W., Wang, H., Yu, H., Guan, Y., Wang, S., Tao, N.: Imaging action potential in single mammalian neurons by tracking the accompanying sub-nanometer mechanical motion. ACS Nano p. acsnano.8b00867 (2018). https://doi.org/10.1021/acsnano.8b00867

Chapter 11
Operator Splits and Multiscale Methods in Computational Dynamics

Harm Askes, Dario De Domenico, Mingxiu Xu, Inna M. Gitman, Terry Bennett, and Elias C. Aifantis

Abstract Gradient enriched continua are an elegant and versatile class of material models that are able to simulate a variety of physical phenomena, ranging from singularity-free descriptions of crack tips and dislocations, via size dependent mechanical response, to dispersive wave propagation. However, the increased order of the governing partial differential equations has historically complicated analytical and numerical solution methods. Inspired by the work of Ru and Aifantis (Acta Mech. **101**(1-4), 59–68 (1993)), this contribution focusses on operator split methods that allow to reduce the order of the governing equations. It will be shown that this order reduction leads to multiscale reformulations of the original equations in which the macrolevel unknowns are fully coupled to the microlevel unknowns. As a first example, gradient enriched equations of elastodynamics are considered with second order and fourth order microinertia terms. The second example concerns

H. Askes (✉)
Department of Civil and Structural Engineering, University of Sheffield, Sheffield, UK
e-mail: h.askes@sheffield.ac.uk

D. De Domenico
Department of Engineering, University of Messina, Messina, Italy
e-mail: dario.dedomenico@unime.it

M. Xu
Department of Applied Mechanics, University of Science and Technology Beijing, Beijing, China
e-mail: xumx@ustb.edu.cn

I. M. Gitman
Department of Mechanical Engineering, University of Sheffield, Sheffield, UK
e-mail: i.gitman@sheffield.ac.uk

T. Bennett
School of Civil, Environmental and Mining Engineering, University of Adelaide, Adelaide, SA, Australia
e-mail: terry.bennett@adelaide.edu.au

E. C. Aifantis
Laboratory of Mechanics and Materials, Aristotle University of Thessaloniki, Thessaloniki, Greece
e-mail: mom@mom.gen.auth.gr

A. Berezovski, T. Soomere (eds.), *Applied Wave Mathematics II*, Mathematics of Planet Earth 6, https://doi.org/10.1007/978-3-030-29951-4_11

dynamic piezomagnetics with gradient enrichment of both the mechanical fields and the magnetic fields.

11.1 Introduction

Generalised continua are suitable modelling tools to simulate the behaviour of materials and structures in cases where the underlying microstructure of the material has a significant effect on the macroscopic behaviour. There are many different types of generalised continua, but in this contribution we will focus on gradient enriched continua, in which higher order spatial derivatives of the standard terms are added to the governing equations. In particular, we will explore gradient enrichments that have been promulgated by Aifantis and coworkers since the 1980s, namely those whereby the higher order terms are the Laplacian of the standard terms (Aifantis, 1984, 1987; Mühlhaus and Alfantis, 1991; Aifantis, 1992; Altan and Aifantis, 1992; Ru and Aifantis, 1993; Vardoulakis and Aifantis, 1994).

Gradient enriched continua have been used successfully in situations where standard continua fail, such as the description of finite width shear bands and damage zones, singularity-free crack tip and dislocation fields, size dependent mechanical behaviour, and dispersive wave propagation. These cases have in common that the macroscale behaviour is partly driven by certain microstructural material parameters. These parameters are lacking in standard continua but appear naturally, as the coefficients of the higher order spatial derivatives, in gradient enriched continua. The physical interpretation of these higher order constitutive coefficients has been the subject of intense research in the community. An overview of approaches and results can be found in (Askes and Aifantis, 2011, Section 4).

Another area of research into gradient enriched continua is their numerical implementation. This is not trivial, since the introduction of higher order spatial derivatives increases the order of the partial differential equations (PDEs). This typically precludes implementation with standard finite element technology and has, thus, inspired a number of alternative solution methods — see (Askes and Aifantis, 2011, Section 5) for a review.

One relatively straightforward approach to finite element implementation of gradient elasticity theories utilises the mathematical format of Laplacian type theories. Namely, the various derivatives may be factorised such that the fourth order PDE may be rewritten into a set of two second order PDEs (Ru and Aifantis, 1993; Lurie et al., 2003). Depending on the format of the original fourth order PDE, the second order PDEs resulting from such an operator split may be either coupled or decoupled; examples of either case are discussed below.

An interesting and, perhaps, unanticipated additional outcome enabled by the operator split is the physical interpretation of the primary unknowns. Whereas the original fourth order PDE is typically a monoscale equation, the primary unknowns of the resulting second order PDEs can be understood as microscale and macroscale variables. Thus, the operator split naturally leads to a multiscale reformulation of the relevant gradient theory (Bennett et al., 2007; Askes et al., 2008a).

Below, we review the main ideas of operator splits in Sect. 11.2 and their multiscale interpretation in Sect. 11.3. Next, we review earlier work in gradient elastodynamics in Sect. 11.4, focussing on two different truncations of microinertia contributions. A new gradient enriched dynamic model for piezomagnetics, drawing on the results for gradient elastodynamics, is presented in Sect. 11.5. A generic variational formulation of all discussed models is given in Sect. 11.6.

11.2 Gradient Elasticity in Statics

In the early 1990s, Aifantis and coworkers developed a theory of gradient enriched elasticity (Aifantis, 1992; Altan and Aifantis, 1992; Ru and Aifantis, 1993) that can be written as

$$\sigma_{ij} = C_{ijkl}\left(\varepsilon_{kl} - \ell^2 \varepsilon_{kl,mm}\right). \tag{11.1}$$

Here, the stresses σ_{ij} are not only related to the strains ε_{kl} but also to their Laplacian. The standard elastic material parameters are contained in the stiffness tensor C_{ijkl}. The higher order term is accompanied by an additional material parameter ℓ which represents the microstructural properties. Together with the usual equilibrium equations and kinematic relations, (11.1) leads to following governing equation:

$$C_{ijkl}\left(u_{k,jl} - \ell^2 u_{k,jlmm}\right) + b_i = 0, \tag{11.2}$$

where u_k are the displacements and b_i are the body forces. The inclusion of this higher order term has been demonstrated to be effective in the removal of singularities (Altan and Aifantis, 1992; Unger and Aifantis, 1995; Gutkin and Aifantis, 1996, 1997) and the description of size dependent mechanical response (Aifantis, 1999; Askes and Aifantis, 2002).

Formally, the Aifantis model of (11.1) can be thought of as a special case of the earlier Mindlin model (Mindlin, 1964, 1965) which is equipped with a much larger number of higher order terms and higher order constitutive constants. However, this observation obfuscates a significant advantage of the Aifantis model: its mathematical structure greatly facilitates analytical and numerical solution methods which are not available for the more general Mindlin model. As was observed and explored by Ru and Aifantis (1993) as well as Lurie et al. (2003), it is possible to factorise the various derivatives in (11.2) as

$$C_{ijkl}\frac{\partial^2}{\partial x_j \partial x_l}\left(u_k - \ell^2 u_{k,mm}\right) + b_i = 0. \tag{11.3}$$

If the factor in parentheses is identified as auxiliary displacements u_k^c, the differential operators can be split and (11.3) can be rewritten as

$$C_{ijkl}u_{k,jl}^c + b_i = 0, \tag{11.4a}$$

$$u_k^g - \ell^2 u_{k,mm}^g = u_k^c. \tag{11.4b}$$

Equation (11.4a) represents the classical equations of elasticity, hence the associated displacements have been indicated with a superscript c. Conversely, the gradient enriched displacements of (11.3) appear with superscript g in (11.4b). Substitution of (11.4b) into (11.4a) yields (11.3), so that the solution of (11.4) should in principle be identical to that of (11.3) — however, whether this holds depends on the boundary conditions used to solve the boundary value problems (Askes et al., 2008b).

The boundary conditions associated with (11.4b) can be manipulated, to a certain extent, by taking derivatives followed by premultiplication with the constitutive tensor C_{ijkl}. This leads to strain based and stress based variants of (11.4b) (Gutkin and Aifantis, 1997, 1999), that is

$$\varepsilon_{kl}^g - \ell^2 \varepsilon_{kl,mm}^g = \frac{1}{2}\left(u_{k,l}^c + u_{l,k}^c\right), \tag{11.4c}$$

and

$$\sigma_{ij}^g - \ell^2 \sigma_{ij,mm}^g = C_{ijkl}u_{k,l}^c. \tag{11.4d}$$

The use of either (11.4c) or (11.4d) leads to singularity-free strains and stresses, whereas some singularities remain if (11.4b) is used (Askes et al., 2008b).

Reducing the order of the governing differential equations from four to two greatly simplifies numerical solution strategies. For instance, within the context of finite element discretisations one can resort to standard $\mathcal{C}^0$-continuous shape functions (Tenek and Aifantis, 2002; Askes et al., 2008b). Subsequent extensions to nonlinear material behaviour (such as plasticity and damage) are also relatively straightforward and allow for generic implementations of nonlinearity (Jirásek and Marfia, 2005; Rodríguez-Ferran et al., 2011), thus demonstrating the versatility of operator split approaches.

11.3 Micro-macro Coupling

The meaning of u_k^c, ε_{kl}^c, and σ_{ij}^c versus those of u_k^g, ε_{kl}^g, and σ_{ij}^g above is clearly that of classic versus gradient enriched fields, respectively. However, when the arguments of nonlocal mechanics are combined with those of homogenisation theory, an alternative interpretation of the two sets of fields may be identified. In addition, homogenisation principles can be used to derive a physical interpretation of the gradient elasticity length scale ℓ.

11.3.1 Linking Nonlocal Mechanics to Homogenisation Theory

In nonlocal continuum mechanics, nonlocal field variables are defined in terms of their local counterparts via an integral equation. For instance, nonlocal stresses $\overline{\sigma}_{ij}$ at a position $\mathbf{x}$ can be written as

$$\overline{\sigma}_{ij}(\mathbf{x}) = \frac{\int_V w(\mathbf{s})\, \sigma_{ij}(\mathbf{x}+\mathbf{s})\, \mathrm{d}\Omega}{\int_V w(\mathbf{s})\, \mathrm{d}\Omega}, \tag{11.5}$$

where σ_{ij} are the standard (or “local”) stresses and w is a nonlocal weight function that typically decreases with increasing distance $\mathbf{s}$ away from $\mathbf{x}$. The differential equation (11.4d) can be derived from the integral equation (11.5) either approximatively using Taylor series expansions (Huerta and Pijaudier-Cabot, 1994; Peerlings et al., 1996) or identically using a Green’s function as the nonlocal weight function (Eringen, 1983; Peerlings et al., 2001).

In homogenisation, upscaling takes place by defining macrolevel stresses σ_{ij}^{M} as the volume average of the microlevel stresses σ_{ij}^{m} via

$$\sigma_{ij}^{\mathrm{M}} = \frac{1}{V_{\mathrm{RVE}}} \int_{V_{\mathrm{RVE}}} \sigma_{ij}^{\mathrm{m}}\, \mathrm{d}\Omega, \tag{11.6}$$

where the integration takes place over a Representative Volume Element (RVE). It can be seen that (11.5) and (11.6) are identical in case the nonlocal weight function is taken as $w = 1$ inside the RVE and $w = 0$ beyond. Thus, a relation is established between gradient models, nonlocal models, and micro-macro coupling (Bennett et al., 2007; Askes et al., 2008a). In particular, (11.4d) can be written in terms of micro- and macroscale variables as

$$\sigma_{ij}^{\mathrm{M}} - \ell^2 \sigma_{ij,kk}^{\mathrm{M}} = \sigma_{ij}^{\mathrm{m}}. \tag{11.7}$$

A similar exercise may be carried out in terms of the strains ε_{kl}.

11.3.2 Linking the Gradient Elasticity Length Scale to the Size of the RVE

The homogenisation principle expressed in (11.6) can also be used as the starting point of another useful derivation. First, we write

$$\sigma_{ij}^{\mathrm{m}} = C_{ijkl}^{\mathrm{m}} \varepsilon_{kl}^{\mathrm{m}}. \tag{11.8}$$

Next, C^{m}_{ijkl} and $\varepsilon^{\mathrm{m}}_{kl}$ are written as a series around their values at the centre of a cubic RVE, which are also identified as the macroscopic values. That is,

$$C^{\mathrm{m}}_{ijkl} = C^{\mathrm{M}}_{ijkl} + C^{\mathrm{M}}_{ijkl,m} x_m \,, \tag{11.9a}$$

$$\varepsilon^{\mathrm{m}}_{kl} = \varepsilon^{\mathrm{M}}_{kl} + \varepsilon^{\mathrm{M}}_{kl,n} x_n \,. \tag{11.9b}$$

Substituting (11.8) and (11.9) into (11.6) leads to

$$\sigma^{\mathrm{M}}_{ij} = \frac{1}{V_{\mathrm{RVE}}} \int_{V_{\mathrm{RVE}}} \left(C^{\mathrm{M}}_{ijkl} + C^{\mathrm{M}}_{ijkl,m} x_m \right) \left(\varepsilon^{\mathrm{M}}_{kl} + \varepsilon^{\mathrm{M}}_{kl,n} x_n \right) \, \mathrm{d}\Omega \,. \tag{11.10}$$

The terms that are linear in x_m are odd functions integrated over a symmetric domain and therefore they vanish. The term proportional to $x_m x_n$ is integrated by parts, i.e.,

$$\int_{V_{\mathrm{RVE}}} C^{\mathrm{M}}_{ijkl,m} \varepsilon^{\mathrm{M}}_{kl,n} x_m x_n \, \mathrm{d}\Omega = \oint_{S_{\mathrm{RVE}}} C^{\mathrm{M}}_{ijkl} \varepsilon^{\mathrm{M}}_{kl,n} n_m x_m x_n \, \mathrm{d}\Gamma =$$

$$= - \int_{V_{\mathrm{RVE}}} C^{\mathrm{M}}_{ijkl} \varepsilon^{\mathrm{M}}_{kl,mn} x_m x_n \, \mathrm{d}\Omega \,. \tag{11.11}$$

Since the origin of the coordinate system is positioned in the centre of the RVE, the boundary integral cancels. With some straightforward manipulations Eq. (11.10) can then be written as

$$\sigma^{\mathrm{M}}_{ij} = C^{\mathrm{M}}_{ijkl} \left(\varepsilon^{\mathrm{M}}_{kl} - \frac{1}{12} L^2_{\mathrm{RVE}} \varepsilon^{\mathrm{M}}_{kl,mm} \right) , \tag{11.12}$$

where L_{RVE} is the size of the RVE. Comparing (11.12) with (11.1), it follows that the length scale parameter of gradient elasticity ℓ is related to the size of the RVE L_{RVE} via (Gitman et al., 2004, 2005; Kouznetsova et al., 2004)

$$\ell^2 = \frac{1}{12} L^2_{\mathrm{RVE}} \,, \tag{11.13}$$

which may be used for identification and quantification of the gradient length scale parameter — see (Askes and Aifantis, 2011, Section 4) for an overview.

11.4 Gradient Elasticity in Dynamics

It is well known that a straightforward extension of (11.2) to dynamics displays some anomalies. In particular, unbounded phase velocities are found, which is not physically realistic. To overcome this, gradient enrichment of the kinetic energy as well as the strain energy can be used (Mindlin, 1964; Georgiadis et al., 2000; Metrikine and Askes, 2002; Engelbrecht et al., 2005; Metrikine and Askes, 2006; Papargyri-Beskou et al., 2009).

11.4.1 Dynamically Consistent Gradient Elasticity

Metrikine and Askes used the term "dynamic consistency" to indicate a simultaneous gradient enrichment of the stress-strain relation and the inertia terms (Metrikine and Askes, 2002). This gives

$$C_{ijkl}\left(u_{k,jl} - \ell^2 u_{k,jlmm}\right) = \rho\left(\ddot{u}_i - \alpha\ell^2\ddot{u}_{i,mm}\right), \tag{11.14}$$

where α is a dimensionless parameter that may be obtained as a result of continualisation (Metrikine and Askes, 2002) or asymptotic analysis (Pichugin et al., 2008; Bennett et al., 2007).

The mathematical structure of (11.14) prohibits a decoupling in the spirit of (11.4), but the observations of Sect. 11.3 can be used to rewrite (11.14) in terms of micro and macroscale displacements. In particular, the template of (11.7) is used for the displacements on the left-hand side of (11.14), so that this expression can be rewritten as

$$C_{ijkl}u^{\mathrm{m}}_{k,jl} = \rho\left(\ddot{u}^{\mathrm{M}}_i - \alpha\ell^2\ddot{u}^{\mathrm{M}}_{i,mm}\right), \tag{11.15a}$$

$$u^{\mathrm{M}}_i - \ell^2 u^{\mathrm{M}}_{i,mm} = u^{\mathrm{m}}_i. \tag{11.15b}$$

A symmetric formulation, in the sense of matching the coefficient of u^{M}_i in (11.15a) with that of u^{m}_i in (11.15b), can be obtained as explained in (Askes and Aifantis, 2006; Askes et al., 2007). First, the second time derivative of Eq. (11.15b) is taken. We may then replace $\ddot{u}^{\mathrm{M}}_{i,mm}$ in (11.15a) with $\ddot{u}^{\mathrm{M}}_i - \ddot{u}^{\mathrm{m}}_i$. Next, the acceleration format of (11.15b) is premultiplied with $\rho(\alpha - 1)$. This yields

$$C_{ijkl}u^{\mathrm{m}}_{k,jl} = \rho\left(\alpha\ddot{u}^{\mathrm{m}}_i - (\alpha - 1)\ddot{u}^{\mathrm{M}}_i\right), \tag{11.16a}$$

$$\rho\left(\alpha - 1\right)\left(\ddot{u}^{\mathrm{M}}_i - \ddot{u}^{\mathrm{m}}_i - \ell^2\ddot{u}^{\mathrm{M}}_{i,mm}\right) = 0. \tag{11.16b}$$

This symmetric formulation facilitates the identification of the underlying energy functionals and, in turn, the correct format of the variationally consistent boundary conditions — see Sect. 11.6 for details. Since the partial differential equations of (11.16) are second order in space, standard $\mathcal{C}^0$-continuous finite element technology can be applied (Bennett and Askes, 2009), and the extension to nonlinear material behaviour is effective in predicting damage zones of finite width (Bennett et al., 2012).

The formulation above is in terms of macroscale and microscale displacements. A related formulation is due to Engelbrecht and coworkers (Engelbrecht et al., 2005; Berezovski et al., 2011, 2013; Engelbrecht and Berezovski, 2015) who use the macroscale displacement in conjunction with the microscale deformation or other internal variables. Similar observations apply regarding ease of implementation, while their approach also opens routes to modelling including multiple levels of microstructure.

11.4.2 Dispersion Correction: A Fourth Order Model

Although the dynamically consistent model of (11.14) is able to describe dispersive wave propagation without anomalies such as infinite or imaginary phase velocities, it is unable to predict the inflection point in the relation between angular frequency and wave number that is observed in certain nano-scale experiments — see for instance Fig. 11.1 or the range of experimental results modelled in (De Domenico and Askes, 2018). This has motivated the introduction of a higher order inertia term (Domenico and Askes, 2016; De Domenico and Askes, 2017). The equations of motion of the resulting model are

$$C_{ijkl}\left(u_{k,jl} - \ell^2 u_{k,jlmm}\right) = \rho\left(\ddot{u}_i - \alpha\ell^2\ddot{u}_{i,mm} + \beta\ell^4\ddot{u}_{i,mmnn}\right), \tag{11.17}$$

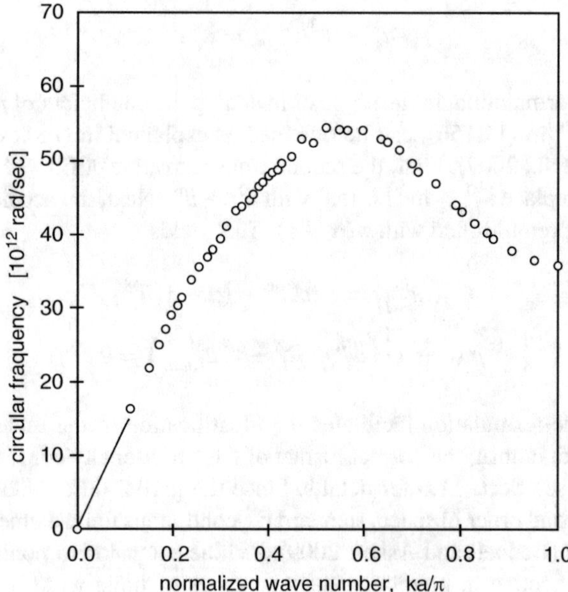

Fig. 11.1 Phonon dispersion observed experimentally in aluminium (Yarnell et al., 1965).

where β is another dimensionless parameter that may be obtained from asymptotic analysis (Domenico and Askes, 2016; De Domenico et al., 2019) or experimental validation (De Domenico et al., 2019). Although the orders of ℓ are unbalanced between the left and right-hand side of (11.17), the highest order of spatial derivatives is the same on both sides and an operator split can be performed. First, the micro-macro relations of Sect. 11.3 are applied to the left-hand side of (11.17), leading to

$$C_{ijkl}u^{\mathrm{m}}_{k,jl} = \rho\left(\ddot{u}^{\mathrm{M}}_i - \alpha\ell^2\ddot{u}^{\mathrm{M}}_{i,mm} + \beta\ell^4\ddot{u}^{\mathrm{M}}_{i,mmnn}\right), \tag{11.18a}$$

$$u^{\mathrm{M}}_i - \ell^2 u^{\mathrm{M}}_{i,mm} = u^{\mathrm{m}}_i . \tag{11.18b}$$

Next, the acceleration format of (11.18b) is used to write

$$\alpha\ell^2\ddot{u}^{\mathrm{M}}_{i,mm} = \alpha\left(\ddot{u}^{\mathrm{M}}_i - \ddot{u}^{\mathrm{m}}_i\right), \tag{11.19a}$$

as well as

$$\begin{aligned}\beta\ell^4\ddot{u}^{\mathrm{M}}_{i,mmnn} &= \beta\ell^2\left(\ddot{u}^{\mathrm{M}}_{i,mm} - \ddot{u}^{\mathrm{m}}_{i,mm}\right) = \\ &= \beta\left(\ddot{u}^{\mathrm{M}}_i - \ddot{u}^{\mathrm{m}}_i\right) - \beta\ell^2\ddot{u}^{\mathrm{m}}_{i,mm} .\end{aligned} \tag{11.19b}$$

Substituting (11.19) into (11.18a) and premultiplying the acceleration format of (11.18b) with $\rho\,(\alpha - \beta - 1)$ results in a symmetric system according to

$$C_{ijkl}u^{\mathrm{m}}_{k,jl} = \rho\left[(\alpha-\beta)\,\ddot{u}^{\mathrm{m}}_i - (\alpha-\beta-1)\,\ddot{u}^{\mathrm{M}}_i - \beta\ell^2\ddot{u}^{\mathrm{m}}_{i,mm}\right], \tag{11.20a}$$

$$\rho\,(\alpha-\beta-1)\left(\ddot{u}^{\mathrm{M}}_i - \ddot{u}^{\mathrm{m}}_i - \ell^2\ddot{u}^{\mathrm{M}}_{i,mm}\right) = 0 . \tag{11.20b}$$

It can be verified that (11.16) are retrieved from (11.20) by taking $\beta = 0$. Although a higher order inertia term has been included in the formulation, the partial differential equations after application of the operator split are again second order in space. Thus, standard finite element methodology can be used (De Domenico and Askes, 2017).

11.4.3 Dispersion Analysis

To investigate the ability of the models of Sects. 11.4.1 and 11.4.2 to describe wave dispersion, their behaviour is compared to that of a one-dimensional discrete model of masses and springs. It is well known that the angular frequency ω of such a model is given in nondimensionalised form through

$$\frac{\omega d}{c_e} = 2d\sin\left(\frac{kd}{2}\right), \tag{11.21}$$

where d is the particle spacing, $c_e \equiv \sqrt{E/\rho}$ is the elastic bar velocity and k is the wave number. For such a periodic system, the Representative Volume Element is exactly equal to one spring. From (11.13) we then have $\ell^2 = L_{\mathrm{RVE}}^2/12 = d^2/12$. Following the asymptotic analysis of Domenico and Askes (2016), a gradient enriched model of the format of Sect. 11.4.2 then leads to

$$\frac{\omega d}{c_e} = dk \sqrt{\frac{1 + \frac{1}{12}d^2k^2}{1 + \frac{1}{6}d^2k^2 + \frac{1}{90}d^4k^4}}, \tag{11.22}$$

which yields $\alpha = 2$ and $\beta = 8/5$. The results are plotted in Fig. 11.2, where the scaling of the axes is adjusted such as to comply with (11.21). It can be seen that the model of Sect. 11.4.2 provides a much improved approximation of the discrete model, compared to the model of Sect. 11.4.1. However, the asymptotic analysis has been based on series expansions around $k \to 0$, thus the accuracy of these particular approximations deteriorate for increasing values of the wave number.

Alternatively, values for α and β may be obtained via a best fit across a range of wave numbers, rather than for the limit case $k \to 0$. Here, we merely wish to give some indications of possible model behaviour. Figure 11.3 shows the dispersion

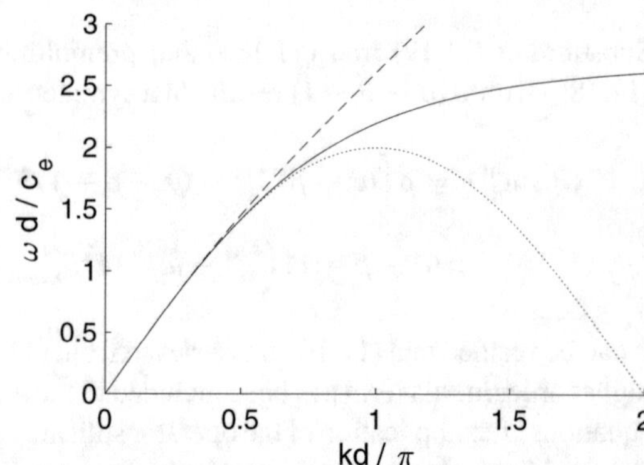

Fig. 11.2 Dispersion behaviour of a discrete model (dotted), model of Sect. 11.4.1 (dashed) and model of Sect. 11.4.2 (solid).

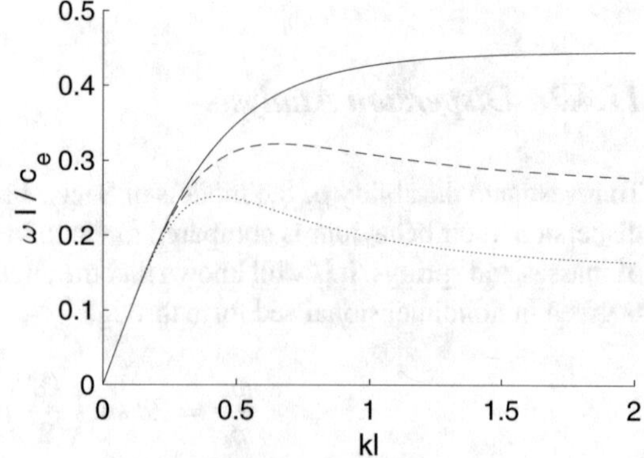

Fig. 11.3 Dispersion behaviour of model of Sect. 11.4.2 with $\alpha = 5$ and $\beta = 5$ (solid), $\beta = 15$ (dashed) and $\beta = 45$ (dotted).

curves for a number of values of β. Depending on the relative magnitudes of α and β, an inflection point may be obtained. Best values for a range of materials and tests have been found and reported in (De Domenico et al., 2019) and, for a slightly simplified variant of the model, in (De Domenico and Askes, 2018).

11.5 Gradient Piezomagnetic Coupling in Dynamics

Next, an operator split approach will be developed for a multiphysics framework. Inspired by the gradient enriched electromechanical model discussed in (Yue et al., 2014), a static gradient enriched piezomagnetic model was developed in (Xu et al., 2019) that can be written as

$$C_{ijkl}\left(u_{k,jl}-\ell_1^2 u_{k,jlmm}\right)+q_{ijk}\left(\varphi_{,jk}-\ell_2^2\varphi_{,jkmm}\right)=0, \tag{11.23a}$$

$$q_{ijk}\left(u_{i,jk}-\ell_2^2 u_{i,jkmm}\right)-\mu_{ij}\left(\varphi_{,ij}-\ell_3^2\varphi_{,ijmm}\right)=0, \tag{11.23b}$$

where the displacements u_i and the magnetic potential φ are the primary unknowns. The material parameters are collected in the elastic stiffness tensor C_{ijkl}, the piezomagnetic coupling tensor q_{ijk} and the magnetic permeability tensor μ_{ij}. Furthermore, the model is equipped with three distinct length scales ℓ_1, ℓ_2 and ℓ_3. It was demonstrated that ℓ_1 and ℓ_3 are required for the removal of singularities from the mechanical fields and the magnetic fields, respectively, whereas ℓ_2 has some quantitative effects but does not lead to a qualitatively different response (Xu et al., 2019).

For a dynamic extension of (11.23) the different wave speeds of the two physical fields must be considered. Loosely speaking, mechanical fields travel with the "speed of sound" whereas magnetic fields travel with the "speed of light". It is well known that these two wave speeds differ by several orders of magnitude. As an approximation, it will be assumed that the magnetic response is instantaneous and thus only the mechanical equation (11.23a) is equipped with inertia terms to simulate transient mechanical effects. Furthermore, the assumption $\ell_1=\ell_2=\ell_3\equiv\ell$ is made. Thus, we have

$$C_{ijkl}\left(u_{k,jl}-\ell^2 u_{k,jlmm}\right)+q_{ijk}\left(\varphi_{,jk}-\ell^2\varphi_{,jkmm}\right)= \\ =\rho\left(\ddot{u}_i-\alpha\ell^2\ddot{u}_{i,mm}+\beta\ell^4\ddot{u}_{i,mmnn}\right), \tag{11.24a}$$

$$q_{ijk}\left(u_{i,jk}-\ell^2 u_{i,jkmm}\right)-\mu_{ij}\left(\varphi_{,ij}-\ell^2\varphi_{,ijmm}\right)=0, \tag{11.24b}$$

where the right-hand side of (11.24a) has been enriched in the spirit of Sect. 11.4.2. Micro-macro relations for the two fields of primary unknowns can be postulated according to

$$u_i^{\mathrm{M}} - \ell^2 u_{i,mm}^{\mathrm{M}} = u_i^{\mathrm{m}}, \tag{11.25a}$$

$$\varphi^{\mathrm{M}} - \ell^2 \varphi_{,mm}^{\mathrm{M}} = \varphi^{\mathrm{m}}, \tag{11.25b}$$

by which (11.24) can be rewritten as

$$C_{ijkl} u_{k,jl}^{\mathrm{m}} + q_{ijk} \varphi_{,jk}^{\mathrm{m}} = \rho \left(\ddot{u}_i^{\mathrm{M}} - \alpha \ell^2 \ddot{u}_{i,mm}^{\mathrm{M}} + \beta \ell^4 \ddot{u}_{i,mmnn}^{\mathrm{M}} \right), \tag{11.26a}$$

$$q_{ijk} u_{i,jk}^{\mathrm{m}} - \mu_{ij} \varphi_{,ij}^{\mathrm{m}} = 0. \tag{11.26b}$$

Next, the methodology of Sect. 11.4.2 is used to rewrite (11.25a) and the right-hand side of (11.26a) to arrive at the following symmetric formulation:

$$C_{ijkl} u_{k,jl}^{\mathrm{m}} + q_{ijk} \varphi_{,jk}^{\mathrm{m}} = \rho \left[(\alpha - \beta)\, \ddot{u}_i^{\mathrm{m}} - (\alpha - \beta - 1)\, \ddot{u}_i^{\mathrm{M}} - \beta \ell^2 \ddot{u}_{i,mm}^{\mathrm{m}} \right], \tag{11.27a}$$

$$\rho\, (\alpha - \beta - 1) \left(\ddot{u}_i^{\mathrm{M}} - \ddot{u}_i^{\mathrm{m}} - \ell^2 \ddot{u}_{i,mm}^{\mathrm{M}} \right) = 0, \tag{11.27b}$$

$$q_{ijk} u_{i,jk}^{\mathrm{m}} - \mu_{ij} \varphi_{,ij}^{\mathrm{m}} = 0. \tag{11.27c}$$

Note that the macroscale magnetic potential φ^{M} does not appear in (11.27). Thus, after the microscale magnetic potential φ^{m} is obtained from (11.27), expression (11.25b) can be used to compute φ^{M}. In other words, (11.25b) is decoupled from (11.27).

11.6 Generic Variational Formulation

The models of Sects. 11.4.1 and 11.4.2 can be retrieved from that of (11.27) by disregarding the piezomagnetic coupling and, in case of Sect. 11.4.1, setting $\beta = 0$. Thus, in what follows we will take (11.27) as the basis for further development. With the symmetric formulation provided in (11.27), it is straightforward to identify the underlying Lagrangian density $\mathcal{L}$ as

$$\mathcal{L} = \frac{1}{2}\rho \left(\dot{u}_i^{\mathrm{m}}\right)^2 + \frac{1}{2}\rho\beta\ell^2 \left(\dot{u}_{i,j}^{\mathrm{m}}\right)^2 + \frac{1}{2}\rho\,(\alpha - \beta - 1) \left[\left(\dot{u}_i^{\mathrm{m}} - \dot{u}_i^{\mathrm{M}}\right)^2 + \ell^2 \left(\dot{u}_{i,j}^{\mathrm{M}}\right) \right] - \\ - \frac{1}{2} u_{i,j}^{\mathrm{m}} C_{ijkl} u_{k,l}^{\mathrm{m}} - u_{i,j}^{\mathrm{m}} q_{ijk} \varphi_{,k}^{\mathrm{m}} + \frac{1}{2} \varphi_{,i}^{\mathrm{m}} \mu_{ij} \varphi_{,j}^{\mathrm{m}}. \tag{11.28}$$

Taking the variation of the Lagrangian density, integrated in space and time, yields

$$\int\!\!\int_{V\,t_1}^{t_2} \left(\delta \dot{u}_i^{\mathrm{m}} \frac{\partial \mathcal{L}}{\partial \dot{u}_i^{\mathrm{m}}} + \delta \dot{u}_{i,j}^{\mathrm{m}} \frac{\partial \mathcal{L}}{\partial \dot{u}_{i,j}^{\mathrm{m}}} + \delta \dot{u}_i^{\mathrm{M}} \frac{\partial \mathcal{L}}{\partial \dot{u}_i^{\mathrm{M}}} + \delta \dot{u}_{i,j}^{\mathrm{M}} \frac{\partial \mathcal{L}}{\partial \dot{u}_{i,j}^{\mathrm{M}}} + \right.$$

$$\left. + \delta u_{i,j}^{\mathrm{m}} \frac{\partial \mathcal{L}}{\partial u_{i,j}^{\mathrm{m}}} + \delta \varphi_{,i}^{\mathrm{m}} \frac{\partial \mathcal{L}}{\partial \varphi_{,i}^{\mathrm{m}}} \right) \mathrm{d}t \; \mathrm{d}\Omega = 0. \quad (11.29)$$

Integration by parts then results in

$$\int\!\!\int_{V\,t_1}^{t_2} \delta u_i^{\mathrm{m}} \left(-\frac{\partial}{\partial t} \frac{\partial \mathcal{L}}{\partial \dot{u}_i^{\mathrm{m}}} + \frac{\partial^2}{\partial t \partial x_j} \frac{\partial \mathcal{L}}{\partial \dot{u}_{i,j}^{\mathrm{m}}} - \frac{\partial}{\partial x_j} \frac{\partial \mathcal{L}}{\partial u_{i,j}^{\mathrm{m}}} \right) \mathrm{d}t \; \mathrm{d}\Omega +$$

$$+ \oint\!\!\int_{S\,t_1}^{t_2} \delta u_i^{\mathrm{m}} n_j \left(-\frac{\partial}{\partial t} \frac{\partial \mathcal{L}}{\partial \dot{u}_{i,j}^{\mathrm{m}}} + \frac{\partial \mathcal{L}}{\partial u_{i,j}^{\mathrm{m}}} \right) \mathrm{d}t \; \mathrm{d}\Gamma +$$

$$+ \int\!\!\int_{V\,t_1}^{t_2} \delta u_i^{\mathrm{M}} \left(-\frac{\partial}{\partial t} \frac{\partial \mathcal{L}}{\partial \dot{u}_i^{\mathrm{M}}} + \frac{\partial^2}{\partial t \partial x_j} \frac{\partial \mathcal{L}}{\partial \dot{u}_{i,j}^{\mathrm{M}}} \right) \mathrm{d}t \; \mathrm{d}\Omega -$$

$$- \oint\!\!\int_{S\,t_1}^{t_2} \delta u_i^{\mathrm{M}} n_j \frac{\partial}{\partial t} \frac{\partial \mathcal{L}}{\partial \dot{u}_{i,j}^{\mathrm{M}}} \mathrm{d}t \; \mathrm{d}\Gamma -$$

$$- \int\!\!\int_{V\,t_1}^{t_2} \delta \varphi^{\mathrm{m}} \frac{\partial}{\partial x_i} \frac{\partial \mathcal{L}}{\partial \varphi_{,i}^{\mathrm{m}}} \mathrm{d}t \; \mathrm{d}\Omega + \oint\!\!\int_{S\,t_1}^{t_2} \delta \varphi^{\mathrm{m}} n_i \frac{\partial \mathcal{L}}{\partial \varphi_{,i}^{\mathrm{m}}} \mathrm{d}t \; \mathrm{d}\Gamma = 0, \quad (11.30)$$

where n_i is the outward normal vector to the surface S of the domain V. Substituting (11.28) into (11.30), it can be verified that the volume integrals lead to (11.27). The surface integrals provide the format of the essential and natural boundary conditions, namely

$$\text{either prescribe } u_i^{\mathrm{m}} \text{ or prescribe } n_j \left(C_{ijkl} u_{k,l}^{\mathrm{m}} + q_{ijk} \varphi_{,k}^{\mathrm{m}} + \rho \beta \ell^2 \ddot{u}_{i,j}^{\mathrm{m}} \right), \quad (11.31\text{a})$$

$$\text{either prescribe } u_i^{\mathrm{M}} \text{ or prescribe } n_j \left(\rho \left(\alpha - \beta - 1 \right) \ell^2 \ddot{u}_{i,j}^{\mathrm{M}} \right), \quad (11.31\text{b})$$

$$\text{either prescribe } \phi^{\mathrm{m}} \text{ or prescribe } n_k \left(q_{ijk} u_{i,j}^{\mathrm{m}} - \mu_{ik} \phi_{,i}^{\mathrm{m}} \right). \quad (11.31\text{c})$$

In addition, boundary conditions must be formulated for (11.25b): essential boundary conditions are expressed in terms of φ^{M} and natural boundary conditions in terms of $n_j \varphi_{,j}^{\mathrm{M}}$.

Equations (11.30) and (11.31) provide a unified framework for the three models discussed earlier. It is worth remarking that the operator splits procedures explored above are *not* a particular case of mixed formulations. In developing the various models, we have taken time derivatives of the relevant micro-macro relations, for instance in progressing from (11.15b) to (11.16b), which means the resulting formulations are no longer reducible to the original formulations.

This is confirmed by inspection of the Lagrangian density of (11.28): there are no terms that couple a primary unknown to one of its derivatives, and there are no terms that represent constraints with Lagrange multipliers. Thus, the typical constraints of mixed formulations on finite element implementations, such as different orders of interpolation for the various fields, do not apply here. Instead, choices for the discretisation can be guided by convenience and user preference, such as selecting the same finite element mesh and the same interpolation functions for all fields of primary unknowns. This has been demonstrated to be appropriate and effective for the model of Sect. 11.4.1 in (Bennett and Askes, 2009), the model of Sect. 11.4.2 in (De Domenico and Askes, 2017) and two special cases of (11.23) in (Xu et al., 2019).

11.7 Closing Remarks

In this contribution, we have provided an overview of a special class of gradient models that lend themselves to differential operator splits. By factorising higher order derivatives, a higher order partial differential equation (PDE) can be split into multiple lower order PDEs. In certain cases this leads to multiscale formulations with a full coupling between micro and macroscale fields, whereas in other cases the operator split leads to decoupled lower order PDEs — see (Xu et al., 2019) for two particular simplifications of (11.23), one of which leading to a coupled system with the other resulting in a decoupled system.

An interesting example whereby both coupling and decoupling occur is the gradient enriched dynamic piezomagnetic model of Sect. 11.5. For reasons given earlier, only the mechanical equation is extended with inertia. As a consequence, the micro- and macroscale displacements are fully coupled. The magnetic equation is not extended with inertia, which results in the macroscale magnetic potential being decoupled from the microscale magnetic potential.

For the models discussed, at most second order spatial derivatives appear in the final field equations. Because of this, and because the resulting formulations are not mixed formulations, standard finite element implementations can be used. We have suggested to use the same mesh and the same interpolation functions for all fields of primary unknowns, but in our earlier work we have also observed that the microscale fields are typically more fluctuating and less smooth than the corresponding macroscale fields (Askes et al., 2007; Bennett and Askes, 2009; De Domenico and Askes, 2017). Thus, different resolutions of the two sets of fields could be pursued if desired or required.

Since the static side of the equations inherently has an additional two spatial derivatives compared to the transient side of the equations, operator split methods typically lead to gradient terms moving from the static side to the transient side (or, in other words, from the potential energy to the kinetic energy). Consequently, the implementation of a stiffness matrix is simplified, which means that addition of nonlinear material behaviour (such as plasticity or damage) can be carried out with relative ease. This has been explored already for the models of Sect. 11.2 (Jirásek and Marfia, 2005; Rodríguez-Ferran et al., 2011) and Sect. 11.4.1 (Bennett et al., 2012), and could be extended to the models of Sects. 11.4.2 and 11.5. Similarly, the operator split approach developed for the multiphysics model of Sect. 11.5 could be expanded to other multiphysics models.

Acknowledgements HA, IMG and ECA gratefully acknowledge support of the EU RISE project FRAMED-734485. MX gratefully acknowledges financial support from the China Scholarship Council and the Fundamental Research Funds for the Central Universities (FRFBR-16-017A).

References

Aifantis, E.: Strain gradient interpretation of size effects. Int. J. Fract. **95**(1-4), 299–314 (1999). https://doi.org/10.1007/978-94-011-4659-3_16

Aifantis, E.C.: On the microstructural origin of certain inelastic models. J. Engng. Mater. Technol. **106**(4), 326–330 (1984). https://doi.org/10.1115/1.3225725

Aifantis, E.C.: The physics of plastic deformation. Int. J. Plasticity **3**(3), 211–247 (1987). https://doi.org/10.1016/0749-6419(87)90021-0

Aifantis, E.C.: On the role of gradients in the localization of deformation and fracture. Int. J. Engng Sci. **30**(10), 279–1299 (1992). https://doi.org/10.1016/0020-7225(92)90141-3

Altan, S., Aifantis, E.: On the structure of the mode III crack-tip in gradient elasticity. Scripta Metallurgica et Materialia **26**(2), 319–324 (1992). https://doi.org/10.1016/0956-716x(92)90194-j

Askes, H., Aifantis, E.C.: Numerical modeling of size effects with gradient elasticity – Formulation, meshless discretization and examples. Int. J. Fract. **117**(4), 347–358 (2002). https://doi.org/10.1023/A:102222552

Askes, H., Aifantis, E.C.: Gradient elasticity theories in statics and dynamics – A unification of approaches. Int. J. Fract. **139**(2), 297–304 (2006). https://doi.org/10.1007/s10704-006-8375-4

Askes, H., Aifantis, E.C.: Gradient elasticity in statics and dynamics: an overview of formulations, length scale identification procedures, finite element implementations and new results. Int. J. Solids Struct. **48**(13), 1962–1990 (2011). https://doi.org/10.1016/j.ijsolstr.2011.03.006

Askes, H., Bennett, T., Aifantis, E.C.: A new formulation and C^0-implementation of dynamically consistent gradient elasticity. Int. J. Numer. Methods Engng. **72**(1), 111–126 (2007). https://doi.org/10.1002/nme.2017

Askes, H., Bennett, T., Gitman, I., Aifantis, E.: A multi-scale formulation of gradient elasticity and its finite element implementation. In: Papadrakakis, M., Topping, B. (eds.) Trends in Engineering Computational Technology, pp. 189–208. Saxe-Coburg Publications, Stirlingshire (2008a). https://doi.org/10.4203/csets.20.10

Askes, H., Morata, I., Aifantis, E.C.: Finite element analysis with staggered gradient elasticity. Computers & Structures **86**(11-12), 1266–1279 (2008b). https://doi.org/10.1016/j.compstruc.2007.11.002

Bennett, T., Askes, H.: Finite element modelling of wave dispersion with dynamically consistent gradient elasticity. Comput. Mech. **43**(6), 815–825 (2009). https://doi.org/10.1007/s00466-008-0347-2

Bennett, T., Gitman, I.M., Askes, H.: Elasticity theories with higher-order gradients of inertia and stiffness for the modelling of wave dispersion in laminates. Int. J. Fract. **148**(2), 185–193 (2007). https://doi.org/10.1007/s10704-008-9192-8

Bennett, T., Rodríguez-Ferran, A., Askes, H.: Damage regularisation with inertia gradients. Eur. J. Mech. A/Solids **31**(1), 131–138 (2012). https://doi.org/10.1016/j.euromechsol.2011.08.005

Berezovski, A., Engelbrecht, J., Berezovski, M.: Waves in microstructured solids: a unified viewpoint of modeling. Acta Mech. **220**(1–4), 349–363 (2011). https://doi.org/10.1007/s00707-011-0468-0

Berezovski, A., Engelbrecht, J., Salupere, A., Tamm, K., Peets, T., Berezovski, M.: Dispersive waves in microstructured solids. Int. J. Solids Struct. **50**(11-12), 1981–1990 (2013). https://doi.org/10.1016/j.ijsolstr.2013.02.018

De Domenico, D., Askes, H.: Computational aspects of a new multi-scale dispersive gradient elasticity model with micro-inertia. Int. J. Numer. Methods Engng. **109**(1), 52–72 (2017). https://doi.org/10.1002/nme.5278

De Domenico, D., Askes, H.: Nano-scale wave dispersion beyond the first Brillouin zone simulated with inertia gradient continua. J. Appl. Phys. **124**(20), 205107 (2018). https://doi.org/10.1063/1.5045838

De Domenico, D., Askes, H., Aifantis, E.C.: Gradient elasticity and dispersive wave propagation: Model motivation and length scale identification procedures in concrete and composite laminates. Int. J. Solids Struct. **158**, 176–190 (2019). https://doi.org/10.1016/j.ijsolstr.2018.09.007

De Domenico, D., Askes, H.: A new multi-scale dispersive gradient elasticity model with micro-inertia: Formulation and-finite element implementation. Int. J. Numer. Methods Engng. **108**(5), 485–512 (2016). https://doi.org/10.1002/nme.5222

Engelbrecht, J., Berezovski, A.: Reflections on mathematical models of deformation waves in elastic microstructured solids. Math. Mech. Complex Systems **3**(1), 43–82 (2015). https://doi.org/10.2140/memocs.2015.3.43

Engelbrecht, J., Berezovski, A., Pastrone, F., Braun, M.: Waves in microstructured materials and dispersion. Phil. Mag. **85**(33-35), 4127–4141 (2005). https://doi.org/10.1080/14786430500362769

Eringen, A.C.: On differential equations of nonlocal elasticity and solutions of screw dislocation and surface waves. J. Appl. Phys. **54**(9), 4703–4710 (1983). https://doi.org/10.1063/1.332803

Georgiadis, H., Vardoulakis, I., Lykotrafitis, G.: Torsional surface waves in a gradient-elastic half-space. Wave Motion **31**(4), 333–348 (2000). https://doi.org/10.1016/s0165-2125(99)00035-9

Gitman, I., Askes, H., Sluys, L.: Representative volume size as a macroscopic length scale parameter. In: Proceedings of 5th International Conference on Fracture Mechanics of Concrete and Concrete Structures, vol. 1, pp. 483–491 (2004)

Gitman, I.M., Askes, H., Aifantis, E.C.: The representative volume size in static and dynamic micro-macro transitions. Int. J. Fract. **135**(1-4), L3–L9 (2005). https://doi.org/10.1007/s10704-005-4389-6

Gutkin, M.Y., Aifantis, E.: Screw dislocation in gradient elasticity. Scripta Mater. **35**(11), 1353–1358 (1996). https://doi.org/10.1016/1359-6462(96)00295-3

Gutkin, M.Y., Aifantis, E.: Edge dislocation in gradient elasticity. Scripta Mater. **36**(1), 129–135 (1997). https://doi.org/10.1016/s1359-6462(96)00352-1

Gutkin, M.Y., Aifantis, E.: Dislocations in the theory of gradient elasticity. Scripta Mater. **5**(40), 559–566 (1999). https://doi.org/10.1016/s1359-6462(98)00424-2

Huerta, A., Pijaudier-Cabot, G.: Discretization influence on regularization by two localization limiters. J. Engng. Mech. **120**(6), 1198–1218 (1994). https://doi.org/10.1061/(asce)0733-9399(1994)120:6(1198)

Jirásek, M., Marfia, S.: Non-local damage model based on displacement averaging. Int. J. Numer. Methods Engng. **63**(1), 77–102 (2005). https://doi.org/10.1002/nme.1262

Kouznetsova, V., Geers, M., Brekelmans, W.: Size of a representative volume element in a second-order computational homogenization framework. Int. J. Multiscale Comput. Engng. **2**(4), 575–598 (2004). https://doi.org/10.1615/intjmultcompeng.v2.i4.50

Lurie, S., Belov, P., Volkov-Bogorodsky, D., Tuchkova, N.: Nanomechanical modeling of the nanostructures and dispersed composites. Comput. Mater. Sci. **28**(3-4), 529–539 (2003). https://doi.org/10.1016/j.commatsci.2003.08.010

Metrikine, A., Askes, H.: An isotropic dynamically consistent gradient elasticity model derived from a 2D lattice. Phil. Mag. **86**(21-22), 3259–3286 (2006). https://doi.org/10.1080/14786430500197827

Metrikine, A.V., Askes, H.: One-dimensional dynamically consistent gradient elasticity models derived from a discrete microstructure: Part 1: Generic formulation. Eur. J. Mech. A/Solids **21**(4), 555–572 (2002). https://doi.org/10.1016/s0997-7538(02)01218-4

Mindlin, R.D.: Micro-structure in linear elasticity. Arch. Rat. Mech. Anal. **16**(1), 51–78 (1964). https://doi.org/10.1007/bf00248490

Mindlin, R.D.: Second gradient of strain and surface-tension in linear elasticity. Int. J. Solids Struct. **1**(4), 417–438 (1965). https://doi.org/10.1016/0020-7683(65)90006-5

Mühlhaus, H.B., Alfantis, E.: A variational principle for gradient plasticity. Int. J. Solids Struct. **28**(7), 845–857 (1991). https://doi.org/10.1016/0020-7683(91)90004-y

Papargyri-Beskou, S., Polyzos, D., Beskos, D.: Wave dispersion in gradient elastic solids and structures: a unified treatment. Int. J. Solids Struct. **46**(21), 3751–3759 (2009). https://doi.org/10.1016/j.ijsolstr.2009.05.002

Peerlings, R., Geers, M., De Borst, R., Brekelmans, W.: A critical comparison of nonlocal and gradient-enhanced softening continua. Int. J. Solids Struct. **38**(44-45), 7723–7746 (2001). https://doi.org/10.1016/s0020-7683(01)00087-7

Peerlings, R.H., De Borst, R., Brekelmans, W.A., Vree, J.H., Spee, I.: Some observations on localization in non-local and gradient damage models. Eur. J. Mech. A/Solids **15**(6), 937–953 (1996)

Pichugin, A., Askes, H., Tyas, A.: Asymptotic equivalence of homogenisation procedures and fine-tuning of continuum theories. J. Sound Vibr. **313**(3-5), 858–874 (2008). https://doi.org/10.1016/j.jsv.2007.12.005

Rodríguez-Ferran, A., Bennett, T., Askes, H., Tamayo-Mas, E.: A general framework for softening regularisation based on gradient elasticity. Int. J. Solids Struct. **48**(9), 1382–1394 (2011). https://doi.org/10.1016/j.ijsolstr.2011.01.022

Ru, C., Aifantis, E.: A simple approach to solve boundary-value problems in gradient elasticity. Acta Mech. **101**(1-4), 59–68 (1993). https://doi.org/10.1007/bf01175597

Tenek, L.T., Aifantis, E.: A two-dimensional finite element implementation of a special form of gradient elasticity. Comput. Modeling Engng. Sci. **3**(6), 731–742 (2002)

Unger, D.J., Aifantis, E.C.: The asymptotic solution of gradient elasticity for mode III. Int. J. Fract. **71**(2), R27–R32 (1995). https://doi.org/10.1007/bf00033757

Vardoulakis, I., Aifantis, E.: On the role of microstructure in the behavior of soils: effects of higher order gradients and internal inertia. Mech. Mater. **18**(2), 151–158 (1994). https://doi.org/10.1016/0167-6636(94)00002-6

Xu, M., Gitman, I.M., Askes, H.: A gradient-enriched continuum model for magneto-elastic coupling: Formulation, finite element implementation and in-plane problems. Comput. Struct. **212**, 275–288 (2019). https://doi.org/10.1016/j.compstruc.2018.11.004

Yarnell, J.L., Warren, J.L., Koenig, S.H.: Experimental dispersion curves for phonons in aluminum. In: Wallis, R. (ed.) Lattice Dynamics, pp. 57–61. Pergamon, Oxford (1965). https://doi.org/10.1016/b978-1-4831-9838-5.50014-5

Yue, Y., Xu, K., Aifantis, E.: Microscale size effects on the electromechanical coupling in piezoelectric material for anti-plane problem. Smart Mater. Struct. **23**(12), 125043 (2014). https://doi.org/10.1088/0964-1726/23/12/125043

Chapter 12
An Explicit Finite Volume Numerical Scheme for 2D Elastic Wave Propagation

Mihhail Berezovski and Arkadi Berezovski

Abstract The construction of a two-dimensional finite volume numerical scheme based on the representation of computational cells as thermodynamic systems is presented explicitly. The main advantage of the scheme is an accurate implementation of conditions at interfaces and boundaries. It is demonstrated that boundary conditions influence the wave motion even in the simple case of a homogeneous waveguide.

12.1 Introduction

Problems in wave propagation in elastic solids can be formulated in terms of hyperbolic conservation laws. Due to the great importance of conservation laws (Dafermos, 2010), numerous numerical methods have been applied to their solution: finite difference methods (Godlewski and Raviart, 1996; Trangenstein, 2009), finite element methods (Cohen, 2002; Kampanis et al., 2008), discontinuous Galerkin methods (Hesthaven and Warburton, 2007; Cohen and Pernet, 2017), finite volume methods (LeVeque, 2002; Guinot, 2003), spectral methods (Hesthaven et al., 2007; Gopalakrishnan et al., 2007) etc.

A comprehensive survey of numerical methods for conservation laws has been presented recently (Hesthaven, 2018). Nevertheless, problems still exist with interface and boundary conditions in multidimensional cases (Gao et al., 2015). In this paper we focus on the construction of a two-dimensional (2D) explicit finite volume numerical scheme with the special attention to the implementation of boundary conditions.

M. Berezovski
Embry–Riddle Aeronautical University, Daytona Beach, FL, USA
e-mail: mihhail.berezovski@erau.edu

A. Berezovski (✉)
Department of Cybernetics, School of Science, Tallinn University of Technology, Tallinn, Estonia
e-mail: arkadi.berezovski@cs.ioc.ee

A. Berezovski, T. Soomere (eds.), *Applied Wave Mathematics II*, Mathematics of Planet Earth 6, https://doi.org/10.1007/978-3-030-29951-4_12

12.1.1 Finite Volume Methods

Finite volume schemes are powerful numerical methods for solving nonlinear conservation laws and related equations. Such methods are locally conservative and based on cell averages. The numerical solution of systems of hyperbolic conservation laws is dominated by Riemann-solver-based schemes (Godlewski and Raviart, 1996; Toro, 1997; LeVeque, 2002; Guinot, 2003). An upgrade of the solution in a given cell is determined by the exchanges (via fluxes) at the interfaces with the neighbouring cells. The fluxes are computed by solving Riemann problems at the interfaces between neighbouring cells.

Computing an exact solution to the Riemann problem can be a very time-consuming task because an iterative procedure is needed. Therefore, approximate Riemann solvers are often preferred because they provide satisfactory solutions while using faster algorithms. Two broad families of solvers can be distinguished: (i) solvers where the Riemann problem is simplified (e.g., by linearising the equations), and (ii) solvers where simplified relationships are used to solve the exact problem. The first family of solver includes Roe's solver (Roe, 1981), where the flux at the location of the initial discontinuity is calculated via a wave decomposition under the assumption of a constant Jacobian matrix.

The Jacobian matrix is approximated in such a way that consistency and conservation conditions are satisfied. An entropy fix is needed when a rarefaction wave extends over the location of the initial discontinuity. From another side, primitive variable Riemann solvers (Toro, 1997) use a linearisation of the hyperbolic system with a constant Jacobian matrix in combination with the Rankine–Hugoniot conditions across each wave. This allows a simplified system of equations to be solved for the unknown variables. The Riemann invariants can also be used along characteristics to obtain the simplified system (Lhomme and Guinot, 2007).

12.1.2 Higher Order Accuracy and Higher Dimensions

The cell average of a solution in a cell contains too little information. In order to obtain higher order accuracy, neighbouring cell averages are used to reconstruct an approximate polynomial solution in each cell. This reconstruction procedure is the key step for many high-resolution schemes (Liu et al., 2007). For example, in the Advection-Diffusion-Reaction (ADER) approach (Titarev and Toro, 2002), the numerical flux function is based on the solution of generalised Riemann problems, where the initial data on both sides of the element interfaces are no longer piecewise constant. Here the initial data is a piecewise polynomial, in general separated by a jump at the interface.

The fundamental idea behind the generalised Riemann problem solvers is a temporal Taylor series expansion of the state at the interface, where time derivatives are then replaced by space derivatives using repeatedly the governing conservation law in differential form, which is the so-called Cauchy–Kovalewski or Lax–Wendroff procedure (Dumbser and Käser, 2007).

When extending the flux difference schemes to multidimensional problems, the so-called grid aligned finite volume approach or dimensional splitting method is adopted traditionally using one-dimensional (1D) Riemann solvers. However, for multidimensional problems, there is in general no longer a finite number of directions of information propagation. It has been pointed out (Roe, 1986) that the Riemann solver is applied in a grid- rather than in the flow-direction, which may lead to a misinterpretation of the local wave structure of the solution.

To overcome the drawbacks of methods based on dimensional splitting, there has been considerable efforts to develop so-called genuinely multidimensional schemes for solving hyperbolic conservation laws (Colella, 1990; Billett and Toro, 1997; LeVeque, 2002; Guinot, 2003).

12.1.3 Discontinuities

While the abovementioned numerical methods have been successfully applied to the solution of problems with smoothly varying fields, they cannot readily handle evolving discontinuities like cracks or martensitic phase transition fronts inside bodies (de Borst, 2008). The reason is the absence of the constitutive information to specify the velocity of the discontinuity uniquely (Abeyaratne and Knowles, 2006).

In the series of papers (Berezovski and Maugin, 2004, 2005a, 2007; Maugin and Berezovski, 2009; Berezovski and Maugin, 2010), it is shown how the additional constitutive information can be extracted from the analysis of the nonequilibrium interaction between two discrete thermodynamic systems. Moreover, this additional information has been successfully embedded into a finite volume algorithm for thermoelastic wave and front propagation represented in terms of averaged and excess quantities (Berezovski et al., 2000; Berezovski and Maugin, 2001, 2002; Berezovski et al., 2003; Berezovski and Maugin, 2005b; Berezovski et al., 2006, 2008; Berezovski, 2011).

It should be noted that in the 1D case, this algorithm can be identified with the conservative wave propagation algorithm (Bale et al., 2003) for smooth solutions (Berezovski, 2011). This means that the splitting of 1D fluxes in the transverse directions in the spirit of the wave propagation algorithm (LeVeque, 1997) is still possible, but only for smooth solutions.

To obtain the multidimensional description of evolving discontinuities, we need to extend the algorithm in terms of averaged and excess quantities onto at least two dimensions. This is the main aim of this chapter, which is devoted to the derivation and the application of the two-dimensional (2D) finite volume numerical scheme for elastic wave propagation.

The chapter is organised as follows. In Sect. 12.2 the governing equations of linear elasticity are presented in the plane strain approximation. Then the finite volume numerical scheme is deduced in Sect. 12.3 on the regular Cartesian grid.

The algebraic relations for excess quantities are written down explicitly in Sect. 12.4. Boundary conditions are formulated for the example of the test problem of wave propagation in an homogeneous waveguide. Results of calculations of the test problem are presented in Sect. 12.5. Conclusions are given in the last section.

12.2 Governing Equations

Elastic solids are characterised by the Hooke law which can be represented in the isotropic case in the form of the stress-strain relation (Mase et al., 2009)

$$\sigma_{ij} = \lambda \delta_{ij} \varepsilon_{kk} + 2\mu \varepsilon_{ij}, \tag{12.1}$$

with the Cauchy stress tensor σ_{ij}, the strain tensor ε_{ij}, and the Lamé parameters λ and μ. In the linear elasticity theory, a motion is governed by the local balance of linear momentum at each regular material point (Achenbach, 1973)

$$\rho \frac{\partial v_i}{\partial t} = \frac{\partial \sigma_{ij}}{\partial x_i} + f_i, \tag{12.2}$$

where ρ is the matter density, v_i is the particle velocity, t is time, f_i is a body force, and x_i are spatial coordinates.

We consider the plane strain situation which means that a body is extremely thick along one coordinate, say, z, and where all applied forces are uniform in the z-direction. Since all derivatives with respect to z vanish, all fields can be viewed as functions of x and y alone. In the plane strain case in the absence of body force, the governing equations for wave motion (12.2) are reduced to

$$\rho \frac{\partial v_1}{\partial t} = \frac{\partial \sigma_{11}}{\partial x} + \frac{\partial \sigma_{12}}{\partial y}, \tag{12.3}$$

$$\rho \frac{\partial v_2}{\partial t} = \frac{\partial \sigma_{21}}{\partial x} + \frac{\partial \sigma_{22}}{\partial y}. \tag{12.4}$$

Stress-strain relations (12.1) are reformulated accordingly

$$\sigma_{11} = (\lambda + 2\mu)\varepsilon_{11} + \lambda \varepsilon_{22}, \tag{12.5}$$

$$\sigma_{12} = \sigma_{21} = 2\mu \varepsilon_{12}, \tag{12.6}$$

$$\sigma_{22} = (\lambda + 2\mu)\varepsilon_{22} + \lambda \varepsilon_{11}. \tag{12.7}$$

Time derivatives of stress-strain relations (12.5)–(12.7) can be represented in terms of velocities

$$\frac{\partial \sigma_{11}}{\partial t} = (\lambda + 2\mu)\frac{\partial v_1}{\partial x} + \lambda \frac{\partial v_2}{\partial y}, \tag{12.8}$$

$$\frac{\partial \sigma_{22}}{\partial t} = \lambda \frac{\partial v_1}{\partial x} + (\lambda + 2\mu)\frac{\partial v_2}{\partial y}, \tag{12.9}$$

$$\frac{\partial \sigma_{12}}{\partial t} = \frac{\partial \sigma_{21}}{\partial t} = \mu \left(\frac{\partial v_1}{\partial y} + \frac{\partial v_2}{\partial x} \right), \tag{12.10}$$

because strains and velocities are connected by compatibility conditions

$$\frac{\partial \varepsilon_{11}}{\partial t} = \frac{\partial v_1}{\partial x}, \tag{12.11}$$

$$\frac{\partial \varepsilon_{12}}{\partial t} = \frac{1}{2}\left(\frac{\partial v_1}{\partial y} + \frac{\partial v_2}{\partial x} \right), \tag{12.12}$$

$$\frac{\partial \varepsilon_{22}}{\partial t} = \frac{\partial v_2}{\partial y}. \tag{12.13}$$

Equations (12.8)–(12.10) together with the balance of linear momentum (12.3)–(12.4) form a closed system of equations, which is convenient for a numerical solution. In the following it will be demonstrated how an explicit finite volume numerical scheme can be constructed which is suitable for the implementation of boundary and interface conditions in a natural way.

12.3 Discretisation

12.3.1 Averaged and Excess Quantities

The first step in the construction of the numerical algorithm is the spatial discretisation of the computational domain. Let us introduce a Cartesian grid of cells $C_{nm} = [x_n, x_{n+1}] \times [y_m, y_{m+1}]$ with interfaces $x_n = n\Delta x$, $y_m = m\Delta y$, and time levels $t_k = k\Delta t$. For simplicity, the grid size Δx, Δy and time step Δt are assumed to be constant. The values of wanted fields are somehow distributed across the cells.

The main idea in the construction of the algorithm is the consideration of every computational cell as a thermodynamic system (Muschik and Berezovski, 2004). Since we cannot expect that such a thermodynamic system is in equilibrium, its local equilibrium state is described by averaged values of field quantities. The use of cell averages is the standard procedure in finite volume methods. What is nonstandard is the introduction into consideration of so-called "excess quantities" in the spirit of the thermodynamics of discrete systems (Muschik and Berezovski, 2004).

The excess quantities represent the difference between values of true and averaged quantities (Berezovski et al., 2008):

$$v_i = \overline{v}_i + V_i , \quad \sigma_{ij} = \overline{\sigma}_{ij} + \Sigma_{ij} . \tag{12.14}$$

Here bars above some symbols denote the averaged quantities and capital letters relate to excess quantities.

12.3.2 Integration Over the Cell

Keeping in mind the representation of field quantities mentioned above, we integrate the governing equations over the computational cell. The integration of the first component of the balance of linear momentum reads

$$\begin{aligned}\int_{\Delta x}\int_{\Delta y} \rho \dot{v}_1 \, \mathrm{d}x \, \mathrm{d}y &= \int_{\Delta y} \left(\sigma_{11}^r - \sigma_{11}^l\right) \mathrm{d}y + \int_{\Delta x} \left(\sigma_{12}^t - \sigma_{12}^b\right) \mathrm{d}x = \\ &= \int_{\Delta y} \left(\overline{\sigma}_{11} + \Sigma_{11}^r - \overline{\sigma}_{11} - \Sigma_{11}^l\right) \mathrm{d}y + \int_{\Delta x} \left(\overline{\sigma}_{12} + \Sigma_{12}^t - \overline{\sigma}_{12} - \Sigma_{12}^b\right) \mathrm{d}x = \\ &= \int_{\Delta y} \left(\Sigma_{11}^r - \Sigma_{11}^l\right) \mathrm{d}y + \int_{\Delta x} \left(\Sigma_{12}^t - \Sigma_{12}^b\right) \mathrm{d}x ,\end{aligned} \tag{12.15}$$

where superscripts r, l, u, and b denote the "right side, "left side", "top", and "bottom", respectively (Fig. 12.1).

Fig. 12.1 Notation for a cell

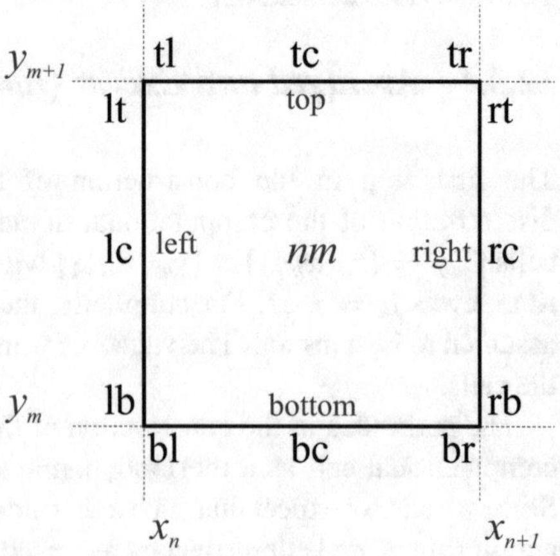

As can be seen from (12.15), the result of the integration is expressed in terms of excess quantities at the boundaries of the cell. These quantities, however, are not constants but vary along the corresponding boundary.

12.3.2.1 Parabolic Approximation at Cell Boundaries

To proceed further, we need to approximate the unknown functions $\Sigma_{11}^{r}(y)$, $\Sigma_{11}^{l}(y)$, $\Sigma_{12}^{t}(x)$, and $\Sigma_{12}^{b}(x)$. Suppose that we know the values Σ_{11}^{rt}, Σ_{11}^{rb} at the right top and right bottom corners and the value Σ_{11}^{rc} at the middle point for the right boundary of the cell numbered by (n, m). Then we can approximate the function $\Sigma_{11}^{r}(y)$ by a quadratic dependence. According to the Simpson rule, we can compute the first integral in $(12.15)_3$ (the third line in this equation) as follows:

$$\int_{\Delta y} \Sigma_{11}^{r}\, \mathrm{d}y = \Delta y \left[\frac{2}{3}\Sigma_{11}^{rc} + \frac{1}{6}\left(\Sigma_{11}^{rt} + \Sigma_{11}^{rb}\right)\right].$$

A similar procedure can be applied for the calculation of all integrals in $(12.15)_3$. Therefore, the integration of the first component of the balance of linear momentum results in the relationship for each computational cell:

$$\int_{\Delta x}\int_{\Delta y} \rho \dot{v}_1\, \mathrm{d}x\, \mathrm{d}y \approx \frac{2}{3}\Delta y\left(\Sigma_{11}^{rm} - \Sigma_{11}^{lm}\right) + \frac{1}{6}\Delta y\left(\Sigma_{11}^{rt} + \Sigma_{11}^{rb} - \Sigma_{11}^{lt} - \Sigma_{11}^{lb}\right) + \\ + \frac{2}{3}\Delta x\left(\Sigma_{12}^{tc} - \Sigma_{12}^{bc}\right) + \frac{1}{6}\Delta x\left(\Sigma_{12}^{tr} + \Sigma_{12}^{tl} - \Sigma_{12}^{br} - \Sigma_{12}^{bl}\right), \tag{12.16}$$

where a combination of two upper indices means the value of the excess quantity at the corresponding corner or middle point (Fig. 12.1). Accordingly, for the second component of the balance of linear momentum we have:

$$\begin{aligned}
\int_{\Delta x}\int_{\Delta y} \rho \dot{v}_2\, \mathrm{d}x\, \mathrm{d}y &= \int_{\Delta y}\left(\sigma_{21}^{r} - \sigma_{21}^{l}\right)\mathrm{d}y + \int_{\Delta x}\left(\sigma_{22}^{t} - \sigma_{22}^{b}\right)\mathrm{d}x = \\
&= \int_{\Delta y}\left(\overline{\sigma}_{21} + \Sigma_{21}^{r} - \overline{\sigma}_{21} - \Sigma_{21}^{l}\right)\mathrm{d}y + \int_{\Delta x}\left(\overline{\sigma}_{22} + \Sigma_{22}^{t} - \overline{\sigma}_{22} - \Sigma_{22}^{b}\right)\mathrm{d}x = \\
&= \int_{\Delta y}\left(\Sigma_{21}^{r} - \Sigma_{21}^{l}\right)\mathrm{d}y + \int_{\Delta x}\left(\Sigma_{22}^{t} - \Sigma_{22}^{b}\right)\mathrm{d}x \approx \\
&\approx \frac{2}{3}\Delta y\left(\Sigma_{21}^{rc} - \Sigma_{21}^{lc}\right) + \frac{1}{6}\Delta y\left(\Sigma_{21}^{rt} + \Sigma_{21}^{rb} - \Sigma_{21}^{lt} - \Sigma_{21}^{lb}\right) + \\
&+ \frac{2}{3}\Delta x\left(\Sigma_{22}^{tc} - \Sigma_{22}^{bc}\right) + \frac{1}{6}\Delta x\left(\Sigma_{22}^{tr} + \Sigma_{22}^{tl} - \Sigma_{22}^{br} - \Sigma_{22}^{bl}\right).
\end{aligned} \tag{12.17}$$

Similarly, the integration of time derivatives of stress-strain relations (12.8)–(12.10) leads to

$$\begin{aligned}
\int_{\Delta x}\int_{\Delta y} \dot{\sigma}_{11}\,\mathrm{d}x\,\mathrm{d}y &= (\lambda+2\mu)\int_{\Delta y}\left(v_1^r - v_1^l\right)\mathrm{d}y + \lambda\int_{\Delta x}\left(v_2^t - v_2^b\right)\mathrm{d}x = \\
&= (\lambda+2\mu)\int_{\Delta y}\left(\overline{v}_1 + V_1^r - \overline{v}_1 - V_1^l\right)\mathrm{d}y + \lambda\int_{\Delta x}\left(\overline{v}_2 + V_2^t - \overline{v}_2 - V_2^b\right)\mathrm{d}x = \\
&= (\lambda+2\mu)\int_{\Delta y}\left(V_1^r - V_1^l\right)\mathrm{d}y + \lambda\int_{\Delta x}\left(V_2^t - V_2^b\right)\mathrm{d}x \approx \\
&\approx \frac{2}{3}(\lambda+2\mu)\Delta y\left(V_1^{rc} - V_1^{lc}\right) + (\lambda+2\mu)\frac{1}{6}\Delta y\left(V_1^{rt} + V_1^{rb} - V_1^{lt} - V_1^{lb}\right) + \\
&+ \frac{2}{3}\lambda\Delta x\left(V_2^{tc} - V_2^{bc}\right) + \lambda\frac{1}{6}\Delta x\left(V_2^{tr} + V_2^{tl} - V_2^{br} - V_2^{bl}\right)
\end{aligned} \tag{12.18}$$

for the first normal component of the stress tensor,

$$\begin{aligned}
\int_{\Delta x}\int_{\Delta y} \dot{\sigma}_{12}\,\mathrm{d}x\,\mathrm{d}y &= \frac{\mu}{2}\int_{\Delta y}\left(v_2^r - v_2^l\right)\mathrm{d}y + \frac{\mu}{2}\int_{\Delta x}\left(v_1^t - v_1^b\right)\mathrm{d}x = \\
&= \frac{\mu}{2}\int_{\Delta y}\left(\overline{v}_2 + V_2^r - \overline{v}_2 - V_2^l\right)\mathrm{d}y + \frac{\mu}{2}\int_{\Delta x}\left(\overline{v}_1 + V_1^t - \overline{v}_1 - V_1^b\right)\mathrm{d}x = \\
&= \frac{\mu}{2}\int_{\Delta y}\left(V_2^r - V_2^l\right)\mathrm{d}y + \frac{\mu}{2}\int_{\Delta x}\left(V_1^t - V_1^b\right)\mathrm{d}x \approx \\
&\approx \frac{\mu}{3}\Delta y\left(V_2^{rc} - V_2^{lc}\right) + \frac{\mu}{12}\Delta y\left(V_2^{rt} + V_2^{rd} - V_2^{lt} - V_2^{lb}\right) + \\
&+ \frac{\mu}{3}\Delta x\left(V_1^{tc} - V_1^{bc}\right) + \frac{\mu}{12}\Delta x\left(V_1^{tr} + V_1^{tl} - V_1^{br} - V_1^{bl}\right)
\end{aligned} \tag{12.19}$$

for the shear stress, and

$$\begin{aligned}
\int_{\Delta x}\int_{\Delta y} \dot{\sigma}_{22}\,\mathrm{d}x\,\mathrm{d}y &= \lambda\int_{\Delta y}\left(v_1^r - v_1^l\right)\mathrm{d}y + (\lambda+2\mu)\int_{\Delta x}\left(v_2^t - v_2^b\right)\mathrm{d}x = \\
&= \lambda\int_{\Delta y}\left(\overline{v}_1 + V_1^r - \overline{v}_1 - V_1^l\right)\mathrm{d}y + (\lambda+2\mu)\int_{\Delta x}\left(\overline{v}_2 + V_2^t - \overline{v}_2 - V_2^b\right)\mathrm{d}x = \\
&= \lambda\int_{\Delta y}\left(V_1^r - V_1^l\right)\mathrm{d}y + (\lambda+2\mu)\int_{\Delta x}\left(V_2^t - V_2^b\right)\mathrm{d}x \approx \\
&\approx \frac{2}{3}\lambda\Delta y\left(V_1^{rc} - V_1^{lc}\right) + \lambda\frac{1}{6}\Delta y\left(V_1^{rt} + V_1^{rb} - V_1^{lt} - V_1^{lb}\right) + \\
&+ \frac{2}{3}(\lambda+2\mu)\Delta x\left(V_2^{tc} - V_2^{bc}\right) + (\lambda+2\mu)\frac{1}{6}\Delta x\left(V_2^{tr} + V_2^{tl} - V_2^{br} - V_2^{bl}\right)
\end{aligned} \tag{12.20}$$

for the second normal component of the stress tensor, respectively.

Defining averaged values for velocities and stresses

$$\overline{v}_i = \frac{1}{\Delta x \Delta y}\int_{\Delta x}\int_{\Delta y} v_i \, \mathrm{d}x \, \mathrm{d}y, \quad \overline{\sigma}_{ij} = \frac{1}{\Delta x \Delta y}\int_{\Delta x}\int_{\Delta y} \sigma_{ij} \, \mathrm{d}x \, \mathrm{d}y, \tag{12.21}$$

we are ready to formulate a numerical scheme in terms of averaged and excess quantities.

12.3.3 Numerical Scheme

The numerical scheme follows the standard approximation of time derivatives

$$\dot{f} \approx \frac{f^{k+1} - f^k}{\Delta t} \quad \forall f, \tag{12.22}$$

which, accounting for (12.16) and (12.17), results in

$$\begin{aligned}(\rho\overline{v}_1)^{k+1} - (\rho\overline{v}_1)^k =& \frac{2}{3}\frac{\Delta t}{\Delta x}\left(\Sigma_{11}^{rc} - \Sigma_{11}^{lc}\right) + \frac{1}{6}\frac{\Delta t}{\Delta x}\left(\Sigma_{11}^{rt} + \Sigma_{11}^{rb} - \Sigma_{11}^{lt} - \Sigma_{11}^{lb}\right) + \\ &+ \frac{2}{3}\frac{\Delta t}{\Delta y}\left(\Sigma_{12}^{tc} - \Sigma_{12}^{bc}\right) + \frac{1}{6}\frac{\Delta t}{\Delta y}\left(\Sigma_{12}^{tr} + \Sigma_{12}^{tl} - \Sigma_{12}^{br} - \Sigma_{12}^{bl}\right),\end{aligned} \tag{12.23}$$

$$\begin{aligned}(\rho\overline{v}_2)^{k+1} - (\rho\overline{v}_2)^k =& \frac{2}{3}\frac{\Delta t}{\Delta x}\left(\Sigma_{21}^{rc} - \Sigma_{21}^{lc}\right) + \frac{1}{6}\frac{\Delta t}{\Delta x}\left(\Sigma_{21}^{rt} + \Sigma_{21}^{rb} - \Sigma_{21}^{lt} - \Sigma_{21}^{lb}\right) + \\ &+ \frac{2}{3}\frac{\Delta t}{\Delta y}\left(\Sigma_{22}^{tc} - \Sigma_{22}^{bc}\right) + \frac{1}{6}\frac{\Delta t}{\Delta y}\left(\Sigma_{22}^{tr} + \Sigma_{22}^{tl} - \Sigma_{22}^{br} - \Sigma_{22}^{bl}\right)\end{aligned} \tag{12.24}$$

for the averaged velocities for each cell n, m. Here all the quantities in the right-hand side are given at time step k and the matter density ρ is assumed to be constant inside each computational cell.

Similarly, for averaged stress components in each cell n, m we have:

$$\begin{aligned}&(\overline{\sigma}_{11})^{k+1} - (\overline{\sigma}_{11})^k = (\lambda + 2\mu)\frac{2}{3}\frac{\Delta t}{\Delta x}\left(V_1^{rc} - V_1^{lc}\right) + \\ &+ (\lambda + 2\mu)\frac{1}{6}\frac{\Delta t}{\Delta x}\left(V_1^{rt} + V_1^{rb} - V_1^{lt} - V_1^{lb}\right) + \lambda\frac{2}{3}\frac{\Delta t}{\Delta y}\left(V_2^{tc} - V_2^{bc}\right) + \\ &+ \lambda\frac{1}{6}\frac{\Delta t}{\Delta y}\left(V_2^{tr} + V_2^{tl} - V_2^{br} - V_2^{bl}\right),\end{aligned} \tag{12.25}$$

$$(\overline{\sigma}_{22})^{k+1} - (\overline{\sigma}_{22})^k = \lambda \frac{2}{3}\frac{\Delta t}{\Delta x}\left(V_1^{rc} - V_1^{lc}\right) + \lambda \frac{1}{6}\frac{\Delta t}{\Delta x}\left(V_1^{rt} + V_1^{rb} - V_1^{lt} - V_1^{lb}\right) +$$
$$+ (\lambda + 2\mu)\frac{2}{3}\frac{\Delta t}{\Delta y}\left(V_2^{tc} - V_2^{bc}\right) + (\lambda + 2\mu)\frac{1}{6}\frac{\Delta t}{\Delta y}\left(V_2^{tr} + V_2^{tl} - V_2^{br} - V_2^{bl}\right), \tag{12.26}$$

$$(\overline{\sigma}_{12})^{k+1} - (\overline{\sigma}_{12})^k = \frac{1}{3}\mu\frac{\Delta t}{\Delta x}\left(V_2^{rc} - V_2^{lc}\right) + \frac{1}{12}\mu\frac{\Delta t}{\Delta x}\left(V_2^{rt} + V_2^{rb} - V_2^{lt} - V_2^{lb}\right) +$$
$$+ \frac{1}{3}\mu\frac{\Delta t}{\Delta y}\left(V_1^{tc} - V_1^{bc}\right) + \frac{1}{12}\mu\frac{\Delta t}{\Delta y}\left(V_1^{tr} + V_1^{tl} - V_1^{br} - V_1^{bl}\right). \tag{12.27}$$

The numerical scheme (12.23)–(12.27) is written down in terms of excess quantities. Therefore, the necessary step is to determine the values of these quantities. Numerical scheme (12.23)–(12.27) uses the values of excess quantities at the middle points of cell boundaries. It is reasonable to identify these values with the values of average excess quantities (i.e., $V_1^{rc} = \overline{V}_1^r$, etc.).

12.4 Determination of Excess Quantities

12.4.1 Averaged Excess Quantities

12.4.1.1 Normal Components

The averaged values of excess quantities are determined exactly by means of jump relations at boundaries between computational cells, which express the continuity of true stresses and velocities (Berezovski et al., 2008)

$$[\![\overline{\sigma}_{ij} + \Sigma_{ij}]\!] = 0, \quad [\![\overline{v}_i + \mathcal{V}_i]\!] = 0. \tag{12.28}$$

In terms of normal components these jump relations for each time step result in

$$(\overline{\sigma}_{11})_{n-1\,m} + \left(\overline{\Sigma}_{11}^r\right)_{n-1\,m} = (\overline{\sigma}_{11})_{n\,m} + \left(\overline{\Sigma}_{11}^l\right)_{n\,m}, \tag{12.29}$$

$$(\overline{\sigma}_{22})_{n\,m-1} + \left(\overline{\Sigma}_{22}^t\right)_{n\,m-1} = (\overline{\sigma}_{22})_{n\,m} + \left(\overline{\Sigma}_{22}^b\right)_{n\,m} \tag{12.30}$$

for stresses and

$$(\overline{v}_1)_{n-1\,m} + \left(\overline{V}_1^r\right)_{n-1\,m} = (\overline{v}_1)_{n\,m} + \left(\overline{V}_1^l\right)_{n\,m}, \tag{12.31}$$

$$(\overline{v}_2)_{n\,m-1} + \left(\overline{V}_2^t\right)_{n\,m-1} = (\overline{v}_2)_{n\,m} + \left(\overline{V}_2^b\right)_{n\,m} \tag{12.32}$$

for velocities. In both cases we have only two equations for four averaged excess quantities. The closure of these systems of equations is achieved by means of conditions of the conservation of analogues of Riemann invariants for excess quantities (Berezovski, 2011)

$$\left(\rho c_p \overline{V}_1^l\right)_{n\,m} + \left(\overline{\Sigma}_{11}^l\right)_{n\,m} = 0, \tag{12.33}$$

$$\left(\rho c_p \overline{V}_1^r\right)_{n-1\,m} - \left(\overline{\Sigma}_{11}^r\right)_{n-1\,m} = 0, \tag{12.34}$$

$$\left(\rho c_p \overline{V}_2^b\right)_{n\,m} + \left(\overline{\Sigma}_{22}^b\right)_{n\,m} = 0, \tag{12.35}$$

$$\left(\rho c_p \overline{V}_2^t\right)_{n\,m-1} - \left(\overline{\Sigma}_{22}^t\right)_{n\,m-1} = 0, \tag{12.36}$$

which leads to closed systems of equations for averaged excess velocities

$$(\overline{\sigma}_{11})_{n-1\,m} + \left(\rho c_p \overline{V}_1^r\right)_{n-1\,m} = (\overline{\sigma}_{11})_{n\,m} - \left(\rho c_p \overline{V}_1^l\right)_{n\,m}, \tag{12.37}$$

$$(\overline{v}_1)_{n-1\,m} + \left(\overline{V}_1^r\right)_{n-1\,m} = (\overline{v}_1)_{n\,m} + \left(\overline{V}_1^l\right)_{n\,m} \tag{12.38}$$

and

$$(\overline{\sigma}_{22})_{n\,m-1} + \left(\rho c_p \overline{V}_2^t\right)_{n\,m-1} = (\overline{\sigma}_{22})_{n\,m} - \left(\rho c_p \overline{V}_2^b\right)_{n\,m}, \tag{12.39}$$

$$(\overline{v}_2)_{n\,m-1} + \left(\overline{V}_2^t\right)_{n\,m-1} = (\overline{v}_2)_{n\,m} + \left(\overline{V}_2^b\right)_{n\,m}. \tag{12.40}$$

Both systems of Eqs. (12.37)–(12.38) and (12.39)–(12.40) have explicit exact solutions. The solution of the first system of equations reads

$$\left(\overline{V}_1^r\right)_{n-1\,m} = \frac{(\overline{\sigma}_{11})_{n\,m} - (\overline{\sigma}_{11})_{n-1\,m} + (\rho c_p)_{n\,m}\left[(\overline{v}_1)_{n\,m} - (\overline{v}_1)_{n-1\,m}\right]}{\left[(\rho c_p)_{n-1\,m} + (\rho c_p)_{n\,m}\right]} \tag{12.41}$$

and

$$\left(\overline{V}_1^l\right)_{n\,m} = \frac{(\overline{\sigma}_{11})_{n\,m} - (\overline{\sigma}_{11})_{n-1\,m} - (\rho c_p)_{n-1\,m}\left[(\overline{v}_1)_{n\,m} - (\overline{v}_1)_{n-1\,m}\right]}{\left[(\rho c_p)_{n-1\,m} + (\rho c_p)_{n\,m}\right]}. \tag{12.42}$$

The corresponding excess values of normal components of the stress tensor follow from the conservation of Riemann invariants

$$\left(\overline{\Sigma}_{11}^{l}\right)_{n\,m} = -\left(\rho c_p \overline{V}_1^{l}\right)_{n\,m}, \tag{12.43}$$

$$\left(\overline{\Sigma}_{11}^{r}\right)_{n-1\,m} = \left(\rho c_p \overline{V}_1^{r}\right)_{n-1\,m}. \tag{12.44}$$

Accordingly, the solution of the second system of Eqs. (12.39)–(12.40) has the form

$$\left(\overline{V}_2^{t}\right)_{n\,m-1} = \frac{(\overline{\sigma}22)_{n\,m} - (\overline{\sigma}22)_{n\,m-1} + \left(\rho c_p\right)_{n\,m}\left[(\overline{v}2)_{n\,m} - (\overline{v}2)_{n\,m-1}\right]}{\left[\left(\rho c_p\right)_{n\,m-1} + \left(\rho c_p\right)_{n\,m}\right]}, \tag{12.45}$$

$$\left(\overline{V}_2^{b}\right)_{n\,m} = \frac{(\overline{\sigma}22)_{n\,m} - (\overline{\sigma}22)_{n\,m-1} - \left(\rho c_p\right)_{n\,m-1}\left[(\overline{v}2)_{n\,m} - (\overline{v}2)_{n\,m-1}\right]}{\left[\left(\rho c_p\right)_{n\,m-1} + \left(\rho c_p\right)_{n\,m}\right]}, \tag{12.46}$$

with excess values of normal components of the stress tensor

$$\left(\overline{\Sigma}_{22}^{b}\right)_{n\,m} = -\left(\rho c_p \overline{V}_2^{b}\right)_{n\,m}, \tag{12.47}$$

$$\left(\overline{\Sigma}_{22}^{t}\right)_{n\,m-1} = \left(\rho c_p \overline{V}_2^{t}\right)_{n\,m-1}. \tag{12.48}$$

12.4.1.2 Shear Components

The values of excess quantities for shear components of the stress tensor are still determined by means of jump relations at boundaries between computational cells (12.28). In terms of shear components these jump relations result in

$$(\overline{\sigma}21)_{n-1\,m} + \left(\overline{\Sigma}_{21}^{r}\right)_{n-1\,m} = (\overline{\sigma}21)_{n\,m} + \left(\overline{\Sigma}_{21}^{l}\right)_{n\,m}, \tag{12.49}$$

$$(\overline{\sigma}12)_{n\,m-1} + \left(\overline{\Sigma}_{12}^{t}\right)_{n\,m-1} = (\overline{\sigma}12)_{n\,m} + \left(\overline{\Sigma}_{12}^{b}\right)_{n\,m} \tag{12.50}$$

for stresses and

$$(\overline{v}1)_{n\,m-1} + \left(\overline{V}_1^{t}\right)_{n\,m-1} = (\overline{v}1)_{n\,m} + \left(\overline{V}_1^{b}\right)_{n\,m}, \tag{12.51}$$

$$(\overline{v}2)_{n-1\,m} + \left(\overline{V}_2^{r}\right)_{n-1\,m} = (\overline{v}2)_{n\,m} + \left(\overline{V}_2^{l}\right)_{n\,m} \tag{12.52}$$

for the corresponding velocities. The closure of systems of equations is again achieved by means of conditions of the conservation of Riemann invariants for corresponding excess quantities

$$\left(\overline{\Sigma}_{12}^{b}\right)_{n\,m} = -\left(\rho c_s \overline{V}_1^{b}\right)_{n\,m}, \tag{12.53}$$

$$\left(\overline{\Sigma}_{12}^{t}\right)_{n\,m-1} = \left(\rho c_s \overline{V}_1^{t}\right)_{n\,m-1}, \tag{12.54}$$

$$\left(\overline{\Sigma}_{21}^{l}\right)_{n\,m} = -\left(\rho c_s \overline{V}_2^{l}\right)_{n\,m}, \tag{12.55}$$

$$\left(\overline{\Sigma}_{21}^{r}\right)_{n-1\,m} = \left(\rho c_s \overline{V}_2^{r}\right)_{n-1\,m}. \tag{12.56}$$

Explicit exact solutions for excess quantities of shear components still exist in the form

$$\left(\overline{V}_2^{r}\right)_{n-1\,m} = \frac{(\overline{\sigma}_{21})_{n\,m} - (\overline{\sigma}_{21})_{n-1\,m} + (\rho c_s)_{n\,m}\left[(\overline{v}_2)_{n\,m} - (\overline{v}_2)_{n-1\,m}\right]}{\left[(\rho c_s)_{n-1\,m} + (\rho c_s)_{n\,m}\right]}, \tag{12.57}$$

$$\left(\overline{V}_2^{l}\right)_{n\,m} = \frac{(\overline{\sigma}_{21})_{n\,m} - (\overline{\sigma}_{21})_{n-1\,m} - (\rho c_s)_{n-1\,m}\left[(\overline{v}_2)_{n\,m} - (\overline{v}_2)_{n-1\,m}\right]}{\left[(\rho c_s)_{n-1\,m} + (\rho c_s)_{n\,m}\right]} \tag{12.58}$$

and

$$\left(\overline{V}_1^{t}\right)_{n\,m-1} = \frac{(\overline{\sigma}_{12})_{n\,m} - (\overline{\sigma}_{12})_{n\,m-1} + (\rho c_s)_{n\,m}\left[(\overline{v}_1)_{n\,m} - (\overline{v}_1)_{n\,m-1}\right]}{\left[(\rho c_s)_{n\,m-1} + (\rho c_s)_{n\,m}\right]}, \tag{12.59}$$

$$\left(\overline{V}_1^{b}\right)_{n\,m} = \frac{(\overline{\sigma}_{12})_{n\,m} - (\overline{\sigma}_{12})_{n\,m-1} - (\rho c_s)_{n\,m-1}\left[(\overline{v}_1)_{n\,m} - (\overline{v}_1)_{n\,m-1}\right]}{\left[(\rho c_s)_{n\,m-1} + (\rho c_s)_{n\,m}\right]}. \tag{12.60}$$

The values of averaged excess quantities for shear components of the stress tensor follow from the conservation of Riemann invariants mentioned above. Now all averaged excess quantities are determined. It remains to calculate the values of excess quantities at the corners of computational cells.

12.4.2 Excess Quantities at Vertices

Suppose that the values of velocity $(V_1^{tr})_{nm}$ at upper right corner of the cell numbered (n, m) are known (Fig. 12.2).

Each corner of the computational cell $C_{nm} = [x_n, x_{n+1}] \times [y_m, y_{m+1}]$ can be considered as the central point of one of the corresponding four virtual cells $C_{n\pm1/2\, m\pm1/2} = [x_{n\pm1/2}, x_{n+1\pm1/2}] \times [y_{m\pm1/2}, y_{m+1\pm1/2}]$. To the first approximation, the value of every field quantity at the corners of computational cells can be represented as the simple average of the corresponding values in neighbouring cells (see the notation in Fig. 12.2). This means that the value of the horizontal excess velocity at the right upper corner $(V_1^{rt})_{nm}$ is calculated by the expression

$$(\overline{v}_1)_{nm} + (V_1^{rt})_{nm} = \frac{1}{4}\Bigg[(\overline{v}_1)_{nm} + \left(\overline{V}_1^{r}\right)_{nm} + (\overline{v}_1)_{n,\,m+1} + \left(\overline{V}_1^{r}\right)_{n\,m+1} + \\ + (\overline{v}_1)_{n+1\,m} + \left(\overline{V}_1^{l}\right)_{n+1\,m} + (\overline{v}_1)_{n+1,\,m+1} + \left(\overline{V}_1^{l}\right)_{n+1\,m+1}\Bigg], \tag{12.61}$$

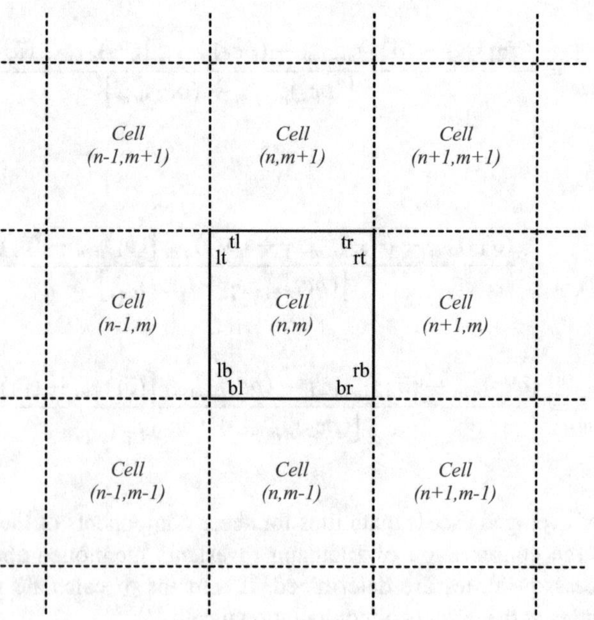

Fig. 12.2 Notation for neighbouring cells

which results in

$$(V_1^{rt})_{n\,m} = \frac{1}{4}\Bigg[-3\,(\overline{v}_1)_{n\,m} + \left(\overline{V}_1^{r}\right)_{n\,m} + (\overline{v}_1)_{n,\,m+1} + \left(\overline{V}_1^{r}\right)_{n\,m+1} + \\ + (\overline{v}_1)_{n+1\,m} + \left(\overline{V}_1^{l}\right)_{n+1\,m} + (\overline{v}_1)_{n+1,\,m+1} + \left(\overline{V}_1^{l}\right)_{n+1\,m+1}\Bigg]. \tag{12.62}$$

It is clear that due to symmetry of averaging the value of another velocity component $(V_1^{tr})_{n\,m}$ is the same

$$(V_1^{tr})_{n\,m} = (V_1^{rt})_{n\,m}. \tag{12.63}$$

Similarly, for $(V_1^{rb})_{n\,m}$ we have:

$$(V_1^{rb})_{n\,m} = \frac{1}{4}\Bigg[-3\,(\overline{v}_1)_{n\,m} + \left(\overline{V}_1^{r}\right)_{n\,m} + (\overline{v}_1)_{n,\,m-1} + \left(\overline{V}_1^{r}\right)_{n\,m-1} + \\ + (\overline{v}_1)_{n+1\,m} + \left(\overline{V}_1^{l}\right)_{n+1\,m} + (\overline{v}_1)_{n+1,\,m-1} + \left(\overline{V}_1^{l}\right)_{n+1\,m-1}\Bigg], \tag{12.64}$$

with

$$(V_1^{br})_{n\,m} = (V_1^{rb})_{n\,m}. \tag{12.65}$$

The remaining values of excess velocities at the corners of computational cells are determined in the same way. Namely,

$$(V_1^{lt})_{n\,m} = \frac{1}{4}\Bigg[-3\,(\overline{v}_1)_{n\,m} + \left(\overline{V}_1^{l}\right)_{n\,m} + (\overline{v}_1)_{n,\,m+1} + \left(\overline{V}_1^{l}\right)_{n\,m+1} + \\ + (\overline{v}_1)_{n-1\,m} + \left(\overline{V}_1^{r}\right)_{n-1\,m} + (\overline{v}_1)_{n-1,\,m+1} + \left(\overline{V}_1^{r}\right)_{n-1\,m+1}\Bigg], \tag{12.66}$$

$$(V_1^{lb})_{n\,m} = \frac{1}{4}\Bigg[-3\,(\overline{v}_1)_{n\,m} + \left(\overline{V}_1^{l}\right)_{n\,m} + (\overline{v}_1)_{n,\,m-1} + \left(\overline{V}_1^{l}\right)_{n\,m-1} + \\ (\overline{v}_1)_{n-1\,m} + \left(\overline{V}_1^{r}\right)_{n-1\,m} + (\overline{v}_1)_{n-1,\,m-1} + \left(\overline{V}_1^{r}\right)_{n-1\,m-1}\Bigg]. \tag{12.67}$$

The same rules of averaging are used for excess stress components. For instance, we have for $(\Sigma_{11}^{rt})_{n\,m}$:

$$(\Sigma_{11}^{rt})_{n\,m} = \frac{1}{4}\Big[-3\,(\overline{\sigma}_{11})_{n\,m} + \left(\overline{\Sigma}_{11}^{r}\right)_{n\,m} + (\overline{\sigma}_{11})_{n,\,m+1} + \left(\overline{\Sigma}_{11}^{r}\right)_{n\,m+1} + \\ + (\overline{\sigma}_{11})_{n+1\,m} + \left(\overline{\Sigma}_{11}^{l}\right)_{n+1\,m} + (\overline{\sigma}_{11})_{n+1,\,m+1} + \left(\overline{\Sigma}_{11}^{l}\right)_{n+1\,m+1}\Big] \tag{12.68}$$

and, consecutively,

$$(\Sigma_{11}^{rb})_{n\,m} = \frac{1}{4}\Big[-3\,(\overline{\sigma}_{11})_{n\,m} + \left(\overline{\Sigma}_{11}^{r}\right)_{n\,m} + (\overline{\sigma}_{11})_{n,\,m-1} + \left(\overline{\Sigma}_{11}^{r}\right)_{n\,m-1} + \\ + (\overline{\sigma}_{11})_{n+1\,m} + \left(\overline{\Sigma}_{11}^{l}\right)_{n+1\,m} + (\overline{\sigma}_{11})_{n+1,\,m-1} + \left(\overline{\Sigma}_{11}^{l}\right)_{n+1\,m-1}\Big], \tag{12.69}$$

$$(\Sigma_{11}^{lt})_{n\,m} = \frac{1}{4}\Big[-3\,(\overline{\sigma}_{11})_{n\,m} + \left(\overline{\Sigma}_{11}^{l}\right)_{n\,m} + (\overline{\sigma}_{11})_{n,\,m+1} + \left(\overline{\Sigma}_{11}^{l}\right)_{n\,m+1} + \\ + (\overline{\sigma}_{11})_{n-1\,m} + \left(\overline{\Sigma}_{11}^{r}\right)_{n-1\,m} + (\overline{\sigma}_{11})_{n-1,\,m+1} + \left(\overline{\Sigma}_{11}^{r}\right)_{n-1\,m+1}\Big], \tag{12.70}$$

$$(\Sigma_{11}^{lb})_{n\,m} = \frac{1}{4}\Big[-3\,(\overline{\sigma}_{11})_{n\,m} + \left(\overline{\Sigma}_{11}^{l}\right)_{n\,m} + (\overline{\sigma}_{11})_{n,\,m-1} + \left(\overline{\Sigma}_{11}^{l}\right)_{n\,m-1} + \\ (\overline{\sigma}_{11})_{n-1\,m} + \left(\overline{\Sigma}_{11}^{r}\right)_{n-1\,m} + (\overline{\sigma}_{11})_{n-1,\,m-1} + \left(\overline{\Sigma}_{11}^{r}\right)_{n-1\,m-1}\Big]. \tag{12.71}$$

All other excess quantities at the corners of computational cells are calculated algebraically in the same way. This finalises the procedure of the determination of excess quantities. The substitution of the values of excess quantities into numerical scheme (12.23)–(12.27) allows us to perform calculations of two-dimensional problems.

12.5 Test Problem

As an example, a stress pulse propagation in a waveguide depicted in Fig. 12.3 is considered. The length of the waveguide is 250 mm and its thickness is 100 mm. Calculations are performed for Al 6061 alloy characterised by the density 2700 kg/m^3, the Young modulus 68.9 GPa, and the Poisson ratio 0.33. This corresponds to the longitudinal wave velocity 5092 m/s. Choosing the space step equal to 1 mm, we have the time step 0.196 µs.

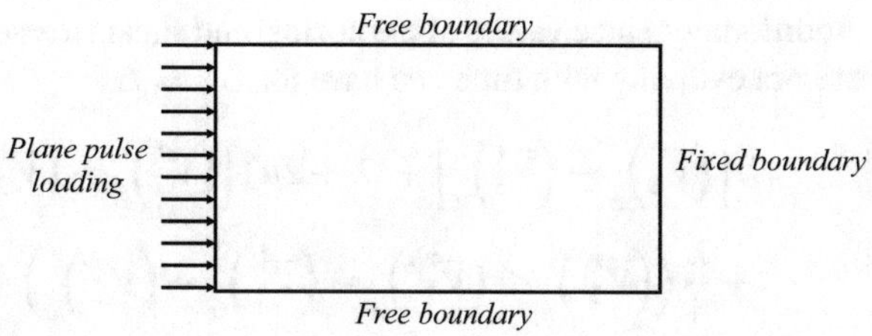

Fig. 12.3 Boundary conditions

To be able to perform the calculation of a particular problem we need to specify initial and boundary conditions. Initial conditions fix the state of each cell at a chosen time instant. We suppose that initially the waveguide is at rest, which assumes zero values for all wanted fields. Boundary conditions should be expressed in terms of averaged and excess quantities used in the numerical scheme. We expect that the state of cells adjacent to each boundary is known (at least partly). For the proper computing, we need to know in advance as many values of averaged and excess quantities as possible.

12.5.1 Upper Boundary: Stress Free

We start with the stress-free upper boundary conditions. The implementation of stress-free boundary conditions in the numerical scheme is not an easy task. The best progress in this field is achieved by the geophysical community (Moczo et al., 2014). The free surface boundary treatment is performed within the finite difference framework because of the efficiency of wave motion simulation in comparison with finite element or discontinuous Galerkin methods (Gao et al., 2015).

In the proposed numerical procedure, however, all the boundary conditions need to be formulated in terms of averaged and excess quantities. As we know, jump relations at boundaries between computational cells express continuity of true stresses and velocities (12.28). It follows that at the stress-free upper boundary the value of the normal stress is zero, yielding (for each time step):

$$(\overline{\sigma}_{22})_n + \left(\overline{\Sigma}^t_{22}\right)_n = 0. \tag{12.72}$$

A similar relationship holds for the shear stress:

$$(\overline{\sigma}_{12})_n + \left(\overline{\Sigma}^t_{12}\right)_n = 0. \tag{12.73}$$

Additionally, since values of the normal and shear stresses at the stress-free boundary are not evolving with time, we have for $\Delta x = \Delta y$:

$$\lambda\left[\left(\overline{V}_1^r\right)_n - \left(\overline{V}_1^l\right)_n\right] + (\lambda+2\mu)\left[\left(\overline{V}_2^t\right)_n - \left(\overline{V}_2^b\right)_n\right] + \\ + \frac{\lambda}{4}\left(\left(\overline{V}_1^{rt}\right)_n + \left(\overline{V}_1^{rb}\right)_n - \left(\overline{V}_1^{lt}\right)_n - \left(\overline{V}_1^{lb}\right)_n\right) + \\ + \frac{(\lambda+2\mu)}{4}\left[\left(\overline{V}_2^{tr}\right)_n + \left(\overline{V}_2^{tl}\right)_n - \left(\overline{V}_2^{br}\right)_n - \left(\overline{V}_2^{bl}\right)_n\right] = 0, \tag{12.74}$$

$$\left(\overline{V}_2^r\right)_n - \left(\overline{V}_2^l\right)_n + \left(\overline{V}_1^t\right)_n - \left(\overline{V}_1^b\right)_n + \\ + \frac{1}{4}\left[\left(\overline{V}_2^{rt}\right)_n + \left(\overline{V}_2^{rb}\right)_n - \left(\overline{V}_2^{lt}\right)_n - \left(\overline{V}_2^{lb}\right)_n\right] + \\ + \frac{1}{4}\left[\left(\overline{V}_1^{tr}\right)_n + \left(\overline{V}_1^{tl}\right)_n - \left(\overline{V}_1^{br}\right)_n - \left(\overline{V}_1^{bl}\right)_n\right] = 0. \tag{12.75}$$

Equations (12.72)–(12.75) allow us to calculate values of four excess quantities, namely, $\left(\overline{V}_1^t\right)_n$, $\left(\overline{V}_2^t\right)_n$, $\left(\overline{\Sigma}_{12}^t\right)_n$, and $\left(\overline{\Sigma}_{22}^t\right)_n$ at the upper stress-free boundary. Fortunately, it is sufficient to update the averaged values of all fields at the upper layer of the computational domain.

A similar consideration is valid for the stress-free bottom boundary with the corresponding transformation of indices.

12.5.2 Left Boundary: Loading; Right Boundary: Fixed

At the loading left boundary the value of the normal stress in each cell and at each time step is given in advance:

$$(\overline{\sigma}_{11})_m + \left(\overline{\Sigma}_{11}^l\right)_m = \sigma_{11}(t), \tag{12.76}$$

where $\sigma_{11}(t)$ is a known function. A similar relationship holds for the shear stress:

$$(\overline{\sigma}_{12})_m + \left(\overline{\Sigma}_{12}^l\right)_m = \sigma_{12}(t). \tag{12.77}$$

Since values of the normal and shear stresses at the loading boundary are evolving with time, we have additionally for each boundary cell and at each time step:

$$\Delta x \dot{\sigma}_{11} = \frac{2}{3}(\lambda+2\mu)\left[\left(\overline{V}_1^r\right)_m - \left(\overline{V}_1^l\right)_m\right] + \frac{2}{3}\lambda\left[\left(\overline{V}_2^t\right)_m - \left(\overline{V}_2^b\right)_m\right] + \\ + (\lambda+2\mu)\frac{1}{6}\left[\left(\overline{V}_1^{rt}\right)_m + \left(\overline{V}_1^{rb}\right)_m - \left(\overline{V}_1^{lt}\right)_m - \left(\overline{V}_1^{lb}\right)_m\right] + \\ + \lambda\frac{1}{6}\left[\left(\overline{V}_2^{tr}\right)_m + \left(\overline{V}_2^{tl}\right)_m - \left(\overline{V}_2^{br}\right)_m - \left(\overline{V}_2^{bl}\right)_m\right], \tag{12.78}$$

$$
\begin{aligned}
\Delta x \dot{\sigma}_{12} = & \frac{1}{3}\mu \left[\left(\overline{V}_2^{r}\right)_m - \left(\overline{V}_2^{l}\right)_m \right] + \frac{1}{3}\mu \left[\left(\overline{V}_1^{t}\right)_m - \left(\overline{V}_1^{b}\right)_m \right] + \\
& + \frac{1}{12}\mu \left[\left(\overline{V}_2^{rt}\right)_m + \left(\overline{V}_2^{rb}\right)_m - \left(\overline{V}_2^{lt}\right)_m - \left(\overline{V}_2^{lb}\right)_m \right] + \\
& + \frac{1}{12}\mu \left[\left(\overline{V}_1^{tr}\right)_m + \left(\overline{V}_1^{tl}\right)_m - \left(\overline{V}_1^{br}\right)_m - \left(\overline{V}_1^{bl}\right)_m \right].
\end{aligned} \tag{12.79}
$$

By means (12.76)–(12.79) we can calculate values of four excess quantities, namely, $\left(\overline{V}_1^{l}\right)_m$, $\left(\overline{V}_2^{l}\right)_m$, $\left(\overline{\Sigma}_1^{l}\right)_m$, and $\left(\overline{\Sigma}_2^{l}\right)_m$, at the left boundary. As mentioned previously, it is sufficient to update the averaged values of all fields at the left boundary of the computational domain.

At a fixed right boundary the values of velocities are zero, i.e.,

$$
(\overline{v}_i)_m + \left(\overline{V}_i^{r}\right)_m = 0 \quad \forall m . \tag{12.80}
$$

Besides, since the values of velocities at the fixed boundary are not evolving with time, we have $\forall\, m$:

$$
\begin{aligned}
& \left(\overline{\Sigma}_{11}^{r}\right)_m - \left(\overline{\Sigma}_{11}^{l}\right)_m + \left(\overline{\Sigma}_{12}^{t}\right)_m - \left(\overline{\Sigma}_{12}^{b}\right)_m + \\
& + \frac{1}{4}\left[\left(\overline{\Sigma}_{11}^{rt}\right)_m + \left(\overline{\Sigma}_{11}^{rb}\right)_m - \left(\overline{\Sigma}_{11}^{lt}\right)_m - \left(\overline{\Sigma}_{11}^{lb}\right)_m \right] + \\
& + \frac{1}{4}\left[\left(\overline{\Sigma}_{12}^{tr}\right)_m + \left(\overline{\Sigma}_{12}^{tl}\right)_m - \left(\overline{\Sigma}_{12}^{br}\right)_m - \left(\overline{\Sigma}_{12}^{bl}\right)_m \right] = 0,
\end{aligned} \tag{12.81}
$$

$$
\begin{aligned}
& \left(\overline{\Sigma}_{21}^{r}\right)_m - \left(\overline{\Sigma}_{21}^{l}\right)_m + \left(\overline{\Sigma}_{22}^{t}\right)_m - \left(\overline{\Sigma}_{22}^{b}\right)_m + \\
& + \frac{1}{4}\left[\left(\overline{\Sigma}_{21}^{rt}\right)_m + \left(\overline{\Sigma}_{21}^{rb}\right)_m - \left(\overline{\Sigma}_{21}^{lt}\right)_m - \left(\overline{\Sigma}_{21}^{lb}\right)_m \right] + \\
& + \frac{1}{4}\left[\left(\overline{\Sigma}_{22}^{tr}\right)_m + \left(\overline{\Sigma}_{22}^{tl}\right)_m - \left(\overline{\Sigma}_{22}^{br}\right)_m - \left(\overline{\Sigma}_{22}^{bl}\right)_m \right] = 0.
\end{aligned} \tag{12.82}
$$

Relationships (12.80)–(12.82) are used to obtain values of four excess quantities, namely, $\left(\overline{V}_1^{r}\right)_m$, $\left(\overline{V}_2^{r}\right)_m$, $\left(\overline{\Sigma}_{11}^{r}\right)_m$, and $\left(\overline{\Sigma}_{21}^{r}\right)_m$ at the fixed right boundary. As before, it is sufficient to update the averaged values of all fields at the right layer of the computational domain.

12.5.3 Results of Computations

The problem of a pulse propagation in an homogeneous waveguide is solved by means of a wave propagation algorithm (LeVeque, 1997) and by means of the proposed numerical scheme. The shape of the plane loading pulse at the left boundary is prescribed by the dependence $\sigma_{11}(t) = \sin^2(\pi t/80)$ for the first 80 time steps. After that, the left boundary is stress-free. Calculations were performed with the Courant number equal to 0.91. The main focus is on the influence of the lateral boundaries. The distribution of longitudinal stress at 220 time steps presented in Fig. 12.4 shows a similarity of results obtained by using the two numerical methods. Here only a small difference is observed on the rear side of the pulse.

The distinction in the distribution of the longitudinal stress becomes more evident after reflection at the fixed right boundary as one can see in Fig. 12.5.

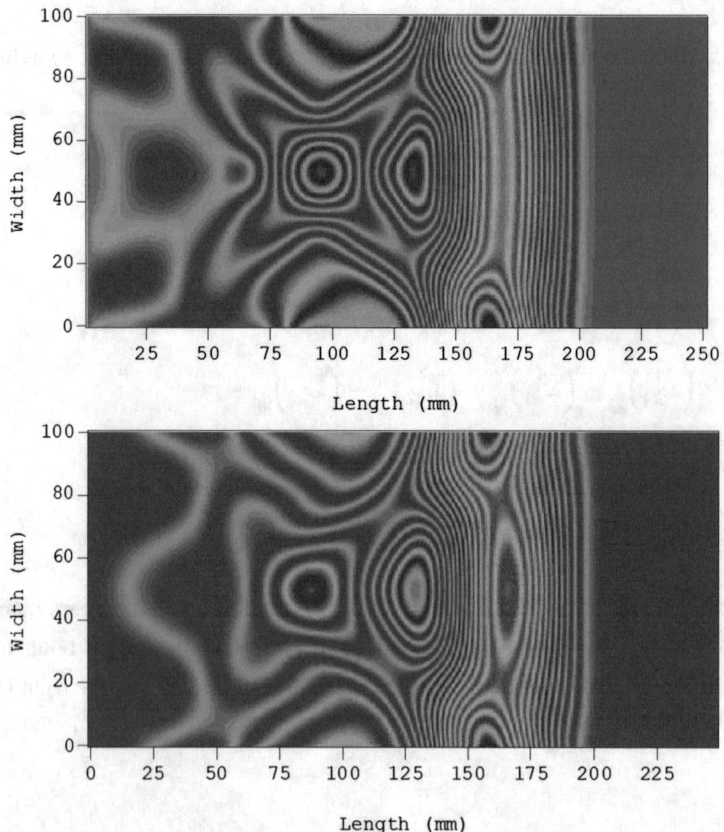

Fig. 12.4 Contour plot for the longitudinal stress distribution at 220 time steps. The upper panel corresponds to results of the standard wave propagation algorithm (LeVeque, 2002) and the bottom panel shows the outcome of the proposed numerical scheme.

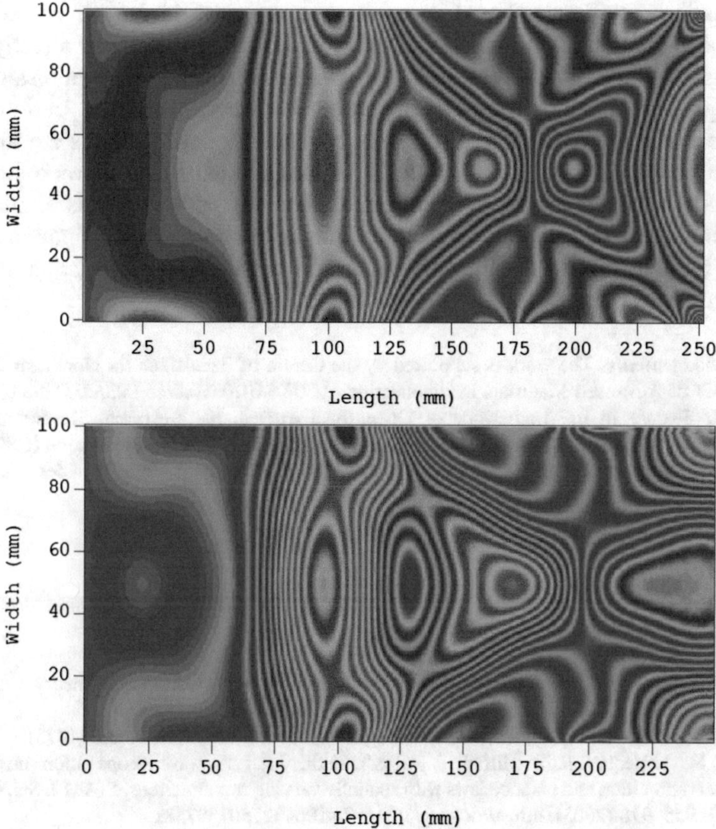

Fig. 12.5 Contour plot for the longitudinal stress distribution at 480 time steps. The upper panel corresponds to results of the standard wave propagation algorithm (LeVeque, 2002). The bottom panel shows the outcome of the proposed numerical scheme.

The main difference in these two approaches is in the implementation of boundary conditions. In the wave propagation algorithm (LeVeque, 2002), boundary conditions are satisfied using the additional "ghost cells". In the proposed thermodynamically consistent scheme, boundary conditions are imposed in terms of excess quantities at the boundaries.

12.6 Conclusions

The propagation of a pulse in elastic waveguides displays the result of interactions of distinct modes. Theoretically, only certain first modes are taken into account. Direct numerical simulation combines all of them by default. However, the implementation of boundary conditions should be as accurate as possible. In this chapter, such an implementation is proposed in terms of excess quantities taken directly at the boundaries. Simulations were performed by means of a wave propagation algorithm (LeVeque, 1997) and by means of the proposed numerical scheme. In the case of a plane wave, the results of calculations obtained by both methods are identical. For the nonplane wave, the distribution of longitudinal stress shows a similarity of results obtained by the two numerical methods. However, this similarity is not complete, especially after reflection. The details of fields distribution depend on the implementation of boundary conditions of the pulse propagation in elastic waveguides.

Acknowledgements The work is supported by the Centre of Excellence for Nonlinear Dynamic Behaviour of Advanced Materials in Engineering CZ.02.1.01/0.0/0.0/15 003/0000493 (Excellent Research Teams) in the framework of Operational Programme Research, Development and Education, by the Grant project with No. 19-04956S of the Czech Science Foundation (CSF) within institutional support RVO:61388998, and by the Estonian Research Council under institutional block grant IUT33-24.

References

Abeyaratne, R., Knowles, J.K.: Evolution of Phase Transitions: A Continuum Theory. Cambridge University Press (2006). https://doi.org/10.1017/CBO9780511547133

Achenbach, J.: Wave Propagation in Elastic Solids. North-Holland, Amsterdam (1973)

Bale, D.S., LeVeque, R.J., Mitran, S., Rossmanith, J.A.: A wave propagation method for conservation laws and balance laws with spatially varying flux functions. SIAM J. Sci. Comput. **24**(3), 955–978 (2003). https://doi.org/10.1137/s106482750139738x

Berezovski, A.: Thermodynamic interpretation of finite volume algorithms. J. Struct. Mech. (Rakenteiden Mekaniikka) **44**(3), 156–171 (2011)

Berezovski, A., Maugin, G.A.: Simulation of thermoelastic wave propagation by means of a composite wave-propagation algorithm. J. Comput. Phys. **168**(1), 249–264 (2001). https://doi.org/10.1006/jcph.2001.6697

Berezovski, A., Maugin, G.A.: Thermoelastic wave and front propagation. J. Thermal Stresses **25**(8), 719–743 (2002). https://doi.org/10.1080/01495730290074504

Berezovski, A., Maugin, G.A.: On the thermodynamic conditions at moving phase-transition fronts in thermoelastic solids. J. Non-Equilibr. Thermodyn. **29**(1), 37–51 (2004). https://doi.org/10.1515/jnetdy.2004.004

Berezovski A., Maugin G.A.: On the velocity of a moving phase boundary in solids. Acta Mech. **179**(3-4), 187–196 (2005a). https://doi.org/10.1007/s00707-005-0251-1

Berezovski, A., Maugin, G.A.: Stress-induced phase-transition front propagation in thermoelastic solids. Eur. J. Mech.-A/Solids **24**(1), 1–21 (2005b). https://doi.org/10.1016/j.euromechsol.2004.09.004

Berezovski, A., Maugin, G.A.: Moving singularities in thermoelastic solids. Int. J. Fract. **147**(1-4), 191–198 (2007). https://doi.org/10.1007/s10704-007-9159-1

Berezovski, A., Maugin, G.A.: Jump conditions and kinetic relations at moving discontinuities. ZAMM-J. Appl. Math. Mech. **90**(7-8), 537–543 (2010). https://doi.org/10.1002/zamm.200900306

Berezovski, A., Engelbrecht, J., Maugin, G.A.: Thermoelastic wave propagation in inhomogeneous media. Arch. Appl. Mech. **70**(10), 694–706 (2000). https://doi.org/10.1007/s004190000114

Berezovski, A., Engelbrecht, J., Maugin, G.A.: Numerical simulation of two-dimensional wave propagation in functionally graded materials. Eur. J. Mech. A/Solids **22**(2), 257–265 (2003). https://doi.org/10.1016/s0997-7538(03)00029-9

Berezovski, A., Berezovski, M., Engelbrecht, J.: Numerical simulation of nonlinear elastic wave propagation in piecewise homogeneous media. Mater. Sci. Engng. A **418**(1), 364–369 (2006). https://doi.org/10.1016/j.msea.2005.12.005

Berezovski, A., Engelbrecht, J., Maugin, G.A.: Numerical Simulation of Waves and Fronts in Inhomogeneous Solids. World Scientific, Singapore (2008). https://doi.org/10.1142/6931

Billett, S., Toro, E.: On WAF-type schemes for multidimensional hyperbolic conservation laws. J. Comput. Phys. **130**(1), 1–24 (1997). https://doi.org/10.1006/jcph.1996.5470

de Borst, R.: Challenges in computational materials science: Multiple scales, multi-physics and evolving discontinuities. Comput. Mater. Sci. **43**(1), 1–15 (2008). https://doi.org/10.1016/j.commatsci.2007.07.022

Cohen, G.: Higher-Order Numerical Methods for Transient Wave Equations. Springer, Berlin (2002). https://doi.org/10.1007/978-3-662-04823-8

Cohen, G., Pernet S.: Finite Element and Discontinuous Galerkin Methods for Transient Wave Equations. Springer, Dordrecht (2017). https://doi.org/10.1007/978-94-017-7761-2

Colella, P,: Multidimensional upwind methods for hyperbolic conservation laws. J. Comput. Phys. **87**(1), 171–200 (1990). https://doi.org/10.1016/0021-9991(90)90233-q

Dafermos, C.M.: Hyperbolic Conservation Laws in Continuum Physics. Springer, Berlin (2010). https://doi.org/10.1007/978-3-642-04048-1

Dumbser, M., Käser, M.: Arbitrary high order non-oscillatory finite volume schemes on unstructured meshes for linear hyperbolic systems. J. Comput. Phys. **221**(2), 693–723 (2007). https://doi.org/10.1016/j.jcp.2006.06.043

Gao, L., Brossier, R., Pajot, B., Tago, J., Virieux, J.: An immersed free-surface boundary treatment for seismic wave simulation. Geophysics **80**(5), T193–T209 (2015). https://doi.org/10.1190/geo2014-0609.1

Godlewski, E., Raviart, P.A.: Numerical Approximation of Hyperbolic Systems of Conservation Laws. Springer, New York (1996). https://doi.org/10.1007/978-1-4612-0713-9

Gopalakrishnan, S., Chakraborty, A., Mahapatra, D.R.: Spectral Finite Element Method: Wave Propagation, Diagnostics and Control in Anisotropic and Inhomogeneous Structures. Springer, London (2007)

Guinot, V. Godunov-type Schemes: An Introduction for Engineers. Elsevier, Amsterdam (2003)

Hesthaven, J.S.: Numerical Methods for Conservation Laws: From Analysis to Algorithms. SIAM, Philadelphia (2018)

Hesthaven, J.S., Warburton, T.: Nodal Discontinuous Galerkin Methods: Algorithms, Analysis, and Applications. Springer Science & Business Media (2007). https://doi.org/10.1007/978-0-387-72067-8

Hesthaven, J.S., Gottlieb, S., Gottlieb, D.: Spectral Methods for Time-dependent Problems. Cambridge University Press (2007). https://doi.org/10.1017/cbo9780511618352

Kampanis, N.A., Dougalis, V., Ekaterinaris, J.A.: Effective Computational Methods for Wave Propagation. Chapman and Hall/CRC, Boca Raton (2008). https://doi.org/10.1201/9781420010879

LeVeque, R.J.: Wave propagation algorithms for multidimensional hyperbolic systems. J. Comput. Phys. **131**(2), 327–353 (1997). https://doi.org/10.1006/jcph.1996.5603

LeVeque, R.J.: Finite Volume Methods for Hyperbolic Problems. Cambridge University Press (2002). https://doi.org/10.1017/CBO9780511791253

Lhomme, J., Guinot, V.: A general approximate-state Riemann solver for hyperbolic systems of conservation laws with source terms. Int. J. Numer. Methods Fluids **53**(9), 1509–1540 (2007). https://doi.org/10.1002/fld.1374

Liu, Y., Shu, C.W., Tadmor, E., Zhang, M.: Non-oscillatory hierarchical reconstruction for central and finite volume schemes. Commun. Comput. Phys. **2**(5), 933–963 (2007)

Mase, G.T., Smelser, R.E., Mase, G.E. Continuum Mechanics for Engineers. CRC Press, Boca Raton (2009)

Maugin, G.A., Berezovski, A.: On the propagation of singular surfaces in thermoelasticity. J. Thermal Stresses **32**(6-7), 557–592 (2009). https://doi.org/10.1080/01495730902848631

Moczo, P., Kristek, J., Gális, M.: The Finite-Difference Modelling of Earthquake Motions: Waves and Ruptures. Cambridge University Press (2014). https://doi.org/10.1017/cbo9781139236911

Muschik, W., Berezovski, A.: Thermodynamic interaction between two discrete systems in non-equilibrium. J. Non-Equilibr. Thermodyn. **29**(3), 237–255 (2004). https://doi.org/10.1515/jnetdy.2004.053

Roe, P.L.: Approximate Riemann solvers, parameter vectors, and difference schemes. J. Comput. Phys. **43**(2), 357–372 (1981). https://doi.org/10.1016/0021-9991(81)90128-5

Roe, P.L.: Discrete models for the numerical analysis of time-dependent multidimensional gas dynamics. J. Comput. Phys. **63**(2), 458–476 (1986). https://doi.org/10.1016/0021-9991(86)90204-4

Titarev, V.A., Toro, E.F.: ADER: Arbitrary high order Godunov approach. J. Sci. Comput. **17**(1-4), 609–618 (2002)

Toro, E.F.: Riemann Solvers and Numerical Methods for Fluid Dynamics: A Practical Introduction. Springer, Dordrecht (1997). https://doi.org/10.1007/978-3-662-03490-3

Trangenstein, J.A.: Numerical Solution of Hyperbolic Partial Differential Equations. Cambridge University Press (2009)

Chapter 13
A Design of a Material Assembly in Space-Time Generating and Storing Energy

Mihhail Berezovski, Stan Elektrov, and Konstantin Lurie

Abstract The paper introduces the theoretical background of the mechanism of electromagnetic energy and power accumulation and its focusing in narrow pulses travelling along a transmission line with material parameters variable in one space dimension and time. This mechanism may be implemented due to a special material geometry, namely, a distribution of two different dielectrics in a spatiotemporal checkerboard. We concentrate on practically reasonable means to bring this mechanism into action in a device that may work both as energy generator and energy storage. The basic ideas discussed below appear to be fairly general; we have chosen their electromagnetic implementation as an excellent framework for the entire concept.

Professor Jüri Engelbrecht is widely known as an outstanding scholar in various branches of mechanics and mathematical physics. Through many years, he has been one of the main contributors to the intensive development of science and academic education in his homeland Estonia and far beyond it, on the European scale. His public service is commensurate with his scientific merits. Jüri's sphere of interests is vast, and it always belonged to the cutting edge of mechanics and thermodynamics. This is the reason why we decided to submit our work to Jüri's jubilee collection of papers; we dedicate it to our dear colleague and friend with the hope that it may get along with his long-time professional aspirations.

M. Berezovski (✉)
Embry–Riddle Aeronautical University, Daytona Beach, FL, USA
e-mail: mihhail.berezovski@erau.edu

S. Elektrov · K. Lurie
Worcester Polytechnic Institute, Worcester, MA, USA
e-mail: klurie@wpi.edu

A. Berezovski, T. Soomere (eds.), *Applied Wave Mathematics II*, Mathematics of Planet Earth 6, https://doi.org/10.1007/978-3-030-29951-4_13

13.1 Introduction

This paper is about a physical implementation of a novel concept in material science and technology termed *dynamic materials* (DMs) (Blekhman and Lurie, 2000; Lurie, 2007). DM are not materials in a traditional sense. They are defined as material substances with properties that may change in *space and time*. In particular, we may speak about formations assembled from traditional material constituents with variable space-time properties. This understanding is unusual because DMs should be thought of as substances that are not fabricated once and for all, but rather brought into the scene and maintained by being properly operated in space and time. Specifically, the temporal property change is possible only due to an exchange of mass, momentum, and energy between the DM and the environment.

To put it briefly, a DM is a thermodynamically open system; only its union with the environment may be considered as closed. The very notion of a DM makes sense only when it is perceived as a spatial temporal entity, and this is the main difference between DMs and ordinary materials that exist as dead substances, i.e., their properties do not change in time. Contrary to DMs, ordinary materials appear to be thermodynamically closed.

Living tissue represents a natural example of a DM. As to a manmade implementation, this has become possible due to recent progress in technology that allows for the property tuning, specifically on a micro- and nano-scale. Some technical means towards such tuning are detailed in (Pelesko and Bernstein, 2002; Oohira et al., 2004; Rozenberg et al., 2006; Krylov et al., 2010, 2008; Krylov and Lurie, 2011). At this point we only mention that the most practical way towards it will be a *material switching*, i.e., a transfer from one material to another implemented by a suitable technical means whenever and wherever necessary.

This transfer is accompanied by energy flux brought into the system (or released from it) by an external agent. When we work with a DM, the presence of an external agent is both inevitable and critical. It serves as a necessary link with the environment, maintaining the dynamic nature of a material formation.

Once implemented, the concept of DMs offers a diversity of effects unthinkable with ordinary materials. Among them are efficient ways to control wave propagation and, in particular, options for introducing novel principles of power generation, transport, and storage of energy. DMs are also of universal significance for optimisation because they may be adjusted to fit the environment, changing in space and time, and therefore they offer a great amount of resources never stored in ordinary materials. In this capacity, they represent a natural material arena for the purposes addressed in dynamics.

These ideas belong with a core of material science today. They can be realised and verified by building dynamic materials structured as large arrays (transmission lines) of coupled and actively controlled oscillatory elements. For this purpose the LC cells in the electromagnetic implementation could be used, and the mass spring elements in mechanical implementation. In either case, the material parameters of the elements are subjected to active control by an external agent.

The wave propagation along the DM arrays presents a number of distinguishing features. As demonstrated by theoretical analysis, immediate consequences of that will be the ability of such structures to dynamically amplify, tune, and compress travelling disturbances, as well as store them, over a wide range of carrier frequencies. Once the typical space and time scales of microfabricated elements are significantly smaller than the spatial wavelength and the temporal period of the disturbance, these structures can be effectively considered as continuous dynamic materials.

This paper offers a fabrication oriented design of a device accumulating and storing the electromagnetic energy in travelling radio frequency waves. The device represents a transmission line controlled by an external energy source that maintains a checkerboard material geometry in space-time.

We first describe the basic principles and provide a mathematical formulation for the functioning of such devices. In subsequent sections, an engineering realisation is described at some length, by using the parts currently available from the market. The robustness of the device as well as its effectiveness, accounting for unavoidable losses and parasitic effects, is given a numerical evaluation.

13.2 General Principles of Power Generation and Energy Storage in DM Structures

Following is an example of a dynamic material that reveals some of special features of such environment. Specifically, it shows how the proper control over travelling waves may produce a new mechanism of power generation and power storage incorporated in the same device.

Assume that we have two conventional isotropic dielectrics: material 1 and material 2, with the wave impedance $\gamma = \sqrt{\mu/\varepsilon}$ and phase velocity $a = 1/\sqrt{\mu\varepsilon}$ that take the values (γ_1, a_1) and (γ_2, a_2), respectively.

Assume that $\gamma_1 = \gamma_2$ and $a_2 > a_1$, so the wave impedances match, and material 2 is "fast" and material 1 is "slow". Consider a periodic material laminate in one-dimensional (1D) space and time (z, t), with materials 1 and 2 occupying the alternating layers with volume fractions m_1 and m_2, respectively, in every period. The laminate will be called static (temporal) if the layers are perpendicular to the z-axis (t-axis). A static laminate is an ordinary material assemblage in space, while a temporal laminate is already a DM produced by a temporal property switching.

A plane electromagnetic wave travelling through the laminate in the z-direction exhibits no reflection on the material interfaces in both static and temporal versions (Morgenthaler, 1958; Lurie and Weekes, 2006; Lurie, 2007; Lurie et al., 2009; Berezovski and Berezovski, 2018). When the wave travels through a static laminate, its energy remains constant in time. When the wave travels through a temporal laminate, its energy increases by the factor a_2/a_1 each time the wave enters "fast" material 2 from "slow" material 1. Similarly, its energy decreases by the factor a_1/a_2 when it leaves material 2 and enters material 1. The net energy gain over a temporal period is therefore zero. The energy gain (loss) occurs due to the work pumped into the wave (or taken out of it) by an external agent that changes the material properties at each temporal property switching.

If the aim is to ensure the accumulation of energy in a travelling wave, laminates do not provide a feasible solution. To guarantee energy accumulation, we have to avoid the moments of energy loss by finding some other way for the wave to enter the "slow" material 1. This may be across the spatial (instead of temporal) interface, because when the wave crosses a spatial interface it does not lose energy since the energy flux remains continuous. We arrive at the idea of testing a rectangular material structure in space-time as a possible energy accumulator, because it offers both spatial and temporal interfaces.

Specifically we try a "checkerboard" structure (Fig. 13.1). This is a double periodic structure in the (z, t)-plane, with periods δ and τ, in z and t, respectively. It can be implemented as a static periodic laminate, with period δ, existing from $t = 0$ to $t = n\tau$ (Fig. 13.1), at which point its material properties "flip over", so that where there was material 1 there now appears material 2, and *vice versa*. This pattern is maintained from $t = n\tau$ to $t = \tau$, and the procedure repeats periodically in time, with period τ. Parameters m and n denote the fraction of material 1 in the periods δ and τ . We expect that this structure will be able to support the wave routes with no energy loss.

A remarkable property of the checkerboard is that it meets these expectations. In Fig. 13.1 the wave enters the "fast" material 2 across the temporal interface where the

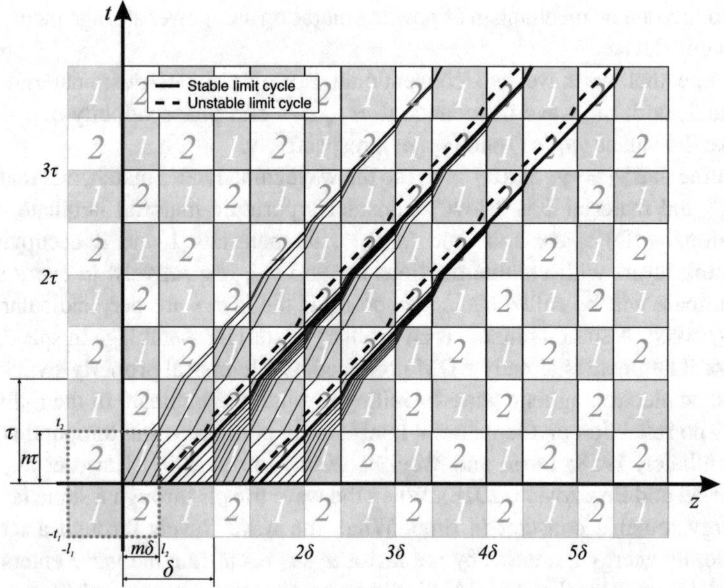

Fig. 13.1 A checkerboard structure in space-time with wave routes and limit cycles for $a_1 = 0.6$, $a_2 = 1.1$, $m = 0.4$, and $n = 0.5$

wave gains energy, and leaves it (i.e., enters material 1) across the spatial interface where no energy loss occurs. The waves travel with no reflections, and consequently the system is not oscillatory and requires no frequency adjustment.

In this framework, the phenomenon of energy accumulation stays in effect for all frequencies. Also, the accumulation is robust because it occurs in some continuous intervals of parameters a_1, a_2, m, and n. Particularly their values in Fig. 13.1 fall into such intervals. As shown in (Lurie et al., 2009), there are given sharp bounds for such intervals, and their attainability is confirmed numerically (so called "plateau effect").

The waves come into distinct groups that approach some selected characteristics called limit cycles. Energy is accumulated in these groups in the form of narrow pulses that carry high power. The cycles are stable, because they attract the neighbouring wave routes. In Fig. 13.1 we have one stable (boldface curve) limit cycle per period. Any two consecutive stable limit cycles are separated by an unstable limit cycle (dashed curve) that repels the neighbouring characteristics. For one period, the energy of the wave increases by a factor $(a_2/a_1)^2$ because it is pumped twice into the wave. As a result, the energy of the wave grows exponentially in time.

The DM checkerboard may be miniaturised by placing two mirrors (ideal reflectors) at the opposite ends of the spatial interval occupied by the system. Reflected waves are amplified as they travel through the checkerboard DM in opposite directions. This will cause nonstop amplification as the waves travel back and forth. When the energy supply from an external agent stops, the temporal switches are no longer maintained, and the system becomes a static laminate serving as a battery storing the energy already accumulated in the waves that still continue travelling between the mirrors but with no energy gain.

13.3 Mathematical Formulation

Consider a plane electromagnetic wave propagating along the z-axis. For such a wave, the electromagnetic field **E, B, D, H** is defined as

$$\mathbf{E} = E\mathbf{j}, \quad \mathbf{B} = B\mathbf{i}, \quad \mathbf{H} = H\mathbf{i}, \quad \mathbf{D} = D\mathbf{j}, \tag{13.1}$$

and Maxwell's equations take the form

$$E_z = B_t\,, \quad H_z = D_t\,, \tag{13.2}$$

with

$$\mathbf{D} = \varepsilon\mathbf{E}, \quad \mathbf{B} = \mu\mathbf{H}. \tag{13.3}$$

We assume that ε and μ (termed, respectively, the dielectric permittivity and the magnetic permeability of a material) as well as E, B, D, and H depend on (z, t). As mentioned in Sect. 13.2, we allow for two immovable material constituents, material 1 with properties ε_1, μ_1, and material 2 with properties ε_2, μ_2. Below we use, instead of ε and μ, an alternative pair of material parameters, namely, wave impedance

$$\gamma = \sqrt{\mu/\varepsilon} \tag{13.4}$$

and phase velocity

$$a = 1/\sqrt{\mu\varepsilon}\,. \tag{13.5}$$

Assume that $\gamma_1 = \gamma_2 = \gamma$ and $a_2 > a_1$, so the wave impedances match, and the material 2 is "fast" and material 1 is "slow". Equations (13.2) are satisfied if E ,B D, and H are expressed through potentials u, v by setting

$$E = u_t\,, \quad B = u_z\,, \quad H = v_t\,, \quad D = v_z\,. \tag{13.6}$$

Equations (13.3) are then reduced to

$$R_t + aR_z = 0\,, \quad L_t - aL_z = 0\,, \tag{13.7}$$

where

$$R = u - \gamma v\,, \quad L = u + \gamma v\,, \tag{13.8}$$

are the Riemann invariants related to the right- and left-going waves, respectively.

Below we discuss the wave propagation through a checkerboard structure introduced in Sect. 13.2. Along the static interfaces $z =$ const, we require continuity of E and H, i.e., of u and v. Along the temporal interfaces $t =$ const, B and D should be continuous, which, again, is equivalent to the continuity of u and v. These variables should be continuous across any interface in a checkerboard. An alternative will be the continuity of Riemann invariants R and L.

The wave propagation along the z-axis can be imitated in a transmission line assembled from the LC cells, with the material switching in space-time from L_1, C_1 to L_2, C_2, and *vice versa*, in each cell (Fig. 13.2). We shall use the same symbols to designate the linear inductance and capacitance in the line.

The equations governing the voltage V, magnetic flux Φ, the charge Q, and the current I are identical to (13.2) and (13.3), with substitutions

$$E \to V\,, \quad B \to \Phi\,, \quad D \to Q\,, \quad H \to I\,, \tag{13.9}$$

and

$$\varepsilon \to C\,, \quad \mu \to L\,. \tag{13.10}$$

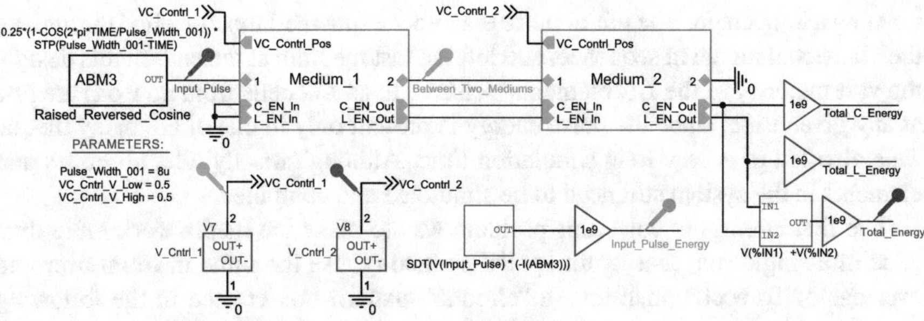

Fig. 13.2 Overall schematics of the *LC* arrangement (ideal elements).

The energy accumulation occurring at the moments of temporal switching in a properly designed checkerboard (Sect. 13.2) appears to be a direct consequence of (13.2), 13.3) or their transmission line counterparts.

In the following sections, we discuss an attempt to produce a reasonable engineering design of a material *LC* assembly that maintains a checkerboard property pattern in space-time and demonstrates the special effects. This effort had to overcome several technical difficulties. The model used a very large number (almost 6000 of each) of *L* and *C* components. This feature may lead to prohibitively expensive (and large, among other issues) implementation. Although the values of *L* were reasonable to achieve, the values of *C* were so low that no physical device could reliably reproduce it in large numbers.

All components in the model were "ideal", that is, not only they did not represent any "actual" physical parts, but they also did not account for such things as losses, parasitic and other effects that "real" parts would have. Finally it remained unclear how temporal change of the values of *C* and *L* would be done in practical implementation.

13.4 Material Design: Stage 1

The goal of Stage 1 is to deal with the first set if the issues formulated above. Still using "ideal" components, we make an attempt to find a solution that contains a reasonable amount of parts, while still exposing the desired effect. All simulation for this Stage was done in OrCAD Pspice (Cadence), which is more powerful and flexible than LTspice. The "ideal" *L* and *C* components were recreated in OrCAD and a circuit consisting of two identical media was constructed. The term "medium" refers here to a "transmission line" simulated with a limited number of lumped *LC* circuits. Since OrCAD allows the creation of hierarchical designs, it is fairly easy to change the amount of *LC* circuits in a medium for a different simulation run.

One of the reasons for the very large number of *LC* circuits is that the design was assumed to be a linear assembly of *LC* cells. Many media were connected in series, forming a long chain. The initial pulse entered the first medium, travelled through all the connected media in sequence, and left the last medium at the end. In this design, the vast majority of the LC elements (referred to as LC cells from now on) are idle at any given time, since the pulse energy is present only in one of the many media. This also led to a very long simulation time. Although mostly idle, all nodes and elements in the system still need to be simulated and computed.

The first attempt to solve this problem was to make the media work more than once in a single run, that is, for spatial switching, use the same medium over and over again. To accomplish this, a "circular" pattern was created in the following manner. The two media are connected in series. The initial pulse enters the first medium. As soon as the pulse is contained entirely within the first medium, the input to this medium is disconnected from the pulse source and connected to the output of the second medium, thus forming a loop. The pulse will then experience spatial switching between the two media indefinitely (in theory, of course).

This approach required that the entire pulse has to "fit" in the medium as it travels across, so that the medium temporal switching can occur at that exact moment. Also, the number of LC cells may be cut in half (with the same result) if we notice that when a transmission line is shorted at its end, the total reflection of the incident wave occurs. The reflected wave will have an opposite polarity and travel in the opposite direction. In the underlying theory it does not matter what direction of travel and what polarity the pulse has to be for the energy accumulation to work, as long as the temporal property switchings of the mediums occur at the right times.

Figure 13.2 shows the overall schematics of such an arrangement. Here, each of the two mediums consists of 32 LC cells. The initial pulse source, ABM3, is connected to terminal 1 of the first medium. Terminal 2 of the second medium is shorted to ground. After generating the initial pulse, the output voltage of ABM3 is held at the ground level, hence terminal 1 of the first medium is automatically shorted to ground for the rest of the simulation. On the schematics there are also some additional elements that control temporal switching of the mediums (VC_{Contrl1} and VC_{Contrl2}), as well as elements for calculating total energy contained within the system.

Figure 13.3 shows an individual LC cell. As well as the actual ideal variable L and C components ($L1$ and $C1$), it contains some parasitic components also. The values were chosen so that they are reasonably close to what might be encountered in the "real" $L1$ and $C1$ with the chosen values and within the frequency range of interest. The quantity $C2$ represents parasitic capacitance of $L1$. The parasitic inductance of $C1$ is negligible in our case and therefore omitted. $R1$, $R2$, and $R3$ represent series resistances of $L1$ and $C1$ to account for losses in these elements. It also contains additional components that allow for the calculation of instantaneous energy contained in $L1$ and $C1$.

The system without temporal switchings was first examined. In other words, only spatial switchings occur in this system. This simulation is useful in observing how the initial pulse travels through both media (that are exactly the same in this case) multiple times, how it "distorts" due to the limited "bandwidth" of the media, and

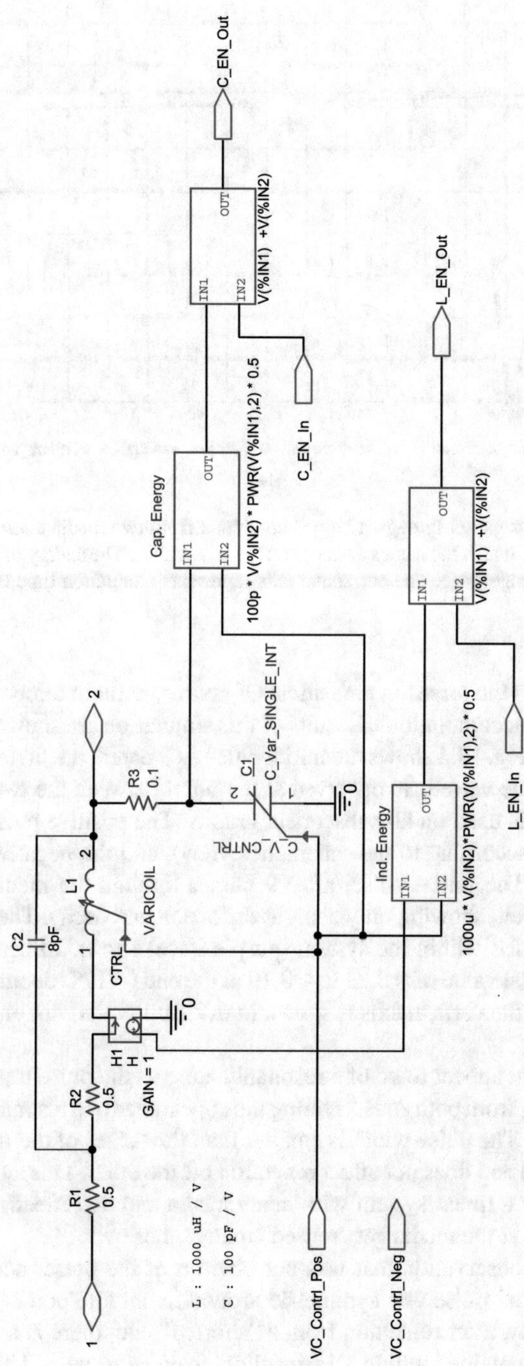

Fig. 13.3 Schematics of an individual *LC* cell (ideal elements).

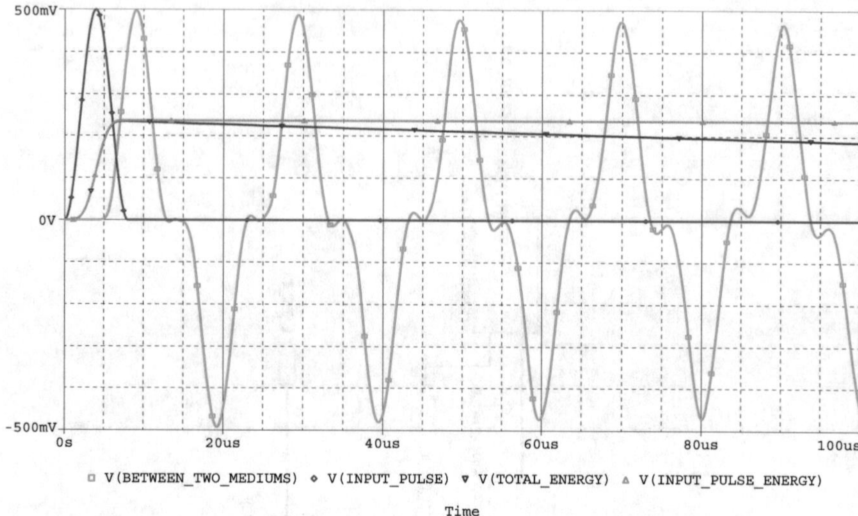

Fig. 13.4 A pulse propagation through a "loop" combined from two media alternating in space. The names of the traces match the names of nodes on the schematics. The energy of the initial pulse is shown in dashed orange trace. The horizontal axis represents simulation time in microseconds (μs).

how it decays due to the losses in parasitics. Of course, without temporal switching there is no "energy accumulation". Results of this simulation are shown in Fig. 13.4.

The red trace in Fig. 13.4 shows the initial pulse as it enters the first medium. The green trace shows the waveform observed at the point between the two media. The same green colour is used on all subsequent graphs. The positive pulses propagate from left to right (according to the schematics view), and the negative ones travel from right to left. The initial pulse, after 9 passes through the media, retains its shape reasonably well, showing only a slight distortion and decay. The trace for the total energy contained within the system (purple trace) shows a linear decline (as expected) from initial value of ≈ 0.23 to ≈ 0.19 at the end ($\approx 17\%$ decline; values are relative). Although the vertical axis is shown in mV (millivolts), for energy traces it is in relative joules.

The system might appear to be of a resonant nature as the pulse travels back and forth fully reflecting from both ends, creating the appearance of a resonator. However, this is not the case. The pulse width is smaller than the "size" of the resonator, and reflection off of one end does not affect reflection off the other. This system behaves exactly the same as a linear system with many media and no reflections at all; the only difference is that the media get "reused" many times over.

One interesting observation that was not obvious at the outset occurred in the cases when the initial pulse was symmetric around its middle point (which it was in our case). Namely, after reflection from a "shorted" end, there is a point in time where the energy contained within all capacitors is equal to zero. This means that voltages across all capacitors are all equal to zero. This corresponds to the time when

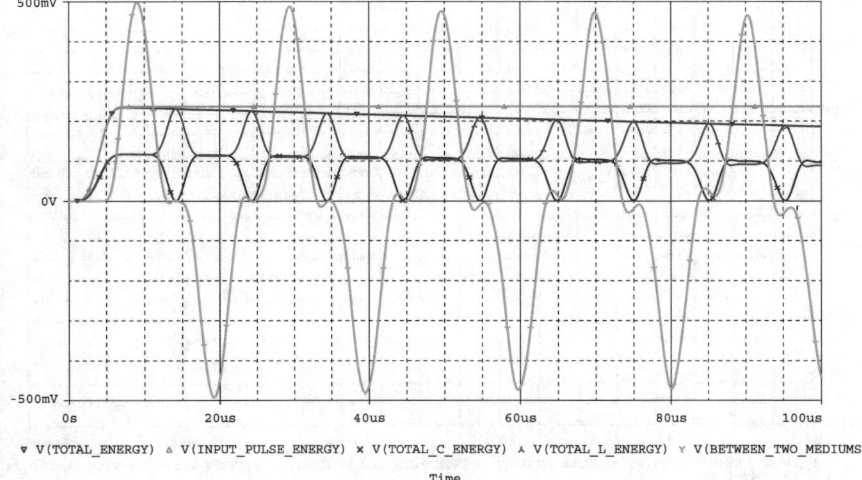

Fig. 13.5 Energy stored in capacitors in the absence of temporal switchings.

exactly half of the pulse is reflected from the "shorted" end. Since the reflected half is exactly the same as the "incoming" half but with an opposite polarity, they cancel each other.

In Fig. 13.5, blue trace shows the total energy stored in all capacitors in the system according to the formula: $E_c = \Sigma_{k=1}^{n}(C_k U_{C_k}^2/2)$, where k is the sequential number of an LC cell, U_{C_k} is the voltage across capacitor in that LC cell, and n is the total number of LC cells in the system. The times when E_c is equal to zero correspond to the moments when the pulse energy is entirely contained within one of the two media: one half incoming and the other half reflected. This provides a convenient way to determine the exact timing for temporal switching of the media. For a reference, the red trace shows the total energy stored in all inductors in the system: $E_L = \Sigma_{k=1}^{n}(L_k I_{L_k}^2/2)$. As in Fig. 13.4, the purple trace represents the total energy.

There is a different termination method for the media that we could have used. Rather than shorting the ends, we could leave them unconnected (floating). In this case there will still be a total reflection from the ends, but without voltage phase reversal (there will be, of course, current phase reversal instead). The boundary condition will change from $V = 0$ to $I = 0$. This means that the same "zeros" will occur in inductor currents, rather than in capacitor voltages. This observation may be beneficial for Stage 2, when we will have to find a real solution for the temporal switching.

Using "zeros" in capacitor voltages as reference points, the timing of the control signals can be specified so that the temporal switching does not happen when the spatial does, and *vice versa* (Fig. 13.6). Here, the temporal switching does occur. The values of L and C are determined by the magnitudes of VC_{Contrl1} (for medium 1) and VC_{Contrl2} (for medium 2) as blue and red traces. The values of the capacitors

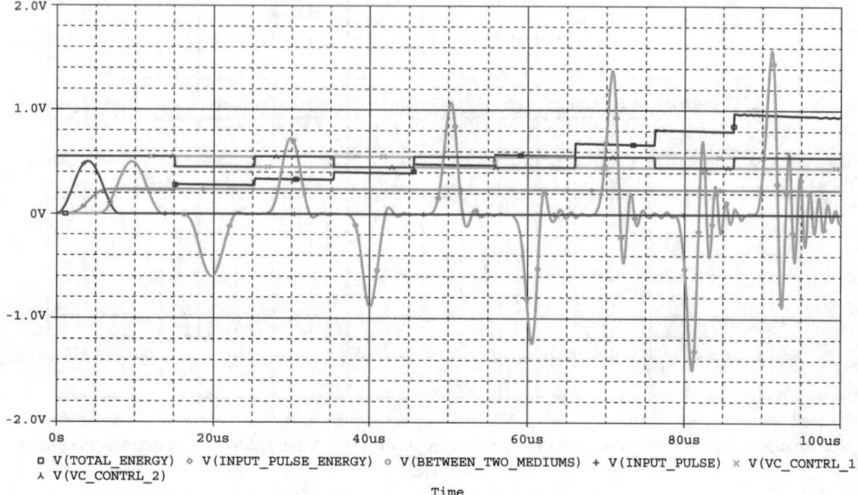

Fig. 13.6 Evolution of pulses in the presence of temporal switchings.

are defined as $C = VC_{\text{Contrl}} \times 100$ pF, and inductors as $L = VC_{\text{Contrl}} \times 1000$ µH. The scaling was chosen for convenience of simulations and for better displaying the control voltages on the same graph with other traces. In Fig. 13.6, the control voltages change from 0.45 V to 0.55 V and back, which correspond to L:C-pairs switching from 45 pF:450 µH to 55 pF:550 µH.

Since both L and C change their values proportionally together, the "characteristic impedance" of the media remains the same (disregarding losses):

$$\gamma = \sqrt{\frac{L}{C}} = \sqrt{\frac{450 \times 10^{-6}}{45 \times 10^{-12}}} = \sqrt{\frac{550 \times 10^{-6}}{55 \times 10^{-12}}} \approx 3160\,\Omega .$$

However, the "phase velocity", $a = 1/\sqrt{LC}$, does change between the values of

$$a_{\text{slow}} = \frac{1}{\sqrt{550 \times 10^{-6} \times 55 \times 10^{-12}}} \approx 5.75 \times 10^{6},$$

and

$$a_{\text{fast}} = \frac{1}{\sqrt{450 \times 10^{-6} \times 45 \times 10^{-12}}} \approx 7.03 \times 10^{6}.$$

The absolute values of phase velocity do not have any physical meaning (as in $\text{m} \cdot \text{s}^{-1}$) in our case. We just model a transmission line with lumped components, which may be large or small, as well as spaced apart differently, but electrically still being the same. The parameter of interest here is the ratio between the two values, which determines the energy accumulation factor. In simulation shown in Fig. 13.6 it is $7.03/5.75 \approx 1.22$.

There were eight temporal switchings in Fig. 13.6, so the total energy accumulation should amount to about $1.22^8 \approx 4.9$. The total energy trace (purple), at the end of the graph, shows the value of ≈ 0.94. As we recall from Fig. 13.5, the initial pulse energy without temporal switching had been reduced to $\approx$0.19 due to losses. That gives us total accumulation of $0.94/0.19 \approx 4.95$, which is very close to the expected value of 4.9 (which, actually should be $\approx$4.98 if we increase the precision of the above calculations).

The total energy accumulation matches very well with theoretical predictions, as does the overall pulse shape. The amplitude increases and the width narrows. However, the pulse exhibits some secondary oscillations that grow with each spatial and temporal switching (unlike in Fig. 13.4, where there is virtually no oscillations throughout the simulation). This is due to a limited "bandwidth" of the lumped-components nature of the media.

The initial pulse shape was chosen to contain as few high frequency spectrum components as possible (for a given pulse width). Numerous profiles were simulated and the "raised inversed cosine" one chosen as the best in this regard. But shaping of the pulse due to the energy accumulation (desired effect) causes the spectrum to contain more and more high frequency components that lead to oscillations and distortions (undesired effect).

These distortions may not present a problem since it is understood that this is just a realistic approximation of a theoretical principle. However, it is desirable to reduce them as much as possible. One possible solution lies in the fact that energy accumulation is proportional to the ratio of $a_{\text{fast}}/a_{\text{slow}}$. If we increase this ratio, we would need fewer temporal and spatial switchings to achieve the same overall gain.

Figure 13.7 shows what happens with the increased energy gain. Here, everything is the same as in Fig. 13.6 except that the control voltages switch between 0.4 to 0.6 V, which corresponds to $a_{\text{fast}}/a_{\text{slow}} = 1.5$. After just four temporal switchings (at $\approx$50 µs), the total energy reaches $\approx$1.06 compared with $\approx$0.21 at the same moment in Fig. 13.4, at the gain of $\approx$5.05 (theoretical gain is about $1.5^4 \approx 5.06$). Not only is the energy accumulation larger here, the pulse shape at this point is better than that at the end of simulation in Fig. 13.6. After the fifth temporal switching in Fig. 13.7 (at $\approx$56 µs), however, the pulse shape deteriorates rapidly. By the end of the simulation it is hard to see the where the pulse is in between the oscillations.

To summarise, the results of Stage 1 reduce to the following. It is possible to demonstrate the principle of energy accumulation with a reasonable number of "ideal" LC cells (even with a close-to-real losses and parasitics present). It is going to be a tradeoff between the $a_{\text{fast}}/a_{\text{slow}}$ value and the number of spatial and temporal switchings, depending on the requirements for the end customer. Finally, while energy accumulation is clearly seen in all simulations, it is just a "convenience" of numerical simulation. In real circuitry, the pulse shape can be observed with an oscilloscope, but it is unclear how to characterise the energy gain so it is obvious to a casual observer.

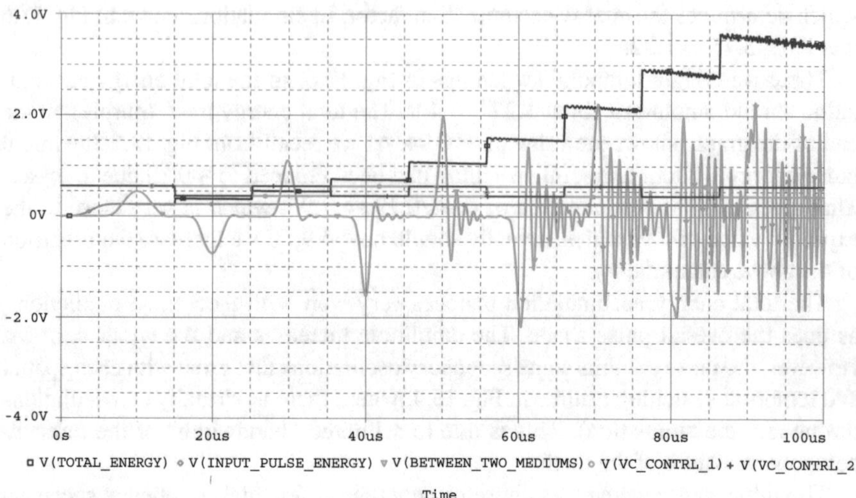

Fig. 13.7 Progressive energy accumulation in pulses travelling through a dynamic material checkerboard assembly.

13.5 Material Design: Stage 2

In Stage 1 (Sect. 13.4), it has been demonstrated, by means of a Pspice simulation, that there may be a viable solution to the practical implementation of a circuit, that is, part of the issues formulated at the end of Sect. 13.3 have been resolved. This Stage 2 deals with two remaining issues. Namely, all components in the model were "ideal" and therefore did not account for such things as losses, parasitic and other effects that "real" parts would have. Also, it remained unclear how temporal change of the values of C and L would be done in practical implementation.

Before we model the system with "actual" L and C components, we have to select them first. These components have to be capable of temporal change in their values in a controlled manner. On the one hand, it is quite easy to realise this for the C components. We can use readily available variable capacitance diodes, i.e., varicaps or varactors. They are semiconductor diodes specifically manufactured so that their reverse biased capacitance is not only well defined, but also changes significantly with applied voltage. There are many types available on the market from various manufacturers. Their primary use is in radio frequency circuits for frequency tuning, and the absolute capacitance values are generally in the range of $\approx$10–$\approx$100 pF range. With that in mind the values of the "ideal" capacitors were chosen for Stage 1 simulations (Sect. 13.4).

Inductor choices, on the other hand, are not that obvious. There are plenty of inductors available for various purposes; however, the vast majority of them are designed with the goal of maintaining the L-value at a constant level under all operating conditions, a situation quite the opposite to ours.

There are also a limited number of inductors for "niche" applications, such as magnetic amplifiers which change their L-value abruptly when current through them reaches a certain level. These inductors tend to be targeted for high power electronics (e.g., power supplies). They are very limited in choice and bulky. Also, it will be difficult to control their L-value since it changes too substantially with a little change in the current. Even if it were possible to control these devices, it is not clear how it can be accomplished in practice, since the inductors are not "ground referenced" like the capacitors but rather connected in series with each other.

Other approaches were also considered. One of them was the possibility of using "gyrators" which are active circuits (consisting of operational amplifiers, resistors and capacitors) that mimic behaviour of real inductances according to the general formula $I = (1/L) \int V dt$. This approach was dismissed due to two factors. Firstly, while single gyrator circuits are generally stable when component values are chosen right, many gyrators connected together might not be stable at all. Secondly, the presence of amplifiers (or any active components, for that matter) in the media might give an impression that it is these components that are responsible for the energy gain, and not the underlying principle under study here.

Another possibility considered was the use of inductors with multiple windings (transformers), where currents through one or more windings control inductance in another winding (since control currents change permeability of the core material). However, these transformers are not designed for this particular purpose. Like regular inductors, they maintain their L-value within wide range of currents. Manufacturing specialised transformers with specifications tailored for this project would be cost prohibitive.

However, the solution for temporal switching of inductors was hinted at in Sect. 13.4. It was noted that if the initial pulse is symmetric around its middle point and a total reflection occurs at an open medium's end, there is a time instant when currents through all inductors are equal to zero. This is exactly the time when temporal switching should take place, since all of the pulse energy is contained within the medium. Since at this time there is no energy stored in the inductors ($E = LI^2/2$), rather than trying to change the value of an inductor itself, we can add (or subtract) another inductor of a fixed value so that total inductance changes to a predetermined value. The principle of conservation of magnetic flux is satisfied, since it remains the same (zero) before and after temporal switching. The same is true when this switching occurs in the "empty" medium for the same reason. When temporal switchings occur, all energy accumulation is due to capacitors' change in value; the only role of the inductors' switching is to maintain the same characteristic impedance of the media.

When two inductors are connected in series, the total inductance $L_{\mathrm{TOT}} = L1+L2$ is the sum of the two (assuming there is no magnetic coupling between the two inductors). So, if we have a switch across $L2$ that opens and closes at appropriate times, the total inductance will change from $L1$ when the switch is closed ($L2$ is shorted), to $L1 + L2$ when the switch is open.

This solution was chosen to be implemented in practice. It requires just two small, off-the-shelf inductors and a simple switch. The only disadvantage of this approach is that once the values of the inductors are chosen and they are physically installed on a board, they cannot be easily changed (e.g., for a different value ratio). This limitation was not considered significant enough for this study, since the goal here is to demonstrate the principle, not to engage in detailed studies of the effects of various values on the outcome.

For Stage 2, the configuration of the media was changed from "shorted ends" to "floating ends" (Fig. 13.8). Here, the sequence of events is as follows. The initial pulse is generated and enters the first medium at the left (as seen on the schematics). When the pulse has entered the media completely, the pulse source is disconnected from the first medium, thus leaving this node floating for the rest of the simulation. The right side of the second medium is floating as well, for now. The simulation is run for a given number of reflections off of both floating ends of the media. After a given number of temporal and spatial switchings, the right side of the second medium is connected to a load resistor with the same value as the characteristic impedance of the media, thus "dumping" the accumulated energy into this resistor.

An initial attempt was made using real components with values close to those used in Stage 1 (Sect. 13.4) with ideal components. However, after running simulations with these values it became obvious that losses in the LC cells (there are 32 of them in each of the two media) outweighed the energy gain from temporal switchings. The amplitude of the pulses had been diminishing over time instead of increasing.

After analysing these results it became clear that majority of the losses occur in inductors. There are very little losses associated with varactors at our frequencies, voltages and currents. Simulations done with "ideal" inductors and "real" varactors show virtually no energy losses. Generally, inductors with larger L-values tend to have larger parasitic properties primarily due to longer windings and/or higher core material volume.

A logical path to try to remedy large losses is to decrease the values of inductors. To maintain timing of the system (i.e., to try not to affect the "phase velocity"), values of the capacitors will have to increase proportionally. This is not a problem for an implementation in a real system. We can connect multiple varactors in parallel to achieve desired value. Varactors are small and inexpensive devices. Increasing the quantity of them in the system will not be detrimental to the size and the cost of the implementation.

Multiple simulations were run with a choice of inductors with a lower L-value and a corresponding increase in number of varactors. The results, indeed, showed that the system performance improved significantly as the inductors' L-value decreased. The final choice of the LC cells was based on a trade-off between performance and number of varactors in each cell.

Figure 13.9 shows the schematics of the individual LC cells. Instead of ideal variable L and C components, they contain "real" components $L1$ and $L2$ with a switch $S1$ for "variable inductors", and $D1$–$D4$ for varactors (each of which consists of two varactors with a common cathode for a total of eight). The passive components that were present in the "ideal" simulation to account for "reasonable" losses were removed, since now we have "real" L–C components with losses represented within their corresponding Pspice models. As before, there are also additional,

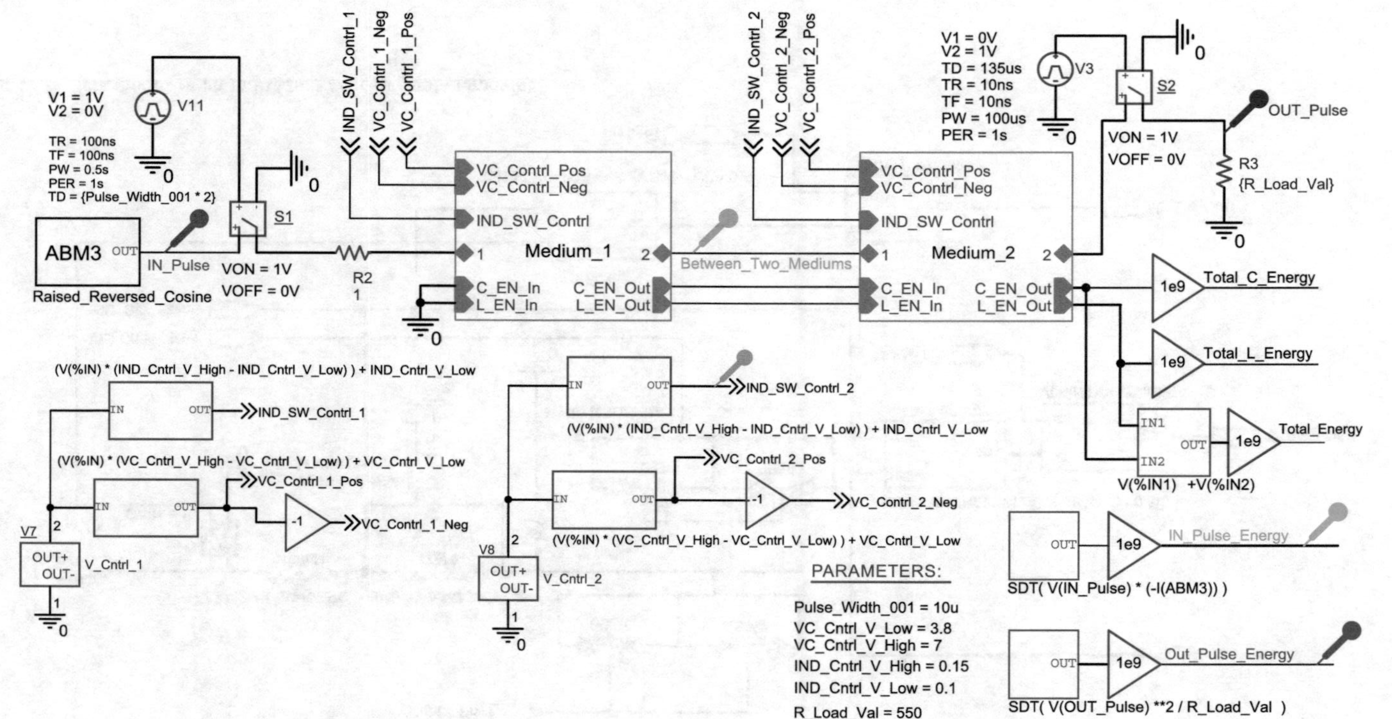

Fig. 13.8 Overall schematics of the LC–arrangement (real elements).

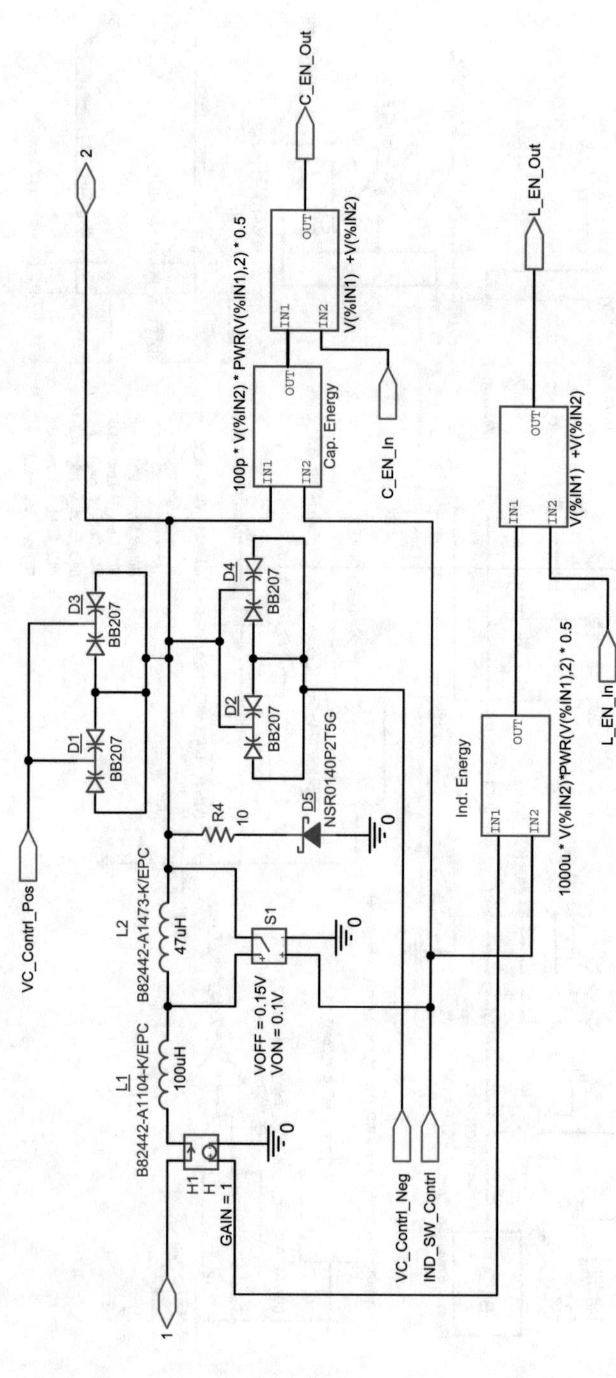

Fig. 13.9 Schematics of an individual *LC* cell (real elements).

purely mathematical, components that allow for calculation of instantaneous energy contained within the LC cells.

On the LC cells schematics there are also two passive elements: resistor $R4$ and diode $D5$. Their presence warrants some additional explanation. As it was discussed in Stage 1 (Sect. 13.4), when a pulse reflects off of a shorted end of a medium, the reflected pulse will have opposite polarity from an "incoming" one (voltage-wise). In the opposite case, when reflection occurs from an open end, the polarity will remain the same (this is the case at this stage). Since our initial pulse has positive polarity, then, in theory, it will remain positive throughout the entire experiment. However, as was seen in Stage 1 (Sect. 13.4), distortions and oscillations cause spikes in voltages that go in the opposite polarity. The presence of $R4$ and $D5$ does not allow these spikes to go in the negative polarity, thus preventing the distortions to grow more and more as the energy amplification takes place. Of course, this "clipping" wastes some of the energy in the system, but it helps in preserving the pulse shape in exchange.

As seen on the LC cells schematics, the values of inductors are 100 μH and 47 μH, which gives the ratio of $L1 + L2/L1 = 147/100 = 1.47$. To match this ratio on the varactor side (to maintain the same characteristic impedance of the media), the control voltages were chosen so that capacitance alternates between 30 pF and 45 pF for each of the varactors. This corresponds to control voltages being 7 V and 3.8 V, respectively. Since there are eight varactors connected in parallel in each cell, the total capacitance alternates between 240 pF and 360 pF for the ratio of 1.5.

An important note regarding simulations with "real" parts: while in simulations with "ideal" components (as in Stage 1, Sect. 13.4) we can control all parameters (including values of L and C) exactly, this is not the case with "real" parts. Real parts have nominal values, usually measured at certain conditions. Depending on stated tolerances, the actual values will vary from part to part. In addition, the values themselves depend on other factors such as currents, voltages, frequency of interest, etc. As such, all values that pertain to real parts should be treated as approximate.

The multiple simulations were run with the parameters outlined above, with varying amplitude of the initial pulse. The result of one of them is shown on Fig. 13.10. The results of other simulations looked similar. Here, the red trace is the initial pulse, teal is the initial pulse's energy, green is the voltage observed between the two media as the pulse travels "back and forth", orange is one of the control voltages (to show when temporal switchings occur), blue is the voltage of the output pulse (as it is being "dumped" into the load resistor), and pink is the total energy out of the system.

Results of these simulations (Table 13.1) show that, after 10 temporal switchings, the overall energy gain is as high as 9.8. This means that, on average, each temporal switching creates $\sqrt[10]{9.8} \approx 1.26$ times energy gain. This is good compared to an "ideal", "lossless" value of $\approx$1.5. Even small reduction in gain (due to losses and other factors) at each switching will lead to a large reduction in overall result (compare 9.8 times for a "theoretical" gain of $1.5^{10} \approx 58$). It does not mean, of course, that the overall system efficiency is $9.8/58 \approx 17\%$. It is also not $100 \times 1.26/1.5 = 84\%$. To find true energy efficiency we have to know how much of the energy was supplied to the media to facilitate the temporal switchings.

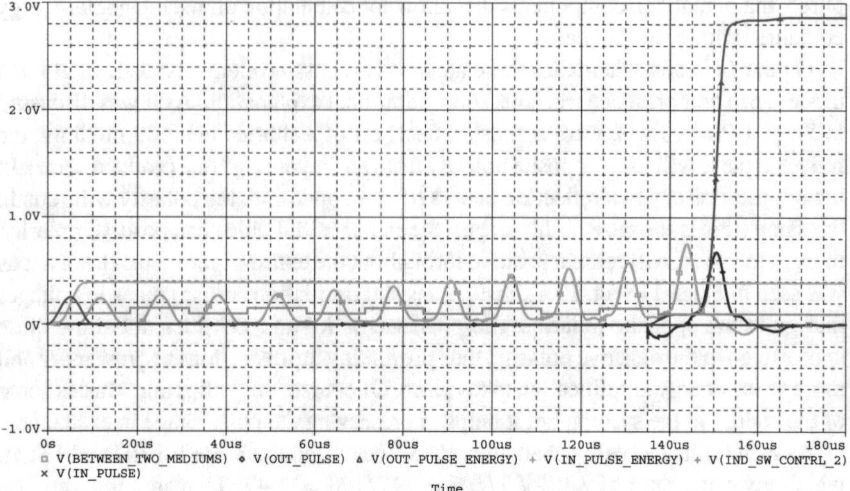

Fig. 13.10 Energy accumulation in a transmission line with real elements.

Table 13.1 Main parameters of simulations. In all cases, the initial pulse shape was one full period of inversed raised cosine with width of 10 μs. Each of the two media contained 32 LC cells. The characteristic impedance of the media was approximately 640 Ω.

Initial pulse amplitude (V)	Initial pulse energy (nJ)	Energy supplied by switching voltage sources (nJ)	Total energy input into the system (nJ)	Output pulse energy (nJ)	Power gain of the initial pulse (times)	Overall power efficiency of the system
0.25	0.39	4.95	5.34	2.87	7.4	54%
0.5	1.56	22.4	23.96	14.4	9.2	60%
0.75	3.5	53	56.5	34.4	9.8	61%
1	6.25	93.6	99.85	60.2	9.6	60%

Since inductors do not contribute to energy gain in our simulations (they are simply switched in and out when their currents are close to zero), the only source of the gain are the varactors, and, more specifically, voltage sources that control the values of the varactors. To evaluate the energy from these sources, we calculate instantaneous values of $P = VI$, integrate them over time, and then add the results from all sources together. This can be seen on the overall schematics in Fig. 13.8, at node VC SW Energy.

True energy efficiency of the system is calculated by taking the energy of the output pulse, as it is seen on the load resistor, and dividing it by the sum of energies of the input pulse and from voltage sources. The results are shown in Table 13.1. It can be seen that this efficiency can be as high as 61%, which is rather impressive. Table 13.1 also shows that there is a limit on the voltages in the pulse. Namely, an increase in the initial pulse amplitude eventually leads to deterioration of the performance due to the fact that we deal with "real" components. Varactors start exhibiting their "diode-like" behaviour with larger pulse voltages, for example.

To summarise the Stage 2 simulations, it is safe to say that a solution has been presented that allows the use of standard off-the-shelf components in LC cells, and still maintains a requirement to preserve the characteristic impedance of the media. Further, it is possible to demonstrate the principle of energy accumulation with a reasonable number of "real" LC cells without the active inductor equivalents or custom-made components. Finally, the results of the simulations show impressive overall energy efficiency – as high as 61%.

As a side note, this process of energy accumulation is akin to a class of amplifiers known as parametric amplifiers. As the name implies, these amplifiers work on the principle of changing a parameter in the system, rather than the use of active components (e.g. transistors). Varactors are one example of devices used in such amplifiers (the parameter being the capacitance of the varactor). As in our system, parametric amplifiers rely on specific timing as it relates to the signal being amplified. One of the main reasons for the use of these amplifiers is the fact that the path of the amplified signal does not contain any active components and resistors required for proper operation, both of which generate various kinds of noise. As a result, parametric amplifiers are known to have very low noise compared to "regular" amplifiers, albeit their use is limited to a few, very specific, areas.

Appendix

Although there is no specific "phase velocity" value in a lumped-element implementation, we can still assess it by assuming a typical LC cell size (including all components and interconnects). With the real components used in this simulation, it is reasonable to assume that a cell will occupy a square with a side of approximately 20 mm. Given the values of L and C, we can say that our "media" have the following "phase velocities":

$$a_{\text{slow}} = \left(\sqrt{\frac{147\,\mu\text{H}}{0.02\,\text{m}} \times \frac{360\,\text{pF}}{0.02\,\text{m}}}\right)^{-1} \approx 0.87 \times 10^5\,\text{m} \cdot \text{s}^{-1},$$

$$a_{\text{fast}} = \left(\sqrt{\frac{100\,\mu\text{H}}{0.02\,\text{m}} \times \frac{240\,\text{pF}}{0.02\,\text{m}}}\right)^{-1} \approx 1.29 \times 10^5\,\text{m} \cdot \text{s}^{-1}.$$

As a point of reference, an optical would have been required to have its index of refraction of the order $3 \times 10^8 \times 10^{-5} \approx 3000$ to achieve similar phase velocity.

Temporal switches occur every $\approx$13.2 µs, which corresponds to a switching frequency of $\approx$76 kHz. This is dictated by the particular values of L and C, as well as the number of the LC cells in the media.

This system was designed with a specific goal: to develop an approach that can be physically implemented with reasonable size and cost. As such, it has inherent limitations as compared to a "limitless and ideal" simulation. One of these limitations has to do with the shape and, more importantly, width of the initial pulse.

It is unadvisable to try it with the width of the initial pulse that does not "fit" entirely within one of the two media due to the following. Firstly, as it was mentioned earlier, it is important that currents through all inductors in both media are near zero when a temporal switching is occurring (to satisfy magnetic flux preservation). If the pulse is present in both media at these moments, this condition may not be satisfied everywhere. In addition, if there are currents flowing through inductors during the switchings, large voltage spikes will occur (the "flyback" effect, according to the formula: $V(t) = LdI/dt$, which will have to be controlled to avoid damage to components. Secondly, due to the limited nature of the system (having only two media "reused" over and over again), it will be almost impossible to interpret the results with wider pulses when different parts of the pulse energy overlap in space-time.

These additional simulations show that, despite being restricted in nature due to design goals, this system is robust and shows consistent performance with respect to varying amplitudes of the initial pulse, as well as to the number of space-time switches.

References

Berezovski, M., Berezovski, A.: Numerical simulation of energy localization in dynamic materials. In: dell'Isola, F., Eremeyev, V.A., Porubov A. (eds.) Advances in Mechanics of Microstructured Media and Structures, pp. 75–83. Springer, Cham (2018). https://doi.org/10.1007/978-3-319-73694-5_5

Blekhman, I.I., Lurie, K.A.: On dynamic materials. Proc. Russian Acad. Sci. (Doklady) **37**, 182–185 (2000). https://doi.org/10.1134/1.171720

Krylov, S., Ilic, B.R., Schreiber, D., Seretensky, S., Craighead, H.: Pull-in behavior of electrostatically actuated bistable microstructures. J. Micromech. Microeng. **18**(5), 055026 (2008). https://doi.org/10.1088/0960-1317/18/5/055026

Krylov, S., Gerson, Y., Nachmias, T., Keren, U.: Excitation of large amplitude parametric resonance by the mechanical stiffness modulation of a microstructure. J. Micromech. Microeng. **20**(1), 015041 (2010). https://doi.org/10.1088/0960-1317/20/1/015041

Krylov, S., Lurie, K.A.: Compliant structures with time-varying moment of inertia and non-zero average momentum and their applications in angular rate microsensors. J. Sound. Vib. **330**(20), 4875–4895 (2011). https://doi.org/10.1016/j.jsv.2011.04.032

Lurie, K.A.: An Introduction to the Mathematical Theory of Dynamic Materials. Advances in Mechanics and Mathematics, vol. 15. Springer, New York (2007, Second Edition 2017). https://doi.org/10.1007/978-3-319-65346-4

Lurie, K.A., Weekes, S.L.: Wave propagation and energy exchange in a spatio-temporal material composite with rectangular microstructure. J. Math. Anal. Appl. **314**(1), 286–310 (2006). https://doi.org/10.1016/j.jmaa.2005.03.093

Lurie, K.A., Weekes, S.L., Onofrei, D.: Mathematical analysis of the waves propagation through a rectangular material structure in space-time. J. Math. Anal. Appl. **355**(1), 180–194 (2009). https://doi.org/10.1016/j.jmaa.2009.01.031

Morgenthaler, F.R.: Velocity modulation of electromagnetic waves. IRE Trans. Microw. Theory Tech. **6**(4), 167–172 (1958). https://doi.org/10.1109/tmtt.1958.1124533

Oohira, F., Iwase, M., Matsui, T., Hosogi, M., Ishimaru, I., Hashiguchi, G., Mihara, Y., Iino, A.: Self-hold and precisely controllable optical cross-connect switches using ultrasonic micro motors. IEEE J. Sel. Top. Quantum Electron. **10**(3), 551–557 (2004). https://doi.org/10.1109/jstqe.2004.830618

Pelesko, J.A., Bernstein, D.H.: Modeling of MEMS and NEMS. CRC Press, New York (2002). https://doi.org/10.1201/9781420035292

Rozenberg, Y.I., Rosenberg, Y., Krylov, V., Belitsky, G., Shacham-Diamand, Y.: Resin-bonded permanent magnetic films with out-of-plane magnetization for MEMS applications. J. Magn. Magn. Mater. **305**(2), 357–360 (2006). https://doi.org/10.1016/j.jmmm.2006.01.026

Xiao, Y., Maywar, D.N., Agrawal, G.P.: Reflection and transmission of electromagnetic waves at a temporal boundary. Opt. Lett. **39**(3), 574–577 (2014). https://doi.org/10.1364/ol.39.000574

Chapter 14
Solving Nonlinear Parabolic Equations by a Strongly Implicit Finite Difference Scheme

Applications to the Finite Speed Spreading of Non-Newtonian Viscous Gravity Currents

Aditya A. Ghodgaonkar and Ivan C. Christov

Abstract We discuss the numerical solution of nonlinear parabolic partial differential equations, exhibiting finite speed of propagation, via a strongly implicit finite difference scheme with formal truncation error $\mathcal{O}\left[(\Delta x)^2 + (\Delta t)^2\right]$. Our application of interest is the spreading of viscous gravity currents in the study of which these type of differential equations arise. Viscous gravity currents are low Reynolds number (viscous forces dominate inertial forces) flow phenomena in which a dense, viscous fluid displaces a lighter (usually immiscible) fluid. The fluids may be confined by the sidewalls of a channel or propagate in an unconfined two-dimensional (or axisymmetric three-dimensional) geometry. Under the lubrication approximation, the mathematical description of the spreading of these fluids reduces to solving the so-called thin film equation for the current's shape $h(x, t)$. To solve such nonlinear parabolic equations we propose a finite difference scheme based on the Crank–Nicolson idea. We implement the scheme for problems involving a single spatial coordinate (i.e., two-dimensional, axisymmetric or spherically symmetric three-dimensional currents) on an equispaced but staggered grid. We benchmark the scheme against analytical solutions and highlight its strong numerical stability by specifically considering the spreading of non-Newtonian power law fluids in a variable width confined channel like geometry (a "Hele-Shaw cell") subject to a given mass conservation/balance constraint. We show that this constraint can be implemented by reexpressing it as nonlinear flux boundary conditions on the domain's endpoints. Then, we show numerically that the scheme achieves its full second order accuracy in space and time. We also highlight through numerical simulations how the proposed scheme accurately respects the mass conservation/balance constraint.

A. A. Ghodgaonkar · I. C. Christov (✉)
School of Mechanical Engineering, Purdue University, West Lafayette, IN, USA
e-mail: christov@purdue.edu

A. Berezovski, T. Soomere (eds.), *Applied Wave Mathematics II*, Mathematics of Planet Earth 6, https://doi.org/10.1007/978-3-030-29951-4_14

14.1 Introduction

In his lucid 2015 book *Questions About Elastic Waves* (Engelbrecht, 2015), Engelbrecht asks "What is a wave?" and answers "As surprising as it may sound, there is no simple answer to this question." Indeed, the definition of 'wave' depends on the physical context at hand (Christov, 2014). Although most wave phenomena in classical continuum mechanics are described by hyperbolic (wave) equations, one of the surprises of 20th century research into nonlinear partial differential equations (PDEs) is that *certain* parabolic (diffusion) equations also yield structures with finite speed of propagation. Two examples are (i) a linear diffusion equation with a nonlinear reaction term (Fisher, 1937; Kolmogorov et al., 1991), and (ii) a diffusion equation that is nonlinear due to a concentration dependent diffusivity (Barenblatt, 1952).[1]

Indeed, it is known that certain aspects of wave phenomena can be reduced to a problem of solving a parabolic PDE, as gracefully illustrated by Engelbrecht (Engelbrecht, 1997, Ch. 6) through a series of selected case studies. Further examples include, but are not limited to, electromagnetic waves propagating along the Earth's surface (Vlasov and Talanov, 1995), seismic waves (Jerzak et al., 2002), underwater acoustics (Jensen et al., 2011), and the classic theory of nerve pulses (Engelbrecht, 1997, §6.4.2), which nowadays has been updated by Engelbrecht et al. (2018a) to a nonlinear hyperbolic (wave) model in the spirit of the Boussinesq paradigm (Engelbrecht et al., 2018b; Christov et al., 2007).

Of special interest to the present discussion are physical problems that are modelled by nonlinear parabolic PDEs. These nonlinear problems lack general, all-encompassing solution methodologies. Instead, finding a solution often involves methods that are specific to the nature of the governing equation or the physical problem that it describes (Evans, 2010, Ch. 4) (see also the discussion in (Christov, 2014) in the context of heat conduction). The classical examples of nonlinear parabolic PDEs admitting travelling wave solutions come from heat conduction (Zel'dovich and Raizer, 2002, Ch. X) (see also (Straughan, 2011)) and thermoelasticity (Straughan, 2011; Berezovski and Ván, 2017). The sense in which these nonlinear parabolic PDEs admit travelling wave and 'wavefront' solutions now rests upon solid mathematical foundations (Gilding and Kersner, 2004; Vázquez, 2007), including the case of gradient dependent nonlinearity (Tedeev and Vespri, 2015) (e.g., the last case in Table 14.1 to be discussed below).

A classic example of a nonlinear parabolic PDE governing the finite speed wave like motion of a substance arises in the study of an ideal gas spreading in a uniform porous medium (Barenblatt, 1952). A similar nonlinear parabolic equation was derived for the interface between a viscous fluid spreading horizontally underneath another fluid of lower density ($\Delta\rho > 0$ between the fluids) (Huppert, 1982). The

[1] More specifically, Barenblatt (1952) (see also (Ostriker et al., 1992, p. 13)) credits the observation of finite speed of propagation in a nonlinear diffusion equation to a difficult-to-find 1950 paper by Zeldovich and Kompaneets.

Table 14.1 Selected models of the propagation of viscous gravity currents herein simulated by a finite difference scheme.

Case/Variable	A [$m^{1-p+q} \cdot s^{-1}$]	p [–]	q [–]	ψ [m]
Newtonian fluid, fixed width HS cell: $b(x) = b_1$. (See Huppert and Woods (1995))	$\frac{\Delta\rho g b_1^2}{12\mu}$	0	0	h
Newtonian fluid, variable width HS cell: $b(x) = b_1 x^n$. (See Zheng et al. (2014))	$\frac{\Delta\rho g b_1^2}{12\mu}$	n	$3n$	h
Power law fluid: $\mu = \mu_0 \dot{\gamma}^{r-1}$, variable width HS cell: $b(x) = b_1 x^n$. (See Di Federico et al. (2017); S (2017))	$\left(\frac{r}{2r+1}\right)\left(\frac{\Delta\rho g}{\mu_0}\right)^{1/r}\left(\frac{b_1}{2}\right)^{(r+1)/r}$	n	$n\left(\frac{2r+1}{r}\right)$	$h\left\lvert\frac{\partial h}{\partial x}\right\rvert^{(1-r)/r}$
Newtonian fluid, 2D porous medium, variable porosity: $\phi(x) = \phi_1 x^m$, variable permeability: $k(x) = k_1 x^n$. (See Zheng et al. (2014))	$\frac{\Delta\rho g k_1}{\mu\phi_1}$	m	n	h
Power law fluid: $\mu = \mu_0 \dot{\gamma}^{r-1}$, 2D porous medium, variable porosity: $\phi(x) = \phi_1 x^m$, variable permeability: $k(x) = k_1 x^n$. (See Ciriello et al. (2016))	$2^{(3r+1)/2}\left(\frac{r}{3r+1}\right)^{1/r} \times \left(\frac{k_1}{\phi_1}\right)^{(r+1)/2r}\left(\frac{\Delta\rho g}{\mu_0}\right)^{1/r}$	m	$\frac{m(r-1)+n(r+1)}{2r}$	$h\left\lvert\frac{\partial h}{\partial x}\right\rvert^{(1-r)/r}$

motion of the denser fluid is dictated by a balance of buoyancy and viscous forces at a low Reynolds number (viscous forces dominate inertial forces).

Such viscous gravity current flows are characterised by 'slender' fluid profiles i.e., they have small aspect ratios ($h/L \ll 1$, where h and L are the typical vertical and horizontal length scales, respectively). Therefore, these flows can be modelled by lubrication theory (Leal, 2007, Ch. 6). Generically, one obtains a nonlinear parabolic equation for the gravity current's shape h as a function of the flow-wise coordinate x and time t. The case of the spreading of a fixed mass of Newtonian fluid was originally explored contemporaneously in (Didden and Maxworthy, 1982; Huppert, 1982).

Being governed by a parabolic (irreversible) equation, these currents 'forget' their initial conditions after some time has elapsed. This is Barenblatt's concept of *intermediate asymptotics* (Barenblatt and Zel'dovich, 1972; Barenblatt, 1996). Moreover, the PDE (14.1) can be reduced to an ordinary differential equation (ODE) through a self-similarity transformation. If the similarity variable can obtained by a scaling (dimensional) analysis, then the solution is termed a self-similar solution of the *first* kind (Barenblatt, 1996, Ch. 3).

Specifically, the transformation is $h(x,t) = \mathfrak{C} t^{\beta} f(\zeta)$ (h in units[2] of metres), where $\zeta = x/(\eta_N t^{\delta})$ is the similarity variable (dimensionless), $f(\zeta)$ is the self-similar profile to be determined by solving an ODE, and $\mathfrak{C}$ and η_N are dimensional consistency constants. The exponents β and δ are obtained through scaling (dimensional analysis) of the governing PDE.

As a representative example, consider the one-dimensional (1D) spreading of a fixed mass of fluid having an arbitrary 'wavy' initial shape (Fig. 14.1). Suppose the fluid's shape $h(x,t)$ is governed by the linear diffusion equation $\partial h/\partial t = A \partial^2 h/\partial x^2$ (taking $A = 1\ \mathrm{m^2 s^{-1}}$ in this example without loss of generality) subject to $(\partial h/\partial x)|_{x=\pm L} = 0$. The initial condition (IC) is quickly 'forgotten,' and the ultimate asymptotic state (here, flat) is achieved after passing through an intermediate asymptotic regime.

It is straightforward to determine the self-similarity transformation: $\beta = -\delta = -1/2$ and $f(\zeta) = \mathrm{e}^{-\zeta^2}$ (see, e.g., (Barenblatt, 1996) and the Appendix). Here, $\mathfrak{C}$ depends on the initial condition and $\eta_N = 2\,\mathrm{m} \cdot \mathrm{s}^{-1/2}$. The convergence of the rescaled $h(x,t)$ profiles towards $f(\zeta)$ can be clearly observed in Fig. 14.1b. The IC is forgotten, and the profile converges onto the Gaussian intermediate asymptotic shape (Barenblatt, 1996). The profile $f(\zeta)$ is termed 'universal' because it is independent of $h(x,0)$.

Having illustrated the notion of first kind self-similarity as intermediate asymptotics, let us summarise its use in studying the gravitational spreading of Newtonian viscous fluids in a variety of physical scenarios. For example, gravity currents arise in geophysical applications associated with flows through porous rocks (Woods, 2015) such as in ground water extraction (Bear, 1988), during oil recovery (Huppert, 2000; Felisa et al., 2018), and during CO_2 sequestration

[2]Throughout the chapter, we use SI units for all dimensional quantities.

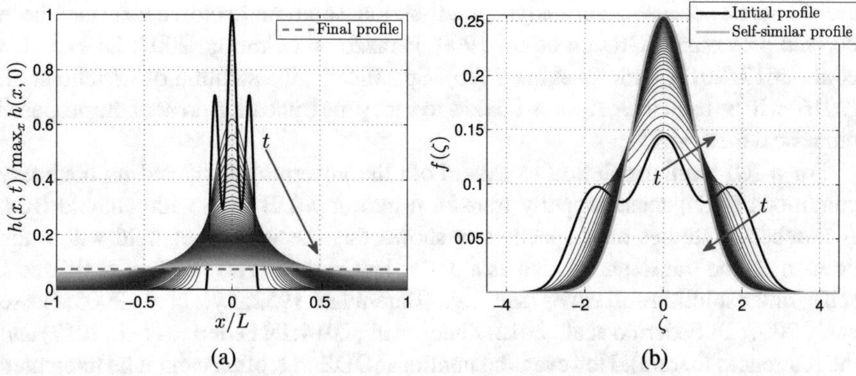

Fig. 14.1 Spreading via 1D linear diffusion and the approach to the universal intermediate self-similar asymptotics. (a) An arbitrary wavy IC $h(x, 0)$ spreads and levels until reaching a flat steady state h_∞. (b) A first kind self-similar transformation (obtained from dimensional analysis) yields a *universal* profile $f(\zeta)$ (highlighted in gold) towards which the solution $h(x, t)$ evolves in the intermediate period after the IC is forgotten (but prior to levelling).

(Huppert and Neufeld, 2014). In these examples, $h(x, t)$ represents an interface between two immiscible fluids in the limit of large Bond number (gravity dominates surface tension). There is now an extensive literature featuring a wealth of exact and approximate analytical self-similar solutions for gravity currents in porous media, (e.g., Barenblatt 1952; Huppert and Woods 1995; Anderson et al. 2003; Lyle et al. 2005; Vella and Huppert 2006; Hesse et al. 2007; Anderson et al. 2010; De Loubens and Ramakrishnan 2011; Ciriello et al. 2013; Huppert et al. 2013; Zheng et al. 2014) amongst many others.

In this chapter, we focus on the propagation of non-Newtonian gravity currents, specifically ones for which the denser fluid obeys a *power law* rheology. This tractable model of non-Newtonian rheological response is also known as the Oswald–de Weale fluid (Bird et al., 1987). In unidirectional flow, the power law model simply dictates that fluid's viscosity depends upon a power of the velocity gradient. Di Federico et al. (2006) generalised Huppert's problem (Huppert, 1982) to power law fluids, although (Gratton et al., 1999; Perazzo and Gratton, 2003) had also considered some related problems.

Even earlier, Kondic et al. (1996, 1998) derived the governing equations for power law fluids under confinement (i.e., in Hele-Shaw cells) using the lubrication approximation. These works have contributed to the use of a modified Darcy law to model the flow of non-Newtonian fluids in porous media using the analogy to flow in Hele-Shaw cells. Aronsson and Janfalk (1992) were perhaps the first to combine a Darcy law for a power law fluid with the continuity equation to obtain a single PDE, of the kind studied herein, governing the gravity current's shape.

Recently, Lauriola et al. (2018) highlighted the versatility of this approach by reviewing the existing literature and extending it to two-dimensional axisymmetric spreading in media with uniform porosity but variable permeability. All these flows are of interest because exact analytical self-similar solutions in closed form have been derived previously (Gratton et al., 1999; Perazzo and Gratton, 2005; Di Federico et al., 2012, 2017; Ciriello et al., 2016). Specifically, the solution of Ciriello et al. (2016) will be used in Sect. 14.4.1 below to verify the truncation error of the proposed numerical method.

For a self-similar solution to exist, both the governing PDE and its boundary conditions (BCs) must properly transform into an ODE in ζ with suitable BCs. A number of studies have specifically shown that the volume of fluid within the domain can be transient, varying as a power law in time, $\mathcal{V}(t) \propto t^{\alpha}$ ($\alpha \geq 0$), and a self-similar solution still exists (see, e.g., (Barenblatt, 1952; Lyle et al., 2005; Hesse et al., 2007; Di Federico et al., 2012; Zheng et al., 2014; Di Federico et al., 2017) and the references therein). However, the nonlinear ODE in ζ often cannot be integrated exactly in terms of known function, except for $\alpha = 0$. In Sect. 14.2.2 below, we discuss how a constraint of the form $\mathcal{V}(t) \propto t^{\alpha}$ can be implemented numerically through flux BCs at the computational domain's ends.

With increasing complexity of the flow physics incorporated in the model, finding a self-similarity transformation may no longer be possible simply by scaling (dimensional) arguments. Gratton and Minotti (1990) classified a number of such situations, including the so-called 'focusing' flows involving fluid axisymmetrically flowing towards the origin on a flat planar surface. Further examples involving confined currents in channels with variable width, and/or in porous media whose permeability and porosity are functions of x, were proposed by Zheng et al. (2014),

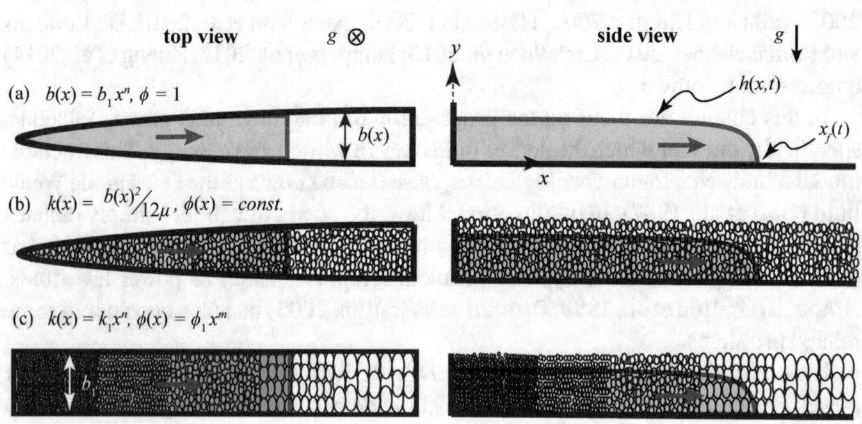

Fig. 14.2 A summary of the gravity current flows and domains considered in this work. (a) Flow away from the origin in a completely porous ($\phi = 1$) HS cell of variable width given by $b(x) = b_1 x^n$ (b_1 = const, $0 \leq n < 1$). (b) Flow in uniformly porous ($\phi = \phi_1 = \text{const} \neq 1$) passage of variable width given by the same $b(x)$ as in (a). (c) Flow in a uniform width slab (i.e., $b(x) = b_1 = \text{const}$) with horizontally heterogeneous porosity and permeability given by $\phi(x) = \phi_1 x^m$ and $k(x) = k_1 x^n$, respectively. The effective permeability of the medium in (a) and (b) is set by the Hele-Shaw analogy via the width: $k(x) = [b(x)]^2/(12\mu)$. Figure reproduced and adapted with permission from (Zheng et al., 2014)© Cambridge University Press 2014.

as illustrated in Fig. 14.2. These gravity currents do enter a self-similar regime, even though a self-similar transformation cannot be obtained by scaling arguments alone.

The exponents β and δ in the transformation are unknown *a priori*, hence this situation represents a self-similarity of the *second* kind (Barenblatt, 1996, Ch. 4). The governing equation can be transformed to an ODE, following which a nonlinear eigenvalue problem must be solved for β and δ through a phase plane analysis (Gratton and Minotti, 1990; Gratton, 1991). Alternatively, experiments or numerical simulations are necessary to determine β and δ. For example, early numerical simulations were performed to this end by Diez et al. (1992). However, a 'prewetting film' ahead of the current's sharp wavefront ($x = x_f(t)$, where $h\big(x_f(t), t\big) = 0$) was required to avoid numerical instabilities. The scheme therein was also first order accurate in time only. In this chapter, we propose a modern, high-order-accurate implicit numerical method for use in such problems.

Specifically, we develop and benchmark a strongly implicit and conservative numerical scheme for 1D nonlinear parabolic PDEs arising in the study of gravity currents. We show how the proposed scheme can be used to simulate (with high accuracy and at low computational expense) the spreading of 1D *non-Newtonian* viscous gravity currents in variable geometries (specifically, Hele-Shaw cells with widths varying as a power law in x). To this end, we build upon the work of Zheng et al. (2014), which introduced this type of finite difference scheme for simulating the spreading of a finite mass of Newtonian fluid in a variable width Hele-Shaw cell. Owing to its accuracy and stability, this finite difference scheme has been recently applied by Alhashim and Koch (2018) to study hydraulic fracturing of low permeability rock.

This chapter is organised as follows. In Sect. 14.2, we briefly summarise existing models describing certain flows of viscous gravity currents. Then, we introduce a convenient general notation for such nonlinear parabolic PDEs. In Sect. 14.3.1, we introduce the 1D equispaced but staggered grid upon which the proposed finite difference scheme is to be implemented. The derivation of the BCs for the PDE, from the mass conservation constraint, is discussed in Sect. 14.2.2. Then, we construct the nonlinear Crank–Nicolson scheme in Sect. 14.3.2 and discuss the discretised form of the nonlinear flux BCs in Sect. 14.3.4.

Continuing, in Sect. 14.4.1, the scheme's accuracy is justified by comparing the numerical solution provided by the finite difference scheme (up to a specified physical time) against an analytical solution obtained through a self-similar transformation of the PDE. Specifically, this approach involves three validation cases: (i) a symmetric (about $x = 0$) lump of fixed fluid mass spreading in two directions (convergence is independent of BCs), (ii) a fixed fluid mass spreading away from the origin ($x = 0$) (requires only no flux BCs), and (iii) a variable fluid mass injected at the origin spreading away from it (requires careful implementation of the nonlinear BCs). In all three cases, the scheme is shown to be capable of accurately computing the evolution of gravity current's shape. In Sect. 14.4.2, we analyse the scheme's conservation properties by verifying numerically that it respects the mass constraint $\mathcal{V}(t) \propto t^{\alpha}$. We consider two validation cases: (i) release of a fixed fluid mass ($\alpha = 0$), and (ii) fluid mass injection into the domain ($\alpha > 0$). In both cases, we specifically focus on the challenging case of a non-Newtonian (power law) displacing fluid in a variable width channel. As a benchmark, we use previously derived first kind self-similar solutions from the literature, which are discussed in the Appendix.

14.2 Preliminaries

In this section, we summarise the mathematical model for viscous gravity currents in a selected set of applications involving Newtonian and non-Newtonian fluids. We study their spreading in a fixed or variable width channel geometry (also known as a "Hele-Shaw cell"), as well as flows in heterogeneous porous media with independently variable permeability and porosity. Our goal is to highlight the fact that all these models can be concisely summarised by a single nonlinear parabolic PDE supplemented with a set of nonlinear Neumann (flux) BCs.

14.2.1 Fluid Domain and Flow Characteristics

The flow domain is assumed to be long and thin. For example, it can be a channel existing in the gap between two impermeable plates, i.e., a Hele-Shaw (HS) cell, which may or may not have variable transverse (to the flow) width (Fig. 14.2a); or, it can be slab of uniform thickness heterogeneously porous material (Fig. 14.2c). The viscous gravity current consists of one fluid displacing another immiscible fluid. Therefore, a sharp interface $y = h(x, t)$ separates the two fluids at all times. The present study considers the limit of negligible surface (interfacial) tension (compared to gravitational forces); this is the limit of large Bond number. The density difference $\Delta\rho$ between the two fluids is large compared to the density of the lighter fluid, and the denser fluid flows along the bottom of the cell, which is a horizontal impermeable surface. In doing so, the denser fluid displaces the lighter fluid out of its way. Here, the geometry is considered to be vertically unconfined so that the details of the flow of the upper (lighter) fluid can be neglected.

We are interested in the evolution of the interface $h(x, t)$ between the two fluids. Owing to the vertically unconfined, long and thin geometry of the flow passage, the denser fluid has a slender profile (small aspect ratio), and the fluid flow can be described by *lubrication theory*. The lubrication approximation also requires that viscous forces dominate inertial forces; this is the limit of small Reynolds number. In this regime of small Reynolds number but large Bond number, the flow is governed by a balance of viscous forces and gravity. Furthermore, the lubrication approximation allows for (at the leading order in the aspect ratio) the variation of quantities across the transverse direction, as well as the vertical velocities of the fluids to be neglected.

As shown in Fig. 14.2a, for the flow in a HS cell, we allow the cell's width to vary as a power law of the streamwise coordinate x, i.e., $b(x) = b_1 x^n$, where $n \geq 0$ is a dimensionless exponent, and $b_1 > 0$ is a dimensional consistency constant having units m^{1-n}. Since the cell has a variable width, it originates from a cell 'origin', which is always taken to be $x = 0$ such that $b(0) = 0$. As discussed in (Zheng et al., 2014), in such a flow geometry, the lubrication approximation may fail when $b(x)$ is an increasing function of x i.e., $\mathrm{d}b/\mathrm{d}x = nb_1 x^{n-1} > 1$. In such quickly widening cells, the transverse variations of properties become significant. We ensure the validity of the lubrication approximation, and models derived on the basis of it, by only considering $n < 1$ such that $\mathrm{d}b/\mathrm{d}x$ remains a decreasing function of x.

The porosity can also be varied by filling the HS cell with beads of fixed diameter, as illustrated in Fig. 14.2b. We also consider a gravity current spreading horizontally in a porous slab of constant transverse width ($b(x) = b_1 = \text{const}$) with heterogeneous porosity $\phi(x) = \phi_1 x^m$ and permeability $k(x) = k_1 x^n$ (Fig. 14.2c). Here, $m, n \geq 0$ are dimensionless exponents and $\phi_1, k_1 > 0$ are dimensional constants needed for consistency with the definitions of porosity and permeability, respectively; specifically ϕ_1 has units of units of m^{-m}, and k_1 has units of m^{2-n}. These variations are illustrated by the streamwise changes of bead radii in Fig. 14.2c. Now, the point at which the porosity and permeability vanish is the origin of the cell.

Another interesting case, that of a medium with vertically heterogeneous porosity, has been explored by Ciriello et al. (2016). In this chapter, we limit our discussion to flow in a completely porous (i.e., unobstructed, $\phi = 1$) HS cell of variable width as in Fig. 14.2a. However, the numerical scheme developed herein can readily treat any of these cases, taking the appropriate parameter definitions from Table 14.1.

We allow the denser fluid to be non-Newtonian. Specifically, it obeys the power law rheology. In unidirectional flow, the one unique nontrivial shear stress component is given by $\tau = \mu(\dot{\gamma})\dot{\gamma}$, where the dynamic viscosity μ depends on the shear rate $\dot{\gamma}$ as $\mu(\dot{\gamma}) = \mu_0 \dot{\gamma}^{r-1}$. Here, μ_0 is the flow consistency index (units of $\mathrm{Pa{\cdot}s}^r$), and r (> 0) is the fluid's rheological index. Fluids having $r < 1$ are termed shear thinning (e.g., blood), and fluids with $r > 1$ are termed shear thickening (e.g., dense particulate suspensions). In the special case $r = 1$, the power law model reduces to the Newtonian fluid. As stated above, the flow of the displaced fluid is immaterial to the dynamics of the gravity current, as long as the viscosity and density contrasts are large. This condition is satisfied, e.g., by assuming (for the purposes of this chapter) the displaced fluid is air.

Finally, the volume of the fluid in the cell itself may be either fixed (constant mass) or vary with time (injection). Consistently with the literature, we consider the instantaneous volume of fluid in the cell to increase as a power law in t: $\mathcal{V}(t) = \mathcal{V}_0 + \mathcal{V}_{\mathrm{in}} t^{\alpha}$, where $\mathcal{V}_0$ is the initial volume of fluid in the HS cell (measured in m^3), $\alpha \geq 0$ is a dimensionless exponent, and $\mathcal{V}_{\mathrm{in}}$ is an injection pseudorate (in units $\mathrm{m}^3\mathrm{s}^{-\alpha}$), becoming precisely the injection rate for $\alpha = 1$. Next, we discuss how this assumption leads to BCs for the physical problem and for the numerical scheme.

14.2.2 Governing Equation, Initial and Boundary Conditions

The propagation of a viscous gravity current is described by a diffusion equation for the interface $h(x, t)$, which is the shape of profile of the denser fluid. The models are derived either from porous medium flow under Darcy's law and the Dupoit approximation (Bear, 1988, Ch. 8) or using lubrication theory with no-slip along the bottom of the cell and zero shear stress at the fluid–fluid interface (Leal, 2007, Ch. 6-C). The resulting velocity field is combined with a depth averaged continuity equation to derive the nonlinear parabolic PDE for $h(x, t)$.

We propose to summarise all gravity current propagation along horizontal surfaces through a single 'thin film' (Oron et al., 1997) equation

$$\frac{\partial h}{\partial t} = \frac{A}{x^p}\frac{\partial}{\partial x}\left(x^q \psi \frac{\partial h}{\partial x}\right). \tag{14.1}$$

According to Engelbrecht (2015, Ch. 5), (14.1) can be classified as an 'evolution equation'. The term in the parentheses on the right-hand side of (14.1), roughly, represents a fluid flux balanced by the change in height on the left-hand side. The multiplicative factor A/x^p arises due to (i) geometric variations of the flow passage in the flow-wise direction, (ii) porosity variations in the flow-wise direction, or (iii) from the choice of coordinate system in the absence of (i) or (ii). Here, A is dimensional constant depending on the flow geometry, the domain, and the fluid properties. Additionally, p and q are dimensionless exponents that depend on the flow geometry and fluid rheology. The quantity denoted by ψ represents specifically the *nonlinearity* in these PDEs. Thus, it is necessarily a function of h, and possibly $\partial h/\partial x$ for a non-Newtonian fluid (as in the third and fifth rows of Table 14.1).[3]

As stated in Sect. 14.1, several versions of (14.1) will be explored herein, incorporating geometric variations, porosity variations, non-Newtonian behaviour. The pertinent physical scenarios that will be tackled herein (using the proposed numerical scheme) are presented in Table 14.1, which lists expressions for A, p, q, and ψ. From a dimensional analysis of the PDE (14.1), it follows that the constant A must have units of $\mathrm{m}^{1+p-q}\mathrm{s}^{-1}$, as long as the nonlinearity ψ has units of length (as is the case for all the models summarised in Table 14.1). It is worth noting that in the case of 1D linear diffusion ($p = q = 0$ and $\psi = 1$), A becomes the 'diffusivity' in units of $\mathrm{m}^2\mathrm{s}^{-1}$.

The PDE (14.1) is solved on the finite space-time interval $(x, t) \in (\ell, L) \times (t_0, t_f]$. Here, t_0 and t_f represent the initial and final times of the numerical simulation's run, respectively. An initial condition (IC) $h_0(x)$ is specified at $t = t_0$, so that $h(x, t_0) = h_0(x)$ is known. Meanwhile, ℓ is a small positive value (close to 0).

[3] Interestingly, an 'r-Laplacian' PDE, similar to (14.1) for a power law fluid in a HS cell (third row of Table 14.1), arises during fluid–structure interaction between a power law fluid and an enclosing slender elastic tube (Boyko et al., 2017). This PDE can also be tackled by the proposed finite difference scheme.

Boundary conditions (BCs) are specified at $x = \ell$ and $x = L$. These involve some combination of h and $\partial h/\partial x$. The reason for taking $x = \ell \neq 0$ becomes clear below.

Thus, let us now discuss such a suitable set of BCs. The BCs are based on the imposed mass conservation/growth constraint. Consider the case of a viscous gravity current in a porous slab with variable porosity $\phi(x) = \phi_1 x^m$, and transverse width $b_1 = const$. Then, the conservation of mass constraint (see Zheng et al. 2014) takes the form

$$\mathcal{V}(t) \equiv \int_\ell^L h(x,t) b_1 \phi(x)\,\mathrm{d}x = \mathcal{V}_0 + \mathcal{V}_{\text{in}} t^\alpha, \tag{14.2}$$

where $\alpha \geq 0$. In the parallel case of a HS cell with variable width $b(x) = b_1 x^n$ and porosity $\phi_1 = \text{const}$, which can either be set to unity or absorbed into $b(x)$ via b_1, the mass constraint becomes

$$\mathcal{V}(t) \equiv \int_\ell^L h(x,t) b(x)\,\mathrm{d}x = \mathcal{V}_0 + \mathcal{V}_{\text{in}} t^\alpha. \tag{14.3}$$

Taking a time derivative of (14.3) and employing (14.1), we obtain:

$$\begin{aligned}\frac{\partial}{\partial t}\int_\ell^L h(x,t) b_1 x^n\,\mathrm{d}x &= \int_\ell^L \frac{\partial h}{\partial t} b_1 x^n\,\mathrm{d}x = \int_\ell^L b_1 x^n \frac{A}{x^n}\frac{\partial}{\partial x}\left(x^q \psi \frac{\partial h}{\partial x}\right)\mathrm{d}x = \\ &= A b_1 \left(x^q \psi \frac{\partial h}{\partial x}\right)\Bigg|_{x=\ell}^{x=L} \overset{\text{by (14.3)}}{=} \frac{\mathrm{d}(\mathcal{V}_{\text{in}} t^\alpha)}{\mathrm{d}t} = \alpha \mathcal{V}_{\text{in}} t^{\alpha-1}.\end{aligned} \tag{14.4}$$

Here, $p = n$ in this case of interest, as described in Table 14.1, and $Ab_1 = const$. Thus, we have obtained conditions relating $x^q \psi \partial h/\partial x$ at $x = \ell$ and $x = L$ to $\alpha \mathcal{V}_{\text{in}} t^{\alpha-1}$. These conditions, if satisfied, automatically take into account the imposed volume constraint from (14.3). The calculation starting with (14.2) is omitted as it is identical, subject to proper choice of p.

For the case of propagation away from the cell's origin (i.e., any injection of mass must occur near $x = 0$, specifically at $x = \ell$), to satisfy (14.4), we can require that

$$\left(x^q \psi \frac{\partial h}{\partial x}\right)\Bigg|_{x=\ell} = \begin{cases} -\frac{\alpha B}{A} t^{\alpha-1}, & \alpha > 0, \\ 0, & \alpha = 0, \end{cases} \tag{14.5a}$$

$$\left(\psi \frac{\partial h}{\partial x}\right)\Bigg|_{x=L} = 0 \quad \Leftarrow \quad \frac{\partial h}{\partial x}\Bigg|_{x=L} = 0, \tag{14.5b}$$

where $B = \mathcal{V}_{\text{in}}/b_1$. Recall, the case of $\alpha > 0$ represents mass injection. Although (14.3) and (14.5) are equivalent, the imposition of the nonlinear BC in (14.5a) must be approached with care. It should be clear that to impose a flux near the origin (at $x = 0$), we need $\left(x^q \psi \partial h/\partial x\right)\big|_{x\to 0}$ to be finite. Then, $\psi \partial h/\partial x = \mathcal{O}(1/x^q)$ as

$x \to 0$. On the spatial domain $x \in (0, L)$, such an asymptotic behaviour is possible for $p = q = 0$. However, in a variable width cell ($p, q \neq 0$), the local profile and slope as $x \to 0$ blow up if they are to satisfy $\psi \partial h/\partial x = \mathcal{O}(1/x^q)$ as $x \to 0$. To avoid this uncomputable singularity issue, we defined the computational domain to be $x \in (\ell, L)$, where ℓ is 'small' but > 0. The BC from (14.5a) at $x = \ell$ can then be rewritten as

$$\left(\psi \frac{\partial h}{\partial x}\right)\bigg|_{x=\ell} = -\frac{\alpha B}{A\ell^q} t^{\alpha-1}, \quad \alpha > 0. \tag{14.6}$$

It may also be of interest to consider the case of a gravity current released a finite distance away from the origin and then spreading towards $x = 0$. In this case, an additional length scale arises in the problem: the initial distance of the current's edge from the origin, say $x_f(0)$. The existence of this extra length scale complicates the self-similarity analysis, leading to solutions of the second kind (Barenblatt, 1996, Ch. 4), as discussed in Sect. 14.1. However, the numerical scheme can handle this case just as well. In fact, it requires no special consideration, unlike spreading away from the origin. Now, we may simply take $\ell = 0$ and consider spreading on the domain $(0, L)$ subject to the following BCs:

$$\left(x^q \psi \frac{\partial h}{\partial x}\right)\bigg|_{x\to 0} = 0 \quad \Leftarrow \quad \frac{\partial h}{\partial x}\bigg|_{x=0} = 0, \tag{14.7a}$$

$$\left(\psi \frac{\partial h}{\partial x}\right)\bigg|_{x=L} = \begin{cases} \frac{\alpha B}{AL^q} t^{\alpha-1}, & \alpha > 0, \\ 0, & \alpha = 0, \end{cases} \tag{14.7b}$$

which together allow us to satisfy (14.4) and, thus, (14.3) for all $t \in (t_0, t_f]$.

The most significant advantage of defining nonlinear flux BCs, such as those in (14.5) or (14.7), is that a nonlinear nonlocal (integral) constraint, such as that in (14.2) or (14.3), no longer has to be applied onto the solution $h(x, t)$. Furthermore, if we start with compact initial conditions, i.e., there exists a nose location $x = x_f(t_0)$ such that $h\big(x_f(t_0), t_0\big) = 0$, then the finite speed of propagation property of the nonlinear PDE (14.1) (Gilding and Kersner, 2004; Vázquez, 2007) ensures that this nose $x_f(t)$ exists for all $t > t_0$ and $h\big(x_f(t), t\big) = 0$ as well.

The proposed fully implicit scheme inherits this property of the PDE. Therefore, we can solve the PDE on the *fixed* domain $x \in (\ell, L)$, without any difficulty, instead of attempting to rescale to a moving domain on which $x_f(t)$ is one of the endpoints with $h = 0$ as the BC applied there. The latter approach proposed by Bonnecaze et al. (1993) (and used in more recent works (Acton et al., 2001) as well) leads to a number of additional variable coefficient terms arising in the PDE (14.1), due to the non-Galilean transformation onto a shrinking/expanding domain. From a numerical methods point of view, having to discretise these additional terms is not generally desirable.

Having defined a suitable set of BCs, the last remaining piece of information required to close the statement of the mathematical problem at hand is the selection of pertinent initial conditions (ICs). For the case of the release of a finite fluid mass ($\alpha = 0$), an arbitrary polynomial IC may be selected, as long as it has zero slope at the origin ($x = 0$), leading to satisfaction of the no-flux boundary condition (14.5a). To this end, let the IC be given by

$$h_0(x) = \begin{cases} a\left(\mathfrak{X}_0^c - x^c\right), & x \le \mathfrak{X}_0, \\ 0, & x > \mathfrak{X}_0, \end{cases} \tag{14.8}$$

where $\mathfrak{X}_0$ is a 'release gate' location defining the initial position of the current's nose, i.e., $\mathfrak{X}_0 = x_f(t_0)$ and $h\big(x_f(t_0), t_0\big) \equiv h_0(\mathfrak{X}_0) = 0$. The constant $c > 1$ is an arbitrary dimensionless exponent. Finally, a (units of m^{1-c}) is set by normalising $h_0(x)$ such that the initial volume of fluid corresponds to the selected initial fluid volume, $\mathcal{V}_0$, via (14.3).

The case of the release of a finite mass of fluid is particularly forgiving in how we set the IC, and its slope at $x = 0$. In fact, we could even take $c = 1$ in (14.8) and the scheme will provide an initial flux of fluid at $t = t_0^+$, with $(\partial h/\partial x)\big|_{x=0} = 0$ thereafter. On the other hand, the case of mass injection ($\alpha > 0$) governed by the nonlinear BCs is not as forgiving. By virtue of the 'point source' mass injection at $x = \ell$, the slope at the origin rises sharply from the moment of mass injection. This very sharp rise has a tendency to introduce unphysical oscillations in the current profile when starting from the IC in (14.8). To avoid this, we must select a 'better' IC, which has a shape more similar to the actual solution's singularity near $x = 0$.

Having tested a few different options, we found that an exponential function works well:

$$h_0(x) = \begin{cases} a\left(-1 + b\mathrm{e}^{cx}\right), & x \le \mathfrak{X}_0, \\ 0, & x > \mathfrak{X}_0. \end{cases} \tag{14.9}$$

Here, b (dimensionless) and c (units of m^{-1}) are positive constants, $\mathfrak{X}_0 = \frac{1}{c}\ln\frac{1}{b}$ ensures that the IC has no negative values and a sharp wavefront, and a (units of m) is set by normalising $h_0(x)$ to the selected initial volume $\mathcal{V}_0$ via (14.3), as above.

Finally, it should be noted that the IC from (14.8) is not used in the convergence studies for finite initial mass (Sects. 14.4.1.1 and 14.4.1.2). Rather, the IC is taken to be the exact self-similar solution of Ciriello et al. (2016) for a power law fluid in a uniform width ($n = 0$) HS cell (see also the Appendix). The reasoning behind this particular choice is further expounded upon in Sect. 14.4.1.

14.3 The Numerical Scheme

The proposed numerical method is a finite difference scheme using the Crank–Nicolson approach toward implicit time stepping. Our presentation follows recent literature, specifically the construction in (Zheng et al., 2014, Appendix B). The proposed scheme's truncation error is formally of second order in both space and time, and we expect the scheme to be unconditionally stable. Furthermore, the scheme is conservative in the sense that it maintains the imposed time dependency of the fluid volume with high accuracy via a specific set of nonlinear BCs. This section is devoted to discussing all these topics one by one.

14.3.1 Notation: Grids, Time Steps, and Grid Functions

The PDE (14.1) is solved on an equispaced 1D grid of $N+1$ nodes with grid spacing $\Delta x = (L-\ell)/(N-1)$. The solution values are kept on a staggered grid of cell centres, which are offset by $\Delta x/2$ with respect to the equispaced grid points. As a result, there is a node lying a half-grid-spacing beyond each domain boundary. It follows that the location of the ith grid point on the staggered grid is $x_i = \ell + (i-1/2)\Delta x$, where $i = 0, 1, 2, \dots, N$. A representative grid with 12 nodes is shown in Fig. 14.3. The use of a staggered grid affords additional stability to the scheme and allows us to evaluate derivatives with second order accuracy via central differences, by default, using only two cell centred values.

As stated in Sect. 14.2.2, the PDE (14.1) is solved over a time period $t \in (t_0, t_f]$, such that $t_f > t_0 \geq 0$, where both the initial time t_0 and the final time t_f of the simulation are user defined. The scheme thus performs M discrete time steps each of size $\Delta t = (t_f - t_0)/(M-1)$. The nth time step advances the solution to $t = t^n \equiv t_0 + n\Delta t$, where $n = 0, 1, \dots, M-1$. Finally, we define the discrete analog ('grid function') to the continuous gravity current shape, which we actually solve for, as $h_i^n \approx h(x_i, t^n)$.

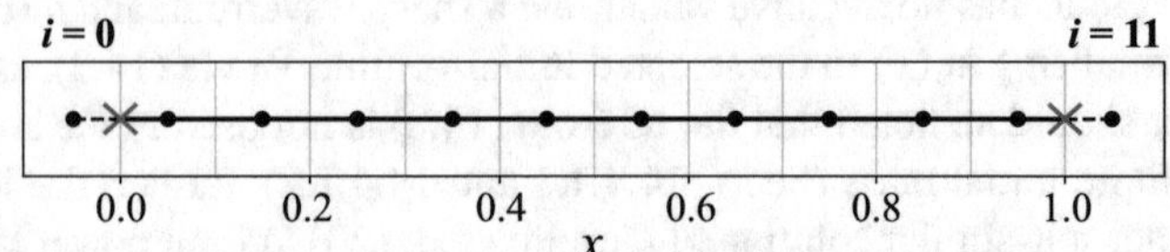

Fig. 14.3 A sample twelve-node equispaced but staggered 1D grid. The grid nodes are staggered by half a grid step $\Delta x/2$ from the cell faces. The boundary conditions are implemented at the 'real' domain boundaries (here marked by **x**). The two grid points *outside* the physical domain (i.e., $i = 0, 11$ or $x_0 = -0.1$ and $x_{11} = 1.1$ in this example) are used to implement the Neumann BCs, which require computing a derivative at the 'real' domain boundaries (i.e., $i = 1/2, 21/2$ or $x_{1/2} \equiv \ell = 0$ and $x_{21/2} \equiv L = 1$ in this example).

14.3.2 The Nonlinear Crank–Nicolson Scheme

Let us denote by $\mathcal{L}$ the continuous spatial operator acting on h on the right-hand side of (14.1), i.e.,

$$\mathcal{L}[h] \equiv \frac{A}{x^p}\frac{\partial}{\partial x}\left(x^q \psi \frac{\partial h}{\partial x}\right). \tag{14.10}$$

Since $\mathcal{L}$ is a second order spatial operator and, thus, (14.1) is a diffusion equation, we are inclined to implement a second order accurate time stepping by the Crank–Nicolson scheme (Crank and Nicolson, 1947). The Crank–Nicolson scheme is fully implicit, which avoids the stringent restriction ($\Delta t \lesssim (\Delta x)^2$) suffered by explicit time discretisations of diffusion equations (Strikwerda, 2004, Ch. 6). Then, the time discrete version of (14.1) is

$$\frac{h_i^{n+1} - h_i^n}{\Delta t} = \frac{1}{2}\left(\mathcal{L}_d\left[h_i^{n+1}\right] + \mathcal{L}_d\left[h_i^n\right]\right), \tag{14.11}$$

where $\mathcal{L}_d$ is the discrete analog to the continuous spatial operator $\mathcal{L}$ defined in (14.10). Based on the approach of Christov and Homsy (2009), the discrete spatial operator is constructed via flux conservative central differencing using two cell-face values, while staggering the nonlinear terms:

$$\mathcal{L}_d\left[h_i^n\right] = \frac{A}{x_i^p}\left[\frac{\left(x_{i+1/2}^q \psi_{i+1/2}^{n+1/2}\right) S_{i+1/2}^n - \left(x_{i-1/2}^q \psi_{i-1/2}^{n+1/2}\right) S_{i-1/2}^n}{\Delta x}\right], \tag{14.12a}$$

$$\mathcal{L}_d\left[h_i^{n+1}\right] = \frac{A}{x_i^p}\left[\frac{\left(x_{i+1/2}^q \psi_{i+1/2}^{n+1/2}\right) S_{i+1/2}^{n+1} - \left(x_{i-1/2}^q \psi_{i-1/2}^{n+1/2}\right) S_{i-1/2}^{n+1}}{\Delta x}\right], \tag{14.12b}$$

where $S \equiv \partial h/\partial x$ is the slope of the gravity current's shape. Note that the nonlinear terms, denoted by ψ, have been evaluated the same way, i.e., at the mid-time step $n + 1/2$, for both $\mathcal{L}_d\left[h_i^n\right]$ and $\mathcal{L}_d\left[h_i^{n+1}\right]$.

Substituting (14.12) into (14.11) results in a system of *nonlinear* algebraic equations because ψ is evaluated at mid-time-step $n + 1/2$ and, thus, depends on both h_i^n (known) and h_i^{n+1} (unknown). This system must be solved for the vector h_i^{n+1} ($i = 0, \ldots, N$), i.e., the approximation to the gravity current's shape at the next time step. Solving a large set of nonlinear algebraic equations can be tedious and computationally expensive. A simple and robust approach to obtaining a solution of the nonlinear algebraic system is through fixed point iterations, or 'the method of internal iterations' (Yanenko, 1971). Specifically, we can iteratively compute approximations to h_i^{n+1}, the grid function at the new time step, by replacing it in (14.11) with $h_i^{n,k+1}$, where $h_i^{n,0} \equiv h_i^n$. Then, the proposed numerical scheme takes the form:

$$\frac{h_i^{n,k+1} - h_i^n}{\Delta t} = \frac{A}{2\Delta x}\left[\frac{x_{i+1/2}^q}{x_i^p}\psi_{i+1/2}^{n+1/2,k} S_{i+1/2}^{n,k+1} - \frac{x_{i-1/2}^q}{x_i^p}\psi_{i-1/2}^{n+1/2,k} S_{i-1/2}^{n,k+1}\right] + \\ + \frac{A}{2\Delta x}\left[\frac{x_{i+1/2}^q}{x_i^p}\psi_{i+1/2}^{n+1/2,k} S_{i+1/2}^{n} - \frac{x_{i-1/2}^q}{x_i^p}\psi_{i-1/2}^{n+1/2,k} S_{i-1/2}^{n}\right]. \tag{14.13}$$

The key idea in the method of internal iterations is to evaluate the nonlinear ψ terms from information known at iteration k and the previous time step n, while keeping the linear slopes S from the next time step $n+1$ at iteration $k+1$. This manipulation linearises the algebraic system, at the cost of requiring iteration over k. Upon convergence of the internal iterations, h_i^{n+1} is simply the last iterate $h_i^{n,k+1}$. Before we can further discuss the iterations themselves or their convergence, we must define our discrete approximations for ψ and S.

The operator $\mathcal{L}_d$ is essentially a second derivative, so we take inspiration from the standard way of constructing the three-point central finite difference formula for the second derivative (Strikwerda, 2004). Therefore, $S_{i\pm1/2}$ can be discretised using a two-point central difference approximation on the staggered grid. For example, at any time step:

$$S_{i+1/2} \equiv \left(\frac{\partial h}{\partial x}\right)_{x=x_{i+1/2}} \approx \frac{h_{i+1} - h_i}{\Delta x}. \tag{14.14}$$

Next, following Zheng et al. (2014) or Christov and Deng (2002), we evaluate ψ at $x_{i\pm1/2}$ by averaging the known values at x_i and x_{i+1} or x_i and x_{i-1}, respectively. Likewise, to approximate $\psi^{n+1/2}$, we average the known values: ψ^n at t^n and $\psi^{n,k}$ at the previous internal iteration. In other words, our approximation of the nonlinear terms is

$$\psi_{i+1/2}^{n+1/2,k} \approx \frac{1}{2}\Bigg[\underbrace{\frac{1}{2}\left(\psi_{i+1}^{n,k} + \psi_i^{n,k}\right)}_{\approx\psi_{i+1/2}^{n,k}} + \underbrace{\frac{1}{2}\left(\psi_{i+1}^{n} + \psi_i^{n}\right)}_{\approx\psi_{i+1/2}^{n}}\Bigg], \tag{14.15a}$$

$$\psi_{i-1/2}^{n+1/2,k} \approx \frac{1}{2}\Bigg[\underbrace{\frac{1}{2}\left(\psi_{i}^{n,k} + \psi_{i-1}^{n,k}\right)}_{\approx\psi_{i-1/2}^{n,k}} + \underbrace{\frac{1}{2}\left(\psi_{i}^{n} + \psi_{i-1}^{n}\right)}_{\approx\psi_{i-1/2}^{n}}\Bigg]. \tag{14.15b}$$

Equations (14.15) afford improved stability for nonlinear PDEs, while preserving the conservative nature of the scheme (as will be shown in Sect. 14.4.2), as discussed

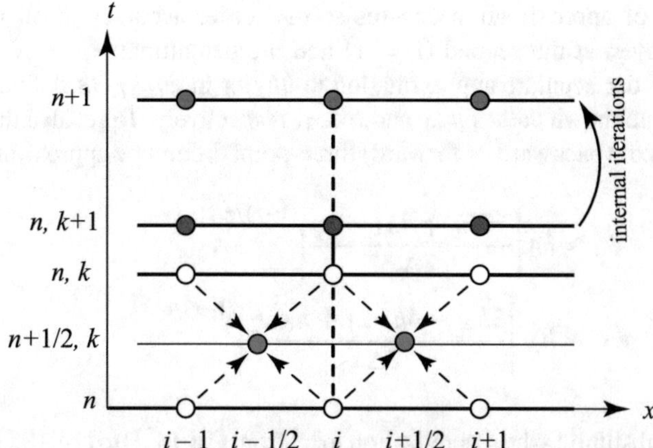

Fig. 14.4 Representative stencil of the proposed scheme. After performing k internal iterations, the nonlinear terms $\psi_{i\pm1/2}$ are computed at the intermediate stage '$n+1/2, k$' (highlighted in blue) from the known quantities h_i^n and $h_i^{n,k}$. The unknown quantity $h_i^{n,k+1}$ at the next internal iteration, stage '$n, k+1$' (highlighted in red), is found by solving the linear system in (14.18). The process continues until the convergence criterion in (14.19) is met, yielding the (initially unknown) solution at $t = t^{n+1}$.

by Von Rosenberg (1975) who credits the idea of averaging nonlinear terms across time stages and staggered grid points to the seminal work of Douglas Jr. et al. (1959, 1962). The scheme thus described is depicted by the stencil diagram in Fig. 14.4.

Here, it is worthwhile noting that, the classical Crank–Nicolson (Crank and Nicolson, 1947) scheme is only *provably* unconditionally stable (Strikwerda, 2004) when applied to a *linear* diffusion equation. It was suggested by Christov and Homsy (2009) that the current approach provides additional stability to this *nonlinear* scheme for large time steps. But, since our problem is nonlinear, some care should be taken in evaluating how large of a time step could be taken. Nevertheless, it is still expected that the largest stable Δt will be independent of Δx.

A complication arising in the present context is that we focus on the case of a power law non-Newtonian viscous gravity current spreading in a variable width cell. As a result, recalling Table 14.1, this model features $\partial h/\partial x$ in ψ, *unlike* the Newtonian case. While the temporal accuracy of the scheme is ensured through the robust implementation of the nonlinear Crank–Nicolson time stepping, the spatial accuracy is contingent upon the discretisation of $\partial h/\partial x$ in ψ.

A further consequence is that, once we discretise $\partial h/\partial x$, the discretisation of ψ becomes *nonlocal* (i.e., it requires information beyond the ith grid point). Nevertheless, the overall scheme still only requires a three-point stencil for $\mathcal{L}_d$. In particular, for interior grid points, we use a central difference formula, giving rise to the expression (at any time step):

$$\psi_i \equiv \left[h \left| \frac{\partial h}{\partial x} \right|^{(1-r)/r} \right]_{x=x_i} \approx h_i \left| \frac{h_{i+1} - h_{i-1}}{2\Delta x} \right|^{(1-r)/r}. \tag{14.16}$$

This choice of approximation ensures second order accuracy at all interior grid nodes. However, at the second ($i = 1$) and the penultimate ($i = N - 1$) nodes, the second order accurate approximation to $\partial h/\partial x$ in $\psi_{i\pm 1/2}$ as defined in (14.15) requires the unknown values h_{-1} and h_{N+1}, respectively. To resolve this difficulty, we use 'biased' (backward or forward) three-point difference approximations:

$$\psi_0 \approx h_0 \left| \frac{-3h_0 + 4h_1 - h_2}{2\Delta x} \right|^{(1-r)/r}, \tag{14.17a}$$

$$\psi_N \approx h_N \left| \frac{3h_N - 4h_{N-1} + h_{N-2}}{2\Delta x} \right|^{(1-r)/r}. \tag{14.17b}$$

Finally, substituting the discretisation for S from (14.14) into (14.13), it is possible to rearrange the scheme into a tridiagonal matrix equation:

$$\underbrace{\left[-\frac{A\Delta t}{2(\Delta x)^2} \frac{x^q_{i-1/2}}{x^p_i} \psi^{n+1/2,k}_{i-1/2} \right]}_{\text{matrix subdiagonal coefficient}} h^{n,k+1}_{i-1} +$$

$$+ \underbrace{\left[1 + \frac{A\Delta t}{2(\Delta x)^2} \left(\frac{x^q_{i+1/2}}{x^p_i} \psi^{n+1/2,k}_{i+1/2} + \frac{x^q_{i-1/2}}{x^p_i} \psi^{n+1/2,k}_{i-1/2} \right) \right]}_{\text{matrix diagonal coefficient}} h^{n,k+1}_i +$$

$$+ \underbrace{\left[-\frac{A\Delta t}{2(\Delta x)^2} \frac{x^q_{i+1/2}}{x^p_i} \psi^{n+1/2,k}_{i+1/2} \right]}_{\text{matrix superdiagonal coefficient}} h^{n,k+1}_{i+1} =$$

$$= h^n_i + \frac{A\Delta t}{2(\Delta x)^2} \left[\frac{x^q_{i+1/2}}{x^p_i} \psi^{n+1/2,k}_{i+1/2} (h^n_{i+1} - h^n_i) - \frac{x^q_{i-1/2}}{x^p_i} \psi^{n+1/2,k}_{i-1/2} (h^n_i - h^n_{i-1}) \right] \tag{14.18}$$

for the interior grid points $i = 1, \dots, N - 1$. In (14.18), the right-hand side and the variable coefficients in brackets on the left-hand side are both known, based on $h^{n,k}_i$, at any given internal iteration k. Then, each internal iteration involves the inversion of a tridiagonal matrix to solve for the grid function $h^{n,k+1}_i$. The inversion of this tridiagonal matrix can be performed efficiently with, e.g., 'backslash' in MATLAB. Subsequently, the coefficient matrix must be recalculated for each internal iteration because of the dependency of $\psi^{n+1/2,k}_{i\pm 1/2}$ on $h^{n,k}_i$ arising from (14.15), (14.16), and (14.17).

The iterations in (14.18) are initialised with $h^{n,0}_i = h^n_i$ ($i = 0, \dots, N$) and continue until an iteration $k + 1 = K$ is reached at which a 10^{-8} relative error tolerance is met. Specifically,

$$\max_{0\leq i\leq N}\left|h_i^{n,K}-h_i^{n,K-1}\right|<10^{-8}\max_{0\leq i\leq N}\left|h_i^{n,K-1}\right|. \tag{14.19}$$

Only a small number (typically, less than a dozen) of internal iterations are required at each time step, making the scheme quite efficient overall.

A detail remains, however. The algebraic system defined in (14.18) applies to all *interior* nodes, i.e., $i = 1, \ldots, N-1$. To complete the system, we must define rows $i = 0$ and $i = N$, which arise from the discretisation of the nonlinear BCs, which comes in Sect. 14.3.4. Upon completing the latter task successfully, $h_i^{n,K}$ becomes the grid function at the next time step h_i^{n+1} upon the completion of the internal iterations, and the time stepping proceeds.

14.3.3 The Special Case of Linear Diffusion

A noteworthy special case of the proposed finite difference scheme arises from setting the dimensionless exponents $p = q = 0$ (i.e., no spatial variation of the diffusivity) and $\psi = 1$ (linear diffusion). Then, (14.18) can be simplified and rearranged in the form ($i = 1, \ldots, N-1$):

$$\left[1+\frac{A\Delta t}{(\Delta x)^2}\right]h_i^{n+1}=\frac{A\Delta t}{(\Delta x)^2}\left(h_{i-1}^{n+1}+h_{i+1}^{n+1}+h_{i-1}^{n}+h_{i+1}^{n}\right)+\left[1+\frac{A\Delta t}{(\Delta x)^2}\right]h_i^n. \tag{14.20}$$

If the grid function $h_i^n \approx h(x_i, t^n)$ represents the temperature field along a 1D rigid conductor situated on $x \in [\ell, L]$, (14.20) is then the original second order (in space and time) numerical scheme proposed by Crank and Nicolson (1947) to solve a linear (thermal) diffusion equation (Strikwerda, 2004, §6.3). As such, this simplification helps illustrate the mathematical roots of the current scheme, and how we have generalised the classical work.

14.3.4 Implementation of the Nonlinear Boundary Conditions

As discussed in Sect. 14.2.2, the boundary conditions are a manifestation of the global mass conservation constraint ((14.2) or (14.3)), imposed on (14.1). The BCs described in (14.5) and (14.7) are defined at the 'real' boundaries of the domain, i.e., at $x = \ell$ and $x = L$. The numerical scheme is implemented over a staggered grid. This allows for derivatives at $x = \ell$ and $x = L$ to be conveniently approximated using central difference formulas using two nearby staggered grid points. In this manner, the BC discretisation maintains the scheme's second order accuracy in space and time. Accordingly, for the case of a current spreading away from the cell's origin, (14.5) are discretised in a 'fully implicit' sense (to further endow numerical stability and accuracy to the scheme (Christov and Deng, 2002)) as follows:

$$\psi_{1/2}^{n+1/2,k}\frac{1}{\Delta x}\left(h_1^{n,k+1}-h_0^{n,k+1}\right)=\begin{cases}-\frac{\alpha B}{A\ell^q}t^{\alpha-1}, & \alpha>0,\\ 0, & \alpha=0,\end{cases} \tag{14.21}$$

$$\frac{1}{\Delta x}\left(h_N^{n,k+1}-h_{N-1}^{n,k+1}\right)=0. \tag{14.22}$$

Within the internal iterations, however, $\psi_{1/2}^{n+1/2,k}$ is known independently of $h_1^{n,k+1}$ and $h_0^{n,k+1}$. Hence, we can express the first ($i=0$) and last ($i=N$) equations, which define the respective rows in the tridiagonal matrix stemming from (14.18), as

$$h_1^{n,k+1}-h_0^{n,k+1}=\begin{cases}-\dfrac{4\alpha Bt^{\alpha-1}\Delta x}{A\ell^q(\psi_0^n+\psi_1^n+\psi_0^{n,k}+\psi_1^{n,k})}, & \alpha>0,\\ 0, & \alpha=0,\end{cases} \tag{14.23a}$$

$$h_N^{n,k+1}-h_{N-1}^{n,k+1}=0. \tag{14.23b}$$

Similarly, we can derive the discretised BCs for spreading towards the origin, upon its release a finite distance away from the origin, from (14.7). Then, the first ($i=0$) and last ($i=N$) equations, which define the respective rows in the tridiagonal matrix, as

$$h_1^{n,k+1}-h_0^{n,k+1}=0, \tag{14.24a}$$

$$h_N^{n,k+1}-h_{N-1}^{n,k+1}=\begin{cases}\dfrac{4\alpha Bt^{\alpha-1}\Delta x}{AL^q(\psi_{N-1}^n+\psi_{N-2}^n+\psi_{N-1}^{n,k}+\psi_{N-2}^{n,k})}, & \alpha>0,\\ 0, & \alpha=0.\end{cases} \tag{14.24b}$$

14.4 Convergence and Conservation Properties of the Scheme

At this point, the numerical scheme and boundary conditions defined in (14.18) and (14.23) or (14.24) form a complete description of the numerical solution to the parabolic PDE (14.1), for a gravity current propagating away from the origin. We have claimed that the finite difference scheme is conservative (i.e., it accurately maintains the imposed time dependency of the fluid volume set by (14.3)) and has second order convergence. These aspects of the scheme will be substantiated in Sects. 14.4.1 and 14.4.2, respectively. The computational domain's dimensions,

Table 14.2 Summary of the simulation parameters used in convergence and conservation studies. The fluid was assumed to be a 95% glycerol-water mixture at 20°C. The width exponent n and fluid's rheological index r were varied on a case-by-case basis to simulate different physical scenarios.

Parameter	Value	Units
Channel length, L	0.75	m
Width coefficient, b_1	0.017390	m^{1-n}
Total released mass, w	0.31550	kg
Density difference, $\Delta\rho$	1250.8	$kg{\cdot}m^{-3}$
Consistency index ($r \neq 1$) or dynamic viscosity ($r = 1$), μ_0	0.62119	$Pa{\cdot}s^r$

which are set by L and b_1, and the properties of fluid being simulated are summarised in Table 14.2. For definiteness, in this chapter we select the fluid properties to be those of a 95% glycerol-water mixture in air at 20°C (Cheng, 2008; Volk and Kähler, 2018).

14.4.1 Estimated Order of Convergence

First, we seek to justify the formal accuracy (order of convergence) of the proposed scheme through carefully chosen numerical examples. To do so, we pursue a series of benchmarks that are successively 'more complicated' (from a numerical perspective). First, we simulate the case of a centrally released fixed mass of fluid propagating in two directions (Sect. 14.4.1.1). Second, we simulate the unidirectional spreading of a fixed mass of fluid (Sect. 14.4.1.2). Last, we simulate the unidirectional spreading of a variable fluid mass (Sect. 14.4.1.3) by taking into account injection of fluid at the boundary.

In each of these three cases, there is a need for a reliable benchmark solution against which the numerical solutions on successively refined spatial grids can be compared. For the case of the release of a fixed mass of fluid, an exact self-similar solution is provided by Ciriello et al. (2016). Specifically the solution is for a power law fluid in uniform HS cell ($n = 0$). The derivation of the self-similar solution is briefly discussed in the Appendix. We use this solution as the benchmark.

As mentioned in Sect. 14.1, parabolic equations 'forget' their IC and the solution becomes self-similar after some time. However, for a general PDE, it is difficult (if not impossible) to estimate how long this process takes. Therefore, to ensure a proper benchmark against the exact self-similar solution, we start the simulation with the exact self-similar solution evaluated at some nonzero initial time ($t_0 > 0$). Then, we let the current propagate up to a final time t_f, with the expectation that the current will remain in the self-similar regime for all $t \in [t_0, t_f]$. Comparing the final numerical profile with the exact self-similar solution at $t = t_f$ then allows for a proper benchmark.

To quantify the error between a numerical solution h_{num} and a benchmark h_{exact} solution at $t = t_f$, we use three standard function space norms (Evans, 2010):

$$\|h_{\text{num}}(x,t_f) - h_{\text{exact}}(x,t_f)\|_{L^\infty} = \max_{x\in[\ell,L]} \left|h_{\text{num}}(x,t_f) - h_{\text{exact}}(x,t_f)\right|, \tag{14.25a}$$

$$\|h_{\text{num}}(x,t_f) - h_{\text{exact}}(x,t_f)\|_{L^1} = \int_\ell^L \left|h_{\text{num}}(x,t_f) - h_{\text{exact}}(x,t_f)\right| \, \mathrm{d}x, \tag{14.25b}$$

$$\|h_{\text{num}}(x,t_f) - h_{\text{exact}}(x,t_f)\|_{L^2} = \sqrt{\int_\ell^L \left|h_{\text{num}}(x,t_f) - h_{\text{exact}}(x,t_f)\right|^2 \, \mathrm{d}x}\,. \tag{14.25c}$$

Using a second order trapezoidal rule for the integrals, the definitions in (14.25) can be expressed in terms of the grid functions to define the 'errors':

$$L^\infty_{\text{error}} \equiv \max_{0\le i\le N} \left|h_i^M - h_{\text{exact}}(x_i,t_f)\right|, \tag{14.26a}$$

$$L^1_{\text{error}} \equiv \Delta x \left\{ \frac{1}{2}\left[\left|h_0^M - h_{\text{exact}}(x_0,t_f)\right| + \left|h_N^M - h_{\text{exact}}(x_N,t_f)\right| \right] + \sum_{i=1}^{N-1} \left|h_i^M - h_{\text{exact}}(x_i,t_f)\right| \right\}, \tag{14.26b}$$

$$L^2_{\text{error}} \equiv \left[\Delta x \left\{ \frac{1}{2}\left[\left|h_0^M - h_{\text{exact}}(x_0,t_f)\right|^2 + \left|h_N^M - h_{\text{exact}}(x_N,t_f)\right|^2 \right] + \sum_{i=1}^{N-1} \left|h_i^M - h_{\text{exact}}(x_i,t_f)\right|^2 \right\} \right]^{1/2}, \tag{14.26c}$$

where M is the time step at which $t^M = t_f$.

Since the solution actually has a corner (derivative discontinuity) at the nose (wavefront) $x_f(t)$ such that $h\big(x_f(t),t\big) = 0$, the propagating gravity current is in fact only a *weak* solution to the PDE (Evans, 2010). Therefore, the L^∞ norm is not a good one to measure the error, as we do not expect the solution to 'live' in this function space. Nevertheless, our numerical results show convergence in the L^∞ norm. The natural functional space for solutions of (14.1) is the space of integrable functions, i.e., L^1. Indeed, we observe excellent second order convergence in this norm. For completeness, the L^2 norm (commonly the function space setting for parabolic equations (Evans, 2010, Ch. 7)) is considered as well. While we observe convergence close to second order in this norm as well, it is clearly not the 'natural' one for these problems either.

For our estimated-order-of-convergence study, Δx is successively halved on a domain of fixed length, such that on the cth iteration of the refinement, the grid spacing is $\Delta x_c = \Delta x_0/2^{c-1}$, where Δx_0 is the initial grid spacing. Doing so ensures a set of common grid points (corresponding to the same physical locations) between successively refined grids. In all studies in this section, we begin with a grid with $N = 101$ nodes, and it becomes the coarsest grid for the refinement study. Given the (formally) unconditionally stable nature of the scheme, we take $\Delta t_c = 2\Delta x_c$ for the refinement studies without loss of generality. From a computational standpoint, it is desirable that time step and grid spacing are of the same order of magnitude in the estimated-order-of-convergence study.

14.4.1.1 Central Release of a Fixed Fluid Mass (No Boundary Effects)

Consider a symmetric domain $x \in [-L, +L]$. Then suppose that a fixed mass of fluid (i.e., $\alpha = 0$ in the volume constraint in (14.3)) is released with an initial shape that is symmetric about $x = 0$. The final simulation time t_f is such that the gravity current does not reach $x = \pm L$ for $t \leq t_f$. Since the fluid mass is constant and the BCs are imposed at $x = \pm L$ (where $h = 0$ initially and remains so for all $t \leq t_f$, by construction), their discretisation simply reduces to the trivial cases, i.e., (14.24a) and (14.23b).

Thus, the BCs for this study are simply linear Neumann (i.e., no flux or homogeneous) BCs, and they do not influence the order of convergence of the overall scheme. Therefore, this study allows us to verify that our approach to the treatment of the nonlinearity ψ, and its weighted averages appearing in the spatially discretised operator $\mathcal{L}_d$ in (14.12), deliver the desired second order of accuracy in space. Coupled with the Crank–Nicolson time stepping's second order accuracy in time, we thus expect second order of convergence in this refinement study.

As stated above, we take the exact self-similar solution to (14.1) provided by Ciriello et al. (2016) (and discussed in the Appendix) evaluated at $t = t_0$ and mirrored about $x = 0$ as the IC. Upon evolving this IC numerically up to $t = t_f$, we compare the numerical profile to the same exact solution now evaluated at $t = t_f$. Hence, in accordance with the assumptions required to obtain this exact solution in (Ciriello et al., 2016), we limit this first convergence study to a uniform width HS cell, i.e., $n = 0$.

Figure 14.5 shows the propagation of constant mass viscous gravity current of three different fluids: (a) Newtonian, (b) shear thinning, and (c) shear thickening power law. The currents propagate symmetrically about the centre of the domain ($x = 0$). The sharp moving front $x_f(t)$ is accurately captured in these simulations on fairly modest (i.e., coarse) grids, without *any* signs of numerical instability or need for special treatment of the derivative discontinuity. Computing the error as a function of Δx during the grid refinement shows second order convergence. This numerical example thus indicates that the proposed approach to treating the implicit nonlinear ψ terms, specifically their evaluation at $n + 1/2$, is consistent with the desired second order accuracy.

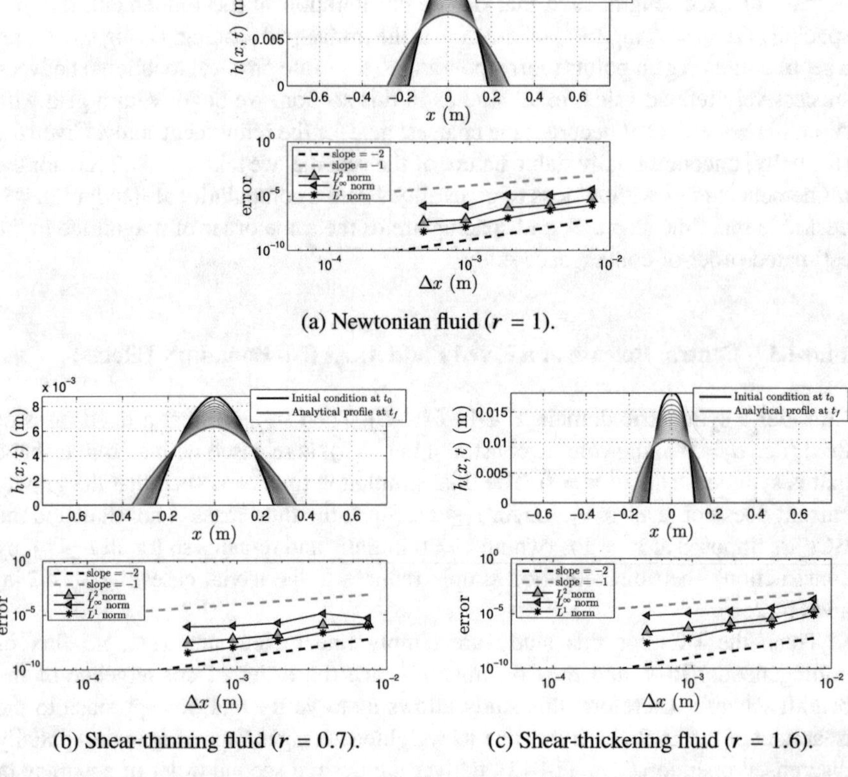

(a) Newtonian fluid ($r = 1$).

(b) Shear-thinning fluid ($r = 0.7$). (c) Shear-thickening fluid ($r = 1.6$).

Fig. 14.5 Estimated order-of-convergence of a 'centrally released' fixed fluid mass propagating in both directions in a uniform width HS cell ($n = 0$). The currents' shapes are plotted from 'early' times (purple/dark) to 'late' times (green/light). In all cases, the volume of fluid is $\mathcal{V}_0 = 2.4902 \times 10^{-5}$ m^3 and $b_1 = 0.01739$ m. The currents are released at $t_0 = 1$ s and spread until $t_f = 3.5$ s.

It should be noted that the restriction on t_f, which is necessary so that the current does not reach the domain boundaries, is critical since the chosen benchmark exact solution only describes the 'spreading' behaviour of the current and not its 'levelling' (once it reaches the no-flux boundaries at $x = \pm L$). Indeed, the levelling regime possesses its distinct self-similar behaviour (see, e.g., Diez et al., 1992; Zheng et al., 2018b), which is beyond the scope of the present work.

14.4.1.2 Propagation of a Fixed Mass of Fluid in a Single Direction

To ascertain the accuracy of our discretisation of the nonlinear BCs, we now return to a one-sided domain $x \in [\ell, L]$ with $\ell = 0$. For the case of a current spreading away from the origin, the BC at the 'left' end of the domain (from which the fluid is released) is nontrivial, and its proper discretisation is the key to the overall order of the convergence of the scheme. Conveniently, for a fixed mass ($\alpha = 0$), the BCs still reduce to homogeneous Neumann conditions (recall Sect. 14.3.4), however, h is no longer zero at the boundary (as was the case in Sect. 14.4.1.1). Thus, this benchmark is our successively 'more complicated' case.

Once again, we ensure that t_f is such that the fluid does not reach the downstream ($x = L$) domain end. Then, as in Sect. 14.4.1.1, we can once again use the exact solution of Ciriello et al. (2016) as the benchmark exact solution; again, this requires restricting to uniform width HS cells (i.e., $n = 0$).

Figure 14.6 shows clear second order estimated order-of-convergence in the L^1 norm. This result indicates the decision to implement the scheme on a staggered grid, in which case the Neumann BCs are conveniently discretised using two-point central differences at the boundary, was indeed correct.

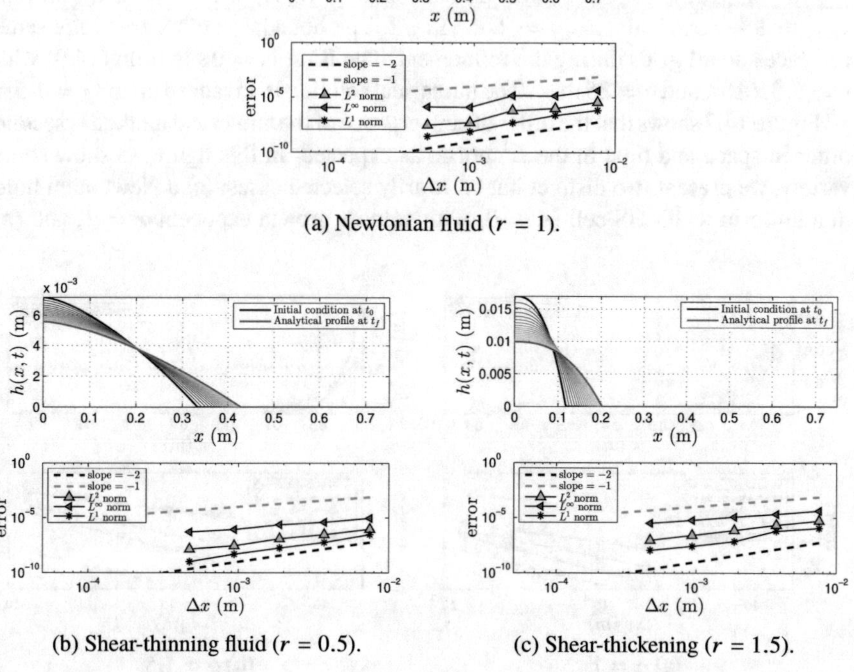

(a) Newtonian fluid (r = 1).

(b) Shear-thinning fluid (r = 0.5).

(c) Shear-thickening (r = 1.5).

Fig. 14.6 Estimated order-of-convergence study for the release of a fixed fluid mass propagating in a single direction (away from cell's origin) in a uniform width HS cell ($n = 0$). Once again, the fluid is released at $t_0 = 1$ s and spreads until $t_f = 3.5$ s. The currents' shapes are plotted from early times (purple/dark) through late times (green/light). The remaining model parameters for these simulations are the same as in Fig. 14.5.

14.4.1.3 Propagation in a Single Direction with Mass Injection

Finally, we subject the numerical scheme to its most stringent test yet. That is, we compute the estimated order of convergence under mass injection conditions ($\alpha > 0$). The injection occurs near the cell's origin and the current propagates away from this location. Since $\alpha > 0$, the fully nonlinear forms of the BCs as given in (14.5) and (14.7) now come into play.

Unlike the previously discussed cases of the release of a fixed fluid mass, a straightforward exact solution to the nonlinear ODE emerging from the self-similar analysis is not possible. For variable mass, obtaining a benchmark solution is now significantly more challenging, given that the nonlinear ODE must be solved *numerically* (see the Appendix). Despite the availability of accurate stiff ODE solvers, such as `ode15s` in MATLAB, it is quite difficult to map the numerical solution of the self-similar ODE onto the selected computational grid, and maintain the desired order of accuracy throughout this procedure. Therefore, for this benchmark, we instead elect to use a 'fine grid' numerical solution as the benchmark solution. This fine grid solution is then compared against the solutions on successively coarser grids to establish the estimated order of convergence of the numerical scheme.

For this study, the simulation domain is $x \in [\ell, L]$ with $\ell = \Delta x_0$, so that $x_{i=0} = \ell - \Delta x_0/2$ and $x_{i=N} = L + \Delta x_0/2$. The boundary points are at the same cell faces on all grids during the refinement. The IC at $t_0 = 0$ s is from (14.9) with $b = 3.5 \times 10^2$, and $c = 25\,\text{m}^{-1}$. The numerical solution is advanced up to $t_f = 1.5$ s.

Figure 14.7 shows that the order of convergence of the numerical method is second order in space and time in the L^1 norm, as expected. In this figure, to show some variety, we present two distinct but arbitrarily selected cases: (a) a Newtonian fluid in a uniform width HS cell ($n = 0$) with volume growth exponent $\alpha = 1$, and (b)

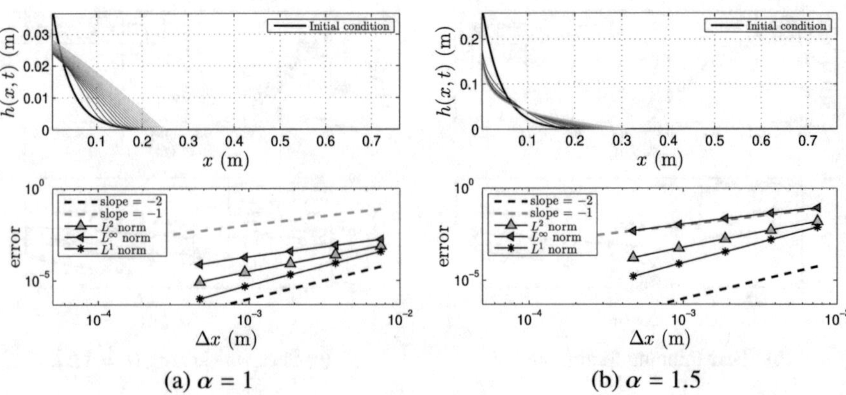

Fig. 14.7 Estimated order-of-convergence study for a variable mass gravity current with injection at $x = \ell$ on the truncated domain $x \in [\ell, L]$ with $\ell = \Delta x_0$. Simulations are shown for the case of (a) a Newtonian fluid in a uniform width HS cell ($r = 1, n = 0$), and (b) a shear thickening fluid in a variable width cell ($r = 0.6, n = 0.6$). The remaining model parameters are as in Fig. 14.5.

a non-Newtonian (shear thinning, $r = 0.6$) fluid in a variable width HS cell (width exponent $n = 0.6$) with a volume growth exponent $\alpha = 1.5$.

With this final numerical test, we have justified the formal truncation error of the proposed finite difference scheme. This result is nontrivial because the PDE and the scheme are both *nonlinear*, requiring subtle approximation on a staggered grid, across half-time steps, and linearisation of an algebraic system via internal iterations.

14.4.2 Satisfaction of the Mass Constraint at the Discrete Level

Since the BCs derived in Sect. 14.2.2 (and discretised in Sect. 14.3.4) stem from the mass conservation constraint (i.e., (14.2) or (14.3)), it is expected that the proposed finite difference scheme should produce a solution $h(x,t)$ that satisfies (14.2) or (14.3) to within $\mathcal{O}\left[(\Delta x)^2 + (\Delta t)^2\right]$ or better, if this constraint is checked independently after computing the numerical solution. To verify this capability of the scheme, in this subsection, we consider two cases: (i) a fixed fluid mass released near the origin ($\alpha = 0$), and (ii) spreading subject to mass injection ($\alpha > 0$) near the origin. Both cases are studied on the domain $x \in [\ell, L]$ with $\ell = \Delta x$. The solution is evolved on the time interval $t \in (t_0, t_f]$, and the volume error for each t is computed as

$$\left| \int_{\ell}^{L} h(x,t) b_1 x^n \, \mathrm{d}x - (\mathcal{V}_0 + \mathcal{V}_{\mathrm{in}} t^{\alpha}) \right|, \tag{14.27}$$

where the x-integration is performed by the trapezoidal rule to $\mathcal{O}\left[(\Delta x)^2\right]$ on the staggered mesh, as before. We expect that the volume error, as defined in (14.27), is $\mathcal{O}\left[(\Delta x)^2 + (\Delta t)^2\right]$, the same as the overall scheme. In this numerical study, the selection of the IC is no longer critical, as we do not compare against an exact self-similar solution. Accordingly, we select generic ICs from (14.8) and (14.9).

In the case of *fixed mass release* ($\alpha = 0$), the IC is a cubic polynomial determined from (14.8) with $c = 3$ and $\mathfrak{X}_0 = 0.25\,\mathrm{m}$. The error in the total fluid volume as a function of t is compared with the initial one. Figure 14.8 shows that, while numerical error does build up in the total volume, the initial volume remains conserved to within (or better than) $(\Delta t)^2 = 10^{-6}$.

A more stringent test of the conservation properties of the proposed scheme is conducted by applying the nonlinear BC associated with *imposed mass injection* at one end. For this case, the IC is taken to be the function in (14.9) with $b = 3.5 \times 10^2$ and $c = 25\,\mathrm{m}^{-1}$ and $\mathfrak{X}_0 = (1/c)\ln(1/b)$. A combination of n, r and α values have been considered to highlight the conservation properties across different physical regimes. Figure 14.9 shows that, in all cases, the volume constraint is properly respected; while the volume error builds up, it remains small (within or better than $(\Delta t)^2 = 10^{-6}$).

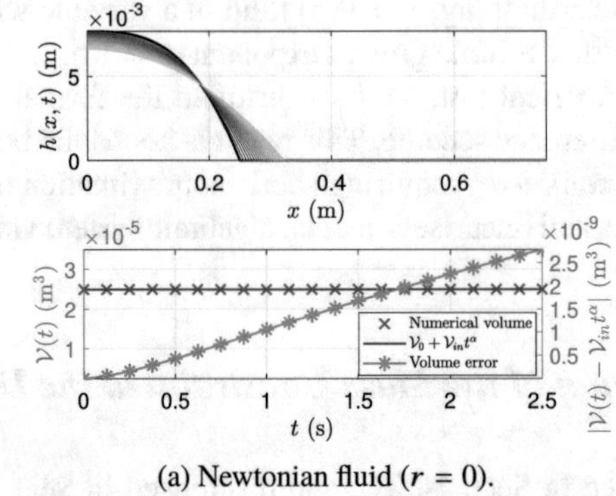

(a) Newtonian fluid (r = 0).

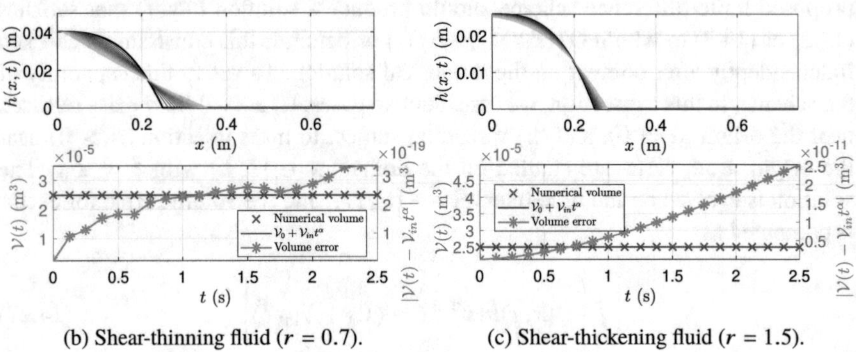

(b) Shear-thinning fluid (r = 0.7). (c) Shear-thickening fluid (r = 1.5).

Fig. 14.8 Results of the conservation study for the release of a fixed fluid mass. To highlight the scheme's capabilities, each case features a different HS cell: (a) $n = 0$, (b) $n = 0.7$, and (c) $n = 0.5$. The currents are allowed to propagate from $t_0 = 0$ s up to $t_f = 2.5$ s, through 2500 time steps ($\Rightarrow \Delta t = 10^{-3}$ s). In all cases, $\alpha = 0$ and $\mathcal{V}(t) = \mathcal{V}_0 = 2.4902 \times 10^{-5}$ m^3. The remaining model parameters for these simulations are the same as in Fig. 14.5.

14.5 Conclusions and Outlook

In this chapter, we developed and benchmarked a finite difference numerical scheme for solving a family of nonlinear parabolic PDEs with variable coefficients given by (14.1). A special feature of these nonlinear PDEs is that they possess solutions that can propagate in a wave like manner with a finite speed of propagation. Our study featured examples from this family of PDEs for modeling the 1D spreading (propagation) of a power law (Oswald–de Weale) fluid in a horizontal, narrow fracture with variable width.

We placed an emphasis on designing a series of numerical tests that show conclusively that the proposed scheme is second order accurate in space and time. Analytical self-similar solutions for special cases of the nonlinear parabolic PDE were used to benchmark the numerical method. Furthermore, we verified that a

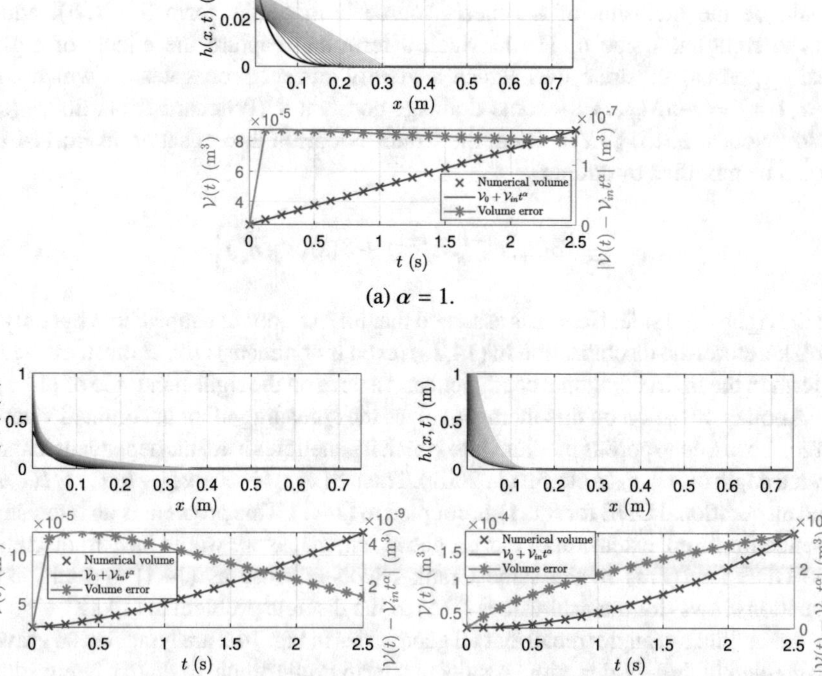

Fig. 14.9 Results of the conservation study for mass injection. Three choices of the volume exponent α are considered. We simulate (a) a Newtonian fluid in uniform HS cell ($r = 1, n = 0$), (b) a shear thinning fluid in a variable width HS cell ($r = 0.7, n = 0.7$), and (c) a shear thickening fluid in a variable width HS cell ($r = 1.5, n = 0.5$). In all cases, t_0, t_f, and Δt are as in Fig. 14.8. Additionally we set $\mathcal{V}_0 = \mathcal{V}_{\text{in}} = 2.4902 \times 10^{-5}\ \text{m}^3$ (see (14.3)), The remaining model parameters for these simulations are the same as in Fig. 14.5.

global mass conservation/injection constraint can be successfully reformulated into a set of nonlinear boundary conditions, which were successfully discretised with second order accuracy as well.

The main advantage of the proposed finite difference scheme is that it is strongly implicit, generalising the time stepping suggested by Crank and Nicolson (1947). Therefore, the proposed scheme does not formally require a time step restriction for stability. By using a staggered grid, along the lines of Christov and Homsy (2009), nonlinear terms were handled within the same three-point stencil as the classical Crank–Nicolson scheme. This choice of grid is particularly convenient for the discretisation of the nonlinear boundary conditions, allowing second order accuracy to be achieved with just a two-point stencil near the domain boundaries. Using fractional steps in time ('internal iterations'), we reformulated the nonlinear algebraic problem at each time step as a fixed point iteration.

In future work, an interesting extension to our proposed numerical scheme could be the inclusion of a generic source term of the form $\mathcal{S}(x, t, h)$, added to the right-hand side of (14.1). Such a term can capture the effects of e.g., a leaky (porous) substrate over which a gravity current propagates, in which case $\mathcal{S}(x, t, h) = -\varkappa h(x, t)$ for some drainage constant $\varkappa$ (Pritchard et al., 2001) (see also (Woods, 2015, §9.2)). Then, the Crank–Nicolson discretisation in Eq. (14.11) could be modified by adding

$$\frac{1}{2}\Big[\mathcal{S}(x_i, t^{n+1}, h_i^{n+1}) + \mathcal{S}(x_i, t^n, h_i^n)\Big] \tag{14.28}$$

to the right-hand side. Here, it is assumed that $\partial h/\partial x$ does not appear in $\mathcal{S}$ but only in $\mathcal{L}$. Therefore, the discretisation in (14.28) (even if nonlinear) will, at most, introduce a term in the matrix diagonal coefficient and a term on the right-hand side of (14.18).

Another variation on this theme involves the spreading of an unconfined viscous fluid above a deep porous medium into which it penetrates in a time dependent manner over a depth of $l(x, t)$ (Acton et al., 2001). Then, $\mathcal{S}(x, t, h) = -\kappa[1+h(x, t)/l(x, t)]$ and an additional ODE for $l(x, t)$ is coupled to (14.1). This problem is an interesting avenue for future extension of our proposed scheme, as we would have to discretise the ODE for $l(x, t)$ in the same Crank–Nicolson sense as (14.1) and add extra equations/rows (for the grid values l_i^{n+1}) to the discrete problem in (14.18).

An inclination angle (recall that all geometries in Fig. 14.2 are lying flat, so gravity is directed in the $-y$-direction) results in a term proportional to $\partial h/\partial x$ being added to (14.1) (for the case of a Newtonian fluid, see, e.g., (Huppert and Woods, 1995; Vella and Huppert, 2006)). This additional term changes the nonlinear diffusion equation (14.1) into a nonlinear *advection diffusion* equation. Care must be taken in discretising this new advective term. A similar PDE arises in the segregation of bidisperse granular mixtures (Dolgunin and Ukolov, 1995; Gray et al., 2015). As discussed by Christov (2018), a strongly implicit Crank–Nicolson scheme can be successfully used for these problems. The scheme of Christov (2018) is so robust that it performs well even in the singular vanishing diffusivity limit of the advection diffusion equation. Considering a generic advection term $\partial\Psi(h)/\partial x$, it can be handled analogously to the nonlinearity in the diffusion term. Specifically, we approximate

$$\left(\frac{\partial\Psi}{\partial x}\right)_{x=x_i} \approx \frac{1}{2}\left[\left(\frac{\Psi_{i+1}^{n+1} - \Psi_{i-1}^{n+1}}{2\Delta x}\right) + \left(\frac{\Psi_{i+1}^{n} - \Psi_{i-1}^{n}}{2\Delta x}\right)\right]. \tag{14.29}$$

Here, the advective term is discretised through a central difference formula involving a local three-point stencil on all interior nodes ($i = 1$ to $N-1$). At the boundary nodes ($i = 0$ and $i = N$), one can use a three-point biased (forward or backward) difference formula, as in (14.17). The now well-established idea of staggering the nonlinear term across fractional time steps is carried forward (recall (14.15)). However, to properly linearise the advective term within the internal iterations, we must be able to write $\Psi(h) = \Upsilon(h)h$ (the most obvious way being to let $\Upsilon(h) \equiv \Psi(h)/h$) so that

$$\Psi_{i\pm1}^{n+1} \approx \Upsilon_{i\pm1}^{n+1/2,k} h_{i\pm1}^{n+1,k}. \tag{14.30}$$

Then, inserting (14.30) into (14.29) and adding the result to the left-hand side of (14.18), modifies the tridiagonal system by adding $\Upsilon_{i\pm1}^{n+1/2,k}/(4\Delta x)$ to the superdiagonal $(i+1)$ and subdiagonal $(i-1)$, respectively. The remaining terms from (14.29) are added to the right-hand side of the system.

Any of these potential extensions would have to be benchmarked against available first kind self-similar solutions in (Huppert and Woods, 1995; Acton et al., 2001; Vella and Huppert, 2006; Woods, 2015), however, no particular difficulties are expected to arise.

Another avenue of future work is as follows. Nowadays, higher order (i.e., greater than second order) nonlinear parabolic PDEs are found to describe a wealth of low Reynolds number fluid phenomena. These phenomena range from the spreading and healing (Zheng et al., 2018b,a) to the rupture dynamics (Garg et al., 2017) of thin liquid films dominated by capillary forces (Leal, 2007, Ch. 6-C). Typically, the spatial operator is of fourth order due to the inclusion of surface tension effects (which depend upon the curvature of h), making the PDE more challenging to solve numerically. (Note that this is distinct from the inclusion of capillary effects in the context of gravity currents propagating in porous media, following Golding et al. (2011).)

Even higher (sixth) order thin film equations arise in dynamics of lubricated thin elastic membranes (Hosoi and Mahadevan, 2004; Flitton and King, 2004; Hewitt et al., 2015) dominated by elastic forces. To interrogate these complex interfacial phenomena, there is a current need for a robust and accurate numerical scheme to simulate these flows with low computational overhead (e.g., without the prohibitive time step stability restrictions of explicit schemes). In future work, it would be of interest to generalise the scheme from this chapter to such problems. Additionally, nonuniform (or adaptive) grids, which could be implemented along the lines of Christov and Deng (2002), can be used to capture singularity formation during thin film rupture.

Acknowledgements We dedicate this work to the 80th anniversary of Prof. Jüri Engelbrecht and his contributions to the fields of complexity science, nonlinear wave phenomena and applied mathematical modeling. ICC acknowledges a very productive trip to Tallinn in September 2014 (on invitation of Prof. Andrus Salupere) and fondly recalls meeting and interacting with Prof. Engelbrecht there, at the IUTAM Symposium on Complexity of Nonlinear Waves.

We also thank Profs. Tarmo Soomere and Arkadi Berezovski for their efforts in editing this volume and their kind invitation to contribute a chapter. Finally, ICC acknowledges many helpful conversations on gravity currents with Prof. H.A. Stone, Dr. Z. Zheng and Prof. S. Longo. Specifically, we thank S. Longo for suggesting the form of the governing equation for a power law fluid in a variable width HS cell (third row in Table 14.1).

Appendix

For many natural phenomena, a relatively simple procedure involving a scaling (dimensional analysis) of the governing equations can be used to yield the similarity variable and appropriate rescaling necessary to obtain a first kind self-similar solution (Barenblatt, 1996). Viscous gravity currents exhibit such self-similar propagation, meaning that the solution (at sufficiently 'long' times (Barenblatt, 1996)) depends solely upon a combined variable of x and t, rather than on each independently. Self-similarity allows for the derivation of exact analytical solutions to the governing equation (14.1) against which numerical solutions can be benchmarked. Specifically, for the case of the release of a fixed mass of fluid ($\alpha = 0$ so that $\mathcal{V}(t) = \mathcal{V}_{\text{in}}$ $\forall t \in [t_0, t_f]$), a closed form analytical self-similar solution was used in Sect. 14.4.1 to test the order of convergence of the numerical scheme.

In this Appendix, following Di Federico et al. (2017), we summarise the derivation of said self-similar solution for a power law non-Newtonian fluid spreading *away* from the origin ($x = 0$) of a HS cell of uniform width b_1 ($n = 0$).[4] First, we introduce the following dimensionless variables (with * superscripts) from (Di Federico et al., 2017):

$$x^* = \left(\frac{B}{A^\alpha}\right)^{1/(\alpha-2)} x, \quad t^* = \left(\frac{B}{A^2}\right)^{1/(\alpha-2)} t,$$

$$h^*(x^*, t^*) = \left(\frac{B}{A^\alpha}\right)^{1/(\alpha-2)} h(x, t), \tag{14.31}$$

where A is the constant from (14.1) (defined in Table 14.1) and $B = \mathcal{V}_{\text{in}}/b_1$. Hereafter, we drop the * superscripts. Next, we must select a suitable similarity variable η. As discussed in Sect. 14.1, a scaling analysis of the dimensionless version of (14.1), suggests that the self-similar solution of the first kind has the form

$$h(x, t) = \eta_N^{r+1} t^{F_2} f(\zeta), \quad \zeta = \frac{\eta}{\eta_N}, \quad \eta = \frac{x}{t^{F_1}}. \tag{14.32}$$

It can be shown that the constant η_N specifically corresponds to the value of η at the nose of the current, i.e., $\eta_N = x_f(t)/t^{F_1}$, where $x = x_f(t)$ is such that $h\big(x_f(t), t\big) = 0$. Here, ζ is a convenient rescaled similarity variable, and the exponents $F_{1,2}$ are

$$F_1 = \frac{\alpha + r}{r + 2}, \tag{14.33a}$$

$$F_2 = \alpha - F_1. \tag{14.33b}$$

[4]While self-similar solutions, of course, exist for $n > 0$, they cannot be found in closed form as analytical solutions.

The shape function $f(\zeta)$ represents the (universal) self-similar profile of the gravity current. We must now determine this function, by substituting (14.32) into the dimensionless version of (14.1) to reduce the latter to a nonlinear ODE:

$$\frac{\mathrm{d}}{\mathrm{d}\zeta}\left(f\left|\frac{\mathrm{d}f}{\mathrm{d}\zeta}\right|^{\frac{1}{r}}\right) + F_2 f - F_1\zeta\frac{\mathrm{d}f}{\mathrm{d}\zeta} = 0, \quad \zeta \in [0, 1]. \tag{14.34}$$

The second order ODE in (14.34) can be rewritten as a first order system:

$$\frac{\mathrm{d}f_1}{\mathrm{d}\zeta} = f_2, \quad \frac{\mathrm{d}f_2}{\mathrm{d}\zeta} = \frac{-r}{f_1 f_2|f_2|^{(1-2r)/r}}\Big[f_2|f_2|^{1/r} + F_2 f_1 - F_1\zeta f_2\Big], \tag{14.35}$$

where, for convenience, we have set $f_1 = f$. The system in (14.35) is 'stiff,' and we must use an appropriate ODE solver, such as `ode15s` in MATLAB, subject to appropriate initial and/or boundary conditions at $\zeta = 0, 1$.

A peculiarity of this self-similar analysis is that we have only a single BC for the ODE (14.34), namely $f(1) = 0$, i.e., this is the location of the gravity current's nose $x = x_f(t)$ at which $\zeta = \eta/\eta_N = 1$ and $h\big(x_f(t), t\big) = 0$. Since the ODE in (14.35) requires a second initial or boundary condition, we use the 'backwards-shooting' idea of Huppert (1982) to provide a second condition near $\zeta = 1$. Then, the ODE in (14.35) can be integrated 'backwards' from $\zeta = 1$ to $\zeta = 0$ subject to two 'initial' conditions at $\zeta = 1$.

To this end, consider the asymptotic behaviour of the current near the nose. By assuming that $f \sim \mathfrak{c}_1(1-\zeta)^{\mathfrak{c}_2}$ as $\zeta \to 1^-$ and substituting this expression into (14.34), we obtain $\mathfrak{c}_1 = F_2^r$ and $\mathfrak{c}_2 = 1$ by balancing the lowest order terms. Now, we have two BCs (see also (Di Federico et al., 2017)):

$$f_1(1-\varepsilon) = F_2^r\varepsilon, \tag{14.36a}$$

$$f_2(1-\varepsilon) = -F_2^r, \tag{14.36b}$$

for a sufficiently small $\varepsilon \ll 1$. We can now solve the system (14.35) subject to the 'final' conditions (14.36) on the interval $\zeta \in [0, 1-\varepsilon]$. By convention, an ODE is solved with initial, not final, conditions. Therefore, we perform the transformation $\zeta \mapsto 1-\hat{\zeta}$, which leads to the right-hand sides of the two equations in (14.35) being multiplied by -1. Then, the final conditions in (14.36) become initial conditions at $\hat{\zeta} = \varepsilon$, and the first order system of ODEs is solved on the interval $\hat{\zeta} \in [\varepsilon, 1]$.

For certain special cases, a closed form analytical solution to (14.34) can be obtained. For the case of the release of a fixed mass of fluid ($\alpha = 0$), Ciriello et al. (2016) derived such an exact solution (as can be verified by substitution):

$$f(\zeta) = \frac{r^r}{(r+2)^r(r+1)}\left(1-\zeta^{r+1}\right), \tag{14.37}$$

which we used to benchmark our finite difference scheme in Sect. 14.4.1. Finally, to obtain the viscous gravity current profile given in (14.32) we must compute η_N. This value follows from imposing the mass conservation constraint in dimensionless form:

$$\eta_N = \left[\int_0^1 f(\zeta)\,\mathrm{d}\zeta\right]^{-1/(r+2)} \approx \left[\int_\varepsilon^1 f(\hat{\zeta})\,\mathrm{d}\hat{\zeta}\right]^{-1/(r+2)}, \tag{14.38}$$

where the second (approximate) equality is needed for the case in which (14.34) has to be integrated numerically (no exact solution); $\varepsilon \ll 1$ is chosen sufficiently small, as above.

Finally we can substitute (14.37) and (14.38) into (14.32) to obtain the analytical solution for the profile of the gravity current, as a function of x at some time t. It should be noted, however, that for this solution to apply, the current must have achieved its self-similar asymptotics, having forgotten the initial condition from which it evolved.

References

Acton, J.M., Huppert, H.E., Worster, M.G.: Two-dimensional viscous gravity currents flowing over a deep porous medium. J. Fluid Mech. **440**, 359–380 (2001). https://doi.org/10.1017/S0022112001004700

Alhashim, M.G., Koch, D.L.: The effects of fluid transport on the creation of a dense cluster of activated fractures in a porous medium. J. Fluid Mech. **847**, 286–328 (2018). https://doi.org/10.1017/jfm.2018.313

Anderson, D.M., McLaughlin, R.M., Miller, C.T.: The averaging of gravity currents in porous media. Phys. Fluids **15**(10), 2810–2829 (2003). https://doi.org/10.1063/1.1600733

Anderson, D.M., McLaughlin, R.M., Miller, C.T.: A sharp-interface interpretation of a continuous density model for homogenisation of gravity-driven flow in porous media. Physica D **239**(19), 1855–1866 (2010). https://doi.org/10.1016/j.physd.2010.06.009

Aronsson, G., Janfalk, U.: On Hele-Shaw flow of power-law fluids. Eur. J. Appl. Math. **3**(4), 343–366 (1992). https://doi.org/10.1017/s0956792500000905

Barenblatt, G.I.: On some unsteady fluid and gas motions in a porous medium. Prikl. Mat. Mekh. (PMM) **16**, 67–78 (1952) (in Russian)

Barenblatt, G.I.: Similarity, Self-Similarity, and Intermediate Asymptotics. Cambridge Texts in Applied Mathematics, vol. 14. Cambridge University Press (1996)

Barenblatt, G.I., Zel'dovich, Y.B.: Self-similar solutions as intermediate asymptotics. Annu. Rev. Fluid Mech. **4**, 285–312 (1972). https://doi.org/10.1146/annurev.fl.04.010172.001441

Bear, J.: Dynamics of Fluids in Porous Media. Dover Publications, Mineola, NY (1988)

Berezovski, A., Ván, P.: Internal Variables in Thermoelasticity. Springer, Cham (2017). https://doi.org/10.1007/978-3-319-56934-5

Bird, R.B., Armstrong, R.C., Hassager, O.: Dynamics of Polymeric Liquids, vol. 1, 2nd edn. John Wiley, New York (1987)

Bonnecaze, R.T., Huppert, H.E., Lister, J.R.: Particle-driven gravity currents. J. Fluid Mech. **250**, 339–369 (1993). https://doi.org/10.1017/s002211209300148x

Boyko, E., Bercovici, M., Gat, A.D.: Viscous-elastic dynamics of power-law fluids within an elastic cylinder. Phys. Rev. Fluids **2**(7), 073301 (2017). https://doi.org/10.1103/physrevfluids.2.073301

Cheng, N.S.: (2008) Formula for the viscosity of a glycerol-water mixture. Ind. Eng. Chem. Res. **47**(9), 3285–3288 (2008). https://doi.org/10.1021/ie071349z

Christov, I.C.: Wave solutions. In: Hetnarski, R.B. (ed.) Encyclopedia of Thermal Stresses, pp. 6495–6506. Springer, Dordrecht (2014). https://doi.org/10.1007/978-94-007-2739-7_33

Christov, C.I., Deng, K.: Numerical investigation of quenching for a nonlinear diffusion equation with a singular Neumann boundary condition. Numer. Methods Partial Differential Eq. **18**(4), 429–440 (2002). https://doi.org/10.1002/num.10013

Christov, C.I., Homsy, G.M.: Enhancement of transport from drops by steady and modulated electric fields. Phys. Fluids **21**(8), 083102 (2009). https://doi.org/10.1063/1.3179555

Christov, C.I., Maugin, G.A., Porubov, A.V. On Boussinesq's paradigm in nonlinear wave propagation. C. R. Mecanique **335**(9-10), 521–535 (2007). https://doi.org/10.1016/j.crme.2007.08.006

Christov, I.C.: On the numerical solution of a variable-coefficient Burgers equation arising in granular segregation. Mat. Phys. Mech. **35**(1), 21–27 (2018). https://doi.org/10.18720/MPM.3512018_4

Ciriello, V., Di Federico, V., Archetti, R., Longo, S.: Effect of variable permeability on the propagation of thin gravity currents in porous media. Int. J. Non-Linear Mech. **57**, 168–175 (2013). https://doi.org/10.1016/j.ijnonlinmec.2013.07.003

Ciriello, V., Longo, S., Chiapponi, L., Di Federico, V.: Porous gravity currents: A survey to determine the joint influence of fluid rheology and variations of medium properties. Adv. Water Res. **92**, 105–115 (2016). https://doi.org/10.1016/j.advwatres.2016.03.021

Crank, J., Nicolson, P.: A practical method for numerical evaluation of solutions of partial differential equations of the heat-conduction type. Math. Proc. Camb. Phil. Soc. **43**(1), 50–67 (1947). https://doi.org/10.1017/s0305004100023197

De Loubens, R., Ramakrishnan, T.S.: Analysis and computation of gravity-induced migration in porous media. J. Fluid Mech. **675**, 60–86 (2011). https://doi.org/10.1017/s0022112010006440

Di Federico, V., Malavasi, S., Cintoli, S.: Viscous spreading of non-Newtonian gravity currents on a plane. Meccanica **41**(2), 207–217 (2006). https://doi.org/10.1007/s11012-005-3354-9

Di Federico, V., Archetti, R., Longo, S.: Similarity solutions for spreading of a two-dimensional non-Newtonian gravity current in a porous layer. J. Non-Newtonian Fluid Mech. **177–178**, 46–53 (2012). https://doi.org/10.1016/j.jnnfm.2012.04.003

Di Federico, V., Longo, S., King, S.E., Chiapponi, L., Petrolo, D., Ciriello, V.: Gravity-driven flow of Herschel–Bulkley fluid in a fracture and in a 2D porous medium. J. Fluid Mech. **821**, 59–84 (2017). https://doi.org/10.1017/jfm.2017.234

Didden, N., Maxworthy, T.: Viscous spreading of plane and axisymmetric gravity waves. J. Fluid Mech. **121**, 27–42 (1982). https://doi.org/10.1017/s0022112082001785

Diez, J.A., Gratton, R., Gratton, J.: Self-similar solution of the second kind for a convergent viscous gravity current. Phys. Fluids A **6**(6), 1148–1155 (1992). https://doi.org/10.1063/1.858233

Dolgunin, V.N., Ukolov, A.A.: Segregation modeling of particle rapid gravity flow. Powder Technol. **83**(2), 95–103 (1995). https://doi.org/10.1016/0032-5910(94)02954-m

Douglas, Jr., J., Peaceman, D.W., Rachford Jr, H.H.: A method for calculating multi-dimensional immiscible displacement. Petrol Trans. AIME **216**, 297–308 (1959)

Douglas, Jr., J., Peaceman, D.W., Rachford Jr, H.H.: Numerical calculation of multidimensional miscible displacement. Soc. Petrol Eng. J. **2**(4), 327–339 (1962). https://doi.org/10.2118/471-pa

Engelbrecht, J.: Nonlinear Wave Dynamics: Complexity and Simplicity. Kluwer Texts in the Mathematical Sciences, vol. 17. Springer, Dordrecht (1997). https://doi.org/10.1007/978-94-015-8891-1

Engelbrecht, J.: Questions About Elastic Waves. Springer, Cham (2015). https://doi.org/10.1007/978-3-319-14791-8

Engelbrecht, J., Peets, T., Tamm, K., Laasmaa, M., Vendelin, M.: On the complexity of signal propagation in nerve fibres. Proc. Estonian Acad. Sci. **67**(1), 28–38 (2018a). https://doi.org/10.3176/proc.2017.4.28

Engelbrecht, J., Salupere, A., Berezovski, A., Peets, T., Tamm, K.: On nonlinear waves in media with complex properties. In: Altenbach, H., Pouget, J., Rousseau, M., Collet, B., Michelitsch, T. (eds.) Generalized Models and Non-classical Approaches in Complex Materials 1, Advanced Structured Materials, vol. 89, pp. 248–270. Springer, Cham (2018b). https://doi.org/10.1007/978-3-319-72440-9_13

Evans, L.C.: Partial Differential Equations, Graduate Studies in Mathematics, vol. 19, 2nd edn. American Mathematical Society, Providence, RI, USA (2010)

Felisa, G., Lenci, A., Lauriola, I., Longo, S., Di Federico, V.: Flow of truncated power-law fluid in fracture channels of variable aperture. Adv. Water Res. **122**, 317–327 (2018). https://doi.org/10.1016/j.advwatres.2018.10.024

Fisher, R.A.: The wave of advance of advantageous genes. Ann. Eugenics **7**(4), 353–369 (1937). https://doi.org/10.1111/j.1469-1809.1937.tb02153.x

Flitton, J.C., King, J.R.: Moving-boundary and fixed-domain problems for a sixth-order thin-film equation. Eur. J. Appl. Math. **15**(6), 713–754 (2004). https://doi.org/10.1017/s0956792504005753

Garg, V., Kamat, P.M., Anthony, C.R., Thete, S.S., Basaran, O.A.: Self-similar rupture of thin films of power-law fluids on a substrate. J. Fluid Mech. **826**, 455–483 (2017). https://doi.org/10.1017/jfm.2017.446

Gilding, B.H., Kersner, R.: Travelling Waves in Nonlinear Diffusion-Convection Reaction. Progress in Nonlinear Differential Equations and Their Applications, vol. 60. Birkhäuser Verlag, Basel, Switzerland (2004). https://doi.org/10.1007/978-3-0348-7964-4

Golding, M.J., Neufeld, J.A., Hesse, M.A., Huppert, H.E.: Two-phase gravity currents in porous media. J. Fluid Mech. **678**, 248–270 (2011). https://doi.org/10.1017/jfm.2011.110

Gratton, J. Similarity and self similarity in fluid dynamics. Fund. Cosmic Phys. **15**, 1–106 (1991)

Gratton, J., Minotti, F.: Self-similar viscous gravity currents: phase plane formalism. J. Fluid Mech. **210**, 155–182 (1990). https://doi.org/10.1017/s0022112090001240

Gratton, J., Mahajan, S.M., Minotti, F.: Theory of creeping gravity currents of a non-Newtonian liquid. Phys. Rev. E **60**(6), 6090–6097 (1999). https://doi.org/10.1103/physreve.60.6960

Gray, J.M.N.T., Gajjar, P., Kokelaar, P.: Particle-size segregation in dense granular avalanches. C. R. Phys. **16**(1), 73–85 (2015). https://doi.org/10.1016/j.crhy.2015.01.004

Hesse, M.A., Tchelepi, H.A., Cantwell, B.J., Orr Jr, F.M.: Gravity currents in horizontal porous layers: transition from early to late self-similarity. J. Fluid Mech. **577**, 363–383 (2007). https://doi.org/10.1017/s0022112007004685

Hewitt, I.J., Balmforth, N.J., De Bruyn, J.R.: Elastic-plated gravity currents. Eur. J. Appl. Math. **26**(1), 1–31 (2015). https://doi.org/10.1017/s0956792514000291

Hosoi, A.E., Mahadevan, L.: Peeling, healing, and bursting in a lubricated elastic sheet. Phys. Rev. Lett. **93**(13), 137802 (2004). https://doi.org/10.1103/physrevlett.93.137802

Huppert, H.E.: The propagation of two-dimensional and axisymmetric viscous gravity currents over a rigid horizontal surface. J. Fluid Mech. **121**, 43–58 (1982). https://doi.org/10.1017/s0022112082001797

Huppert, H.E.: Geological fluid mechanics. In: Batchelor, G.K., Moffatt, H.K., Worster, M.G. (eds.) Perspectives in Fluid Dynamics, pp. 447–506. Cambridge University Press (2000)

Huppert, H.E., Neufeld, J.A.: The fluid mechanics of carbon dioxide sequestration. Annu. Rev. Fluid Mech. **46**, 255–272 (2014). https://doi.org/10.1146/annurev-fluid-011212-140627

Huppert, H.E., Woods, A.W.: Gravity driven flows in porous layers. J. Fluid Mech. **292**, 55–69 (1995). https://doi.org/10.1017/s0022112095001431

Huppert, H.E., Neufeld, J.A., Strandkvist, C.: The competition between gravity and flow focusing in two-layered porous media. J. Fluid Mech. **720**, 5–14 (2013). https://doi.org/10.1017/jfm.2012.623

Jensen, F.B., Kuperman, W.A., Porter, M.B., Schimdt, H.: Computational Ocean Acoustics, 2nd edn. Springer, New York (2011). https://doi.org/10.1007/978-1-4419-8678-8

Jerzak, W., Collins, M.D., Evans, R.B., Lingevitch, J.F., Siegmann, W.L.: Parabolic equation techniques for seismic waves. Pure Appl. Geophys. **159**(7-8), 1681–1689 (2002). https://doi.org/10.1007/s00024-002-8702-2

Kolmogorov, A., Petrovskii, I., Piskunov, N.: A study of the diffusion equation with increase in the amount of substance, and its application to a biological problem. In: Tikhomirov, V.M. (ed.) Selected Works of A.N. Kolmogorov, Mathematics and its Applications (Soviet Series), vol. 25, pp. 248–270. Springer, Dordrecht, (1991) translation of 1937 Russian original

Kondic, L., Palffy-Muhoray, P., Shelley, M.J.: Models of non-Newtonian Hele-Shaw flow. Phys. Rev. E **54**, 4536–4539 (1996). https://doi.org/10.1103/physreve.54.r4536

Kondic, L., Shelley, M.J., Palffy-Muhoray, P.: Non-Newtonian Hele-Shaw flow and the Saffman–Taylor instability. Phys. Rev. Lett. **80**(7), 1433–1436 (1998). https://doi.org/10.1103/physrevlett.80.1433

Lauriola, I., Felisa, G., Petrolo, D., Di Federico, V., Longo, S.: Porous gravity currents: Axisymmetric propagation in horizontally graded medium and a review of similarity solutions. Adv. Water Res. **115**, 136–150 (2018). https://doi.org/10.1016/j.advwatres.2018.03.008

Leal, L.G.: Advanced Transport Phenomena: Fluid Mechanics and Convective Transport Processes. Cambridge University Press, New York (2007). https://doi.org/10.1017/CBO9780511800245.001

Longo, S.: Second-kind self-similar solutions for power-law and Herschel–Bulkley gravity currents, unpublished (2017)

Lyle, S., Huppert, H.E., Hallworth, M., Bickle, M., Chadwick, A.: Axisymmetric gravity currents in a porous medium. J. Fluid Mech. **543**, 293–302 (2005). https://doi.org/10.1017/s0022112005006713

Oron, A., Davis, S.H., Bankoff, S.G.: Long-scale evolution of thin liquid films. Rev. Mod. Phys. **69**(3), 931–980 (1997). https://doi.org/10.1103/revmodphys.69.931

Ostriker, J.P., Barenblatt, G.I., Sunyaev, R.A. (eds.): Selected Works of Yakov Borisovich Zeldovich, vol 1. Princeton University Press, Princeton, NJ (1992)

Perazzo, C.A., Gratton, J.: Thin film of non-Newtonian fluid on an incline. Phys. Rev. E **67**(1), 016307 (2003). https://doi.org/10.1103/physreve.67.016307

Perazzo, C.A., Gratton, J.: Exact solutions for two-dimensional steady flows of a power-law liquid on an incline. Phys. Fluids A **17**(1), 013102 (2005). https://doi.org/10.1063/1.1829625

Pritchard, D., Woods, A.W., Hogg, A.J.: On the slow draining of a gravity current moving through a layered permeable medium. J. Fluid Mech. **444**, 23–47 (2001). https://doi.org/10.1017/s002211200100516x

Straughan, B.: Heat Waves. Applied Mathematical Sciences, vol. 117. Springer, New York, NY (2011)

Strikwerda, J.: Finite Difference Schemes and Partial Differential Equations, 2nd edn. SIAM, Philadelphia (2004). https://doi.org/10.1137/1.9780898717938

Tedeev, A., Vespri, V.: Optimal behavior of the support of the solutions to a class of degenerate parabolic systems. Interfaces and Free Boundaries **17**(2), 143–156 (2015). https://doi.org/10.4171/ifb/337

Vázquez, J.L.: The Porous Medium Equation: Mathematical Theory. Oxford University Press, Oxford, UK (2007)

Vella, D., Huppert, H.E.: Gravity currents in a porous medium at an inclined plane. J. Fluid Mech. **555**, 353–362 (2006). https://doi.org/10.1017/s0022112006009578

Vlasov, S.N., Talanov, V.I.: The parabolic equation in the theory of wave propagation. Radiophys. Quantum Electron. **38**(1-2), 1–12 (1995). https://doi.org/10.1007/bf01051853

Volk, A., Kähler, C.: Density model for aqueous glycerol solutions. Exp. Fluids **59**(5), 75 (2018). https://doi.org/10.1007/s00348-018-2527-y

Von Rosenberg, D.U.: Methods for the Numerical Solution of Partial Differential Equations. Modern Analytic and Computational Methods in Science and Mathematics, vol. 16, 3rd edn. Elsevier, New York (1975)

Woods, A.W.: Flow in Porous Rocks: Energy and Environmental Applications. Cambridge University Press, Cambridge, UK (2015)

Yanenko, N.N.: The Method of Fractional Steps. Springer-Verlag, Berlin/Heidelberg (1971). English translation edited by M. Hoult

Zel'dovich, Y.B., Raizer, Y.P.: Physics of Shock Waves and High-Temperature Hydrodynamic Phenomena. Dover Publications, Mineola, NY (2002)

Zheng, Z., Christov, I.C., Stone, H.A.: Influence of heterogeneity on second-kind self-similar solutions for viscous gravity currents. J. Fluid Mech. **747**, 218–246 (2014). https://doi.org/10.1017/jfm.2014.148

Zheng, Z., Fontelos, M., Shin, S., Michael, D., Tseluiko, D., Kalliadasis, S., Stone, H.A.: Healing capillary films. J. Fluid Mech. **838**, 404–434 (2018a). https://doi.org/10.1017/jfm.2017.777

Zheng, Z., Fontelos, M., Shin, S., Stone, H.A.: Universality in the nonlinear leveling of capillary films. Phys. Rev. Fluids **3**(3) 032001 (2018b). https://doi.org/10.1103/physrevfluids.3.032001

Chapter 15
A Parabolic Approach to the Control of Opinion Spreading

Domènec Ruiz-Balet and Enrique Zuazua

This paper is dedicated to Jüri Engelbrecht with gratitude and admiration

Abstract We analyse the problem of controlling to consensus a nonlinear system modelling opinion spreading. We derive explicit exponential estimates on the cost of approximately controlling these systems to consensus, as a function of the number of agents N and the control time horizon T. Our strategy makes use of known results on the controllability of spatially discretised semilinear parabolic equations. Both systems can be linked through time rescaling.

15.1 Introduction

Agent based models (Macy and Miller, 2002; Motsch and Tadmor, 2014) are a common tool to study dynamics in social and biological sciences. Due to the complexity of the phenomena under consideration in these fields, one focuses on the modelling simple but relevant interaction rules among the agents entering in the system.

D. Ruiz-Balet (✉)
Departamento de Matemáticas, Universidad Autónoma de Madrid, Madrid, Spain

Fundación Deusto, Bilbao, Basque Country, Spain
e-mail: domenec.ruizi@uam.es

E. Zuazua
Department of Mathematics, Friedrich-Alexander-Universität Erlangen-Nürnberg, Erlangen, Germany

Fundación Deusto, Bilbao, Basque Country, Spain

Departamento de Matemáticas, Universidad Autánoma de Madrid, Madrid, Spain
e-mail: enrique.zuazua@fau.de

A. Berezovski, T. Soomere (eds.), *Applied Wave Mathematics II*, Mathematics of Planet Earth 6, https://doi.org/10.1007/978-3-030-29951-4_15

Examples of complex phenomena in sociobiological systems, which have been studied from both a mathematical and computational perspective, are the paradigms of flocking, the Cucker–Smale model (Cucker and Smale, 2007), or synchronisation of fireflies, the Kuramoto model (Kuramoto, 1975). Both are examples of agent based models that rely on ordinary differential equations (ODEs).

The number of agents entering in these interactive dynamics is often very large, for example when modelling bacteria motility (Kearns, 2010), and the limit processes when it tends to infinity becomes then relevant. Often a mean field approach is adopted for this purpose, describing the evolution of the density of individuals, leading to systems governed by partial differential equations (PDEs), mainly hyperbolic conservation laws and kinetic models of nonlocal nature (Ha and Tadmor, 2008; Cañizo et al., 2011).

The control of these models has also been subject of investigation (Liu et al., 2011; Liu and Barabási, 2016), together with their mean field limit models (Piccoli et al., 2015). However, we are far from having a complete understanding of this topic and, in particular, about the transition of control properties from finite to infinite dimensional dynamics.

This chapter is a nonlinear complement of the analysis in (Biccari et al., 2019), where linear models were considered. It is devoted to analyse the control properties of an opinion model, understood as the evolution of a population ahead of a dichotomy.

The opinion of an agent with respect to an external proposal is modelled through as a real number $y \in \mathbb{R}$. The sign of y reveals whether the agent is in favour (positive) or against (negative) this proposal. The absolute value of y represents the intensity of the agent's opinion. This opinion can evolve in time by itself, through a maturation process. It can be also influenced by interactions with other individuals.

In this article we consider the following system of ordinary differential equations (ODE), modelling the evolution of the opinion on N interacting agents:

$$\begin{cases} \frac{\mathrm{d}}{\mathrm{d}t}\boldsymbol{y} = L\boldsymbol{y} + \boldsymbol{N}(\boldsymbol{y}), \\ \boldsymbol{y}(0) = \boldsymbol{y}^0, \end{cases} \tag{15.1}$$

where L is a matrix related to the graph associated with the social network of agents and $\boldsymbol{N}$ is a nonlinear interaction term.

Similar models have been treated in the literature in the context of networks control, e.g., (Sorrentino et al., 2007). We shall work under an *awareness assumption*, according to which every agent knows the size of the network and makes use of this information to weight the influence of others. In mathematical terms this means that self-evolution, represented by the nonlinear term $\boldsymbol{N}(\boldsymbol{y})$ in (15.1), might depend on N: $\boldsymbol{N} = \boldsymbol{N}_N$.

We are interested in steering the system to a given consensus steady state, by means of influencing the opinion of some individuals in the social network. We aim at getting estimates of the cost of the control uniformly with respect to the system size N. Classical control techniques for nonlinear ODE systems (quasistatic deformations, Lie brackets, linearisations, etc. (Coron, 2007; Sontag, 1998)) do not provide explicit bounds with respect to N.

However, there is a context in which we naturally find such kind of estimates, namely the numerical discretisation schemes for partial differential equations (PDE) control problems. In particular, finite differences semidiscretisations of PDE systems provide a wide class of large systems of ODEs whose control properties are by now well understood. As the mesh size parameters tend to zero to approximate the PDE, the size N of these systems increases, precisely as in the analysis we aim at developing in here.

This paper focuses on networks with a lattice structure, so to connect the system under consideration with time reparameterisations of finite difference approximations of nonlinear parabolic PDEs. This allows us to use the existing controllability results, that rely on the Carleman estimates, so to derive explicit estimates, as a function of N, on the control of the nonlinear ODE systems under consideration.

We will specifically deal with the following dynamical system modelling opinion spreading in a one-dimensional (1D) chain network with nonlinear self-evolution:

$$\boldsymbol{y}_t = \mathcal{A}\boldsymbol{y} + \boldsymbol{G}_N(\boldsymbol{y}), \tag{15.2}$$

where the N-dimensional state $\boldsymbol{y} = (y_1, y_2, \cdots, y_N)^T \in \mathbb{R}^N$ represents the opinion of N agents, $j = 1, \ldots, N$, $\mathcal{A}$ is the $N \times N$ diffusion matrix

$$\mathcal{A} := \frac{1}{3}\begin{pmatrix} -1 & 1 & 0 & \cdots & \cdots & 0 \\ 1 & -2 & 1 & \cdots & \cdots & 0 \\ 0 & 1 & -2 & 1 & \cdots & 0 \\ \cdots & \cdots & \cdots & \cdots & \cdots & \cdot \\ 0 & 0 & \cdots & 1 & -2 & 1 \\ 0 & 0 & \cdots & 0 & 1 & -1 \end{pmatrix}_{N\times N}, \tag{15.3}$$

and $\boldsymbol{G}_N : \mathbb{R}^N \to \mathbb{R}^N$ is a nonlinear perturbation. For the sake of simplicity in the presentation, $\boldsymbol{G}_N$ is assumed to be of the form

$$\boldsymbol{G}_N[\boldsymbol{y}) = (G_N(y_1), G_N(y_2), \cdots, G_N(y_N)]^T. \tag{15.4}$$

The multiplicative factor $1/3$ in the matrix $\mathcal{A}$ averages the effects of the interacting agents. In the present case each of them is influenced, either by its own opinion or configuration, by those of the neighbouring ones, to the left and right (Fig. 15.1).

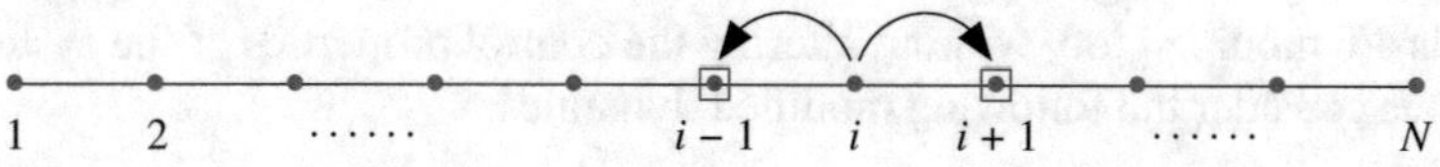

Fig. 15.1 Scheme of interactions of agent i who communicates with agents $j = i \pm 1$.

We assume that $\boldsymbol{G}_N$ is locally a Lipschitz function. Suitable growth conditions will be imposed on $\boldsymbol{G}_N$. Typically, and to avoid technicalities, we shall assume that $\boldsymbol{G}_N$ is globally Lipschitz. A *c*onsensus configuration is the one in which all the components of the state coincide, that is, $y_1 = y_2 = \cdots = y_N = \overline{y}$. It constitutes a steady state of the system if $\boldsymbol{G}(\overline{\boldsymbol{y}}) = 0$ with $\overline{\boldsymbol{y}} = (\overline{y}, \ldots, \overline{y})$. We shall assume, moreover, that $\boldsymbol{G}_N(0) = 0$. In this case $\boldsymbol{y} \equiv 0$ is a steady state, the trivial consensus. If the nonlinearity $\boldsymbol{G}_N$ has another zero, i.e., $\boldsymbol{G}(\overline{\boldsymbol{y}}) = 0$ for some $\overline{y}$, then $\boldsymbol{y} \equiv \overline{\boldsymbol{y}}$ is also a consensus equilibrium state. In this case, a translation of the nonlinear term, $\boldsymbol{G}(\boldsymbol{y}) \to \boldsymbol{G}(\boldsymbol{y} + \overline{\boldsymbol{y}})$, allows us to assume, without loss of generality, that $\overline{\boldsymbol{y}} \equiv 0$. Below we shall call this term simply nonlinearity.

Thus, in the following, we shall assume that $\boldsymbol{G}(0) = 0$ and discuss the problem of control to consensus $\boldsymbol{y} \equiv 0$. This is the so called problem of *null controllability*, in which the goal is to drive the system to the null state. Even if this is not made explicit in the notation, the $N \times N$ matrix $\mathcal{A}$ depends on the dimension N of the system. In our analysis we shall also allow the nonlinearity to be dependent on the number of agents N. We shall analyse how the amplitude of the nonlinearity affects the way it interacts with the diffusion matrix $\mathcal{A}$ at the control level.

Driving the system to consensus (equivalently, to $\overline{\boldsymbol{y}} \equiv 0$ according to the discussion above) in a finite time $0 < T < \infty$, requires acting on some of the agents in the system through the control matrix $\mathcal{B}_N$ which encodes the way the control $\boldsymbol{v} = \boldsymbol{v}(t)$ affects some components of the system

$$\boldsymbol{y}_t = \mathcal{A}\boldsymbol{y} + \boldsymbol{G}_N(\boldsymbol{y}) + \mathcal{B}_N \boldsymbol{v}. \tag{15.5}$$

We assume that the $N \times 2$ control operator $\mathcal{B}_N$

$$\mathcal{B}_N = \begin{pmatrix} 1 & 0 \\ 0 & 0 \\ \cdot & \cdot \\ \cdot & \cdot \\ 0 & 0 \\ 0 & 1 \end{pmatrix}_{N\times 2} \tag{15.6}$$

acts on the two extremal agents corresponding to the states y_1 and y_N. The control $\boldsymbol{v}$ has two components:

$$\boldsymbol{v}(t) = (v_1(t), v_N(t))^T. \tag{15.7}$$

Taking into account that the control acts on the two extremal components of the state, the matrix $\mathcal{A}$ governing the dynamics can be modified at those entries, provided the controls are modified too, without altering the control properties of the system. We shall thus consider the following modified dynamics:

$$\boldsymbol{y}_t = A\boldsymbol{y} + \boldsymbol{G}_N(\boldsymbol{y}) + \mathcal{B}_N \overline{\boldsymbol{v}}, \tag{15.8}$$

with $\overline{v} \in \mathbb{R}^2$ such that $\overline{v} = v + (y_1, y_N)^T/3$, and

$$A = \frac{1}{3}\begin{pmatrix} -2 & 1 & 0 & \cdots & \cdots & 0 \\ 1 & -2 & 1 & \cdots & \cdots & 0 \\ 0 & 1 & -2 & 1 & \cdots & 0 \\ \cdots & \cdots & \cdots & \cdots & \cdots & \cdot \\ 0 & 0 & \cdots & 1 & -2 & 1 \\ 0 & 0 & \cdots & 0 & 1 & -2 \end{pmatrix}_{N\times N}. \tag{15.9}$$

In above discussed problems, the controls enter the extremes of the chain. In the PDE setting, this corresponds to a boundary control problem. The later can be related to the interior control problem, in which the control enters an interior subset by using a classic extension restriction argument. The same occurs for semidiscrete systems, as it does for those we are considering here.

Inspired by this fact and for technical reasons (also in order to use the results of (Boyer and Le Rousseau, 2014)), we will focus on a problem with controls acting on M agents inside the chain of a larger system. Thus, we shall also consider the system

$$y_t = Ay + G_N(y) + B_N v, \tag{15.10}$$

with $v = v(t) \in \mathbb{R}^M$, M being the number of controlled agents, and

$$B_N = \begin{pmatrix} 0\,0 \cdots & 0 & 0 \\ 0\,0 \cdots & 0 & 0 \\ \cdot\ \cdot & \cdot\ \cdot & \cdot\ \cdot \\ 0 \cdot & 1 & 0\ 0 \\ 0 \cdot & 0 & 1\ 0 \\ 0\,0 \cdots & \cdots & 0 \\ 0\,0 \cdots & \cdots & 0 \end{pmatrix}_{N\times M}. \tag{15.11}$$

The number of controlled components M will be chosen so that M/N remains constant as N grows. This corresponds, in the PDE setting, to controlling a one-dimensional heat propagation problem in a bounded interval, with controls supported in a fixed subinterval.

The line of thinking in this chapter is inspired by the same idea as in (Biccari et al., 2019), viewing the systems above as the analogue of the semidiscrete approximation of the semilinear heat equation, the known results in the later setting, and in particular those in (Boyer and Le Rousseau, 2014), can be employed to analyse the controllability of the models under consideration.

As in (Boyer and Le Rousseau, 2014), the controllability results we shall achieve, contrarily to that typically undertaken in the linear and PDE setting (Biccari et al., 2019), will not guarantee that the target is reached in an exact manner but in an approximate manner. This is due to the fact that the discrete Carleman inequalities allowing the observability inequalities (leading to the controllability results in (Boyer and Le Rousseau, 2014) to be achieved), present some exponentially small (with respect to N) remainder terms. Roughly speaking, the three systems presented above can be treated similarly and exhibit the same control properties. For the sake of clarity we present them in the context of the original control system (15.5).

Our main result is as follows:

Theorem 15.1 *Consider the control system (15.5). Assume that the matrices* $(\mathcal{A}, \mathcal{B})$ *governing the controlled system to be as above ((15.3) and (15.6)), so that the Kalman rank condition is satisfied, i.e.,*

$$\operatorname{rank}[\mathcal{B}, \mathcal{A}\mathcal{B}, \dots, \mathcal{A}^{N-1}\mathcal{B}] = N. \tag{15.12}$$

Assume that the nonlinearity is scaled as in

$$\boldsymbol{G}_N(\boldsymbol{y}) = \frac{\boldsymbol{G}(\boldsymbol{y})}{N^2}, \tag{15.13}$$

with $\boldsymbol{G}$ *as above, (15.4), globally Lipschitz continuous and fulfilling* $\boldsymbol{G}(0) = 0$. *We fix* $T > 0$ *and take a control time horizon* $[0, N^2T]$, *which grows quadratically with* N. *Then, there exists a fixed bound for the cost of the control* $K > 0$, *independent of* N, *such that for every* N *large enough and all* $\boldsymbol{y}^0 \in \mathbb{R}^N$, *there exists a control function* $\boldsymbol{v} \in L^2([0, N^2T]; \mathbb{R}^2)$ *steering system* (15.10) *nearly to equilibrium in time* N^2T, *i.e., so that the solution of*

$$\boldsymbol{y}_t = \mathcal{A}\boldsymbol{y} + \frac{\boldsymbol{G}(\boldsymbol{y})}{N^2} + \mathcal{B}_N\boldsymbol{v}, \quad t \in [0, N^2T], \tag{15.14a}$$

$$\boldsymbol{y}(0) = \boldsymbol{y}^0 \in \mathbb{R}^N, \tag{15.14b}$$

reaches an exponentially small ball in the final time $t = N^2T$,

$$\|\boldsymbol{y}(N^2T)\|_2 \leq K \exp\{-C_0 N\} \|\boldsymbol{y}^0\|_2, \tag{15.15}$$

and C_0 *being independent of* N, *by means of uniformly bounded controls*

$$\|\boldsymbol{v}\|_{L^2([0,N^2T];\mathbb{R}^2)} \leq C(T, \|g\|_\infty) \|\boldsymbol{y}^0\|_2, \tag{15.16a}$$

$$C(T, \|g\|) = \frac{C_\beta}{C_\alpha}\left[\left(e^{C_\alpha T} C_\alpha T + 1 + T\right) K^2 + C_\alpha e^{C_\alpha T}\right]^{1/2}, \tag{15.16b}$$

$$C_\alpha = 2\|g\|_\infty + 1, \tag{15.16c}$$

where C_β *is a constant independent of* N, $\|g\|$, T *and* K:

$$K = \exp\left[C_1\left(1 + \frac{1}{T} + T\|g\|_\infty + \|g\|_\infty^{2/3}\right)\right], \tag{15.17}$$

with

$$g(s) = G(s)/s\,, \tag{15.18}$$

and C_1 being independent of N.

Here and in the following $\|\cdot\|_2$ stands for the Euclidean norm; for example:

$$|\boldsymbol{y}|_2^2 := \frac{1}{N}\sum_{j=1}^{N}(y_j)^2\,.$$

15.1.1 Insights of the Contribution

Notice that by a suitable time scaling $t = N^2\tau$, system (15.5), with $\boldsymbol{G}_N = \boldsymbol{G}/N^2$, can be rewritten as

$$\boldsymbol{y}_\tau = N^2 A\boldsymbol{y} + \boldsymbol{G}(\boldsymbol{y}) + N^2 B_N \boldsymbol{v}\,,$$

which can be viewed as a controlled version the semidiscrete free dynamics

$$\boldsymbol{y}_\tau = N^2 A\boldsymbol{y} + \boldsymbol{G}(\boldsymbol{y})\,,$$

which constitutes a N-point finite difference space semidiscretisation of the semilinear heat equation:

$$y_\tau = \frac{1}{3}\partial_{xx} y + G(y)\,,$$

with Neumann boundary controls in the space interval $0 < x < 1$. Similarly, systems (15.8) and (15.10) can be seen as Dirichlet control problems, with boundary and interior controls, respectively.

Theorem 15.1 is based on the finite time controllability of semidiscrete approximations of semilinear parabolic equations, (Boyer and Le Rousseau, 2014), and extends the results of Biccari et al. (2019) on linear systems. Biccari et al. (2019) make use of the spectral properties of linear parabolic semidiscrete systems and classical results on the controllability of heat like equations (Fattorini and Russell, 1971, 1974). In the linear system the null state is reached exactly at the final time.

The extension of those results to nonlinear systems requires making use of Carleman inequalities as in (Boyer and Le Rousseau, 2014) and leads to the exponentially small reminder at the final time. Whether this remainder can be dropped to assure the exact reachability of consensus is an open problem. As pointed out in (Biccari et al., 2019), this reminder can be avoided in the linear setting.

Moreover, the result above holds when the control is active in both extremes. Similarly, for the Dirichlet system (15.8), using (Boyer and Le Rousseau, 2014), one sole boundary control would suffice. But our argument, linking Dirichlet and Neumann boundary conditions, to directly use the results in (Boyer and Le Rousseau, 2014), leads to the Neumann controllability with two controls. The result of Theorem 15.1 is very likely true with one single control but this would require adapting the discrete Carleman inequalities in (Boyer and Le Rousseau, 2014).

Our results apply only for weak normalised nonlinearities of the form $\boldsymbol{G}/N^2$. This normalisation allows us to ensure that the controls are uniformly bounded in the time interval $[0, N^2T]$. In case the nonlinearity were not normalised by the factor N^2, the cost of controlling the system would diverge as $N \to \infty$.

Furthermore, the result above holds in long time horizons of the order of N^2T. This allows controlling the system with uniformly bounded controls. In case the control time horizon $[0, T]$ was fixed, independent of T, as we shall see, the control would grow exponentially in N^2, as it occurs in the linear setting (Biccari et al., 2019).

Similarly, the cost of control at time $t = N^2T$ for a nonlinearity $\boldsymbol{G_N}(\boldsymbol{y}) = (g_N(y_1)y_1, \cdots, g_N(y_N)y_N)^T$, independent of N, would also grow exponentially with N:

$$K_N = \exp\left[C_1\left(1 + \frac{1}{T} + TN^2\|g_N\|_\infty + N^{4/3}\|g_N\|_\infty^{2/3} +\right)\right]. \tag{15.19}$$

From the modelling perspective, the presence of the nonlinearity $\boldsymbol{G}/N^2$ allows a weak nonlinear interaction among all agents of the system, normalised by the multiplicative factor N^{-2}. The main result applies for N large enough but it does not guarantee the controllability of the original control system for small N. Note that this is not expected to be the case in the general setting above, since, when N is not large, the nonlinearity can interact with the matrix governing the dynamics $\mathcal{A}$ in a way that the Kalman rank condition is lost. Dealing with the control of those systems with N fixed would require to use the genuine techniques of nonlinear finite dimensional control systems (Coron, 2007).

In this paper we have assumed the nonlinearity $\boldsymbol{G}$ to be globally Lipschitz or, in other words, g, as in (15.18), to be uniformly bounded. The results could be extended to nonlinearities growing at infinity in a slightly logarithmic superlinear way, but this would require further analysis (Fernández-Cara and Zuazua, 2000b).

However, the result above is limited to the simplest network in which all agents are aligned and interconnected through a 3-point homogeneous interaction rule. Dealing with more general graphs, possibly heterogeneous, is a challenging problem, even in the linear case (Biccari et al., 2019).

The results of this paper could be extended to 1D networks with slightly varying diffusive interactions in the linear component (thus leading to parabolic equations with variable coefficients) and to square grid shaped networks (lattices) in several space dimensions, making use of the results in (Boyer and Le Rousseau, 2014).

Finally, the main result also applies for certain nonlinear nonlocal systems with nonlinearities of the form $\boldsymbol{G}_N(\boldsymbol{y}) = \left(g_{N,1}(\boldsymbol{y})y_1, \cdots, g_{N,N}(\boldsymbol{y})y_N\right)^T$, provided that, for all N and $j \in \{1, \ldots, N\}$, the functions $g_{N,j}$ are uniformly bounded in L^∞.

15.1.2 Organisation of the Work

The structure of the work is the following. In Sect. 15.2, we present some well known results on the null controllability of parabolic equations, linear and semilinear, their finite difference counterparts, and a brief summary of Biccari et al. (2019).

In Sect. 15.3 we analyse the divergent behaviour of the control properties when the nonlinearity is scaled differently. In Sect. 15.4 we prove Theorem 15.1. Section 15.5 presents some numerical experiments. Finally, in Sect. 15.6, we summarise our conclusions and present some open problems for future research.

15.2 Preliminaries

15.2.1 Continuous Models

15.2.1.1 Controllability of the Heat Equation

Let $\Omega \subset \mathbb{R}$ be a bounded interval, ω nonempty subinterval and fix $T > 0$. Consider the following control problem for the heat equation

$$y_t - y_{xx} = \mathbb{1}_\omega u \quad \text{in } \Omega \times (0, T), \tag{15.20a}$$

$$y = 0 \quad \text{on } \partial\Omega \times (0, T), \tag{15.20b}$$

$$y = y^0 \quad \text{on } \Omega \times \{t = 0\}. \tag{15.20c}$$

The following result is classical and well known and can be found, for instance, in (Fernández-Cara and Zuazua, 2000a, Theorem 1.3).

Theorem 15.2 *For any $T > 0$ and any $y_0 \in L^2(\Omega)$ there exists a control $u \in L^2(\omega \times (0, T))$ such that the solution of (15.20) satisfies:*

$$y(T) \equiv 0. \tag{15.21}$$

Furthermore, the cost of control can be estimated as

$$\|u\|_{L^2(\omega\times(0,T))} \leq \exp\left[C\left(1 + \frac{1}{T}\right)\right] \|y^0\|_{L^2(\Omega)},$$

where the constant $C > 0$ depends on Ω and ω but is independent of T.

15.2.1.2 Controllability of the Semilinear Heat Equation

Let us now consider the semilinear control problem for the heat equation:

$$y_t - y_{xx} - G(y) = \mathbb{1}_\omega u \quad \text{in } \Omega \times (0,T), \tag{15.22a}$$

$$y = 0 \quad \text{on } \partial\Omega \times (0,T), \tag{15.22b}$$

$$y = y^0 \quad \text{on } \Omega \times \{t = 0\}. \tag{15.22c}$$

The following result is well known:

Theorem 15.3 (Fernández-Cara and Zuazua, 2000b, Theorem 1.2) *Let $T > 0$. Assume that $G : \mathbb{R} \to \mathbb{R}$ is globally Lipschitz, with Lipschitz constant L and $G(0) = 0$. Then system* (15.22) *is null controllable at any time $T > 0$ with cost:*

$$C(T) = \exp\left[C\left(1 + \frac{1}{T} + TL + L^{2/3}\right)\right]. \tag{15.23}$$

Remark 15.1 System (15.22), in the absence of control (i.e., with $u = 0$), when the nonlinear G is superlinear at infinity, can blow up in finite time. Fernández-Cara and Zuazua (2000b) proved that some blowing up processes for nonlinearities G growing at infinity in a slightly superlinear fashion

$$\limsup_{|s|\to\infty} \frac{|G(s)|}{|s| \log^{3/2}(1 + |s|)} < \infty, \tag{15.24}$$

can be controlled by acting fast enough, i.e., controlling the system in a short enough time horizon $[0, T]$, with small T, before the system blows up.

15.2.2 Uniform Controllability of Semidiscrete Heat Equations

Here we recall the results in (Boyer and Le Rousseau, 2014), concerning a space semidiscrete version on the results in the previous subsection, that hold uniformly on the mesh size parameter $h \to 0$ (in our setting, $h = 1/N$).

Theorem 15.4 (Boyer and Le Rousseau, 2014, Theorem 5.2, Theorem 5.11) *Let $\sigma > 0$ be a constant diffusivity, and the nonlinearity G be as in Theorem 15.1. The system*

$$\partial_\tau \boldsymbol{y} - \sigma N^2 A \boldsymbol{y} - \boldsymbol{G}(\boldsymbol{y}) = B_N \boldsymbol{u}, \tag{15.25a}$$

$$\boldsymbol{y}(0) = \boldsymbol{y}^0, \tag{15.25b}$$

where $B_N \in \mathbb{R}^N \times \mathbb{R}^M$ is as in (15.11) and $M/N > 0$ is kept constant, is uniformly controllable as $N \to \infty$ for any $T > 0$, in the sense that for all initial data there are controls assuring that

$$|\boldsymbol{y}(T)|_2 \leq C(T)e^{-C_0 N}|\boldsymbol{y}^0|_2, \tag{15.26a}$$

$$\|\boldsymbol{u}\|_{L^2((0,T);\mathbb{R}^M)} \leq C(T)|\boldsymbol{y}^0|_2, \tag{15.26b}$$

$$C(T) = \exp\left[C_1\left(1 + 1/T + \|g\|_\infty T + \|g\|_\infty^{2/3}\right)\right], \tag{15.26c}$$

with C_0, $C_1 > 0$ depending on the location of the controlled components and σ, but independent of g and T.

Remark 15.2 Several remarks are in order:

- In agreement with the notation adopted for the weighted euclidean norm $|\cdot|_2$, for $\boldsymbol{y}(t) \in L^2((0,T);\mathbb{R}^N)$ we shall employ the notation

$$\|\boldsymbol{y}(t)\|_{L^2((0,T);\mathbb{R}^N)} = \left[\int_0^T |\boldsymbol{y}(t)|_2^2 dt\right]^{1/2}.$$

- This result is uniform in $N \to \infty$ in the sense that the controls are uniformly bounded, but the state is not guaranteed to reach exactly the null state. An exponentially small rest remains, of the order of $\exp(-C_0 N)$. Obviously, as $N \to \infty$, this reminder vanishes and one recovers the null control of the semilinear heat equation, as in the previous subsection.
- Theorem 15.4 was proved by means of Carleman estimates for semidiscrete parabolic equations. The exponential reminder term in the state at time $\tau = T$ is a consequence of the Carleman inequality.
- According to (Boyer and Le Rousseau, 2014), N ($= 1/h$ in the context of that article, h being the mesh size parameter, $h \to 0$) needs to be large enough:

$$N \geq C\left(1 + \frac{1}{T} + \|g\|_\infty^{2/3}\right). \tag{15.27}$$

- Theorem 15.4 can be extended to the case where g is superlinear too (Boyer and Le Rousseau, 2014).

15.2.3 The Linear Case

Recently, in (Biccari et al., 2019), these issues were addressed in the linear setting ($\boldsymbol{G} = 0$) using spectral techniques (Fattorini and Russell, 1971, 1974). In particular, the following result was obtained:

Theorem 15.5 ((Biccari et al., 2019), Proposition 4.1) *Let us consider the following N-dimensional control problem:*

$$\boldsymbol{y}_t + A\boldsymbol{y} = B\boldsymbol{v}, \tag{15.28}$$

with A given by (15.9) *and*

$$B = (1, 0, ..., 0)^T, \tag{15.29}$$

representing a scalar control entering in one of the extremes of the network.

Then:

- *When the time of control is of the order of* $T \sim N^2$*, null controllability is achievable acting only on one of the extreme agents, with a uniformly bounded (on N) control.*
- *When time T is independent of N, null controllability requires controls of size* $\exp(cN^2)$.

15.3 The Impact of Scaling on the Cost of Control

In this section we discuss in more detail the impact of scaling the nonlinearity and time, as a function of N, on the cost of controlling the systems under consideration.

15.3.1 The Scaling Factor $1/N^2$ *in the Nonlinearity*

We fix $T > 0$ and consider model (15.10) in the time horizon N^2T with a nonlinearity of the following form:

$$\boldsymbol{y}_t = A\boldsymbol{y} + \frac{\boldsymbol{G}(\boldsymbol{y})}{N^2} + B_N\boldsymbol{v}, \quad t \in [0, N^2T]. \tag{15.30}$$

Reparameterising time as $t = N^2\tau$, we get:

$$\boldsymbol{y}_\tau = N^2A\boldsymbol{y} + \boldsymbol{G}(\boldsymbol{y}) + B_N\boldsymbol{u}, \quad \tau \in [0, T], \tag{15.31}$$

with

$$\boldsymbol{u}(\tau) = N^2\boldsymbol{v}(N^2\tau), \quad 0 < \tau < T. \tag{15.32}$$

Note that, in view of (15.32),

$$||\boldsymbol{u}(\cdot)||_{L^2_\tau(0,T)} = N||\boldsymbol{v}(\cdot)||_{L^2_t(0,N^2T)}. \tag{15.33}$$

We can understand (15.31) as a semidiscretisation with mesh size $h = 1/N$ of the following semilinear heat equation in $\Omega = (0, 1)$:

$$y_\tau = \frac{1}{3} y_{xx} + G(y) + \mathbb{1}_\omega u \,, \quad (\tau, x) \in [0, T] \times (0, 1) \,.$$

Applying Theorem 15.4 we obtain that, if N is large enough, the semidiscrete system satisfies the following condition:

$$|\boldsymbol{y}(t = N^2 T)|_2 = |\boldsymbol{y}(\tau = T)|_2 \leq \exp\left[C_1 \left(1 + \frac{1}{T} + T\|g\|_\infty + \|g\|_\infty^{2/3} \right) \right] \times$$
$$\times \exp\left(-C_0 N\right) |\boldsymbol{y}^0|_2 \,,$$

with controls

$$\|\boldsymbol{u}\|_{L^2\left([0,T];\mathbb{R}^M\right)} \leq \exp\left[C_1 \left(1 + \frac{1}{T} + T\|g\|_\infty + \|g\|_\infty^{2/3} \right) \right] |\boldsymbol{y}^0|_2 \,.$$

Taking (15.33) into account, we deduce that

$$\|\boldsymbol{v}\|_{L^2\left([0,N^2T];\mathbb{R}^M\right)} \leq \frac{1}{N} \exp\left[C_1 \left(1 + \frac{1}{T} + T\|g\|_\infty + \|g\|_\infty^{2/3} \right) \right] \|\boldsymbol{y}^0\|_2 \,.$$

This holds under condition (15.27).

15.3.2 *Other Scaling Factors*

We now discuss other two cases where the nonlinearity is scaled differently, namely, $\boldsymbol{G}_N = \boldsymbol{G}$ and $\boldsymbol{G}_N = \boldsymbol{G}/N$.

15.3.2.1 The Homogeneous Case

Consider

$$\boldsymbol{y}_t = A\boldsymbol{y} + \boldsymbol{G}(\boldsymbol{y}) + B_N \boldsymbol{v} \,, \quad t \in [0, N^2 T] \,,$$

which, under time rescaling $N^2 \tau = t$, can be understood as a semidiscretisation of

$$y_\tau = \frac{1}{3} \partial_{xx} y + N^2 G(y) + \mathbb{1}_\omega u \,, \quad \tau \in [0, T] \,,$$

where, as in (15.32), $\boldsymbol{u}(\tau) = N^2 \boldsymbol{v}(N^2 \tau)$.

In this case the cost of control is

$$\|\boldsymbol{v}\|_{L^2\left([0,N^2T];\mathbb{R}^M\right)} \leq \frac{1}{N}\exp\left[C_1\left(1+\frac{1}{T}+TN^2\|g\|_\infty+N^{4/3}\|g\|_\infty^{2/3}\right)\right]\|\boldsymbol{y}^0\|_2,$$

that blows up as $N \to \infty$.

On the other hand, the state at the final time satisfies the following condition:

$$|\boldsymbol{y}(\tau=T)|_2 \leq \exp\left[C_1\left(1+\frac{1}{T}+TN^2\|g\|_\infty+N^{4/3}\|g\|_\infty^{2/3}\right)\right]\exp\{-C_0N\}|\boldsymbol{y}^0|_2.$$

This estimate does not allow to approach the null state at the final time, since the upper bound on the right-hand side term diverges exponentially as $N \to \infty$.

15.3.2.2 An Intermediate Case

Consider now the equation

$$\boldsymbol{y}_t = A\boldsymbol{y} + \frac{\boldsymbol{G}(\boldsymbol{y})}{N} + B_N\boldsymbol{v}, \quad t \in [0, N^2T],$$

which, after scaling, can be understood as a semidiscretisation of

$$y_\tau = \frac{1}{3}\partial_{xx}y + NG(y) + \mathbb{1}_\omega u, \quad \tau \in [0, T].$$

Applying Theorem 15.4 we obtain a cost that blows up exponentially as $N \to \infty$:

$$\|\boldsymbol{v}\|_{L^2\left([0,N^2T];\mathbb{R}^M\right)} \leq \frac{1}{N}\exp\left[C_1\left(1+\frac{1}{T}+TN\|g\|_\infty+N^{2/3}\|g\|_\infty^{2/3}\right)\right]\|\boldsymbol{y}^0\|_2,$$

achieving a target ball

$$|\boldsymbol{y}(t=N^2T)|_2 = |\boldsymbol{y}(\tau=T)|_2 \leq \exp\left[C_1\left(1+\frac{1}{T}+TN\|g\|_\infty+N^{2/3}\|g\|_\infty^{2/3}\right)\right]\times$$
$$\times\exp\left(-C_0N\right)|\boldsymbol{y}^0|_2.$$

This target ball can be assured to shrink as $N \to \infty$ provided the control time $T > 0$ is taken small enough

$$T < \frac{C_0}{C_1\|g\|_\infty}.$$

This argument can be iterated repeatedly to control the system in longer time intervals (corresponding to T large). The smallness of the final target can be enhanced, but the controls diverge.

15.4 Proof of the Main Result

In this section we present the proof of Theorem 15.1. The strategy is as follows:

1. *Step 1.* First, as explained above, we understand system (15.5) as a semidiscretisation of a semilinear heat equation by means of a reparameterisation of the time scale.
2. *Step 2.* By an extension argument we reduce the problem to consider an interior control problem in a larger network, so to apply the results in Theorem 15.4.
3. *Step 3.* By restriction we conclude the controllability results with two controls acting on the extremes of the original network.
4. *Step 4.* We conclude providing precise estimates on the cost of control and the size of the target achieved in the final time.

Step 1. Scaling. Consider the dynamical system (15.5) with $\boldsymbol{G}_N(\boldsymbol{y}) = \boldsymbol{G}/N^2$, where $\mathcal{A}$ is defined as in (15.3) and with a control $\boldsymbol{v} \in L^2\left([0, N^2T]; \mathbb{R}^2\right)$ acting on both extremes of the chain. Reparameterising time the system reads:

$$\boldsymbol{y}_\tau = N^2\mathcal{A}\boldsymbol{y} + \boldsymbol{G}(\boldsymbol{y}) + \mathcal{B}\boldsymbol{u}, \quad \tau \in [0, T], \tag{15.34}$$

with $\boldsymbol{u}$ as in (15.32).

As explained above, this system can be seen as the semidiscretisation of a semilinear heat equation with Neumann boundary conditions in the space domain, N being the number of nodes and $h = 1/N$ the mesh size.

Step 2. Extension. As it is classical in the PDE setting, we relate the boundary control problem with that of interior control in an extended domain.

We thus introduce an extended state $\boldsymbol{y}_E \in \mathbb{R}^{2N+1}$

$$\boldsymbol{y}_E = \left(y_{-\frac{N}{2}}, \cdots, y_0, y_1, \cdots, y_N, y_{N+1}, \cdots, y_{N+\frac{N}{2}}\right)^T,$$

corresponding to the extended network in which the original one, corresponding to the components $j = 1, \ldots, N$, is now extended to $j = -N/2, \ldots, N + N/2$. Of course this construction is valid when N is even, the adaptation to the case where N is odd being straightforward.

We then consider the extended controlled dynamics

$$\boldsymbol{y}_{E,\tau} = N^2 A_{2N+1}\boldsymbol{y}_E + \boldsymbol{G}(\boldsymbol{y}_E) + B_E\boldsymbol{u}, \quad \tau \in [0, T], \tag{15.35a}$$

$$\boldsymbol{y}_E(0) = \boldsymbol{y}_E^0, \tag{15.35b}$$

where A_{2N+1} is the $(2N + 1) \times (2N + 1)$-dimensional version of A in (15.9).

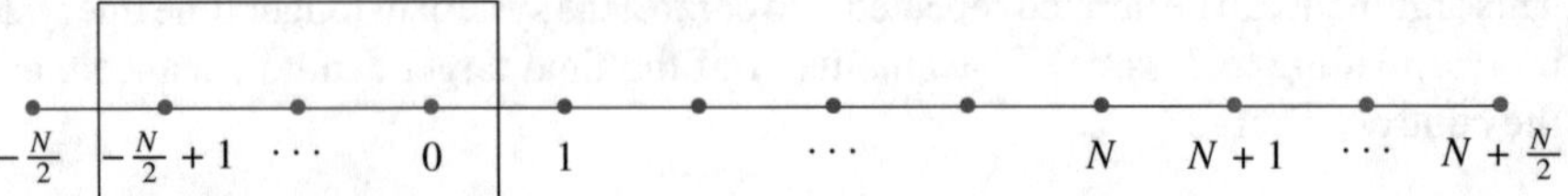

Fig. 15.2 The blue dots represent the original network, before extension. The red dots represent the extended ones. The black square indicates the support of the control operator B_E.

$$
\begin{pmatrix}
-2 & 1 & 0 & & \cdots & & & & 0 \\
1 & -2 & 1 & 0 & \cdots & & & & 0 \\
 & \ddots & \ddots & \ddots & & & & & \\
0 & \cdots & 1 & -2 & 1 & 0 & \cdots & & 0 \\
\vdots & & & & \vdots & & & & \vdots \\
0 & \cdots & & 0 & 1 & -2 & 1 & \cdots & 0 \\
0 & \cdots & & & & \cdots & 0 & 1 & -2
\end{pmatrix}
$$

Fig. 15.3 The $N \times N$ submatrix governing the dynamics of the original nodes $\{1, \ldots, N\}$ extracted from the extended $(2N+1) \times (2N+1)$ matrix A_{2N+1}.

The initial datum $\boldsymbol{y}^0$ can be extended to $\boldsymbol{y}_E^0$ in such a way that

$$
|\boldsymbol{y}^0|_2 \leq |\boldsymbol{y}_E^0|_2 \leq 2|\boldsymbol{y}^0|_2 .
$$

For this extended system the $2N \times M$ control operator B_E is built to be only active on the nodes that fall outside the original network, i.e., with support in the complement of $j = 1, \ldots, N$. To fix ideas, B_E will be active on the $N/2$ nodes corresponding to the indexes $j = -\frac{N}{2} + 1, \ldots, 0$ so that $M = N/2$ (Fig. 15.2).

Applying Theorem 15.4 to the extended system, we get

$$
|\boldsymbol{y}_E(T)|_2 \leq K e^{-C_0 N} |\boldsymbol{y}_E^0|_2 \leq 2K e^{-C_0 N} |\boldsymbol{y}^0|_2 , \tag{15.36}
$$

with uniformly bounded controls

$$
\|\boldsymbol{u}\|_{L^2((0,T);\mathbb{R}^M)} \leq C(T)|\boldsymbol{y}^0|_2 . \tag{15.37}
$$

Step 3. Restriction. The nodes corresponding to the original network fulfill a subsystem of ODEs governed by a $N \times N$ submatrix (Fig. 15.3). The projected N-dimensional state of $\boldsymbol{y}_E$, denoted by $\boldsymbol{y}$, satisfies

$$
\frac{d}{d\tau}\boldsymbol{y} = N^2 \mathcal{A}\boldsymbol{y} + \boldsymbol{G}(\boldsymbol{y}) + \frac{N^2}{3}\begin{pmatrix} y_0 - y_1 \\ 0 \\ \vdots \\ 0 \\ y_{N+1} - y_N \end{pmatrix}_{N\times 1} . \tag{15.38}
$$

Accordingly, the controls we obtain for system (15.5) are:

$$\overline{\boldsymbol{u}}(\tau) = \begin{pmatrix} u_1(\tau) \\ u_2(\tau) \end{pmatrix} = \frac{N^2}{3} \begin{pmatrix} y_0(\tau) - y_1(\tau) \\ y_{N+1}(\tau) - y_N(\tau) \end{pmatrix}, \tag{15.39}$$

$y_0(\tau)$ and $y_{N+1}(\tau)$ being the controlled states for the extended problem, in the nodes immediately close to the extremes $j = 1$ and $j = N$, respectively. Note that, because of the scaling factor in the diffusion matrix, they are also multiplied by N^2.

Obviously, as a consequence of (15.36), the restricted dynamics also fulfills the terminal bound:

$$|\boldsymbol{y}(T)|_2 \le |\boldsymbol{y}_E(T)|_2 \le K e^{-C_0 N} |\boldsymbol{y}_E^0|_2 \le 2K e^{-C_0 N} |\boldsymbol{y}^0|_2 .$$

Step 4. Estimates on the Cost of Control. As we have seen in (15.39), the controls are related to the components $j = 0$ and $j = N + 1$ of the extended state $\boldsymbol{y}_E$.

According to (15.33), in order to have an uniform bound $L^2_t(0, N^2 T)$ for the controls $\boldsymbol{v}$ of the original system, we need to show that

$$\|(u_1, u_2)\|^2_{L^2\tau\left([0,T];\mathbb{R}^2\right)} \le CN. \tag{15.40}$$

This is not completely obvious from (15.39), because of the scaling factor N^2. However, rewriting (15.39) as

$$\overline{\boldsymbol{u}}(\tau) = \begin{pmatrix} u_1(\tau) \\ u_2(\tau) \end{pmatrix} = \frac{N}{3} \begin{pmatrix} N[y_0(\tau) - y_1(\tau)] \\ N[y_{N+1}(\tau) - y_N(\tau)] \end{pmatrix}, \tag{15.41}$$

we see that these controls are the amplification (by a multiplicative factor N) of semidiscrete approximations of the normal derivatives of the 1D heat equation at the boundary points.

In view of (15.33) it is sufficient to show that

$$N\left[||y_0(\tau) - y_1(\tau)||_{L^2_\tau(0,T)} + ||y_{N+1}(\tau) - y_N(\tau)||_{L^2_\tau(0,T)} \right] \le C. \tag{15.42}$$

Standard regularity properties for parabolic equations and their semidiscrete counterparts will then suffice since the expression of the terms in the left-hand side of (15.42) represent semidiscrete approximations of the normal derivatives.

In particular, it is sufficient to show that the solutions $\boldsymbol{y}_E$ of the extended control system (15.35a) satisfy, uniformly on N, the semidiscrete version of the $L^2(0, T; H^2)$-bound of solutions of the heat equation with a right-hand side term in L^2. This can be easily achieved using classical energy estimates applied to the semidiscrete system, and taking into account that the nonlinearity is globally Lipschitz.

Obviously, as in the context of the heat equation, to achieve the $L^2(0, T; H^2)$ on the solutions of the heat equation, the initial data needs to be H^1-smooth. Such a smoothness property was not assumed in our main statements. Indeed we always considered initial data in L^2. But, as in the context of the heat equation, in the present semidiscrete setting, this extra regularity assumption on the initial datum does not impose any restriction since the regularising effect of these models guarantees that the solutions starting from initial data in L^2 automatically enter in H^1, in the absence of controls.

15.5 Numerical Experiments

We consider the control system (15.5) with $N = 45$, the nonlinearity

$$G_N(y) = \frac{1}{N^2} y e^{-y^2},$$

and the time horizon is $N^2 T$ with $T = 2$. Rather than considering the controllability problem, we analyse an optimal control one, minimising the functional

$$J(\boldsymbol{v}) = \sum_{i=1}^{N} |y_i(N^2 T)|^2 + \beta \int_0^{N^2 T} [|v_1(t)|^2 + |v_2(t)|^2]\, \mathrm{d}t,$$

with two controls $\boldsymbol{v} \in L^2\left([0, N^2 T]; \mathbb{R}^2\right)$ acting on the extremes of the chain, i.e., on agents $j = 1$ and $j = N$.

The penalisation parameter is taken as $\beta = 10^{-15}$, so to force the final state towards zero, in analogy with the null controllability problem. We choose the initial datum

$$y_j^0 = \sin(\pi j / N) \quad j \in \{1, \ldots, N\}.$$

The numerical experiments are developed using the DyCon Toolbox (DYC, 2019).

In Figs. 15.4 and 15.5 we can visualise and compare the free dynamics of the system and the controlled dynamics, corresponding to the optimal control minimising the functional J above. It is clearly observed that, while the free solution has the tendency to grow, the controlled one collapses around the null state at the final time.

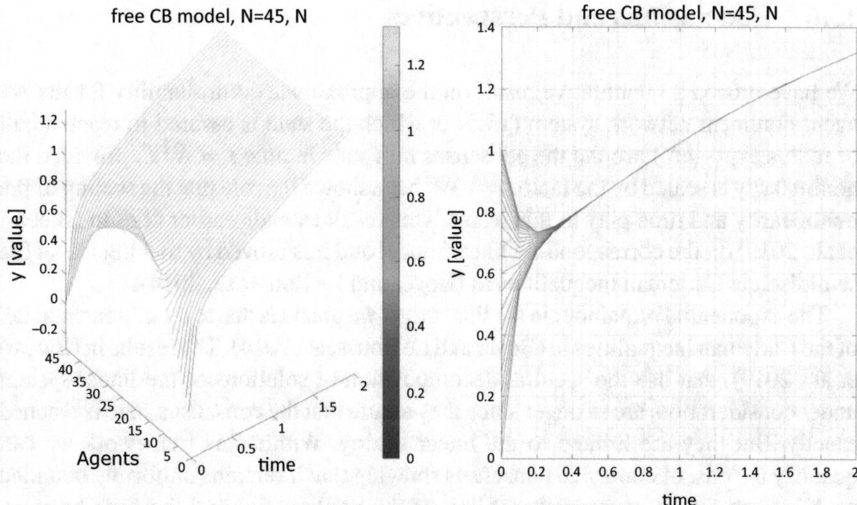

Fig. 15.4 Free dynamics of the system. On the left figure, in the in-plane axes we represent the $N = 45$ agents and time rescaled by the factor $1/N^2$, while in the vertical axis we draw the full state y. In the right picture we represent the same dynamics, drawing the vertical variation of each of the agents positions along the horizontal time variable.

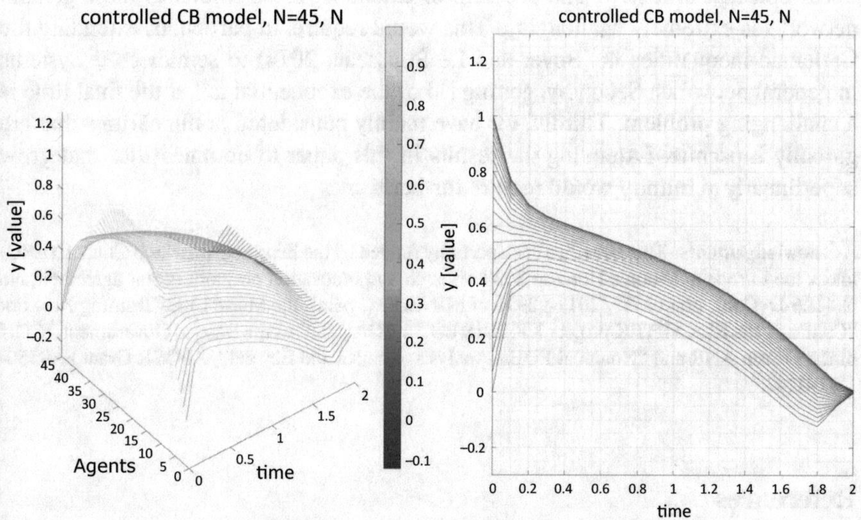

Fig. 15.5 Same as above but for the controlled dynamics.

15.6 Conclusions and Perspectives

We have proved a quantitative result on the approximate controllability for the N-agent nonlinear network system (15.5) in which the state is assured to reach a ball of radius $\exp(-cN)$ around the consensus null state in time $t = N^2 T$, provided the nonlinearity is scaled by the factor N^2. We have shown the role that the scaling of the nonlinearity and time play in this result. This result extends earlier ones in (Biccari et al., 2019) on the corresponding linear model and it is proved by making use of the semidiscrete Carleman inequalities in (Boyer and Le Rousseau, 2014).

The exponential remainder in the final target we obtain is due to the exponential tail of the Carleman inequalities in (Boyer and Le Rousseau, 2014). The results in (Biccari et al., 2019), that use the spectral decomposition of solutions of the linear system under consideration, are stronger since they assure that the consensus state is reached exactly. But they are limited to the linear setting. Within this framework we can quantify the cost of control to consensus showing that it remains uniformly bounded as $N \to \infty$ for an appropriate scaling of the nonlinearity and the time horizon. Our results apply for nonlocal nonlinearities, under the same scaling, provided the nonlinearities are uniform on N.

The presented contents raise some interesting open problems. Among them the following are worth highlighting. Firstly, the network we have considered is particularly simple, all agents being aligned and interconnected by dominant linear constant diffusion. The problem of extending these results to more general networks is extremely challenging. This would require, in particular, extending the Carleman inequalities in (Boyer and Le Rousseau, 2014) to semidiscrete systems in general networks. Secondly, getting rid of the exponential tail at the final time is a challenging problem. Thirdly, we have mainly considered nonlinearities that are globally Lipschitz. Extending the results in this paper to nonlinearities that grow superlinearly at infinity would require further work.

Acknowledgements This work has been partially funded by the European Research Council (ERC) under the European Union's Horizon 2020 research and innovation program (grant agreement No 694126-DyCon), grant MTM2017-92996 of MINECO (Spain), the Marie Curie Training Network "Conflex", the ELKARTEK project KK-2018/00083 ROAD2DC of the Basque Government, ICON of the French ANR and "Nonlocal PDEs: Analysis, Control and Beyond", AFOSR Grant FA9550-18-1-0242.

References

Biccari, U., Ko, D., Zuazua, E.: Dynamics and control for multi-agent networked systems: a finite difference approach. arXiv Math, to appear in M3AS (2019). https://doi.org/10.1142/S0218202519400050

Boyer, F., Le Rousseau, J.: Carleman estimates for semi-discrete parabolic operators and application to the controllability of semi-linear semi-discrete parabolic equations. Annales de l'IHP Analyse non linéaire **31**(5), 1035–1078 (2014). https://doi.org/10.1016/j.anihpc.2013.07.011

Cañizo, J.A,, Carrillo, J.A,, Rosado, J.: A well-posedness theory in measures for some kinetic models of collective motion. Math. Models Methods Appl. Sci. **21**(3), 515–539 (2011). https://doi.org/10.1142/s0218202511005131

Coron, J.M.: Control and Nonlinearity. American Mathematical Society, Boston, MA, USA (2007)

Cucker, F., Smale, S.: Emergent behavior in flocks. IEEE Trans. Automatic Control **52**(5), 852–862 (2007). https://doi.org/10.1109/tac.2007.895842

DyCon Toolbox. Universidad de Deusto & Universidad Autónoma de Madrid, Spain (2019) https://deustotech.github.io/dycon-platform-documentation/. Accessed on 21.06.2019

Fattorini, H., Russell, D.: Uniform bounds on biorthogonal functions for real exponentials with an application to the control theory of parabolic equations. Quarterly Appl. Math. **32**, 45–69 (1974). https://doi.org/10.1090/qam/510972

Fattorini, H.O., Russell, D.L.: Exact controllability theorems for linear parabolic equations in one space dimension. Arch. Rat. Mech. Anal. **43**(4), 272–292 (1971). https://doi.org/10.1007/bf00250466

Fernández-Cara, E., Zuazua, E.: The cost of approximate controllability for heat equations: the linear case. Adv. Differential Equations **5**(4-6), 465–514 (2000a)

Fernández-Cara, E., Zuazua, E.: Null and approximate controllability for weakly blowing up semilinear heat equations. Annales de l'Institut Henri Poincare (C) Non Linear Analysis **17**(5), 583–616 (2000b). https://doi.org/10.1016/s0294-1449(00)00117-7

Ha, S.Y., Tadmor, E.: From particle to kinetic and hydrodynamic descriptions of flocking. Kinetic & Related Models **1**(3), 415–435 (2008). https://doi.org/10.3934/krm.2008.1.415

Kearns, D.: A field guide to bacterial swarming motility. Nature Reviews Microbiology **8**(9), 634–644 (2010). https://doi.org/10.1038/nrmicro2405

Kuramoto, Y.: Self-entrainment of a population of coupled non-linear oscillators. In: Araki, H. (ed.) Mathematical Problems in Theoretical Physics. Lecture Notes in Physics, vol. 39, pp. 420–422, Springer, Berlin (1975). https://doi.org/10.1007/bfb0013365

Liu, Y., Slotine, J.J., Barabási, A.: Controllability of complex networks. Nature **473**, 167–173 (2011). https://doi.org/10.1038/nature10011

Liu, Y.Y., Barabási, A.L.: Control principles of complex systems. Rev. Mod. Phys. **88**, 035006 (2016). https://doi.org/10.1103/revmodphys.88.035006

Macy, M.W., Willer, R.: From factors to actors: Computational sociology and agent-based modeling. Annu. Rev. Sociology **28**, 143–166 (2002). https://doi.org/10.1146/annurev.soc.28.110601.141117

Motsch, S., Tadmor, E.: Heterophilious dynamics enhances consensus. SIAM Review **56**(4), 577–621 (2014). https://doi.org/10.1137/120901866

Piccoli, B., Rossi, F., Trélat, E.: Control to flocking of the kinetic Cucker–Smale model. SIAM J. Math. Anal. **47**(6), 4685–4719 (2015). https://doi.org/10.1137/140996501

Sontag, E.D.: Mathematical Control Theory: Deterministic Finite Dimensional Systems, 2nd edn. Springer, Berlin (1998)

Sorrentino, F., Di Bernardo, M., Garofalo, F., Chen, G.: Controllability of complex networks via pinning. Phys. Rev. E **75**(4), 046103 (2007). https://doi.org/10.1103/physreve.75.046103

Chapter 16
Entropy Production in Phase Field Theories

Peter Ván

Abstract Allen–Cahn (Ginzburg–Landau) dynamics for scalar fields with heat conduction is treated in rigid bodies using a nonequilibrium thermodynamic framework with weakly nonlocal internal variables. The entropy production and entropy flux is calculated with the classical method of irreversible thermodynamics by separating full divergences.

16.1 Introduction

Phase field theories are dissipative. At least they seem to be dissipative, because the governing equations are parabolic. There have been various attempts to characterise the dissipation in the framework of their construction methods. However, they were originally introduced without thermodynamic considerations, with the help of a combination of variational and thermodynamic like methods (Cahn and Hilliard, 1958; Cahn, 1961; Hohenberg and Halperin, 1977). The universal background of these equations is questioned in spite of their widespread applicability and success in modelling several different phenomena (Emmerich, 2008; Hohenberg and Krekhov, 2015).

Theoretically, the most problematic aspect is their incompatibility with classical continuum theories. If they are dissipative, then the unification with classical theories, in particular heat conduction, must be straightforward and important. Several conceptual frames were developed to understand this aspect, among them the most notable are the method of configurational forces (Gurtin, 1996) and General equation for Non-Equilibrium Reversible-Irreversible Coupling (GENERIC) (Öttinger, 2005).

P. Ván (✉)
Department of Theoretical Physics, Wigner Research Centre for Physics, Budapest, Hungary

Department of Energy Engineering, Faculty of Mechanical Engineering, Budapest University of Technology and Economics, Budapest, Hungary
e-mail: van.peter@wigner.mta.hu

A. Berezovski, T. Soomere (eds.), *Applied Wave Mathematics II*, Mathematics of Planet Earth 6, https://doi.org/10.1007/978-3-030-29951-4_16

Nonequilibrium thermodynamics with internal variables (NET-IV) is a natural framework to derive the governing differential equations of continua without variational considerations for both dissipative and nondissipative evolution (Ván, 2018). With dual internal variables, inertial effects can be modelled, without or with dissipation (Ván et al, 2008; Berezovski et al., 2011a,b; Berezovski and Ván, 2017; Berezovski et al., 2018). An advantage of a pure thermodynamic background is the universality of the results. As long as the general conditions of the derivation are fulfilled, the results of the derivation are valid, independently of the micro- or mesoscopic structure of the material.

The thermodynamic compatibility of phase fields have been discussed and researched for a long time mostly with variational techniques. The identification of entropy flux and entropy production is problematic (Penrose and Fife, 1990). The relation to continuum balances is investigated mostly when the gradient extensions of classical fields are considered (see, e.g., (Sekerka, 2011)).

In this short chapter a single scalar internal variable is treated with classical thermodynamic methods in rigid heat conductors. In this case Allen–Cahn-type evolution is a solution of the entropy inequality. The relation of entropic and free energy representations is analysed. It is shown that heat flux is different with and without internal variables.

16.2 Classical Variational-Relaxational Derivation of the Allen–Cahn Equation

What is the evolution equation of a single scalar internal variable field without any constraint in a continuum at rest? First one would look for a variational principle. However, then a second order time derivative and a nondissipative evolution cannot be avoided and we obtain something similar to continuum mechanics. There are several other systems, with diffusive properties and in this case the best evolution equations are obtained by a characteristic mixture of variational and thermodynamical ideas.

Let us denote the scalar field by α. We assume, that the Helmholtz free energy density, F, depends on this variable and its gradient: $F(\alpha, \nabla\alpha)$. For the sake of simplicity we consider the following square gradient form for Ginzburg–Landau free energy function:

$$F(\alpha, \nabla\alpha) = F_0(\alpha) + \frac{\gamma}{2}\nabla\alpha \cdot \nabla\alpha, \tag{16.1}$$

where γ is a nonnegative material parameter, which is scalar for isotropic continua. F_0 is the classical, local part of the free energy, that may be a double well potential, if α is an order parameter of a second order phase transition.

Then, following the usual arguments, one assumes that the rate of α in a body with volume V is negatively proportional to the change of the free energy, denoted by δ:

$$\frac{\mathrm{d}}{\mathrm{d}t}\int_V \alpha \mathrm{d}V = -l\delta \int_V F(\alpha, \nabla\alpha)\mathrm{d}V \,. \tag{16.2}$$

Assuming that this equality is valid for any V, we obtain the general Allen–Cahn (Ginzburg–Landau) equation in the following form:

$$\partial_t \alpha = -l\frac{\delta F}{\delta \alpha} = -l\left[\partial_\alpha F - \nabla \cdot \frac{\partial F}{\partial \nabla \alpha}\right]. \tag{16.3}$$

Here ∂_t is the partial time derivative, $\delta/\delta\alpha$ in the functional derivative, and l is a material parameter. With the square gradient free energy (16.1), the classical form of the equation:

$$\partial_t \alpha = -l\big(\partial_\alpha F_0 - \gamma \triangle \alpha\big) \tag{16.4}$$

is obtained. The question is whether and in what sense the equation is dissipative and how is that related to the second law. In the following we will clarify this question and also show, that the Allen–Cahn equation follows from simple thermodynamics.

16.3 Thermostatics of Internal Variables

In our continuum theory, the entropy is the function of the internal energy and also the scalar field α and its gradient. Therefore, the Gibbs relation for specific quantities is written as

$$\mathrm{d}e = T\mathrm{d}s - \frac{A}{\rho}\mathrm{d}\alpha - \frac{\mathbf{A}}{\rho} \cdot \mathrm{d}\nabla\alpha \,. \tag{16.5}$$

Here e is the specific internal energy, s is the specific entropy, T denotes the temperature, ρ is the density and $\nabla\alpha$ is the gradient of α field. The dot denotes the inner product of the corresponding vectors.

The Gibbs relation is a convenient representation the specific entropy function $s(e, \alpha, \nabla\alpha)$, with the partial derivatives:

$$\frac{\partial s}{\partial e} = \frac{1}{T}, \quad \frac{\partial s}{\partial \alpha} = \frac{A}{\rho T}, \quad \frac{\partial s}{\partial \nabla \alpha} = \frac{\mathbf{A}}{\rho T} \,. \tag{16.6}$$

These partial derivatives define the internal variable related intensive quantities A and $\mathbf{A}$.

By means of the specific Helmholtz free energy, $f(T, \alpha, \nabla\alpha)$, considering the definition by Legendre transformation, $f = e - Ts$, one can see easily that

$$\frac{\partial f}{\partial T} = -s, \quad \frac{\partial f}{\partial \alpha} = -\frac{A}{\rho}, \quad \frac{\partial f}{\partial \nabla\alpha} = -\frac{\mathbf{A}}{\rho}. \tag{16.7}$$

In case of a continuum at rest the specific free energy is related to the free energy density as $F = \rho f$. With these expressions the thermostatics for a classical field theory with a single scalar weakly nonlocal internal variable is defined. The derivation of the corresponding relations for densities and also with global quantities is straightforward, introducing the extensivity of the thermodynamic potentials (Berezovski and Ván, 2017).

16.4 Entropy Production

The substantial balance of internal energy is given as

$$\rho\dot{e} + \nabla\cdot\mathbf{q} = 0, \tag{16.8}$$

where $\mathbf{q}$ is the heat flux, the overdot denotes substantial time differentiation and $\nabla\cdot$ denotes divergence. In our case, for rigid heat conductors, the substantial time derivative is equal to the partial derivative. For the calculation of the entropy balance we follow the classical method of de Groot and Mazur (1962) with the identification of the entropy flux by a convenient separation of full divergences. The method was also applied by Maugin (2006) for internal variables. Therefore, for the time derivative of the entropy we obtain

$$\rho\dot{s}(e, \alpha, \nabla\alpha) = -\frac{1}{T}\nabla\mathbf{q} + \frac{A}{T}\dot{\alpha} + \frac{\mathbf{A}}{T}\cdot\frac{\mathrm{d}}{\mathrm{d}t}(\nabla\alpha), \tag{16.9}$$

where we have substituted the time derivative of the internal energy with (16.8) and represented the overdot notation of the substantial time derivative by $\mathrm{d}/\mathrm{d}t$. In case of rigid bodies the spatial and substantial derivatives commute, therefore $\mathrm{d}\nabla/\mathrm{d}t = \nabla\mathrm{d}/\mathrm{d}t$. In this case, continuing the calculation yields:

$$\rho\dot{s} = -\nabla\cdot\frac{\mathbf{q}}{T} + \mathbf{q}\cdot\nabla\frac{1}{T} + \frac{A}{T}\dot{\alpha} + \nabla\cdot\left(\frac{\mathbf{A}\dot{\alpha}}{T}\right) - \dot{\alpha}\nabla\cdot\frac{\mathbf{A}}{T}. \tag{16.10}$$

Therefore, the complete entropy balance can be written in the following form:

$$\rho\dot{s} + \nabla\cdot\left(\frac{\mathbf{q} - \mathbf{A}\dot{\alpha}}{T}\right) = (\mathbf{q} - \mathbf{A}\dot{\alpha})\cdot\nabla\frac{1}{T} + \frac{\dot{\alpha}}{T}\left(A - \nabla\cdot\dot{\mathbf{A}}\right) \geq 0. \tag{16.11}$$

Here we can identify the entropy flux $\mathbf{J}$ and the entropy production σ as follows:

$$\mathbf{J} = \frac{\mathbf{q} - \mathbf{A}\dot{\alpha}}{T}, \tag{16.12}$$

$$\sigma = (\mathbf{q} - \mathbf{A}\dot{\alpha}) \cdot \nabla \frac{1}{T} + \frac{\dot{\alpha}}{T}\left(A - \nabla \cdot \dot{\mathbf{A}}\right) \geq 0. \tag{16.13}$$

The constitutive relations can be determined by solving the inequality (16.13) and recognising that it leads to the evolution equation of α, and the constitutive function of the heat flux $\mathbf{q}$. In case of linear relationship between the thermodynamic fluxes and forces and for isotropic materials:

$$\dot{\alpha} = l\,(A - \nabla \cdot \mathbf{A}) = l\rho T\left(\frac{\partial s}{\partial \alpha} - \nabla \cdot \frac{\partial s}{\partial(\nabla\alpha)}\right) = -l\rho\left(\frac{\partial f}{\partial \alpha} - \nabla \cdot \frac{\partial f}{\partial(\nabla\alpha)}\right), \tag{16.14}$$

$$\mathbf{q} = \dot{\alpha}\mathbf{A} + \lambda\nabla\frac{1}{T} = \rho T\dot{\alpha}\frac{\partial s}{\partial(\nabla\alpha)} + \lambda\nabla\frac{1}{T} = -\rho\dot{\alpha}\frac{\partial f}{\partial(\nabla\alpha)} + \lambda\nabla\frac{1}{T}. \tag{16.15}$$

Here $\lambda = \lambda_F T^2$, where λ_F is the Fourier heat conduction coefficient. Equation (16.14) is the Cahn–Allen–Ginzburg–Landau equation. The second law requires nonnegative relaxation and heat conduction coefficients l and λ_F. With a given thermodynamic potential (s or f), (16.8) and (16.14)–(16.15) are a closed system of differential equations to be solved for considering phase fields coupled to thermal effects.

16.5 Conclusion

The derivation of entropy production can be extended easily with mechanical interaction for fluids and elastic solids. Then the restriction of constant density must be released with the present thermostatic representation of the internal variable and its gradient (Berezovski and Ván, 2017).

It is also remarkable that the simple and heuristic exploitation method for separation of divergences is compatible with the more rigorous Liu or Coleman–Noll procedures in as it was shown in (Ván, 2018). It was also shown that for a Cahn–Hilliard evolution the rigorous exploitation is more technical.

For phase transition models the concavity of the entropy or/and the proper convexity relations for free energy are important requirements. This property ensures the stability of equilibria and the basin of attraction is related to simply connected concave regions. With additional considerations for boundary conditions the total entropy is a good candidate for a Lyapunov functional of equilibria as it is indicated in (Penrose and Fife, 1990).

Acknowledgements The work was supported by the grants National Research, Development and Innovation Office – NKFIH 116197(116375), NKFIH 124366(124508) and NKFIH 123815.

The paper is dedicated to Jüri Engelbrecht on the occasion of his 80th birthday.

References

Berezovski, A., Ván, P.: Internal Variables in Thermoelasticity. Springer, Cham (2017). https://doi.org/10.1007/978-3-319-56934-5

Berezovski, A., Engelbrecht, J., Berezovski, M.: Waves in microstructured solids: a unified viewpoint of modelling. Acta Mech. **220**(1-4), 349–363 (2011a). https://doi.org/10.1007/s00707-011-0468-0

Berezovski, A., Engelbrecht, J., Maugin, G.A.: Generalized thermomechanics with dual internal variables. Arch. Appl. Mech. **81**(2), 229–240 (2011b). https://doi.org/10.1007/s00419-010-0412-0

Berezovski, A., Yildizdag, M.E., Scerrato, D.: On the wave dispersion in microstructured solids. Contin. Mech. Thermodyn. online first (2018). https://doi.org/10.1007/s00161-018-0683-1

Cahn, J.W.: On spinodal decomposition. Acta Metallica **9**, 795–801 (1961)

Cahn, J.W., Hilliard, J.E.: Free energy of a nonuniform system I. Interfacial free energy. J. Chem. Phys. **28**(2), 258–267 (1958). https://doi.org/10.1063/1.1744102

Emmerich, H.: Advances of and by phase-field modelling in condensed-matter physics. Adv. Phys. **57**(1), 1–87 (2008). https://doi.org/10.1080/00018730701822522

de Groot, S.R., Mazur, P.: Non-equilibrium Thermodynamics. North-Holland, Amsterdam (1962)

Gurtin, M.G.: Generalized Ginzburg–Landau and Cahn–Hilliard equations based on a microforce balance. Physica D **92**(3), 178–192 (1996). https://doi.org/10.1016/0167-2789(95)00173-5

Hohenberg, P.C., Halperin, B.I.: Theory of dynamic critical phenomena. Rev. Mod. Phys. **49**(3), 435–479 (1977). https://doi.org/10.1103/revmodphys.49.435

Hohenberg, P.C., Krekhov, A.: An introduction to the Ginzburg-Landau theory of phase transitions and nonequilibrium patterns. Phys. Rep. **572**, 1–42 (2015). https://doi.org/10.1016/j.physrep.2015.01.001

Maugin, G.A.: On the thermomechanics of continuous media with diffusion and/or weak nonlocality. Arch. Appl. Mech. **75**, 723–738 (2006)

Öttinger, H.C.: Beyond Equilibrium Thermodynamics. Wiley-Interscience, Hoboken, NJ, USA (2005). https://doi.org/10.1002/0471727903

Penrose, O., Fife, P.C.: Thermodynamically consistent models of phase-field type for the kinetics of phase transitions. Physica D **43**(1), 44–62 (1990). https://doi.org/10.1016/0167-2789(90)90015-h

Sekerka, R.F.: Irreversible thermodynamic basis of phase field models. Phil. Mag. **91**(1), 3–23 (2011). https://doi.org/10.1080/14786435.2010.491805

Ván, P.: Weakly nonlocal non-equilibrium thermodynamics: the Cahn–Hilliard equation. In: Altenbach, H., Pouget, J., Rousseau, M., Collet, B., Michelitsch, T. (eds.) Generalized Models and Non-Classical Approaches in Complex Materials, vol. 1, pp. 745–760. Springer, Cham (2018)

Ván, P., Berezovski, A., Engelbrecht, J.: Internal variables and dynamic degrees of freedom. J. Non-Equilibr. Thermodyn. **33**(3), 235–254 (2008). https://doi.org/10.1515/jnetdy.2008.010

Index

A. Berezovski, T. Soomere (eds.), *Applied Wave Mathematics II*, Mathematics of Planet Earth 6, https://doi.org/10.1007/978-3-030-29951-4

Printed in the United States
By Bookmasters

Printed in the United States
By Bookmasters